Contents

SECTION II Spike Timing: Coding, Decoding, and Sensation

Series Preface

The Frontiers in Neuroscience Series presents the insights of experts on emerging experimental technologies and theoretical concepts that are, or will be, at the vanguard of neuroscience.

The books cover new and exciting multidisciplinary areas of brain research, and describe breakthroughs in fields such as visual, gustatory, auditory, olfactory neuroscience, as well as aging biomedical imaging. Recent books cover the rapidly evolving fields of multisensory processing, depression and different aspects of reward.

Each book is edited by experts and consists of chapters written by leaders in a particular field. Books are richly illustrated and contain comprehensive bibliographies. Chapters provide substantial background material relevant to the particular subject.

The goal is that these books will become the references that neuroscientists use to get acquainted with new information and methodologies in brain research. I view my task as series editor as that of producing outstanding products that contribute to the broad field of neuroscience. Now that the chapters are available on line, the effort put in by me, the publisher, the book editors, and individual authors will contribute to the further development of brain research. To the extent that you learn from these books we will have succeeded.

Sidney A. Simon, Ph.D.
Series Editor

Editors

Patricia M. Di Lorenzo is a professor of psychology and the director of the Integrative Neuroscience Program at Binghamton University in Binghamton, New York. She earned a BS in psychology and a PhD in biopsychology from the University of Rochester, Rochester, New York. Her PhD studies under Dr. Jerome S. Schwartzbaum marked the beginning of her career studying the gustatory system in the brainstem. She continued her studies of neural coding of taste as a postdoctoral fellow under Drs. John Garcia and Donald Novin at the University of California at Los Angeles. In 1983, she joined the faculty of Smith College in Northampton, Massachusetts, where she spent two years. In 1985, she accepted a faculty position at Binghamton University. Since then, Dr. Di Lorenzo has continued to probe issues of neural coding in the brainstem gustatory system using behavioral, electrophysiological, and computational techniques.

Dr. Di Lorenzo is a member of the Society for Neuroscience, the American Physiological Society, and the Association for Chemoreception Sciences. She is also an associate editor for *Frontiers in Integrative Neuroscience* and has served on review panels at the National Science Foundation and the National Institute on Deafness and Other Communication Disorders. In 2007, she was awarded the New York State Chancellor's Award for Excellence in Scholarship and Creative Activities. She has authored or coauthored over 55 publications and book chapters on neural coding in the taste system. Her work has been supported by the Whitehall Foundation, the National Science Foundation, and the National Institute on Deafness and Other Communication Disorders.

Jonathan D. Victor is a professor of neurology and neuroscience and the director of the Division of Systems Neurology and Neuroscience at Weill Cornell Medical College of Cornell University in New York City. He received a BA in mathematics from Harvard College and a PhD in neurophysiology from the Rockefeller University, where he studied retinal computations in the horseshoe crab and the cat under the mentorship of Floyd Ratliff, Bruce Knight, and Robert Shapley. Following an MD at Cornell and neurology residency at the New York Presbyterian Hospital, he was appointed assistant professor in the Laboratory of Biophysics at Rockefeller University. He moved to Weill Cornell in 1986, where he continued his work in neural computations and sensory processing, primarily in the visual system. He is also engaged in the development of novel approaches to investigate the dynamics of neural circuits at the systems level, and the application of these tools to normal and abnormal human brain function. In 2006, he was named the Fred Plum Professor of Neurology.

Dr. Victor serves as the coeditor-in-chief of the *Journal of Computational Neuroscience* and a member of the editorial boards of *Vision Research* and *Neural Computation* and has been a member of numerous NIH and NSF panels. He is the author of over 180 peer-reviewed publications, invited articles, and book chapters. His work has been supported by the Klingenstein Foundation, the McKnight Foundation, the Swartz Foundation, the McDonnell Foundation, and the National Institutes of Health.

Contributors

Gautam Agarwal
Redwood Center for Theoretical Neuroscience
Helen Wills Neuroscience Institute
University of California
Berkeley, California

Zane N. Aldworth
National Institute of Child Health and Human Development
National Institutes of Health
Laboratory of Cellular and Synaptic Physiology
Bethesda, Maryland

Asohan Amarasingham
Department of Mathematics
City College of New York
New York, New York

Francesco P. Battaglia
Center for Neuroscience
Swammerdam Institute for Life Sciences
University of Amsterdam
Amsterdam, the Netherlands

Laurel H. Carney
Departments of Biomedical Engineering and Neurobiology and Anatomy
University of Rochester
Rochester, New York

Patricia M. Di Lorenzo
Department of Psychology
Binghamton University
Binghamton, New York

Alexander G. Dimitrov
Department of Mathematics
Washington State University
Vancouver, Washington

Timothy J. Gawne
Department of Vision Sciences
University of Alabama at Birmingham
Birmingham, Alabama

Wulfram Gerstner
School of Computer and Communication Sciences
and
School of Life Sciences, Brain-Mind Institute
Ecole Polytechnique Federale de Lausanne
Lausanne, Switzerland

Ranier Gutierrez
Department of Pharmacology
CINVESTAV-IPN
Mexico City, Mexico

Robert Gütig
Max Planck Institute of Experimental Medicine
Goettingen, Germany

Matthew T. Harrison
Division of Applied Mathematics
Brown University
Providence, Rhode Island

Matthew H. Higgs
Neurology Section
Department of Veterans Affairs
Medical Center
and
Department of Physiology and
Biophysics
University of Washington
Seattle, Washington

Robert E. Kass
Department of Statistics
and
Machine Learning Department
and
Center for the Neural Basis of
Cognition
Carnegie Mellon University
Pittsburgh, Pennsylvania

Saša Koželj
Donders Institute for Brain, Cognition
and Behaviour
Radboud University Nijmegen
Nijmegen, the Netherlands

Paul R. Martin
School of Medical Sciences
and
Save Sight Institute
and
ARC Centre of Excellence
in Vision Science
University of Sydney
New South Wales, Australia

Richard Naud
Department of Physics
Ottawa University
Ottawa, Ontario, Canada

Israel Nelken
Department of Neurobiology
Institute of Life Sciences
and
The Interdisciplinary Center for
Neural Computation
and
The Edmond and Lily Safra Center
for Brain Sciences
Hebrew University
Jerusalem, Israel

Sheila Nirenberg
Department of Physiology and
Biophysics
and
Institute for Computational Biomedicine
Weill Medical College of Cornell
University
New York, New York

Sam Reiter
National Institute of Child Health
and Human Development
National Institutes of Health
Bethesda, Maryland

and

Department of Neuroscience
Brown University
Providence, Rhode Island

Ran Rubin
Racah Institute of Physics
and
The Edmond and Lily Safra Center
for Brain Sciences
Hebrew University
Jerusalem, Israel

Ayelet-Hashahar Shapira
Department of Neurobiology
Institute of Life Sciences
Hebrew University
Jerusalem, Israel

Sidney A. Simon
Department of Neurobiology
Duke University Medical Center
Durham, North Carolina

Samuel G. Solomon
School of Medical Sciences
and
ARC Centre of Excellence
in Vision Science
University of Sydney
New South Wales, Australia

and

Institute of Behavioural
Neuroscience
University College London
London, England

Friedrich T. Sommer
Redwood Center for Theoretical
Neuroscience
Helen Wills Neuroscience Institute
University of California
Berkeley, California

Haim Sompolinsky
Racah Institute of Physics
and
The Edmond and Lily Safra Center
for Brain Sciences
Hebrew University
Jerusalem, Israel

and

Center for Brain Science
Harvard University
Cambridge, Massachusetts

William J. Spain
Neurology Section, Department of
Veterans Affairs Medical Center
and
Department of Physiology and
Biophysics
and
Department of Neurology
University of Washington
Seattle, Washington

Mark Stopfer
National Institute of Child Health
and Human Development
National Institutes of Health
Bethesda, Maryland

Paul H. Tiesinga
Donders Institute for Brain, Cognition
and Behaviour
Radboud University Nijmegen
Nijmegen, the Netherlands

and

Department of Physics and Astronomy
University of North Carolina
Chapel Hill, North Carolina

Jonathan D. Victor
Brain and Mind Research Institute
Division of Systems Neurology and
Neuroscience
and
Department of Neurology
and
Institute for Computational Biomedicine
Weill Cornell Medical College
New York, New York

Introduction

Since the discovery of the action potential in the late nineteenth century, the idea that the timing of spikes in the nervous system can convey information has captivated scientists. The number of studies that have focused on spike timing has increased exponentially since that time, especially over the last decade. Although most early studies were carried out in invertebrates and largely addressed basic biophysical issues, recent investigations have a much more extensive scope, addressing the role of spike timing in coding and decoding, and often with explicit relevance to sensory, motor, and cognitive function in mammalian brains. Emerging from these studies is the fundamental notion that spike timing is a primary mode of communication in nervous systems.

The importance of spike timing for signaling is, of course, the central theme of this book. Our hope is that the variations on this theme represented by the chapters go beyond this broad statement, and give it substantial biological reality. Each contributor was asked not only to review his or her work related to spike timing, but also to speculate. The result, we believe, is a fascinating overview of how spike timing contributes to communication in the nervous system, laid out in two interrelated sections.

Section I describes the foundation for quantitative analyses and theory. We have organized the chapters beginning with studies of spike trains of individual neurons and ending with systems-level analyses of information gleaned from spike timing. Section II can be loosely characterized as data driven, and the component chapters use sensory systems as experimental models. Here the chapters dealing with a given sensory system are sequenced so that studies of spike timing in more peripheral structures precede studies of more central areas. However, we recognize that this organization may not suit the needs or interests of all readers, as the chapters are diverse not only in terms of topic, but also in terms of the degree of each reader's familiarity with the neural systems under study and the mathematical methods that are applied. The reader is, therefore, encouraged to use this organization only as for general orientation, and to pick and choose as his/her interests dictate.

In Section I, "Spike Timing: Tools and Models," the authors lay out the how's and why's of what information is contained in spike timing: how it can be quantified, and how neural systems can extract it. Each chapter attempts to tackle the daunting task of extracting meaning (information) from a temporally unique sequence of spikes. In the first chapter, "Spike Trains as Event Sequences: Fundamental Implications" by Jonathan D. Victor and Sheila Nirenberg, some of the basic challenges of the analyses of spike timing are addressed as well as consideration of some potential solutions. In the next chapter, "Neural Coding and Decoding with Spike Times," Ran Rubin, Robert Gütig, and Haim Sompolinsky describe a general model of spike-timing-based coding and decoding, the Tempotron, and show that it can address many important issues associated with decoding spike trains in a biologically plausible manner. The chapter "Can We Predict Every Spike?" by Richard Naud and Wulfram Gerstner, tackles the question of whether and how spike times can be predicted given some knowledge of spike timing history and the time course of input. Matthew T. Harrison, Asohan Amarasingham, and Robert E. Kass discuss methods for analyzing synchrony across neurons in "Statistical Identification of Synchronous Spiking." Next, the chapter by Ayelet-Hashahar Shapira and Israel Nelken, called "Binless Estimation of Mutual Information in Metric Spaces," develop an approach for analyzing mutual information between spike trains and the stimuli that evoke them that succeeds in the particularly challenging situation presented by the auditory system. The final two chapters in this section, "Measuring Information in Spike Trains about Intrinsic Brain Signals" by Gautam Agarwal and Friedrich T. Sommer and "The Role of Oscillation-Enhanced Neural Precision in Information Transmission between Brain Areas" by Paul H. Tiesinga, Saša Koželj, and Francesco P. Battaglia, address the relationship of spike timing to oscillations in brain activity.

Section II, "Spike Timing: Coding, Decoding, and Sensation," speaks on the issue of how input–output relationships are reflected in spike timing across a range of sensory systems. In the first chapter, "Timing Information in Insect Mechanosensory Systems," Alexander G. Dimitrov and Zane N. Aldworth describe studies of the filiform sensilla of the cricket and the campaniform sensilla of the moth. They show that, while the timing of individual spikes carries information about the stimulus, patterns of spikes are more than the sum of their parts—thus demonstrating, in a very concrete sense, that spike patterns carry information. The next two chapters use the auditory system to study spike timing. First, in "Neural Encoding of Dynamic Inputs by Spike Timing," Matthew H. Higgs and William J. Spain address the problem of how neurons respond to an oscillating input with precisely timed spikes, and focus on how response dynamics are shaped by the biophysics of ion channels. Second, in the chapter "Relating Spike Times to Perception: Auditory Detection and Discrimination," Laurel H. Carney directly addresses the question of whether spike timing can predict auditory perception. In it, she systematically examines several aspects of sound and its perception and relates each to spike timing–based decoding schemes. The next three chapters deal with the role of spike timing in the chemical senses. In the first of these, Sam Reiter and Mark Stopfer in "Spike Timing and Neural Codes for Odors" show that neural coding of olfaction can be understood in terms of transformations of the olfactory-generated signal across populations of neurons that vary in both number and spike timing. Second, in "Spike Timing as a Mechanism for Taste Coding in the Brainstem," Patricia M. Di Lorenzo summarizes studies of spike timing in the perception of taste stimuli in rats, examining both mechanisms for generating spike sequences and the functionality of unique temporal patterns of stimulation. Third, in "Increases in Spike Timing Precision Improves Gustatory Discrimination upon Learning," Ranier Gutierrez and Sidney A. Simon describe evidence on the effects of learning on spike timing across several brain areas and its relationship to behavior in the gustatory system of rats. Finally, the last two chapters of the book deal with the visual system. In their chapter titled, "Spike Timing in Early Stages of Visual Processing," Paul R. Martin and Samuel G. Solomon describe the factors that govern the transmission of timing information between the nonspiking neurons at the earliest stages of visual processing, as well as the transfer of spike timing information at the retinal output in the thalamus. Finally, in his chapter titled, "Cortical Computations Using Relative Spike Timing,"

Timothy J. Gawne develops the hypothesis that much of the cortical computation may rely on the relative timing of spikes between different neurons.

Each chapter brings the authors' unique perspective to bear on the broad issues associated with the study of spike timing. Taken together, we believe these contributions provide a compelling overview of the field as a whole as well as a glimpse of future developments.

Patricia M. Di Lorenzo
Jonathan D. Victor

Section I

Spike Timing

Tools and Models

1 Spike Trains as Event Sequences
Fundamental Implications

Jonathan D. Victor and Sheila Nirenberg

CONTENTS

1.1 INTRODUCTION

The goal of this chapter is to consider some of the fundamental implications of the event-sequence nature of neural activity. We approach this, primarily, from the point of view of function: we will look at some simple, idealized vignettes that highlight the implications of the event-sequence nature of neural activity for representing and processing information. While the issues that these vignettes raise are all related to spike timing, they are also logically distinct from each other, and therefore, deserve separate consideration.

The starting point is the well-known fact that neurons communicate with each other via a sequence of action potentials, a profound observation that has been with us for the better part of a century. The reason that the observation is so profound is that it has implications for how the brain operates at the most fundamental level. This is because action potentials ("spikes") are stereotyped—for a given neuron, each spike is largely identical to every other one. Thus, for a neuron to transmit information, it must vary the timing of the sequence of spikes that it emits. This, of course, is the motivation for the collection of topics discussed in this book: understanding how spike sequences represent information requires understanding the mechanisms for encoding signals into temporal sequences of spikes and the mechanisms for decoding these sequences, as well as having appropriate analytical tools for these investigations.

However, the event-sequence nature of neural activity has a number of distinct implications beyond spike timing per se. Perhaps the most basic is that it forces us to carefully consider the kind of mathematical entity used to represent spike trains. This, in turn, has implications for how we formulate models for the evolution of neural activity over time, and the ways in which we analyze how neural activity can represent information.

The first vignette moves from these abstract considerations to a very concrete aspect of the event sequences: they can contain a positive number of events, or no events, but they cannot contain a negative number of events. This elementary asymmetry has theoretical implications for how increments and decrements can be detected: we illustrate the basic idea and then present some data from a model system (the subdivision of ON and OFF signals in the retina) that suggest that these ideas are in fact relevant to real neural circuits.

The last two vignettes address how a discrete sequence of events can represent a continuous quantity. In broad terms, this can happen two ways: in time and in space. To focus on the temporal strategy, we consider how a single neuron can represent a continuous quantity by varying its spike rate over an extended period of time. To focus on the spatial strategy, we consider how a population of neurons can represent a continuous quantity via a distributed pattern of activity at a single instant. Although we look at highly simplified scenarios in both cases and take an elementary viewpoint, these vignettes nevertheless serve to illustrate the impact of the event-like nature of neural activity on the roles of noise, variability, and neuronal diversity.

Finally, we emphasize that our goal is to describe the range of possibilities afforded by event sequences, not to predict how the brain should operate. While we do not doubt that evolution has pushed the brain toward some kind of optimality, the standards of optimality are very complex, and include many biologic constraints beyond the ones considered here: constraints of chemistry and physics, constraints of the genetic code, constraints of development, the need to adapt successfully to an environment that changes over short and long timescales, and so on. Thus, the simple vignettes we consider make no attempt to deduce optimal strategies, but serve only to illustrate how the event-like nature of neural activity shapes the repertoire of strategies that the brain can use, and, consequently, the ways in which we need to approach experimental data.

1.2 TWO WAYS TO REPRESENT SPIKE TRAINS

Whether one takes an empirical or theoretical approach, working with spike trains requires some kind of framework for describing them. Since the characteristics of this framework shape the resulting analysis, the choice of a framework is a fundamental one that can have pervasive implications.

The simple fact that a spike train is a temporal sequence of events means that two kinds of frameworks are possible. That is, a spike train can be parameterized in two ways (Figure 1.1a): we can specify what happens at each time, or we can specify the times at which the events occur. These descriptions are complementary. In the first strategy, the description consists of a large number of variables (one for each time), but each assumes only a binary value (event vs. no event). In the second strategy, the description consists of a small but indefinite number of variables (one for each event), and each can assume a continuum of values (the time corresponding to each event).

1.2.1 DIMENSIONALITY AND GEOMETRY

Both representations fulfill the need of representing spike trains as mathematical entities, but they have very different characteristics. The first difference relates to the dimensionality of the representation. To see this in a concrete way, consider the mechanics of creating a digital record of a recorded spike train. The fidelity of this record depends, of course, on the temporal resolution chosen by the experimentalist.

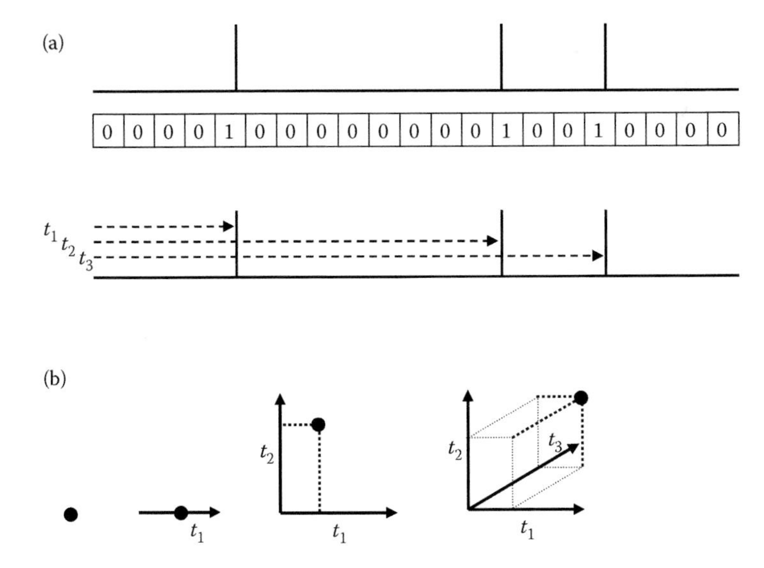

FIGURE 1.1 (a) Two ways to represent spike trains. A spike train can be described by what happens in each time bin (1 = spike, 0 = no spike) or by the times t_i at which each spike occurs. (b) The spike time representation consists of a sequence of components of differing dimensions. Spike trains with n spikes are represented as points in a space of dimension n.

The impact of this choice differs greatly for the two representation strategies. For the first strategy, a 10-fold increase of temporal resolution requires a 10-fold increase in the number of binary values that must be stored since there are 10 times as many bins in which the behavior (spike vs. no spike) must be recorded. For the second strategy, a 10-fold increase in resolution requires only that an additional decimal digit needs to be added to the record of each event's time. Thus, as temporal resolution increases, the event-list representation becomes progressively more efficient.

The point here is not one of practicality, but rather the difference in the geometric characteristics of the representations. The first representation places spike trains into a vector space, with one dimension for each time point. Thus, the representation is intrinsically high-dimensional (e.g., 1000 dimensions for 1 s of data recorded at a resolution of 1 ms). Within this high-dimensional space, actual spike trains occupy only a very sparse portion of the space since the vast majority of coordinate values are zero. The second representation does not have a fixed dimensionality. Instead, it consists of multiple components, each with different dimensions (Figure 1.1b). For example, the spike trains with one spike occupy points on a line segment $0 \leq t_1 < 1$ and thus form a one-dimensional component, and the spike trains with two spikes occupy points in the unit square, with $0 \leq t_1 \leq t_2 < 1$, and thus form a two-dimensional component. Note that within each of these separate components, actual spike trains occupy a relatively large fraction of the space. That is, the event-time representation automatically provides for an efficient representation, but it does so at the expense of complexity: the representation consists of a stack of discrete components with different dimensions, rather than a single high-dimensional space.

This difference in geometry in turn has implications for how we frame the basic questions about neural function: how does neural activity progress over time, and how does neural activity represent things (sensations, motor plans, etc.)? In both cases, we would like to formulate the answers in terms of mappings (or transformations): a mapping from neural activity at one time to activity at the next, or a mapping from neural activity in one brain region to activity in another, or a mapping from a nonneural pattern, such as a sensory stimulus, to a neural one. For a mapping to be viable, it must have at least a minimal element of robustness: that is, a small change in the input should typically lead to a small change in the output. Thus, the viability of a proposal for brain dynamics or for a sensory representation depends, critically, on what we mean by a "small change" in a spike train.

Defining such a notion of a "small change" in a spike train is, in effect, defining the topology of the space in which they are embedded. Importantly, for the two kinds of representations, the natural topologies are quite different. For the first (vector space) representation, the topology is fundamentally discrete, and the continuity of time is irrelevant. A spike train A is represented by quantities a_0, $a_{\Delta t}$, $a_{2\Delta t}$, and so on, one for each time step. These quantities, which are the coordinates of the spike train, count the number of spikes that occur in the corresponding interval. (To keep things simple, we consider time steps that are small enough so that no interval contains more than one spike; thus, each coordinate is either 0 or 1.) The distance between two spike trains, A and B, depends on the differences in their coordinates, that is, the quantities $|a_{k\Delta t} - b_{k\Delta t}|$. Because these differences can only be 0 or 1, the natural notion of distance is discrete. Because the distance considers each instant (each time $k\Delta t$) independently, the continuity of time is ignored.

For the second representation (as event sequences), there is a discrete component, but also a continuous one. The discrete component is that the representation consists of a stack of components, one for each number of spikes. But within these discrete components, the natural notion of distance is one that takes the continuous nature of time into account. That is, the distance between a spike train A with spikes at times $t_1^{[A]}$ and $t_2^{[A]}$, and a spike train B with spikes at $t_1^{[B]}$ and $t_2^{[B]}$ depends on the times between the events, for example, the quantities $|t_i^{[A]} - t_j^{[B]}|$. Crucially, if the events in the two spike trains occur at similar times, these quantities are small, so the spike trains A and B would be considered similar. But in the vector-space representation, this is not the case: as long as the spikes occurred in different time bins, the spike trains would be considered different, and the difference would not depend on how far apart the times are.

Both representations become progressively more exact as one increases temporal resolution (i.e., the precision with which spike times are registered). Although it might at first appear that this results in an eventual diminution of the consequences in the differences in these strategies, the opposite is in fact the case. As temporal resolution increases, more and more dimensions are required for the vector-space representation since each bin requires its own dimension. But there is no increase in the complexity of the spike time representation because this is determined by the number of spikes. Thus, the spike time representation retains its efficiency, even as time resolution increases. Having an efficient representation with this property can be crucial when using experimental data to estimate quantities of central importance, such as stimulus-response probabilities (Jacobs et al. 2009) and information (Victor 2002, 2006).

Finally, we want to emphasize that, although these representations have very different characteristics, they are entirely equivalent: each representation provides a complete description of spike trains. The difference, though, is that some aspects of spike trains may be much more evident in one representation than in the other. For example, since an interspike interval is the difference of a pair of adjacent spike times, it can be calculated by subtracting successive values t_j and t_{j+1} in the event-sequence representation. Interspike intervals can of course also be calculated from the binary representation, but the number of binary coordinate values $a_{k\Delta t}$ that must be inspected to determine each interval can be indefinitely large.

1.2.2 Choosing a Representation Determines the Natural Analysis Strategies

As the preceding remark suggests, the two representations carry with them distinct notions of the "natural" mathematical operations that can be applied to spike trains. That is, while any kind of operation on spike trains can be carried out in either representation, the representations differ in the kinds of operations that are considered to be fundamental—and therefore, suggest different starting points for theories of neural coding and decoding.

In the vector-space representation, it is natural to use vector-space operations, such as addition and dot products,* and this is of course of immense practical value. Addition of spike trains as vectors corresponds to the creation of peristimulus histograms. The dot product (scalar product) between two vectors gives them a Euclidean geometry since it determines the notion of length and angle. Representing spike trains as vector spaces thus makes an implicit statement about whether two spike trains are similar: their natural distance is the length of their difference in the vector space.

Moreover, with one additional ingredient, time translation, the vector-space distance yields the familiar cross-correlogram. Time translation (i.e., changing the recorded spike times from t_k to $t_k + \Delta\tau$) leads to a different set of coordinate values, but one that corresponds to a spike train that has the identical internal structure. The cross-correlogram is simply the set of dot products of one spike train with all time translations of the other.

In contrast, the event-sequence representation focuses on the topology of spike trains, rather than on vector-based operations between them. Thus, it naturally leads to approaches that focus on strategies for comparing and classifying spike trains, rather than arithmetic procedures, such as averages and correlations. These approaches include several kinds of "spike metrics" that can be used to capture different ways in which event sequences can be compared. A metric based on differences in event times (as described above) is perhaps the simplest, but the idea readily extends to metrics based on interspike intervals (Figure 1.2b), or on other features such as bursts (Victor and Purpura 1997).†

Finally, there is an important class of metrics—"kernel-based metrics"—that provide a bridge between these two kinds of representations. To formulate a metric of this class, each spiking event is first replaced by a waveshape—considered, perhaps, to represent the synaptic conductance that it causes (Figure 1.2c). This changes the representation of a spike train A from an event sequence $\{t_1^{[A]}, t_2^{[A]}, \ldots\}$ into a function of time, $A(t)$. In the simplest implementation, each spiking event is replaced by the same waveform (van Rossum 2001), but one can also allow the waveforms to be dependent on spiking history (Houghton 2009). Standard vector-space distances can now be applied to these functions, and this in turn induces a metric on the spike trains. The critical part of this construction is that the waveshape used to represent each spike is a smooth function of time. Because of this, the metric induced by the construction recognizes that a small shift in the time of an event is not as significant a change as a large shift. Thus, these metrics retain a key feature of the geometry of event sequences—the continuity of time—and the vector-space embedding allows for arithmetic operations and computational efficiency.

* In general, vector spaces need not have a dot product (also known as an inner product or scalar product); the presence of an inner product is what makes a vector space into a Hilbert space. However, the vector spaces that are relevant here—and nearly all of those that arise in neural data analysis—are, in fact, Hilbert spaces.

† It is possible to impose vector space operations on a "spike time" representation—for example, by considering a more complex kind of event sequence, in which each event is allowed to have a numerical weight. Such constructs can be useful for understanding the mathematical properties of the spike time representation and spike metrics. Our point, thus, is not about whether the spike time representation can be made compatible with some form of vector space operations, but rather that the spike time representation puts the emphasis on the relationships between spike trains, instead of on how they can be arithmetically combined.

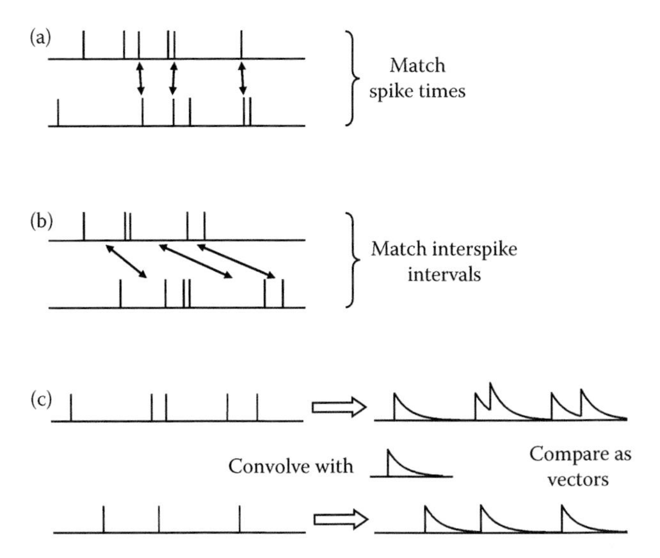

FIGURE 1.2 Quantifying the similarity between spike trains. (a) The "spike time" metric, based on correspondences between spikes that occur at nearby times (Victor and Purpura 1997). (b) The "spike interval" metric, based on correspondences between interspike intervals of matching lengths. (c) A "kernel-based" metric (van Rossum 2001), in which each spike train is first convolved with a waveshape representing synaptic activity, and the resulting time series are compared as vectors.

1.3 DETECTING LARGE CHANGES: A FUNDAMENTAL ASYMMETRY

Here and in the subsequent sections, we consider the implications of the event-like nature of neural activity for signaling. We begin by examining the consequences of an observation so simple as to appear trivial: the number of events cannot be negative. As we describe below, even this basic observation has important consequences for signaling: there can be substantial differences in the speed with which positive and negative changes in firing rates can be detected.

To see this, we analyze the performance of an observer who has access to the firing of a single neuron, and is interested in determining whether the firing rate has increased or decreased from its baseline. The observer's optimal strategy depends on the statistics of the firing process—for example, whether it is regular or irregular; here, we will assume that the firing process is Poisson.* That is, baseline firing

* We assume Poisson firing to keep things simple and generic, and even this most simple model has interesting behavior. We do not intend to imply that Poisson firing is a complete model for real neural firing patterns (although it is often a good first approximation). The Poisson assumption allows us to analyze the ideal observer's performance in complete detail, and avoids having to frame a complex, parametric model for firing statistics. Below, when we analyze the detection of small changes in firing rates, we will consider some simple alternatives to Poisson firing.

events occur randomly at a rate λ, and the observer's task is to determine whether there has been a change in firing by a fraction c, to a rate $\lambda(1 + c)$. Our focus will be on the observer's performance for detecting increments ($c > 0$) versus decrements ($c < 0$), and how this performance depends on the duration of the spike train, T, that the observer has available.

To determine this performance, we examine the statistics of the observer's data on a trial-by-trial basis. If the firing rate is at baseline, the distribution of observations is governed by Poisson statistics, with a mean of λT. Thus, the probability that the observer sees n spikes is

$$q_n = \frac{(\lambda T)^n}{n!} e^{-\lambda T}. \tag{1.1}$$

The gray histograms in Figure 1.3a show examples of this distribution, for a baseline firing rate of $\lambda = 1.0$ spike/s, and observation intervals of $T = 0.25$ s and $T = 1.0$ s.

If the firing rate deviates from baseline by a fraction c, then the mean value of the Poisson distribution changes to $\lambda(1 + c)T$. Consequently, the probability that the observer sees n spikes is

$$p_n = \frac{(\lambda(1 + c)T)^n}{n!} e^{-\lambda(1+c)T}. \tag{1.2}$$

The black and white histograms in Figure 1.3a show examples of this distribution corresponding to the above baseline conditions, with two modulation depths ($c = \pm 0.5$ and $c = \pm 1.0$). To casual observation, the black distributions (rate decreases) deviate more from the baseline (gray) distributions than the white distributions (rate increases), owing in large part to the peak at 0 (i.e., the greater frequency of observing no events). The asymmetry is larger when modulation depth is higher ($c = \pm 1.0$ vs. $c = \pm 0.5$, bottom row vs. top row) and suggests that it is easier to detect a decrement than to detect an increment.

1.3.1 Quantitative Analysis

To make this observation more precise, we quantify how readily the observer can detect positive and negative changes from baseline firing. We use an information-theoretic analysis that provides an analytic expression for the asymmetry between decrements and increments (Pandarinath et al. 2010); for a receiver operating curve analysis, see Thibos et al. (1979).

Assuming that a change from baseline firing has occurred, the probability that the observer sees n spikes is determined by p_n (Equation 1.2). To decide whether a firing rate change has occurred, the observer must ask whether this observation is more likely to have come from this distribution versus the baseline distribution, q_n (Equation 1.1), that is, whether the ratio p_n/q_n is greater than 1. It is, therefore, natural to quantify the discriminability of the p-distribution from the q-distribution by the

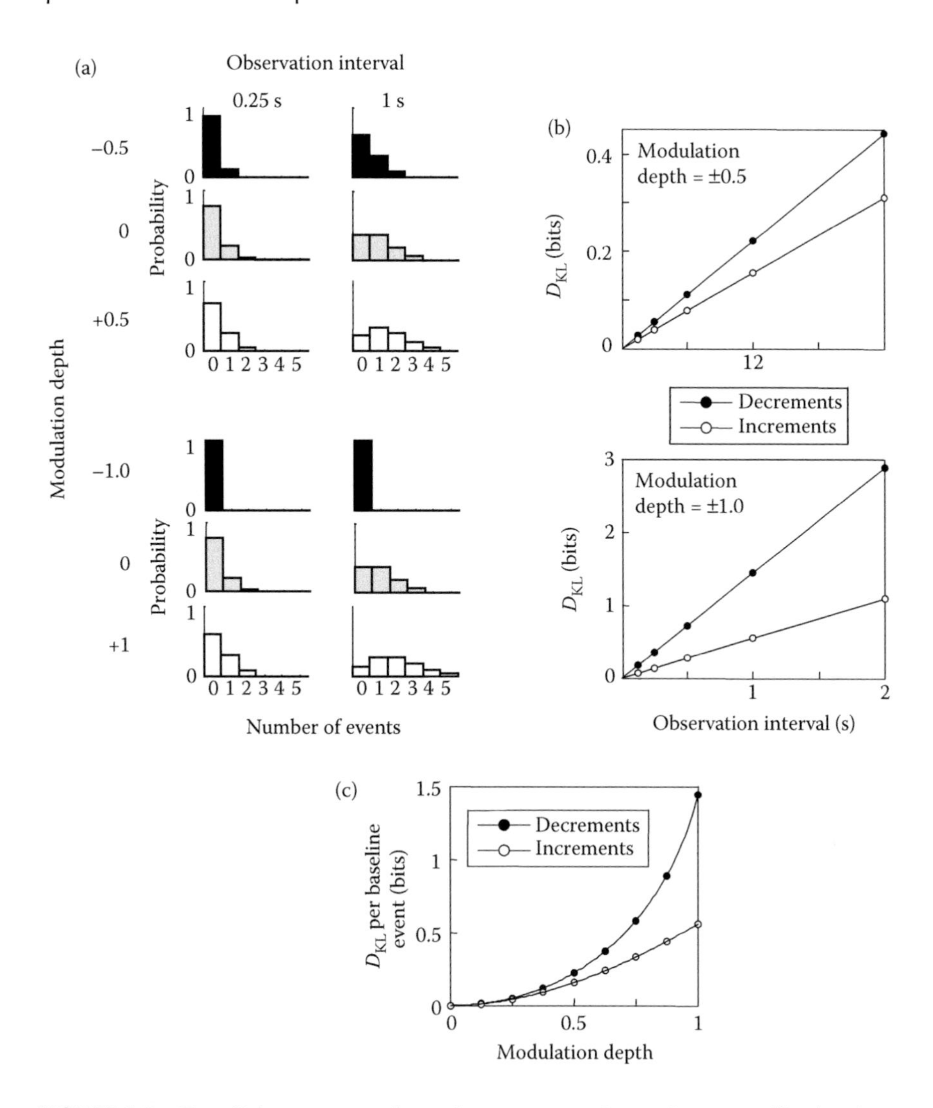

FIGURE 1.3 For a Poisson process, large decrements are detected more readily than large increments. (a) Expected distributions of event counts for a Poisson process whose rate is one event per second at baseline (gray), and for processes whose rates are incremented (white) or decremented (black) from this baseline by a modulation depth (contrast) c. Count distributions are shown for two observation intervals, 0.25 s (left) and 1.0 s (right), and for two modulation depths, $c = \pm 0.5$ (top) and $c = \pm 1.0$ (bottom). (b) Discriminability of increments (white) or decrements (black) from the baseline process, as measured by the Kullback–Leibler divergence (Equation 1.3). There is an advantage for decrements, and the advantage is greater for large modulation depths ($c = \pm 1.0$, lower graph) than for small modulation depths ($c = \pm 0.5$, upper graph). In all cases, the discriminability increases in proportion to the observation time, as expected for a Poisson process. (c) Discriminability per baseline event as a function of modulation depth. At high modulation depths, decrements can be detected almost three times more rapidly than increments.

average of the logarithm of this ratio, weighted by the probability p_n that each observation occurs. This is the Kullback–Leibler divergence:

$$D(P||Q) = \sum_n p_n \log_2 \frac{p_n}{q_n}. \tag{1.3}$$

Thus, $D(P||Q)$ measures the discriminability of the q-distribution from the p-distribution in bits per observation. Intuitively, $\log_2 p_n/q_n$ indicates the informativeness of an observation of n spikes, and p_n indicates how often this happens. (For additional background, see Latham and Nirenberg 2005.)

For Poisson distributions, one can calculate the Kullback–Leibler divergence exactly. After simple algebra, substituting Equations 1.1 and 1.2 into Equation 1.3 yields

$$D(P||Q) = \frac{\lambda T}{\ln 2}((1+c)\ln(1+c) - c) \tag{1.4}$$

This result is shown in Figure 1.3b. As surmised from the histograms of Figure 1.3a, the asymmetry increases with increasing modulation depth: there is about a 30% advantage for decrements when the depth of modulation is 0.5, and more than a 2.5-fold advantage when the depth of modulation is 1.0.

Two facts allow us to summarize this analysis across all values of the baseline firing rate λ and the observation time T. First, we note (either from Equation 1.4 or from Figure 1.3b) that for both increments and decrements, discriminability is proportional to the observation time. (Since we are dealing with Poisson processes, this is not surprising—the occurrence of events in each instant is independent, and therefore, the data from each instant should contribute independently to a decision.) Moreover, changing the baseline event rate by some factor has the same effect as changing the observation time by its reciprocal—since for a Poisson process, the only informative aspect of the data is the total number of events. Thus, the discriminability of an increment or a decrement is proportional to the number of expected baseline events (λT). This proportionality factor, $((1 + c)\ln(1 + c) - c)/(\ln 2)$, shown in Figure 1.3c, summarizes the asymmetry between increments and decrements, and how it becomes progressively more prominent as modulation depth increases.

1.3.2 Implications

The above asymmetry between detection of increments and decrements of equal size arises because detectability depends not only on signal but also on noise (variance). Because event counts cannot be less than 0, very low event counts have a lower variance than high ones. At first glance, this would seem to imply that the asymmetry is only relevant for very low firing rates. But in fact, this is not the case: the asymmetry depends on the number of events in the observation interval, not the firing rate per se. That is, no matter how high the firing rate λ, one can choose a sufficiently short observation interval $T < 1/\lambda$ for which the expected number of events is small.

Thus, no matter how high the firing rate, there is always a regime—the regime of short observation times—in which decrements have an advantage over increments. This is also the regime in which the simplifying assumption of Poisson behavior is effectively irrelevant. For sufficiently short intervals, any spike train can be viewed as approximately Poisson since the only kinds of intervals that occur with significant frequency are the ones with no spikes and the ones with a single spike.

Even when there are a large number of counts per interval, the asymmetry is likely to persist. This is because one expects that when the number of counts is higher, then so is the variance in the number of counts. Since the ability to detect a signal is determined not only by signal size but also by noise, it follows that increments will be more difficult to detect than decrements. However, to quantify the asymmetry in the high-count regime, one must choose a model that specifies how variability depends on event count—such as the Poisson model considered here.

Detection, of course, is only part of what a sensory system must do; for many sensory tasks, discrimination among multiple suprathreshold alternatives is required. Since discrimination also depends on both signal and noise, one expects that the asymmetry will apply to discrimination as well. As is the case for detection, quantification of this asymmetry—how it depends on the number of counts, and the number of stimulus intensities to be distinguished—requires choosing a model for variability. With the Poisson model, a threefold asymmetry (similar to that observed in Figure 1.3c) is present for up to 10^3 counts, when there are 10 intensities to be discriminated (Pandarinath et al. 2010).

Recent evidence suggests that the brain takes advantage of this asymmetry, when it comes to reading out neural activity (Pandarinath et al. 2010). We used the retinal output (the ganglion cells of the mouse) as a model system to study this question for two reasons. First, retinal ganglion cells fall into two categories, ON and OFF: ON cells increase in firing with light increments and OFF cells increase their firing with light decrements. Second, it is straightforward to shift between conditions in which the noise characteristics of the input to ganglion cells are Poisson (in which the above asymmetry is expected), and conditions in which the noise characteristics are likely to be approximately Gaussian in character. The Poisson regime corresponds to low light levels, where the quantal nature of light becomes the dominant source of variability. The approximately Gaussian regime corresponds to high light levels, in which other sources of noise dominate.* The latter regime serves as a control—no asymmetry is expected when variability is Gaussian. Indeed, this is what was found: in the light, OFF and ON cells are closely matched in terms of the duration of signal fluctuations to which they respond (i.e., OFF and ON cells have similar temporal tuning). But when the retina shifts to a dark-adapted state and the quantal nature of light becomes the dominant noise source, this changed: in dim light, OFF cells respond to briefer fluctuations than ON cells (i.e., OFF cells are tuned to higher temporal frequencies than ON cells). The magnitude of the shift in tuning for OFF cells

* Note that the retinal network that conveys signals to ganglion cells consists of *nonspiking* neurons that signal increments and decrements in an analog fashion, rather than in stereotyped events. Because of the central limit theorem, we anticipate that approximately additive combination of noisy signals from many such neurons will produce a signal with nearly Gaussian fluctuations.

compared to ON cells, was approximately 3:1, similar to the magnitude of the asymmetry seen in Figure 1.3c at the highest depths of modulations. Although the analysis in Figure 1.3 is obviously far too simple to be directly applicable to the retina, it is still interesting to note that it accounts for the nature and approximate magnitude of a fundamental asymmetry in retinal processing.

While the above analysis of retinal processing suggests that the decrement/increment asymmetry is something that the brain can take advantage of under appropriate circumstances, we do not mean to imply that it is the sole principle that shapes neural signaling. It is just one of many factors that need to be considered. Other factors include the cost incurred by maintaining populations that respond to decrements as well as increments, the extent to which the large-signal, short-time regime is the relevant one, and the precise nature of variability over extended periods of time. Thus, our point is not that this asymmetry should dominate the design of neural signaling, but rather, that it is a consideration that arises because long-range communication between neurons is based on event sequences. Had we thought of this communication in terms of a continuous real variable—an ongoing "firing rate" that can fluctuate up or down—the consideration would not even have arisen.

1.4 DETECTING SMALL CHANGES: HOW CRUCIAL IS REGULAR FIRING?

In the next vignette, we again examine the task of determining when a neuron's firing rate has changed, but we shift the focus to detecting small changes in firing rate, rather than large ones—and we allow for observation of the neuron's behavior over an extended period of time. Intuitively, this is a regime in which the intrinsic regularity of a neuron's firing pattern has a major impact. If a neuron ordinarily fires in highly regular, clock-like fashion, then even a small change in the interspike interval is a reliable signal that its underlying rate has changed. On the other hand, if a neuron's firing pattern is irregular (such as a Poisson neuron, considered above), then a wide range of interspike intervals can occur, even if the underlying rate does not change.

Before embarking on the analysis, it is worthwhile emphasizing that this issue is not merely a hypothetical one: although a Poisson process is often taken as a default model for neuronal firing, real neurons vary widely along the continuum from highly regular to highly irregular. In the periphery, mammalian retinal ganglion cells are modestly more regular than Poisson (Troy and Robson 1992; Reich et al. 1998), and mechanoreceptive sensory afferents can be nearly clocklike (Werner and Mountcastle 1965). Cortical neurons (also see the review in Maimon and Assad 2009), though widely considered to be approximately Poisson (Tolhurst et al. 1983; Buracas et al. 1998), can range from substantially more regular than Poisson (DeWeese et al. 2003; Gur and Snodderly 2006; Maimon and Assad 2009) to substantially more irregular (Victor and Purpura 1996), depending on cortical area (DeWeese et al. 2003; Maimon and Assad 2009), anesthetic state, and behavioral task (Gur and Snodderly 2006).

As we will see below, the intuition that regularly firing neurons are more efficient in transmitting a change over time is correct. But we will also find that in terms of information throughput, the gains realized by regularity can also be realized by increasing firing rate or adding additional neurons—and this equivalence is precise.

This in turn suggests that a neuron's position on the continuum of irregularity versus regularity is driven by factors other than just information throughput, such as the metabolic cost of spikes, and its participation in a population that needs to function effectively under a wide range of conditions.

1.4.1 A Convenient Set of Distributions, Ranging from Poisson to Regular

To quantify the way that regularity impacts on the ability to detect that a small change in the underlying rate has occurred, we consider a parametric family of neurons, whose behavior ranges from Poisson to perfectly regular. We then quantify the behavior of an observer who is attempting to determine whether the underlying rate has changed, and again use the Kullback–Leibler divergence as a measure of discriminability.

The parametric family of neurons is defined as follows. Each neuron emits spikes according to a "renewal process," that is, a process in which the probability distribution for the time of the next spike ("the renewal density") depends only on the time since the last spike. Thus, each interspike interval represents an independent sample of a probability distribution, and the regularity of their firing is determined by the narrowness of this distribution.

A Poisson neuron with firing rate λ has the familiar renewal density

$$q(t) = \lambda e^{-\lambda t}. \tag{1.5}$$

To model more regular firing patterns, we replace the distribution (1.5) by a gamma distribution

$$q(t) = \frac{(k\lambda)^k}{(k-1)!} t^{k-1} e^{-k\lambda t}. \tag{1.6}$$

The shape of the gamma distribution (and hence, the narrowness of its peak) is determined by the parameter k. For $k = 1$, a Poisson distribution results; as k increases, the distribution becomes progressively narrower (see Figure 1.4a for examples); the standard deviation is proportional to $1/\sqrt{k}$.

We choose the gamma-distribution family to run the gamut from Poisson to clock-like for two reasons. First, it is mathematically simple, yet still provides a reasonable description of neuronal interspike interval distributions during steady firing (Troy and Robson 1992; Reich et al. 1998; McKeegan 2002; Maimon and Assad 2009). Second, it has an intuitive interpretation, which becomes relevant when we consider the implications of the analysis. Specifically, a gamma distribution with parameter k and firing rate λ can be generated in the following fashion. First, create an underlying ("hidden") Poisson process, whose firing rate is given by $k\lambda$. Then, to create the gamma process, select every kth event from this underlying Poisson process. Choosing every kth event reduces the firing rate from $k\lambda$ to λ, and, more importantly, dampens out the variability: multiple, independent intervals from the

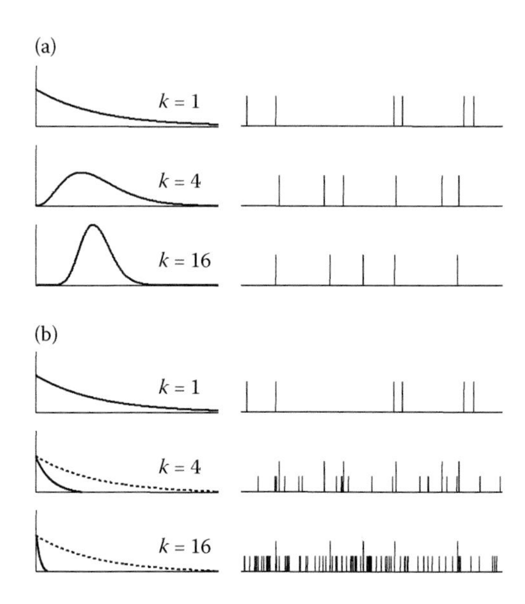

FIGURE 1.4 Gamma processes and their relationship to Poisson processes. (a) The left panel shows interspike interval distributions of gamma processes with identical firing rates and three values of the regularity parameter k ($k = 1$ corresponds to a Poisson process; $k = 4$ and $k = 16$ are processes of progressively greater regularity). The right panel shows typical spike trains generated by these processes. (b) Generation of gamma processes from hidden Poisson processes of a higher rate. The left panel side shows the interspike interval distribution of a Poisson process whose rate is increased by a factor of k (solid) with respect to the baseline rate (dotted). The right panel shows events generated by these faster Poisson processes (short line segments). Every kth event is indicated by a longer line segment, which corresponds to the gamma processes shown in (a).

underlying process Poisson are combined to make a single interval in the gamma process. This relationship is shown in Figure 1.4b.

The implications of the relationship between a gamma process of parameter k and a Poisson process of k-fold higher firing rate are broader than what is directly indicated by the figure. As illustrated, we can think of these underlying "hidden" events as coming from a single neuron but we can also think of them as being distributed across any population of Poisson neurons, as long as the total firing rate is $k\lambda$ (e.g., k independent Poisson neurons, each of firing rate λ). Either way, choosing every kth event results in the same gamma process.

We now consider the task of an observer who attempts to determine whether the underlying rate has changed from λ to some slightly different firing rate λ'. That is, the observer sees a single interspike interval, say of length t, and needs to decide whether it is more likely that the interval t came from the original distribution (1.6), or, alternatively, from a distribution

$$p(t) = \frac{(k\lambda')^k}{(k-1)!} t^{k-1} e^{-k\lambda' t}. \tag{1.7}$$

As in the previous section, the informativeness of each observation can be quantified by the Kullback–Leibler divergence. We slightly modify the definition (1.3) because here the observations form a continuum (the intervals t) rather than a discrete set of spike counts:

$$D(P\|Q) = \int_0^\infty p(t)\log_2 \frac{p(t)}{q(t)}\,dt. \tag{1.8}$$

Substituting Equations 1.6 and 1.7 into Equation 1.8 yields

$$D(P\|Q) = \frac{(k\lambda')^k}{(k-1)!}\int_0^\infty t^{k-1}e^{-k\lambda' t}\log_2\left(\left(\frac{\lambda'}{\lambda}\right)^k e^{-kt(\lambda'-\lambda)}\right)dt. \tag{1.9}$$

Evaluation via standard means leads to

$$D(P\|Q) = \frac{k}{\ln 2}\left(\ln(1+c) - \frac{c}{1+c}\right), \tag{1.10}$$

where $c = (\lambda'/\lambda) - 1$ is the fractional change in firing rate. We note that Equation 1.10 is an expression for discriminability on a per-interval basis, in contrast to Equation 1.4, which quantifies discriminability in an observation period of length T. This is because in the present analysis—which involves non-Poisson processes—it is the lengths of the interspike intervals that are independent, not the spike times themselves. For the Poisson case ($k = 1$), Equations 1.4 and 1.10 are brought into alignment by multiplying Equation 1.10 by the average number of spike intervals per observation period T, namely, $(1 + c)\lambda T$.

1.4.2 An Interesting Equivalence

The above analysis allows us to compare two factors that make a change in firing rate easier to detect: increases in the number of events per unit time, or increases in regularity. Equation 1.10 shows that increasing the regularity parameter by a factor k increases the per-event discriminability by a factor k. Thus, increasing the regularity parameter by a factor k but not changing the baseline firing rate is precisely equivalent to increasing the baseline firing rate by the same factor k but not changing the regularity.

While it is not surprising that higher firing rates and increased regularity both improve the ability to detect firing rate changes, the precise nature of the relationship indicates an interesting equivalence. As mentioned above (and see Figure 1.4), a gamma process with regularity k is equivalent to every kth spike of a "hidden" Poisson process that has a k-fold higher underlying rate. Equation 1.10 shows that these two processes—one that is k times more regular than a Poisson process of rate λ, and one that has k times the firing rate—are precisely equivalent in terms of how readily they allow firing rate changes to be detected.

This equivalence can be taken one step further because the Poisson process of rate $k\lambda$ can be spread across neurons, with no loss (or gain) of information. Thus, from the point of view of enabling a change in firing rate to be identified, a single neuron whose firing pattern is characterized by a gamma process of regularity k and firing rate λ is equivalent not only to a single neuron whose firing rate is $k\lambda$ but also to any population of independent Poisson neurons whose total firing rate is $k\lambda$, including k independent neurons each with firing rate λ. That is, an independent collection of k Poisson neurons of firing rate λ has precisely the same performance as a single neuron of firing rate λ if its interspike intervals have the regularity of a gamma process of order k.

In sum, to build a neural population that attains a criterion degree of signaling fidelity, one can choose two kinds of strategies: strategies based on regularly firing neurons, and strategies based on irregularly firing neurons with a higher total firing rate. The former uses fewer spikes, and might be most efficient from the point of view of the metabolic cost of firing spikes and the number of neurons that need to be supported. The latter strategy, based on large numbers of independent, irregular neurons, has other potential advantages. Although the two strategies are perfectly balanced in terms of information throughput over the long term, a single, regular neuron is uninformative just after it spikes—since it is known that it will not spike again. In contrast, a population of irregular neurons can always transmit at least some information. Moreover, since a stimulus that activates a greater number of neurons will necessarily be transmitted with greater fidelity (because it has a higher overall Poisson firing rate), such a population immediately reconfigures itself to enable more precise signaling of intensity when spatial resolution is not required or available.

Since the position on the regularity-versus-irregularity continuum is perfectly balanced by firing rate from the point of view of overall information throughput, we speculate that this aspect of neural dynamics is driven by other factors, such as the metabolic cost of spikes, the time value of information, and the benefits of on-the-fly reconfiguration.

1.5 POPULATION CODING: VARIABILITY AND DIVERSITY

The final vignette considers the discrimination of graded quantities in a short period of time. This is complementary to the two vignettes considered above: the first considered short periods of time, but primarily examined detection; the second considered discrimination of graded quantities, but extended periods of time. To transmit graded levels of a signal in a short period of time, a population of neurons is required because the information that can be provided by a single neuron is too limited: it can either fire or not.

As we will see, building such a population necessitates a compromise: design strategies that are efficient for single neurons and small populations work poorly for large populations, and vice versa. Moreover, the nature of this trade-off depends on the extent to which the neurons are deterministic, the diversity of the neural population, and whether the neurons are "labeled" so that their individual identities can be taken into account when the population activity is decoded. Interestingly, we will see that one specific trade-off—in which neurons are labeled by their membership in one of two classes that divide the domain into halves (such as "ON" and "OFF" cells) is a favorable one. But as in the previous vignettes, our goal is not to determine the optimal trade-off, but merely to identify considerations that arise specifically because a neuron's output is a stereotyped event, rather than a graded signal.

1.5.1 The Basic Trade-Off

To address this issue, we will examine how the ability of a population of neurons to signal a graded quantity depends on the size of the population, N. To pose the problem precisely, we will assume that the quantity to be transmitted (the "stimulus") is a value anywhere in the range from 0 to 1, and that neural activity is observed for a period that is sufficiently brief so that no neuron can fire twice. Since we are interested in the extent to which a graded quantity can be reliably transmitted, we use information theory to provide a theoretically grounded means to quantify performance (see Rieke et al. 1997 for an introduction to information theory and its application to neural coding).

With this framework, transmitted information is defined as the extent to which observation of the neural response reduces the uncertainty about the value of the stimulus. Uncertainty, in turn, is quantified by entropy; entropy (in bits) is essentially the base-2 logarithm of the number of possible stimulus values. Thus, transmitted information is quantified by the difference between the *a priori* stimulus entropy, and the average entropy of the *a posteriori* stimulus distribution, conditional on observing a sample of neural activity.

If there is only one neuron, there are only two possible outcomes: either it fires or it does not. Since there are only two possibilities for what the neuron can do, the uncertainty about the stimulus can be reduced by at most a factor of two, that is, at most one bit of information can be transmitted. This maximal performance is achieved by setting the neuron's threshold to 0.5, and making it noise-free. For such a neuron, either outcome (whether it fires or not) is a reliable report of which half of the interval contains the stimulus.

However, constructing an entire population out of this kind of neuron (threshold of 0.5 and noiseless) is obviously a very poor strategy for transmitting graded signals. A population containing N such neurons carries no more information than a single neuron: either every neuron in the entire population fires or every neuron does not. The additional neurons lead to no further reduction in uncertainty about the stimulus, and therefore, do not transmit any additional information.

This low ceiling on the information that a population can transmit is clearly undesirable, but there are two obvious ways to fix it. One strategy is to use neurons that are noisy—for example, neurons for which the *probability* of firing is determined by the value to be represented. For a large population built out of such neurons, the fraction of neurons firing gives a reliable estimate of the input quantity. The second strategy is to use neurons that are diverse. For example, if we build a population of neurons whose thresholds are uniformly distributed over the [0,1]-interval, then the fraction of neurons that fire also gives a reliable estimate of the value to be represented.

Both these strategies ensure that the input quantity can be represented to any desired degree of accuracy, provided that the population is sufficiently large, but they do this at the expense of a trade-off. The trade-off is that the performance of a single neuron drawn from this population is no longer optimal. In the "noisy neuron" scenario, the response of a single neuron is no longer a reliable indicator of which half of the interval contains the signal. In the "diverse neuron" scenario, the response of a single neuron is largely uninformative if that neuron happens to be one whose threshold is close to 0, or close to 1 because for those neurons, its behavior can often be guessed in advance.

Figure 1.5 quantifies these trade-offs, based on calculations of how transmitted information varies as a function of population size. (The details behind these calculations are provided in the Appendix.) The simplest case is the noiseless neuron of threshold 0.5; as indicated above, the transmitted information is one bit, independent

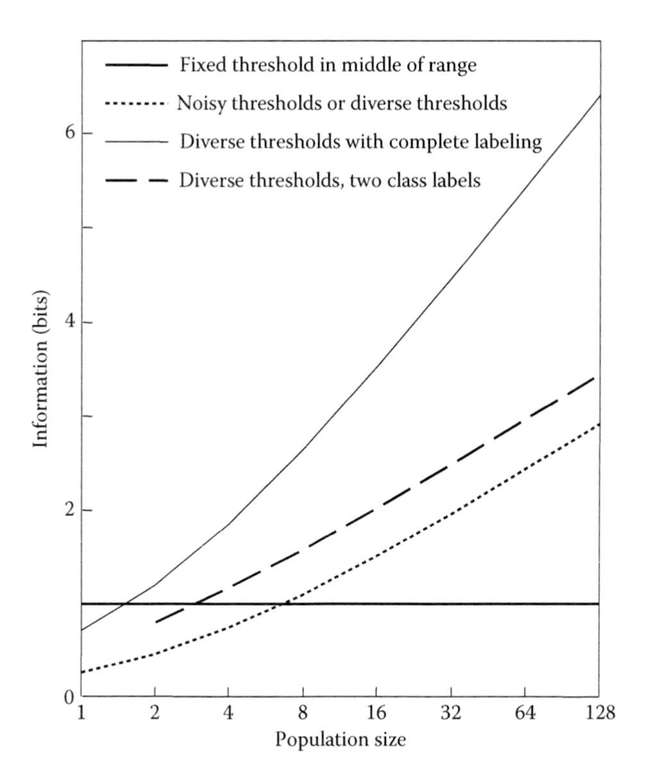

FIGURE 1.5 The optimal strategy for population coding depends on population size. Each curve quantifies how well a snapshot of the activity of neural population can represent a graded quantity, as a function of the number of neurons in the population. Several kinds of populations are considered: (a) noiseless neurons with a threshold in the middle of the range to be represented (heavy solid line); (b) neurons with a noisy threshold, whose firing probability increases in proportion to the quantity to be represented (dotted line); (c) noiseless neurons drawn randomly from an unlabeled population of neurons with diverse thresholds—this curve is identical to (b); (d) diverse thresholds as in (c), but with the neurons labeled so that their thresholds are known to the observer (thin solid line); and (e) diverse thresholds as in (c), but with neurons are drawn equally from one of two labeled classes, with each serving half of the range (dashed line).

of the size of the population. The two scenarios mentioned above—populations built out of noisy neurons whose probability of firing is given by the stimulus value, and populations built out of noiseless neurons, whose thresholds are randomly distributed over the [0,1]-interval—yield much less information (~0.28 bits) for a population size 1 than the noiseless neuron, but for sufficiently large populations, these alternatives perform better. The crossover point is at $N = 7$ neurons, and, as the number of neurons increases, transmitted information grows approximately as $\frac{1}{2}\log_2 N$.

Interestingly, both kinds of populations—identical neurons with noisy thresholds and diverse neurons with noiseless thresholds—have exactly the same performance. The reason is simple: the decoding problem confronting an observer of the population is identical. To see this, we first consider the population of noisy neurons. Since each neuron's firing probability is equal to the stimulus s, we can simulate its behavior by drawing a random number from the [0,1]-interval, and having the neuron fire when this random number is less than s, the value to be transmitted. We then consider the population of diverse neurons with noiseless thresholds. To simulate each neuron in it, we also draw a random number from the [0,1]-interval. This random number represents the neuron's deterministic threshold, and the neuron, therefore, fires if the random number is less than s. Thus, for either population, a simulated neuron fires when a random draw from the [0,1]-interval has a value of s or less. Consequently, for either population, each trial is characterized by a list of $N + 1$ random numbers drawn from the [0,1]-interval: N random numbers that describe the neurons on that trial, along with the random number, s, that describes the stimulus. If k neurons fire, there are k numbers that are less than s (the neurons that fire), and $N - k$ numbers that are larger (the neurons that do not fire). From the observer's point of view, the number of neurons that fire determines the rank order of s in this list—but nothing more. This allows us to rephrase the decoding problem facing the observer: if k neurons fire, the *a posteriori* distribution of s is the distribution of the $(k + 1)$-th ranked number in a list of $N + 1$ numbers drawn uniformly from the [0,1]-interval.* Since the statistical description of each trial is identical for the two kinds of populations, the information transmitted by each kind of population must be identical as well.

1.5.2 Variability, Diversity, Labels, and Classes

A key element of the above analysis is that for the population of diverse neurons with noiseless thresholds, we assumed that the identity of the neurons was not known. Because of this uncertainty, information that might be available to the observer is lost. We quantify this loss by considering another scenario: neurons are again drawn from the same diverse population (thresholds uniformly distributed in the interval from 0 to 1), but now, the neurons are "labeled" by their thresholds, so that the thresholds are known to the observer. As shown in Figure 1.5, this leads to a substantial improvement (see Appendix for details of the calculation): while the noiseless neuron (threshold at 0.5) is still the optimal choice for $N = 1$, the labeled population crosses over at $N = 2$, in contrast to the unlabeled one, which crosses over at $N = 7$. For large populations,

* This is known as an "order statistic"; we use this in the Appendix to carry out the calculations shown in Figure 1.5.

transmitted information grows approximately as $\log_2 N$ when neurons are labeled by their thresholds, rather than as $\frac{1}{2}\log_2 N$ when the thresholds are unknown.

The improvement in performance is dramatic but it may require a substantial biological cost—the cost of labeling every neuron in the population by its individual threshold. This suggests one final strategy: partial labeling, that is, subdividing neurons into classes. More specifically, suppose that neurons belonged to one of two classes (such as ON or OFF), with each class having thresholds that uniformly cover half of the range, and half of the neurons drawn from each class. Results of the information calculation, shown in Figure 1.5, indicate that this "labeling by class" improves performance by about 0.5 bits over the unlabeled case. This translates into a considerable (>50%) improvement for modest populations sizes ($N = 2$ or 4). Thus, for small populations, labeling by class may be a useful trade-off, in that it results in substantially better performance than anonymous neurons but does not require the biological cost of labeling each neuron individually.

In sum, a neuron that is optimal when considered as an individual (a neuron with a reliable threshold at a fixed level) is decidedly suboptimal when used to build a population; conversely, neurons that work well as part of a population are necessarily suboptimal when considered as individuals. A network typically needs to work well under a wide range of conditions—conditions in which only one neuron is stimulated as well as conditions in which a large population is engaged. There is no universal solution that optimizes performance across this gamut of conditions, and factors such as the reliability of a threshold and population diversity lead to trade-offs that improve performance in one range of this gamut, at the expense of performance in another range.

The reason that no universal optimum is possible—and hence, the reason that the above considerations even arise—is that the output of a neuron is stereotyped. If a neuron could produce a graded output, the problem would simply go away: an individual neuron could report the value of the input signal, and there would be no need to add variability or diversity (thus compromising single-neuron performance) to achieve a benefit as the population size increases.

1.6 CONCLUSIONS

In this chapter, we have considered four implications of the spiking nature of neural activity. At the most fundamental level, it shapes the range of mathematical approaches to describing neural activity. We then constructed three vignettes to focus on different aspects of information transmission: signaling of large changes over brief periods of time, signaling of graded quantities over extended periods of time, and signaling of graded quantities over brief periods of time. In each case, the spiking nature of neural activity gives the analysis a distinctive character, and shapes the qualitative conclusions we draw. For large changes over brief periods of time, decrements are signaled several-fold more rapidly than similar increments. For signaling of graded quantities over extended periods of time, there is a precise equivalence between firing regularity and population firing rate. For signaling graded quantities rapidly, intrinsic noise, population diversity, and labeled lines provide for a range of trade-offs that optimize performance in different regimes of population size.

In each of these three examples, our considerations only scratch the surface of the implications of neuronal spiking, and raise further questions about the design strategies and their costs. In the first example—detection of large increments and decrements—our analysis began by assuming a baseline Poisson-like occurrence of neural events. But without an ongoing level of activity, decrements are obviously impossible. The implication is that the nervous system can only take advantage of the superiority of decrements if it is willing to invest in the cost of maintaining ongoing activity.

The second example relates to the equivalence between firing regularity and population rate—that any benefits of regularly firing neurons could be achieved by simply increasing the number of irregularly firing ones. To derive this equivalence, we assumed that the neurons in the population are independent. Any deviation from independence would tend to make the population redundant. Thus, any potential advantages that synchrony may have (from the point of view of reading out the population signal) need to be balanced against their cost in terms of reducing the overall amount of information that the population can carry.

The third example concerns strategies for population-based signaling of graded quantities. We showed that a two-class design for the population conferred some advantages over designs in which neurons had stochastic thresholds, or were noiseless and drawn from a uniform pool. Similar advantages continue to accrue as the number of classes in the population increase. But this leads to further issues: the cost of maintaining multiple subpopulations and with specific connections so that they can be distinguished, and the potential benefits of arranging these classes into orderly mosaics, to increase the likelihood that spatially restricted stimuli engage approximately equal numbers of neurons in each class.

In conclusion, the spiking nature of neural activity has profound implications for understanding the design of the nervous system, and the three vignettes considered here are but starting points for analyzing them.

APPENDIX

This appendix provides details underlying the calculations given in Figure 1.5 and the related asymptotic results in the text.

Our goal is to calculate the information that is carried by a snapshot of the activity of a population of N neurons, in each of several scenarios. In general, this is given by the difference between the *a priori* entropy of the stimulus and its average entropy conditional on a response:

$$I(N) = -\int_0^1 p(s)\log_2 p(s)ds + \left\langle \int_0^1 p_N(s|r)\log_2 p_N(s|r)ds \right\rangle_r, \tag{A.1}$$

where $p(s)$ is the *a priori* distribution of the stimulus, $p_N(s|r)$ is its distribution conditional on observing the response r, and $\langle\ \rangle_r$ indicates an average over all possible responses r, weighed by their frequency of occurrence. Since we view the response as a "snapshot" of activity, the response variable is a specification of which neurons fire and which do not. The integrals in Equation A.1 are differential entropies and the probability distributions are normalized over [0,1].

In each scenario, the stimulus is uniformly distributed in [0,1]. Thus, $p(s) = 1$ and the first term in Equation A.1 is 0. We now proceed to calculate the second term.

Unlabeled Neurons

We first calculate $I^{unlabeled}(N)$, the amount of information about the stimulus contained in a snapshot of N neurons whose thresholds are randomly chosen in the interval [0,1] and not known to the observer. Since the thresholds are unknown, the response variable r is simply the number of neurons that fire (k), which can range from 0 to N.

As mentioned in the main text, the probability $p_N(s|k)$ can be determined by thinking of the stimulus value and the N thresholds as a set of $N+1$ random numbers drawn from the [0,1]-interval: N random numbers that describe the neurons, along with the random number, s, that describes the stimulus. $p_N(s|k)$ is the probability that the stimulus value s is the $(k+1)$-th ranked number on the list (i.e., that it is higher than the descriptors of the k neurons that fired and lower than the descriptors of the $N-k$ neurons that did not fire). This has two consequences. First, each of the $N+1$ possible values of k is equally likely since it is equally likely that any of the $N+1$ random draws is assigned to the stimulus. Since the first term in Equation A.1 is zero, Equation A.1 is therefore equivalent to

$$I^{unlabeled}(N) = \frac{1}{N+1}\sum_{k=0}^{N}\int_{0}^{1} p_N(s|k)\log_2 p_N(s|k)ds. \tag{A.2}$$

The second consequence is that $p_N(s|k)$ is the $(k+1)$-th "order statistic" (David and Nagajara 2003) for $N+1$ draws from the uniform [0,1] distribution. Therefore, its value is given by

$$p_N(s|k) = \frac{1}{B(k+1, N-k+1)} s^k(1-s)^{N-k}, \tag{A.3}$$

where $B(a,b)$ is the beta function

$$B(a,b) = \int_{0}^{1} u^{a-1}(1-u)^{b-1}du. \tag{A.4}$$

As is well known, the beta function can be written in terms of the gamma function

$$B(a,b) = \frac{\Gamma(a)\Gamma(b)}{\Gamma(a+b)}, \tag{A.5}$$

where $\Gamma(a) = \int_0^{\infty} u^{a-1}e^{-u}\, du.$

To evaluate the differential entropy of $p_N(s|k)$ for Equation A.2, we use the following approach, which also proves convenient for the other kinds of populations considered below. For a probability distribution $q(s)$ on the interval [0,1], let

$$M(\alpha) = \int_0^1 (q(s))^\alpha \, ds. \tag{A.6}$$

Since

$$\frac{d}{d\alpha} x^\alpha = \frac{d}{d\alpha} e^{\alpha \ln x} = (\ln x) e^{\alpha \ln x} = x^\alpha \ln x, \tag{A.7}$$

it follows that the differential entropy of q is given by

$$-\int_0^1 q(s) \log_2 q(s) ds = -\frac{1}{\ln 2} \frac{d}{d\alpha} \left(\int_0^1 (q(s))^\alpha \, ds \right) \Bigg|_{\alpha=1} = -\frac{1}{\ln 2} \frac{dM}{d\alpha} \bigg|_{\alpha=1} = -\frac{1}{\ln 2} \frac{d \ln M}{d\alpha} \bigg|_{\alpha=1}. \tag{A.8}$$

The last step follows because

$$\frac{dM}{d\alpha}\bigg|_{\alpha=1} = M(\alpha) \left(\frac{1}{M} \frac{dM}{d\alpha} \right) \bigg|_{\alpha=1} = M(\alpha) \frac{d \ln M}{d\alpha} \bigg|_{\alpha=1}. \tag{A.9}$$

and $M(\alpha) = 1$ at $\alpha = 1$ (via Equation A.6).

We now apply this strategy to $q(s) = p_N(s|k)$, as given by Equation A.3. Using the definition (A.4) of the beta function, Equation A.6 yields

$$M(\alpha) = \int_0^1 \frac{s^{k\alpha} (1-s)^{(N-k)\alpha}}{(B(k+1, N-k+1))^\alpha} ds = \frac{B(k\alpha + 1, (N-k)\alpha + 1)}{(B(k+1, N-k+1))^\alpha}. \tag{A.10}$$

From this, it follows from Equation A.5 that

$$\begin{aligned} \ln M(\alpha) = {} & -\alpha \ln B(k+1, N-k+1) + \ln \Gamma(k\alpha + 1) \\ & + \ln \Gamma((N-k)\alpha + 1) - \ln \Gamma(N\alpha + 2). \end{aligned} \tag{A.11}$$

Consequently, the logarithmic derivative of M, as required by Equation A.8, can be expressed in terms of the digamma function $\psi(u) = d(\ln \Gamma(u))/du$:

$$\begin{aligned} \frac{d}{d\alpha} \ln M(\alpha) \bigg|_{\alpha=1} = {} & -\ln B(k+1, N-k+1) + k\psi(k+1) \\ & + (N-k)\psi(N-k+1) - N\psi(N+2). \end{aligned} \tag{A.12}$$

Since Equation A.12 is related to the differential entropy of $q(s) = p_N(s|k)$ by Equation A.8, it follows from Equation A.2 that

$$I^{unlabeled}(N) = -\frac{1}{(N+1)\ln 2}\cdot$$
$$\sum_{k=0}^{N}(\ln B(k+1, N-k+1) - k\psi(k+1) - (N-k)\psi(N-k+1) + N\psi(N+2)),$$
$$\text{(A.13)}$$

which may be rewritten as

$$I^{unlabeled}(N) = -\frac{N\psi(N+2)}{\ln 2} - \frac{1}{(N+1)\ln 2}\cdot$$
$$\sum_{k=0}^{N}(\ln B(k+1, N-k+1) - k\psi(k+1) - (N-k)\psi(N-k+1)). \qquad \text{(A.14)}$$

This exact result is the quantity plotted as the dotted line in Figure 1.5.

Labeled Neurons

A similar strategy provides for an evaluation of Equation A.1 for the "labeled neuron" case—that is, the information $I^{labeled}(N)$ contained in a snapshot of the response of a population of N neurons whose thresholds, randomly drawn from [0,1], are *known* to the observer.

To calculate $I^{labeled}(N)$, we proceed as follows. First, denote the N randomly drawn thresholds by $0 < y_1 < \cdots < y_N < 1$, and, for convenience, write $y_0 = 0$ and $y_{N+1} = 1$. Since these thresholds are known to the observer, the response variable r is equivalent to a pair of variables $(\vec{y}, k)$, where $\vec{y}$ denotes the set of threshold values, and k (as above) is the number of neurons that fire. Given these data, the observer knows that the stimulus lies in the interval $[y_k, y_{k+1}]$; within this interval, any value is equally likely. Thus, $p_N(s|\vec{y}, k) = 1/(y_{k+1} - y_k)$ if s is in the interval $[y_k, y_{k+1}]$, and zero otherwise. Writing the intervals between the thresholds as $x_k = y_{k+1} - y_k$, the differential entropy of $p_N(s|\vec{y}, k)$, needed for Equation A.1, is

$$-\int_0^1 p_N(s|\vec{y},k)\log_2 p_N(s|\vec{y},k)ds = -\frac{1}{\ln 2}\int_{y_k}^{y_k+x_k}\frac{1}{x_k}\ln\frac{1}{x_k}ds = \frac{\ln x_k}{\ln 2}. \qquad \text{(A.15)}$$

We now find the average of Equation A.15 over all values of $\vec{y}$ (the choices of thresholds) and k. Since each choice of thresholds is equally likely, it follows that each choice of intervals $\vec{x}$ is equally likely, provided that the sum of the interval lengths is 1. Given a choice of intervals $\vec{x}$, the probability that exactly k neurons fire is equal to x_k since x_k is the fraction of the [0,1] range that exceeds the threshold of exactly k neurons. It follows that

$$\left\langle \int_0^1 p_N(s|r)\log_2 p_N(s|r)ds \right\rangle_{\vec{y},k} = -\frac{\Gamma(N+1)}{\ln 2}\sum_{k=0}^{N}\int x_k \ln x_k \delta\left(\sum_{j=0}^{N} x_j - 1\right)d\vec{x}. \quad (A.16)$$

The normalizing factor $\Gamma(N+1)$ follows by evaluating the multivariate beta function

$$\int x_0^{c_0-1}x_1^{c_1-1}\cdot\cdots\cdot x_N^{c_N-1}\delta\left(\sum_{j=0}^{N} x_j - 1\right)d\vec{x} = B(c_0,c_1,\ldots,c_N)$$
$$= \frac{\Gamma(c_0)\Gamma(c_1)\cdot\cdots\cdot\Gamma(c_N)}{\Gamma(c_0+c_1+\cdots+c_N)} \quad (A.17)$$

at $c_0 = c_1 = \cdots = c_N = 1$. To evaluate the integral inside the sum on the right-hand side of Equation A.16, we use the same approach as above (Equations A.6 through A.12) for converting integrals involving logarithms to derivatives of moments. Without loss of generality, one can choose $k = 0$, yielding

$$\int x_0 \ln x_0 \delta\left(\sum_{j=0}^{N} x_j - 1\right)d\vec{x} = \frac{d}{d\alpha}\int x_0^{\alpha}\delta\left(\sum_{j=0}^{N} x_j - 1\right)d\vec{x}\Bigg|_{\alpha=1} = \frac{d}{d\alpha}B(\alpha+1,1,\ldots,1)\Big|_{\alpha=1}. \quad (A.18)$$

There are $N+1$ equal terms (one for each value of k) that contribute to the sum in Equation A.16, so, therefore,

$$\left\langle \int_0^1 p_N(s|r)\log_2 p_N(s|r)ds \right\rangle_{\vec{y},k} = -\frac{(N+1)\Gamma(N+1)}{\ln 2}\frac{d}{d\alpha}B(\alpha+1,1,\ldots,1)\Big|_{\alpha=1}. \quad (A.19)$$

As above (Equation A.9), we calculate the derivative logarithmically. Since $u\Gamma(u) = \Gamma(u+1)$

$$\left\langle \int_0^1 p_N(s|r)\log_2 p_N(s|r)ds \right\rangle_{\vec{y},k}$$
$$= -\frac{\Gamma(N+2)}{\ln 2}\left(B(\alpha+1,1,\ldots,1)\frac{d}{d\alpha}\ln B(\alpha+1,1,\ldots,1)\right)\Bigg|_{\alpha=1} \quad (A.20)$$

from which the definition (A.15) of the beta function implies

$$\left\langle \int_0^1 p_N(s|r)\log_2 p_N(s|r)ds \right\rangle_{\vec{y},k} = -\frac{\Gamma(2)\Gamma(1)^N}{\ln 2}\left(\frac{d}{d\alpha}\ln\frac{\Gamma(\alpha+1)}{\Gamma(\alpha+N+1)}\right)\Bigg|_{\alpha=1}. \quad (A.21)$$

This is can be expressed in terms of the digamma function $\psi(u) = d(\ln \Gamma(u))/du$, yielding

$$I^{labeled}(N) = \left\langle \int_0^1 p_N(s|r)\log_2 p_N(s|r)ds \right\rangle_{\vec{y},k} = \frac{1}{\ln 2}\left(\psi(N+2) - \psi(2)\right), \quad \text{(A.22)}$$

which is the quantity plotted as the thin line in Figure 1.5.

Labeled Classes

The last case we consider is a population that consists of C classes, with N_h neurons ($N_h \geq 1$) in class h. We posit that for each of the C classes, the thresholds of the neurons occupy an equal ($1/C$) and nonoverlapping portion of the [0,1]-interval, and the observer knows the class identity of each neuron but not its threshold. The main text considered only $C = 2$ (each cell class covers half of the interval); we consider the more general case here because it clarifies the structure of the problem. The approach taken extends readily to unequal ranges covered by each class, provided that these ranges do not overlap and together cover the full interval.

In this scenario, the information contained in a snapshot of activity, which we denote $I^{class}(N_1,\ldots,N_C)$, can be readily calculated from the values of $I^{unlabeled}(N_h)$ (Equation A.13). We relate these quantities by breaking down all trials into two kinds: "typical" trials, in which the stimulus is somewhere between the lowest-threshold neuron of some class h, and the highest-threshold neuron of that class, and "atypical" trials, in which the stimulus is between the highest threshold of one class, and the lowest threshold of the next.

In a "typical" trial, the observer knows which of the C intervals contains the stimulus (say, the interval $[(h-1)/C, h/C]$), and the number of neurons within the class h that fire. Knowledge of which interval contains the stimulus yields $\log_2 C$ bits because the C intervals are equally likely. Knowledge of the number of neurons within class h that fire yields, on average, $I^{unlabeled}(N_h)$ bits. This last statement follows from the above "unlabeled neurons" calculation on the interval [0,1] since the numerical labels assigned to the endpoints of the interval are irrelevant.

In an "atypical" trial, the stimulus lies somewhere between the highest-threshold neuron of one class (say, h), and the lowest-threshold neuron of the next class, $h + 1$. Thus, the observer is faced with the ambiguity of which interval (interval h vs. interval $h + 1$) contains the stimulus. Moreover, the observer knows that there is an ambiguity because all neurons of class $\leq h$ are firing, and all neurons of class $\geq h + 1$ are not firing. Once this ambiguity is removed, the atypical trial would yield the same amount of information as a typical one—since then, the relevant interval and class would be known.

The above analysis shows that we can calculate $I^{class}(N_1,\ldots,N_C)$ by first determining the information as if all trials were typical, and then subtracting a term reflecting the information necessary to resolve the ambiguity:

$$I^{class}(N_1,\ldots,N_C) = \log_2 C + \frac{1}{C}\sum_{h=1}^{C} I^{unlabeled}(N_h) - \sum_{h=1}^{C-1} b(h) I^{ambig}(h), \quad \text{(A.23)}$$

where $b(h)$ is the probability that there is an ambiguity between the intervals represented by class h and $h+1$, and $I^{ambig}(h)$ is the information required to resolve this ambiguity.

To determine $b(h)$, we note that this is the sum of two terms: the probability $b_{above}(h)$ that the stimulus s is above the highest threshold in class h but still within the interval that it represents, and the probability $b_{below}(h+1)$ that s is below the lowest threshold in class $h+1$ but still within its interval. As mentioned above in the analysis of the unlabeled population, if a stimulus is drawn randomly within the range that the population represents, all values for the number of neurons that fire are equally likely. Since the chance that the stimulus is in each interval is $1/C$, it follows that

$$b_{above}(h) = b_{below}(h) = \frac{1}{C(N_h + 1)}$$

and

$$b(h) = b_{above}(h) + b_{below}(h+1) = \frac{1}{C}\left(\frac{1}{N_h + 1} + \frac{1}{N_{h+1} + 1}\right). \quad \text{(A.24)}$$

The amount of information required to resolve the ambiguity between $b_{above}(h)$ and $b_{below}(h+1)$ is the amount of information that disambiguates between two alternatives whose probabilities are $b_{above}(h)/b(h)$ and $b_{below}(h+1)/b(h)$, namely,

$$I^{ambig}(h) = -\left(\frac{b_{above}(h)}{b(h)}\log_2\frac{b_{above}(h)}{b(h)} + \frac{b_{below}(h+1)}{b(h)}\log_2\frac{b_{below}(h+1)}{b(h)}\right). \quad \text{(A.25)}$$

Combining Equations A.23, A.24, and A.25 yields the desired result:

$$\begin{aligned} I^{class}(N_1,\ldots,N_C) &= \log_2 C + \frac{1}{C}\sum_{h=1}^{C} I^{unlabeled}(N_h) \\ &+ \frac{1}{C}\sum_{h=1}^{C-1}\left(-\frac{\log_2(N_h+1)}{N_h+1} - \frac{\log_2(N_{h+1}+1)}{N_{h+1}+1}\right. \\ &\left. + \left(\frac{1}{N_h+1} + \frac{1}{N_{h+1}+1}\right)\log_2\frac{(N_h+1)(N_{h+1}+1)}{N_h + N_{h+1} + 2}\right). \end{aligned} \quad \text{(A.26)}$$

When all classes have the same number M of neurons, this simplifies to

$$I^{class}(M,\ldots,M) = \log_2 C + I^{unlabeled}(M) - \frac{2(C-1)}{C(M+1)}. \quad \text{(A.27)}$$

With $C = 2$ and $M = N/2$, this is the quantity plotted as the dashed line in Figure 1.5.

Asymptotic Estimates

To obtain additional insight into population coding, we calculate the asymptotic behavior of $I^{unlabeled}$, $I^{labeled}$, and I^{class} for large populations. We first consider $I^{labeled}$ since it is the most straightforward, and then $I^{unlabeled}$; estimates for I^{class} follow immediately from estimates for $I^{unlabeled}$ via Equation A.27.

Labeled Neurons

For $I^{labeled}(N)$, the asymptotic behavior for large N follows directly from Equation A.22, via the asymptotic behavior of the digamma function (Abramowitz and Stegun 1964), namely, $\psi(N) = \ln N + O(1/N)$:

$$I^{labeled}(N) = \log_2(N) - \frac{\psi(2)}{\ln 2} + O(1/N). \tag{A.28}$$

This yields an interesting comparison between the performance of a population of neurons with random thresholds, and the optimal choice of thresholds (thresholds that are evenly-spaced throughout the interval). The optimal choice yields $\log_2(N+1) = \log_2(N) + O(1/N)$ bits; the random choice (Equation A.28) incurs a penalty of $-\psi(2)/\ln 2$, which is approximately -0.61 bits.

Unlabeled Neurons

For $I^{unlabeled}(N)$, we cannot use the above strategy since the expression (A.14) for $I^{unlabeled}(N)$ includes a sum of terms involving the digamma function with small argument. Instead, we go back to Equation A.2 and approximate the distribution of $p_N(s|k)$ by a Gaussian of matching variance, and we use the differential entropy of this approximating Gaussian to estimate $I^{unlabeled}(N)$.

As noted above, the distribution $p_N(s|k)$ is given by the $(k+1)$-th order statistic for $N+1$ draws from the uniform [0,1] distribution (Equation A.3). Replacing this distribution by a Gaussian will be a good approximation, provided that the number k of neurons that fire is neither very close to 0 nor very close to N. Since this holds for all but $O(1/N)$ of the terms in the sum (A.2), the resulting approximation will be valid up to $O(1/N)$.

To carry out this approximation, we first note that a Gaussian of variance V has differential entropy equal to

$$-\frac{1}{\sqrt{2\pi V}}\int_{-\infty}^{\infty} e^{-x^2/2V}\log_2\left(\frac{1}{\sqrt{2\pi V}}e^{-x^2/2V}\right)dx = \frac{1}{2\ln 2}(1+\ln 2\pi V). \tag{A.29}$$

We next use Equation A.3 to calculate the variance of $p_N(s|k)$:

$$\begin{aligned} V_N(k) &= \int_0^1 s^2 p_N(s|k)ds - \left(\int_0^1 s p_N(s|k)ds\right)^2 \\ &= \frac{B(k+3, N-k+1)}{B(k+1, N-k+1)} - \left(\frac{B(k+2, N-k+1)}{B(k+1, N-k+1)}\right)^2 = \frac{(k+1)(N-k+1)}{(N+2)^2(N+3)}, \end{aligned} \tag{A.30}$$

where the final equality makes use of the relationship (A.5) between the beta function and the gamma function and the fact that $\Gamma(u+1) = u\Gamma(u)$. We now estimate the differential entropy of $p_N(s|k)$ in Equation A.2, by taking $V = V_N(k)$ in Equation A.29. This yields

$$\begin{aligned} I^{unlabeled}(N) &= -\frac{1}{2\ln 2}\frac{1}{N+1}\sum_{k=0}^{N}\left(1+\ln 2\pi\frac{(k+1)(N-k+1)}{(N+2)^2(N+3)}\right)+O(1/N) \\ &= -\frac{1}{2\ln 2}\left(1+\ln 2\pi-\ln N+\frac{1}{N+1}\sum_{k=0}^{N}\ln\left(\frac{k+1}{N+1}\right)\left(\frac{N-k+1}{N+1}\right)\right)+O(1/N) \\ &= -\frac{1}{2\ln 2}\left(1+\ln 2\pi-\ln N+\frac{2}{N+1}\sum_{k=0}^{N}\ln\left(\frac{k+1}{N+1}\right)\right)+O(1/N), \end{aligned} \tag{A.31}$$

where the last step follows because as k runs from 0 to N, the two factors in the logarithm sample the same values but in reverse order. Finally, since

$$\frac{2}{N+1}\sum_{k=0}^{N}\ln\frac{k+1}{N+1} = 2\int_0^1 \ln u\,du + O(1/N) = -2 + O(1/N) \tag{A.32}$$

(A.31) yields

$$I^{unlabeled}(N) = \frac{1}{2}\log_2 N + \frac{1-\ln 2\pi}{2\ln 2} + O(1/N). \tag{A.33}$$

Thus, compared to the case of N labeled neurons, for which the amount of information is asymptotically $\log_2 N$, the lack of a label causes a loss of about half the information.

Labeled Classes

For I^{class}, we first use Equation A.27 to relate it to $I^{unlabeled}$, and estimate the latter with Equation A.33. There are two cases of particular interest. For the two-class case ($C = 2$ and $M = N/2$) considered in the main text and Figure 1.5, Equation A.27 states that

$$I^{class}(N/2, N/2) = 1 + I^{unlabeled}(N/2) + O(1/N), \tag{A.34}$$

which, in conjunction with the asymptotic expression (A.33) for $I^{unlabeled}$, yields

$$I^{class}(N/2, N/2) = 1/2 + I^{unlabeled}(N) + O(1/N). \tag{A.35}$$

While this approximation was developed for $N \to \infty$, it is reasonably accurate across the entire range (even for $N = 2$): as shown in Figure 1.5, separation of the population into two classes provides a half-bit advantage over the unlabeled population.

The second case is that of extreme of class labeling: N separate classes, with one neuron in each. Thus, the threshold of each neuron is known to be somewhere within an interval of length $1/N$ but the precise value of each threshold is not known. In this case ($C = N$, $M = 1$), Equation A.27 reduces to

$$I^{class}(1,\ldots,1) = \log_2 N + I^{unlabeled}(1) - 1 + O(1/N). \qquad \text{(A.36)}$$

We compare this to $\log_2 (N + 1)$, which is the performance of a population of N neurons with optimally placed and known thresholds. The asymptotic difference is $I^{unlabeled}(1) - 1$, which is approximately −0.72—a penalty that is comparable to, but slightly worse than, choosing the thresholds completely at random, and knowing the threshold values once they are chosen (see comments after Equation A.28).

ACKNOWLEDGMENTS

This work is supported in part by the National Eye Institute Grants RO1 EY07977 and RO1 EY09314 to J. Victor and RO1 EY12978 to S. Nirenberg.

REFERENCES

Abramowitz M and Stegun IA. *Handbook of Mathematical Functions*. Reprinted by Dover. New York, 1970: National Bureau of Standards, 1964.

Buracas GT, Zador AM, DeWeese MR and Albright TD. Efficient discrimination of temporal patterns by motion-sensitive neurons in primate visual cortex. *Neuron* 20:959–969, 1998.

David HA and Nagajara HN. *Order Statistics*. Hoboken, NJ: Wiley, 2003.

DeWeese MR, Wehr M and Zador AM. Binary spiking in auditory cortex. *J Neurosci* 23:7940–7949, 2003.

Gur M and Snodderly DM. High response reliability of neurons in primary visual cortex (V1) of alert, trained monkeys. *Cereb Cortex* 16:888–895, 2006.

Houghton C. Studying spike trains using a van Rossum metric with a synapse-like filter. *J Comput Neurosci* 26:149–155, 2009.

Jacobs AL, Fridman G, Douglas RM, Alam NM, Latham PE, Prusky GT, and Nirenberg S. Ruling out and ruling in neural codes. *Proc Natl Acad Sci U S A* 106:5936–5941, 2009.

Latham PE and Nirenberg S. Synergy, redundancy, and independence in population codes, revisited. *J Neurosci* 25:5195–5206, 2005.

Maimon G and Assad JA. Beyond Poisson: Increased spike-time regularity across primate parietal cortex. *Neuron* 62:426–440, 2009.

McKeegan DE. Spontaneous and odour evoked activity in single avian olfactory bulb neurones. *Brain Res* 929:48–58, 2002.

Pandarinath C, Victor JD, and Nirenberg S. Symmetry breakdown in the ON and OFF pathways of the retina at night: Functional implications. *J Neurosci* 30:10006–10014, 2010.

Reich DS, Victor JD, and Knight BW. The power ratio and the interval map: Spiking models and extracellular recordings. *J Neurosci* 18:10090–10104, 1998.

Rieke F, Warland D, Ruyter van Steveninck RR, and Bialek W. *Spikes: Exploring the Neural Code*. Cambridge, MA: MIT Press, 1997.

Tolhurst DJ, Movshon JA, and Dean AF. The statistical reliability of signals in single neurons in cat and monkey visual cortex. *Vision Res* 23:775–785, 1983.
Thibos L, Levick W, and Cohn T. Receiver operating characteristic curves for Poisson signals. *Biol Cybern* 33:57–61, 1979.
Troy JB and Robson JG. Steady discharges of X and Y retinal ganglion cells of cat under photopic illuminance. *Vis Neurosci* 9:535–553, 1992.
van Rossum MC. A novel spike distance. *Neural Comput* 13:751–763, 2001.
Victor JD. Binless strategies for estimation of information from neural data. *Phys Rev E* 66: 51903, 2002.
Victor JD. Approaches to information-theoretic analysis of neural activity. *Biol Theory* 1: 302–316, 2006.
Victor JD and Purpura KP. Nature and precision of temporal coding in visual cortex: A metric-space analysis. *J Neurophysiol* 76:1310–1326, 1996.
Victor JD and Purpura KP. Metric-space analysis of spike trains: Theory, algorithms and application. *Network* 8:127–164, 1997.
Werner G and Mountcastle VB. Neural activity in mechanoreceptive cutaneous afferents: Stimulus-response relations, Weber functions, and information transmission. *J Neurophysiol* 28:359–397, 1965.

2 Neural Coding and Decoding with Spike Times

Ran Rubin, Robert Gütig, and Haim Sompolinsky

CONTENTS

2.1 INTRODUCTION

The question of how neurons encode, store, and process information has challenged systems neuroscience for more than five decades. Approaches to neuronal coding have long been dominated by the rate coding hypothesis, which stipulates that neurons encode information by the number of action potentials generated in response to a stimulus. Indeed, neuronal firing rates have been used as the central measure for characterizing neuronal responses and much of our intuition of how neurons integrate information arriving from afferent populations is based on these measures. However, the observed fine temporal modulation of the neuronal firing in several systems has motivated the study of alternative schemes of neuronal codes, such as synfire chains (Abeles, 1991; Diesmann et al., 1999) that exploit spike synchrony and rank order decoders (Thorpe and Gautrais, 1998) that are based on spike latencies. Nevertheless, the computational capacity of these schemes remained unknown, and more generally it was unclear to what extent neurons are capable of decoding information that is encoded in the relative timing of incoming spikes. To address this question, the Tempotron has been devised as a generic model of spike-timing-based information processing by single neurons. The Tempotron modifies its synaptic efficacies to carry out peform appropriate classification of incoming spatio temporal patterns of spikes. Using this model, we have shown that a simple biologically plausible form of post synaptic integration suffices to decode a broad range of spike-timing-based neuronal codes with high capacity. In addition, the model clarifies the nature of synaptic learning rules that would allow neurons to realize such capabilities.

The goal of this chapter is to review recent studies that explored spike-timing-based computation. These works address several issues: models of encoding information in spike times, estimating the capacity of information embedded in spike times, mechanisms of decoding this information, and models of spike-time-based learning. In Section 2.2, we review some of the main models of neuronal codes and their proposed decoding schemes. In Sections 2.3 and 2.4 we present the Tempotron classification model and characterize its computational properties, highlighting the qualitative difference between rate-based and spike-timing-based processing. The Tempotron learning rule and its possible biological implementation are described in Section 2.5. Section 2.6 extends the model, incorporating synaptic inputs as conductance changes. This section highlights the computational role of synaptic conductances in processing inputs with variable timescales (the "time-warp" problem). We summarize the results and discuss some of the outstanding issues in Section 2.7.

2.2 NEURAL CODES

2.2.1 Biological Context

In most of the vertebrate nervous system, neurons communicate by action potentials, which carry information in their time of arrival. In many systems, changes in sensory stimuli as well as in behavioral and internal states correlate well with modulation in the number of spikes generated in a coarse time window. However, studies of the visual, auditory, olfactory, and somatosensory pathways have revealed precise timing relationships in neuronal firing patterns elicited by sensory stimuli (Carr and Konishi, 1990; Meister et al., 1995; deCharms and Merzenich, 1996; Wehr and Laurent, 1996; Johansson and Birznieks, 2004), suggesting that an important component of stimulus information could be encoded in the timing of individual spikes (Victor, 2000).

Visual processing provides an important biologically relevant example. In humans and most other animals, vision occurs in discrete episodes where the eye is relatively still, interrupted by rapid gaze shifts. During such a saccade, the visual image sweeps rapidly over the retina, and several retinal ganglion cell types are strongly suppressed (Roska and Werblin, 2003). After the image comes to rest, many ganglion cells fire a burst of spikes and these bursts comprise all retinal information about the new image (Noda, 1975; Greschner et al., 2006; Segev et al., 2007). For certain ganglion cells, the timing of the very first spike at the onset of a new image reveals much of the image information (Gollisch and Meister, 2008). This suggests that subsequent brain circuits could indeed begin processing visual information very rapidly, using the spikes generated at the onset of fixation.

Recent experimental evidence has shown that neuronal firing in the cortex can be very sparse and that a small number of action potential generated by some neurons can exert a powerful influence in downstream networks (Wolfe et al., 2010). For example, many neurons in the auditory cortex fire only one or zero spikes in response to a tone (Hromádka and Zador, 2009). When the spike occurs, it is locked in time to the stimulus. Whisker deflections are encoded by a small number of coincident spikes in the somatosensory cortex (Jadhav et al., 2009). A recent study has found that the vast majority of neurons in the vibrissa cortex fire at frequencies of a few Hertz or below (O'Connor et al., 2010). Sparse discharge rates suggest that the system may carry information not only in spike counts but also in the time of occurrence.

2.2.2 Computational Context

The use of spike timing in the neural code has a clear computational advantage, in that it offers the possibility of fast computation on the basis of relatively small number of spikes. For example, the human visual system can make high-level decisions about images just 100 ms after light strikes the retina (Thorpe et al., 1996; Liu et al., 2002). Taking into account retinal delays and the number of synapses in the cortical pathways, it is likely that such decisions rely on just one or at most a few spikes per afferent. In contrast, rate coding requires averaging over time or over a large population of neurons, which limits the temporal and spatial bandwidth of the underlying computation.

Spike-time-based computation faces several challenges. Reliance on small number of event times is potentially sensitive to jitter in these times. Spike times are determined by complex biophysical nonlinear mechanisms. Many of the existing models of learning and computation rely on linear operations or on smooth nonlinearities, such as principal component analysis, Perceptrons, linear support vector machines, and multilayer sigmoidal Perceptrons, all of which are hard to implement in the strongly nonlinear framework of spike timing. Thus, an important challenge is to understand how information encoded in spike times can be efficiently decoded by downstream systems, and how the synaptic weights of these "readout" circuits can be modified through learning to be able to implement specific tasks.

Considerable work has been done on unsupervised spike-time-based learning, primarily in the context of spike-timing-dependent-plasticity (STDP) (Kempter et al., 1999; Song et al., 2000; Gutig et al., 2003; Guyonneau et al., 2005; Legenstein et al.,2005; Worgotter and Porr, 2005). However, it is difficult to gauge the general computational capacity that can be achieved through unsupervised learning such as STDP. Here, we will focus on supervised computation and learning, in which the underlying circuit has to implement certain input–output functions. The supervised framework provides a more concrete setting for elucidating the capacity and limitation of spike-timing-based computation.

2.2.3 Neural Coding and Decoding

The neural code is any feature of the neuronal firing that covaries with changes in the stimulus but is robust and reproducible from trial to trial. Such features can be associated with the firing pattern of a single neuron or more generally with the spatiotemporal spiking response of a population of neurons. Neural decoding is a neuronal mechanism that reads out the neural code. That is, neural decoding extracts information about the stimulus from the relevant features of the firing response of the neuronal population. In general, the neural code is fully specified by $P(T_N|s)$, where T_N is the spike time sequence of N neurons in the relevant time window, s is the stimulus, and the conditional probability takes into account the trial-to-trial variability in the neuronal responses. The information content of the code can be assessed using Shannon's mutual information, maximum likelihood or Bayesian frameworks. These methods can also be used to construct "optimal" readouts. However, in general, these readouts are too complex to be realized by simple neuronal circuits, motivating the search for specific models of coding and decoding that are plausibly implemented by brain systems. Several examples of simple neuronal codes are presented in Figure 2.1.

2.2.4 Rate Codes

Perhaps the simplest form of neuronal code is the rate code, which has been used to describe the stimulus coding in a plethora of neuronal systems (e.g., orientation tuning in the primary visual cortex, frequency tuning in the primary auditory cortex, and place cells in the rat hippocampus). In a static rate code, some features of the stimulus are represented in the number of spikes emitted by the neuron in response to the stimulus. The *timing* of spikes within a trial is considered irrelevant for the

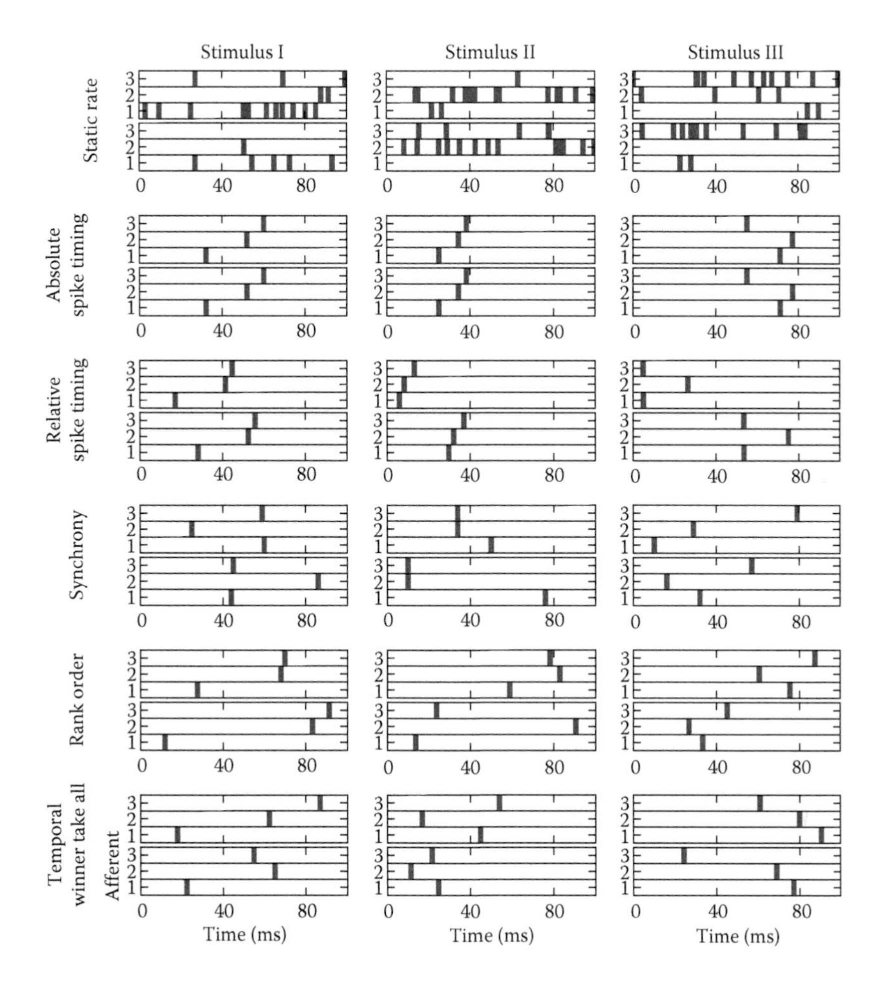

FIGURE 2.1 Examples of neural codes. Spike patterns of three neurons responding to three stimuli (columns) for six different neural codes (rows). In each row, two neural responses to each stimulus are shown to demonstrate trial-to-trial variability. In all models, time $t = 0$ is locked to the onset of the stimulus. *Rate code*: Information about the stimulus is encoded in the firing rates of the neurons. In this example, each neuron is selective to one stimulus and responds to it by firing with an elevated rate while the other two neurons respond with a low-baseline firing rate (neurons 1, 2, and 3 are selective to stimuli I, II, and III, respectively). Note the high variability due to the Poisson statistics used to generate the spikes. *Absolute spike timing*: Information is encoded in the precise spike timing of the neurons. For clarity, no trial-to-trial variability is introduced although in realistic situations, some amount of jitter in the spike timing will exist and define the "temporal resolution" of the code. *Relative spike timing*: Information is encoded only in ISIs and not in the absolute timing and the neuronal response. *Synchrony*: The identity of the synchronous events determines the stimuli identity: neurons 1 and 3 for stimulus I, neurons 2 and 3 for stimulus II, and no synchronous event for stimulus III. *Rank order*: Only the spike firing order carries information about the stimulus. Stimuli I, II, and III are characterized by firing orders (1,2,3), (1,3,2), and (2,1,3), respectively. *Temporal winner take all*: Information is in the identity of the first spiking afferent. The spike firing order of other afferents may vary.

neuronal code. Trial-to-trial variability in the spike count in response to the same stimulus condition is considered as noise, which limits the information content of the rate code. Trial-to-trial variations in spike counts and spike timing are often described by a Poisson statistics.

Within the context of rate codes, temporally varying stimuli can be encoded via modulation of the underlying firing rate of the neuron. This is usually formalized in the framework of inhomogeneous Poisson process (and its extensions) where the stimulus-dependent firing rate determines the underlying rate function of the process. As noted above, a reliable representation of the stimulus in behaviorally relevant time scales, by rate modulation, requires high firing rates in the case of single neurons or averaging over a large population of neurons.

2.2.5 Latency Codes

In a latency code, information is carried in the time of the response onset, for example, the timing of the first spike of individual neurons relative to some external temporal cue such as stimulus onset, or relative to other neurons' spike times. Latency codes are insensitive to trial-to-trial variability in the number or timing of subsequent spikes. In the case of latencies relative to a reference neuronal firing, the code is also invariant to the absolute timing of the population spike responses. These codes are sensitive to noise in the form of jitter in individual spike times as well as to a non zero probability of spike failure and spurious spikes.

2.2.6 Rank Order Codes

Another form of a neural code considers the temporal order of the first spikes of individual neurons (Thorpe and Gautrais, 1998), which in principle can code $N!$ stimulus values for a population of N neurons. Rank order codes are insensitive to the variation in the latency times as long as their order is preserved, or to the occurrence of multiple spikes per neurons. A special case of rank order code is the temporal winner take all (tWTA) model (Shamir, 2009). Here, different stimuli are represented in a labeled line code, where the label is the identity of the neuron that is the first to spike. In tWTA code, N possible stimuli can be coded with N neurons. Shamir (2009) has studied the performance of such coding as well as an n-tWTA code that determines the stimulus by the identity of the group of cells firing the first n spikes. Shamir (2009) showed that in various conditions, fast and accurate coding of the stimulus can be achieved even in the presence of noise on spike times.

2.2.7 Other Spatiotemporal Codes

Spike times can carry information in a variety of additional ways. Interspike interval (ISI) codes embed information in the interval between the first two spikes or alternatively in the aggregate of ISIs of individual neurons. Synchrony code modulates the patterns of spike time synchrony within a population of neurons, for instance, a stimulus may result in synchronous firing of pairs of neurons, while in the "null" state, the firing times of pairs of neurons is uncorrelated.

2.2.8 Models of Decoding of Spatiotemporal Spike Codes

Several neural mechanisms for decoding rate codes for various tasks have been studied extensively. Simple models of rate-based stimulus estimation are the well-known population vector and more general linear readouts, which carry out a weighted sum of the input rates via a single layer of synapses (Seung and Sompolinsky, 1993; Salinas and Abbott, 1994). For binary decision tasks, simple readout models are the majority rule (Shadlen et al., 1996) and the more general Perceptron (Rosenblatt, 1962; Minsky and Papert, 1988), which thresholds a weighted sum of the input rates (Seung and Sompolinsky, 1993). Other readout models include recurrent networks (Deneve et al., 1999) and nonlinear synaptic gating (Burak et al., 2010). Some of these mechanisms exhibit close-to-optimal performance, given a Poisson statistics of firing.

Decoding of synchrony code can be carried out by neuronal coincidence detectors. This can be realized by a leaky integrate-and-fire (IF) model with a short integration time constant, as described in Section 2.3.

Thorpe (Thorpe and Gautrais, 1998) suggested a model for the decoding of rank order code, in which the synapse of a presynaptic neuron is suppressed multiplicatively by the previous firing of other neurons in the afferent population. In this way, the efficacy of a spike from a given neuron on a postsynaptic readout neuron increases with the spike's rank within the arrival order of all spikes. Different rank orders can be distinguished by an appropriate choice of the readout baseline synaptic weights. To implement such a model, the neurons must possess a mechanism for the multiplicative suppression as well as a mechanism to cancel the effect of variations in the spike timing of the inputs within a given rank order. Decoding of tWTA can be implemented by a layer of neurons with strong, all-to-all lateral inhibition (as suggested in Shamir, 2009), such that the firing of one of the readout neurons will silence the others.

A complementary approach to the problem of processing of information embedded in spike times is to explore the decoding power of a generic single-layer network composed of a single spiking neuron receiving inputs from many afferents via a layer of linearly summating synapses. A first step in this direction is the Tempotron model, proposed by Gütig and Sompolinsky (Gütig and Sompolinsky, 2006, 2009; Rubin et al., 2010), which is a leaky IF neuron whose firing and quiescence represents a binary decision. Remarkably, Gütig and Sompolinsky found that the underlying threshold nonlinearity, together with the linear spatiotemporal filter specified by the layer of input synapses, endows the system with powerful capabilities for decoding a broad range of spatiotemporal spike time codes. Furthermore, a relatively simple online learning algorithm allows the Tempotron to learn to carry out the classification of its input patterns, via error-dependent synaptic plasticity rule. This work will be the focus of the following sections.

2.3 TEMPOTRON CLASSIFIER MODEL

2.3.1 Neuron's Input–Output Model

The Tempotron classifier model consists of a leaky IF neuron, with N input synapses of strength ω_i, $i = 1, \ldots, N$, which can be positive or negative. Each input pattern is represented by N sequences of spikes, where the spike timings for the afferent i are denoted by the set $\{t_i\}$. Synaptic currents are modeled as a single exponential with synaptic decay time τ_s. They are filtered by the membrane time constant τ_m,* resulting in a postsynaptic potential

$$U(t) = \sum_{i=1}^{N} \omega_i \sum_{t_i<t} u(t - t_i) \tag{2.1}$$

where $u(t)$ denotes a fixed causal temporal kernel, given by

$$u(t) = u_0 \left[\exp\left(-\frac{t}{\tau_m}\right) - \exp\left(-\frac{t}{\tau_s}\right) \right] \quad t > 0 \tag{2.2}$$

and zero for $t < 0$. An output spike is generated whenever $U(t)$ crosses the threshold, U_{thr}, from below.

The Tempotron carries out a binary classification of its input patterns. When presented with a pattern, the Tempotron classifies it as a "target" pattern by firing one or more output spikes and as a "null" pattern by remaining quiescent. Figure 2.2

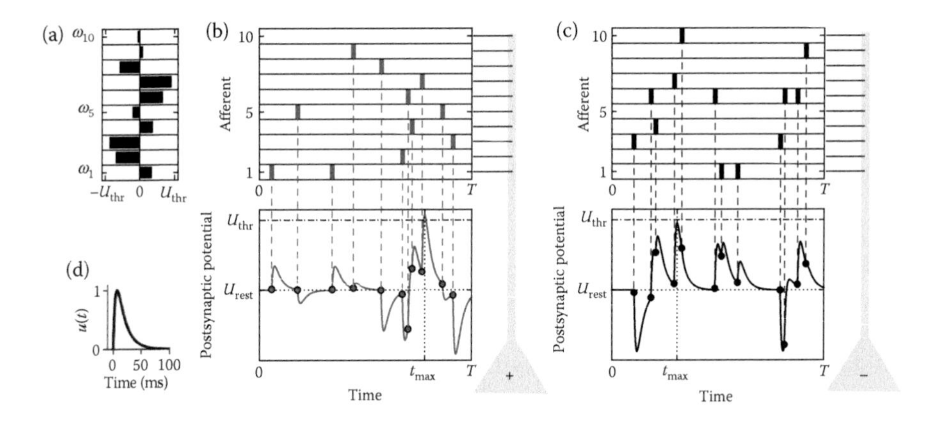

FIGURE 2.2 Tempotron classification. (a) A bar chart of synaptic weights of a Tempotron with 10 input afferents. (b) and (c) show the Tempotron's response to two input spike patterns. The top panels depict the spike arrival times (between 0 and $T = 500$ ms) of the two input patterns. The bottom panels depict the postsynaptic potential induced by each pattern. The Tempotron integrates the incoming spikes with the weights presented in (a) and the postsynaptic kernel presented in (d). The circles on the voltage traces mark the timing of the input spikes. The time of the maximal postsynaptic potential is denoted by t_{max}. In (b), the postsynaptic potential at t_{max} is above U_{thr} representing the classification of inputs as "target." In (c), the potential remains below the firing threshold, corresponding to classification of the pattern as "null." (d) Postsynaptic integration kernel $u(t)$ with membrane time constant $\tau_m = 15$ ms and synaptic time $\tau_s = \tau_m/4$.

* In all the numerical results presented here, we have used $\tau_s = \tau_m/4$.

illustrates the response of the Tempotron to two patterns, gray and black, classified as "target" and "null," respectively.

The simple model described above marks the Tempotron as a spike-timing-based analog of the rate-based Perceptron, which also carries out a binary decision by performing a linear threshold operation on a weighted sum of its static inputs (Rosenblatt, 1962; Minsky and Papert, 1988). Thus, the comparison of the two models would be fruitful in highlighting the qualitative and quantitative differences between rate-based and spike-timing-based computations.

2.3.2 Nonlinear Decision Surfaces: Two- and Three-Dimensional Examples

A useful insight into the decoding capabilities of the Tempotron can be gained by considering the case of two-dimensional (2D) inputs, that is, $N=2$. The left column of Figure 2.3 shows two simple implementations of a rank order code by the Tempotron, one with two excitatory inputs (middle panel) and one with an excitatory and an inhibitory input (bottom panel). Two input patterns ("blue" and "red") each consisting of a pair of spikes are superimposed in the left top panel. The time difference between the spikes of the two afferents is the same for both blue and red patterns, but the spike order is different. In both implementations, the Tempotron fires when the spike from

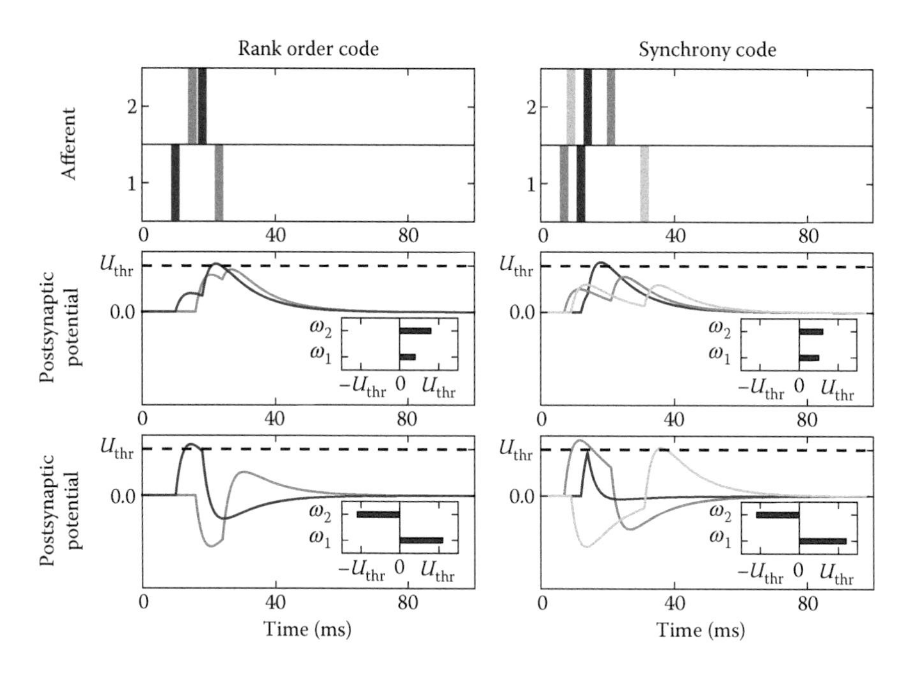

FIGURE 2.3 **(See color insert.)** Implementation of rank order (left column) and synchrony (right column) codes by the Tempotron. The input patterns are presented in different colors in the top row and consist of a single spike arriving from each of the two input afferents with a certain time difference. The bottom two rows depict the postsynaptic potential traces evoked by the different input patterns. Insets depict the synaptic efficacies of the two input afferents.

afferent 1 arrives first ("blue") and does not fire when it arrives last ("red"). This simple example demonstrates the counter intuitive sensitivity of the Tempotron's firing to the *order* of the arrival of two inputs with different synaptic efficacies, even if both inputs are excitatory. The right column of Figure 2.3 illustrates the ability of the Tempotron to selectively fire for a synchronous input ("blue") or for nonsynchronous ones ("red"). Selectivity for the synchronous input (middle panel) is implemented with two excitatory synapses and selectivity for the nonsynchronous input (bottom panel) is implemented by a combination of excitatory and inhibitory synapses.

The general classification capabilities of the Tempotron with two inputs, each with a single spike, can be described as a simple function of the two synaptic weights. Since the Tempotron is insensitive to the absolute timing of the input spikes, all possible input patterns can be described by time difference between the two input spikes

$$\Delta t_{12} = t_1 - t_2, \tag{2.3}$$

which can be either positive or negative. Only four classifications of the input are realizable by the Tempotron. The Tempotron can either classify the input patterns as "target" patterns for all Δt_{12}, as "null" patterns for all Δt_{12}, or, more interestingly, it can segment the Δt_{12} line into three segments with either target, null, target classification, (+ − +) or null, target, null classification (− + −). The location of the decision boundaries on the Δt_{12} axis depends on both the synaptic weights and the neuron's

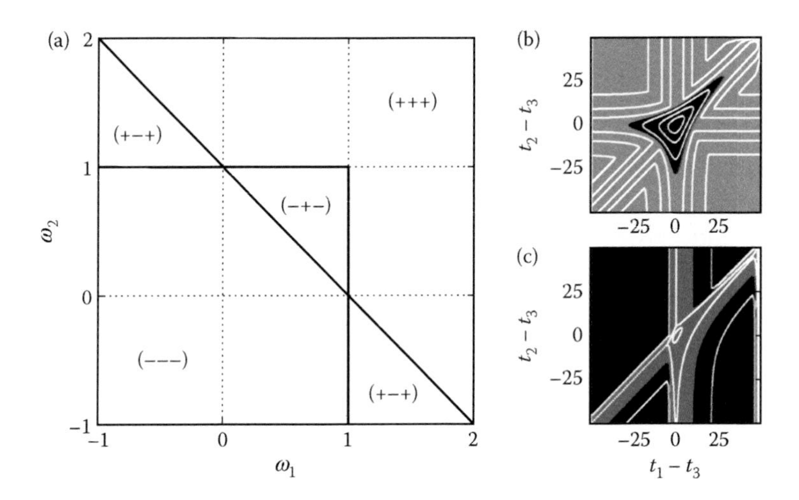

FIGURE 2.4 (a) Classification-type diagram for a Tempotron with two inputs. Each region in the ω_1, ω_2 space is marked with one classification type. (b) and (c) depict examples of possible classifications for a Tempotron with three inputs. The input space is now two dimensional, $(t_1 - t_3)$, $(t_2 - t_3)$, and each set of weights induces a specific partitioning of the input space into regions classified as "target" (marked in black) and regions classified as "null" (marked in gray). The lines depict curves of constant maximal postsynaptic potential. (b) The classification induced by $\omega_1 = \omega_2 = \omega_3 = 0.48$. (c) The classification induced by $\omega_1 = -0.18$, $\omega_2 = 1.08$, and $\omega_3 = -0.18$. Synaptic weights are given in units of the neuron's threshold, U_{thr}.

time constants. Note that these decision boundaries translate into two straight lines in the 2D t_1, t_2 input space. The type of classification is solely determined by the synaptic weights and divides the 2D ω_1, ω_2 weight space into four regions corresponding to the four possible classifications (Figure 2.4a).

The decision surface of the Tempotron with three inputs can be much more complicated than the 2D case, as illustrated by the examples of Figure 2.4b and c. Comparing these surfaces to the linear decision surface of the Perceptron illustrates the role of the Tempotron's integration time. Although, at any given time, the decision to fire or not is a simple linear threshold decision, the overall decision of the Tempotron to fire involves an OR operation on the potentials at all times. This can also be formalized as a MAX operation, that is, the decision of the Tempotron is based on whether the maximal postsynaptic the potential, $U(t_{\max})$, is above or below threshold, where $t_{\max} = \text{argmax}_t\, U(t)$.

2.4 TEMPOTRON PERFORMANCE IN LATENCY CODE

Although the neural codes described above (Section 2.2) and their implementation by the Tempotron with a small number of synapses (Section 2.3) are very intuitive, the possible coding and decoding schemes that can be implemented by a Tempotron with a large number of synapses (as is the case in the brain) can be considerably more complicated and less tractable. The ultimate test of any candidate model of neural information processing would be its performance on interesting challenging natural tasks that the nervous system learns. However, to assess the computational capacity of the model and to compare it to other models, it is useful to quantify the performance of the model on a set of generic high-dimensional synthetic benchmark tasks. Random input–output tasks have often been used as generic tasks. This is justified since in many high-dimensional problems, the distribution of performance is concentrated around the average (Gardner, 1987; Monasson and O'Kane, 1994; Seung et al., 1992; Amit et al., 1985). An additional important advantage of random sets of tasks is that in many cases, performance can be assessed by an analytical theory in the asymptotic limit of large N.

We describe here a similar approach in the case of the Tempotron. The capacity of the Tempotron to decode spike time information has been studied systematically using random latency tasks as a benchmark (Gütig and Sompolinsky, 2006; Rubin et al., 2010). Consider a set of $P = \alpha N$ random spike-timing patterns, of duration T, divided into two classes. For each pattern, the times of the input spikes from each input neuron are randomly chosen from independent Poisson processes with rate $1/T$. For each pattern, the desired output, $y = \pm 1$, is independently chosen with equal probability. A solution to the classification problem is a set of synaptic weights $\{\omega_i\}$ such that the Tempotron classifies patterns labeled with $y = +1$ as "target" patterns and classifies patterns labeled with $y = -1$ as "null" patterns. We are interested in the statistical properties of the solutions, such as the distribution of synaptic weights, the distribution of the neuron's synaptic potential, and the classification capacity of the Tempotron. The latter is defined as the maximal number of random patterns per synapse, α_c, which can be correctly classified with high probability. These can be studied numerically using, for example, the Tempotron learning algorithm, discussed in

Section 2.5, and by analytical methods based on statistical mechanics and extreme value theory (EVT) of Gaussian processes (Leadbetter et al., 1983).

Timescales: Before describing the performance of the Tempotron on random latency tasks, we note that this performance depends on the time constants, τ_m and τ_s, relative to the input firing rates, which is equal to $1/T$. The properties of the Tempotron can be most easily understood, when $N \gg T/\max\{\tau_m, \tau_s\} \gg 1$. In this regime, the theory predicts that the system's properties depend on the dimensionless pattern duration time

$$K = \frac{T}{\sqrt{\tau_s \tau_m}}. \tag{2.4}$$

2.4.1 Statistics of Postsynaptic Potential

We first describe the statistics of the solution in the regime $\alpha < \alpha_c$. The Tempotron's decision is based on the maximal postsynaptic potential occurring in each trial, U_{max}. Therefore, the distribution of U_{max} will be concentrated around the neuron's threshold and the average postsynaptic potential will lie below it. To understand the typical distribution of potentials generated by the different learned patterns, we note a few observations from the statistical mechanical theory of the Tempotron. First, the individual synaptic weights associated with a typical solution of the random latency classification tasks are of the order $1/\sqrt{N}$, whereas their spatial mean is only of the order $1/N$. Thus, for this task, the Tempotron works in the balanced regime where the excitation and inhibition nearly cancel each other. In addition, in large N, the spatial mean of the weights and the norm of the weight vector are not uniquely specified by the task, but are related linearly. Choosing the mean of the weights to be zero (which fixes the weight vector norm) results in a distribution of the potential $U(t)$ after learning (defined by sampling over different times and different patterns), which is approximately Gaussian with zero mean. The standard deviation of the potential's distribution is proportional to U_{thr} in a manner that is consistent with the required output statistics of the Tempotron, that is, that the probability of the maximal potential to cross the threshold is equal to the fraction of patterns labeled with $y = +1$ (Figure 2.5a). The full trace of the potential is approximately a temporally correlated Gaussian with a correlation function that is determined by the postsynaptic temporal kernels.

The Gaussian nature of the postsynaptic potential implies that its maximum (over time) has approximately a Gumbel-distribution shape (Figure 2.5a). Note that by the definition of the task, the median of the distribution of U_{max} of learned patterns is at the neuron's threshold. Furthermore, the standard deviation of U_{max} is small compared to the threshold in the large K limit and scales according to

$$\mathrm{std}(U_{max}) \propto \frac{U_{thr}}{\ln K}, \tag{2.5}$$

which is narrower than the typical fluctuations in $U(t)$ by a factor $1/\ln K$. This decrease of the fluctuations in U_{max} results in an increase in the Tempotron's capacity

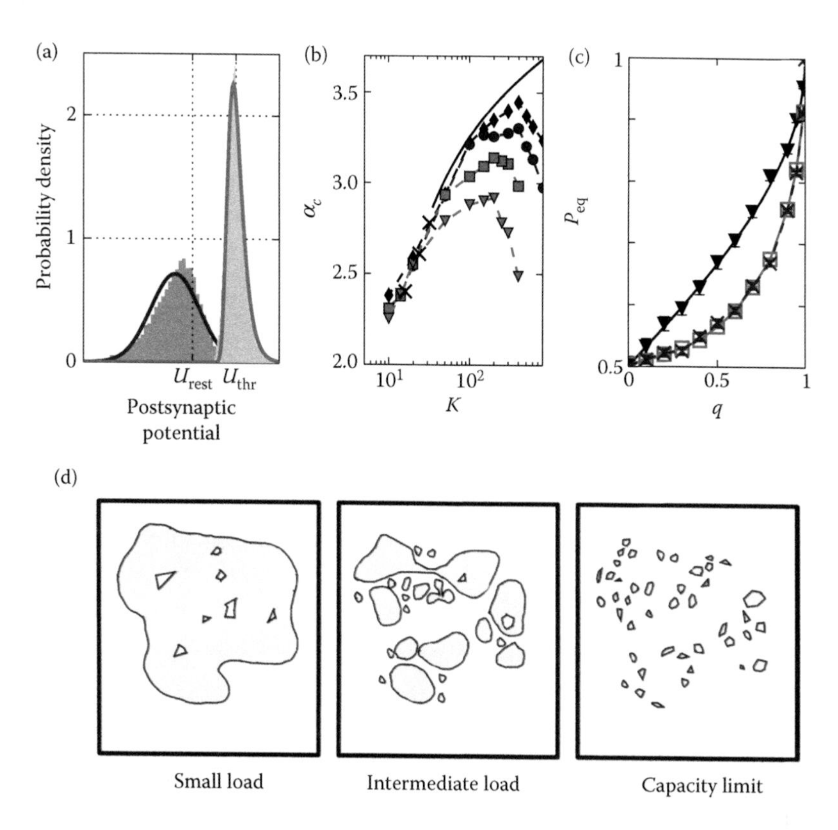

FIGURE 2.5 **(See color insert.)** (a) Probability density of $U(t)$ (blue) and probability density of U_{max} (red). The lines correspond to a Gaussian with the mean and variance of the postsynaptic potential, $U(t)$, and a Gumbel law with median at U_{thr} and a scale parameter that was fitted to the data. The data were measured with $K = 400$, $\alpha = 1.68$, $N = 500$, and 34 samples. (b) Capacity α_c of the Tempotron versus K. Lines with symbols show results of the Tempotron learning algorithm for networks of different size ($N = 250$, triangles; $N = 500$, squares; $N = 1000$, circles; $N = 2000$, diamonds). The solid line and × symbols depict fitted theoretical predictions with one free parameter. (c) Probability that two neurons will classify a random pattern in the same manner, P_{equal}, versus the correlation coefficient between their weight vectors, q, for the Perceptron (theory and simulations in black), the Tempotron (blue × symbols, $K = 100$), and the Hodgkin–Huxley (red squares, input pattern duration of $T = 1.5$ s) models. See Rubin et al. (2010) for details of the Hodgkin–Huxley model. (d) The evolution of solution space with increasing load. Each square represents the space of all weight vectors and every point represents one N-dimensional vector. Gray regions represent solutions to the classification task. The solution space is connected only for a small number of patterns, and breaks to many clusters as the load increases. At the capacity limit, the solution space consists of many small clusters spread evenly throughout the weight space.

with K (see below). Another interesting statistical feature is the distribution of the number of threshold crossings in a single pattern of duration T, N_{spikes} In this large K regime, this number obeys a Poisson distribution with a mean rate $r = \ln 2/T$, which is consistent with a 1/2 probability of firing within time T (see Figure 1 in Rubin et al., 2010).

2.4.2 Capacity

EVT and statistical mechanics tools allow us to evaluate the capacity of the Tempotron for the random patterns classification task. In the $K \gg 1$ limit, the leading order term for the capacity, α_c, is given by (Rubin et al., 2010)

$$\alpha_c = \frac{\ln\ln K}{2\ln 2} \tag{2.6}$$

Hence, the capacity of the Tempotron is not bounded as K increases, and may exceed the capacity ($\alpha_c = 2$) of the Perceptron model (Gardner, 1987). Note that for a fixed finite N and K of $O(N)$, the few input spikes that arrive within a single decision time window, T/K, do not carry sufficient information to classify the patterns. We therefore expect that for any fixed N, α_c is a nonmonotonic function of K while the value of K that maximizes the capacity increases with N, as implied by Equation 2.6. These results are corroborated by numerical simulations in Figure 2.5b. Note that even in large K, the predicted increase in α_c is rather weak, and for a broad range of K, its value is around 3.

2.4.3 Geometry of the Solution Space

Below capacity, the solution to the classification task is not unique. The Tempotron is free to choose the timing of its output spike for patterns labeled with $y = +1$ and, therefore, many different solutions exist to the same classification problem, each implementing the required classification with different timings of the output spikes. The statistical mechanics calculation provides us with a qualitative description of the evolution of the solution space with increasing load, α. For a very small number of patterns, $\alpha < 1/\ln K$, the solutions space is connected. Above a critical load, the solutions space breaks into many different clusters that continue to divide, shrink, and disappear as the load increases. Finally close to the capacity limit, the solution space consists of many small clusters spread evenly throughout the entire weight space. The typical correlation coefficient (or overlap) of a pair of solutions inside a single cluster, q, approaches 1 in the $K \to \infty$ limit and scales according to (Rubin et al., 2010)

$$1 - q \simeq O\left(\frac{1}{\ln K}\right) \tag{2.7}$$

Thus, the typical size of these clusters diminishes as K increases. Furthermore, these clusters abruptly disappear at the capacity limit. On the other hand, the typical correlation coefficient of solution pairs belonging to different clusters vanish in the large K limit. The illustration in Figure 2.5d demonstrates this qualitative picture. The geometrical structure of the solution space of the spike-time-based Tempotron is dramatically different from that of the rate-based Perceptron. The solution space of the Perceptron consists of one convex domain that shrinks gradually as the load increases and reduces to a single point at the capacity limit.

The fractured nature of the Tempotron's solution space also highlights another important difference between the Tempotron and the Perceptron. In the case of the Perceptron, a pair of solutions to a specific classification problem have a significant overlap, and in addition, a pair of Perceptrons with correlated synaptic weights will yield correlated classification for any given set of inputs. In contrast and somewhat counterintuitively, the above results for the Tempotron imply that even close to capacity, IF neurons with very different synaptic weights can carry out exactly the same classification, whereas IF neurons with significant degree of similarity in their weight vectors will typically fail to solve the same task. This behavior can be traced to the balanced nature of a typical solution, which implies that the decisions depend on the precise cancellations of excitatory and inhibitory inputs induced by each input pattern.

To show this property, we measured the probability of a pair of neurons with random synaptic weights to classify a given random pattern in the same manner, P_{equal}, as a function of the correlation coefficient between the two weight vectors, q (Figure 2.5c). The standard deviation and mean of the synaptic weights were chosen to ensure that both neurons have an overall 1/2 probability of spiking for a random pattern; thus, we expect that $P_{\text{equal}} = 1/2$ for completely independent weight vectors ($q = 0$) and that $P_{\text{equal}} = 1$ for identical weights ($q = 1$). As expected, two Tempotrons are likely to agree on their classifications of a random pattern only if the correlation coefficient between their synaptic weights is close to 1. We also present the simulation results for the Hodgkin–Huxley neuron, a classical biophysical model for spike generation. Interestingly, despite its nonlinear spike generation dynamics, the classification pattern of a pair of Hodgkin–Huxley neurons is similar to that of the Tempotron, indicating that this behavior does not depend on the details of the spike generation but on the summation of input spikes within temporal windows. In contrast, in the case of the Perceptron, which lacks temporal windows, the probability that two weight vectors agree on their classification increases roughly linearly with their correlation coefficient, q (Figure 2.5c).

It is interesting to note the performance–stability trade-off that exists due to (2.7) and the capacity (2.6). Even though the capacity monotonically increases with K, the stability of the solution to perturbations in the synaptic weights decreases. Thus, we expect that under realistic conditions where jitter in the synaptic weights is present, the system will function robustly only when the duration of input patterns (relevant to the decision) is not too long relative to the neuron's integration time.

Despite the simplicity of its architecture and dynamics, the nonlinearity of the Tempotron decision rule yields a rather complex structure of the solution space. Importantly, the Tempotron is not constrained to fire at a given time in response to a "target" pattern. Thus, by adjusting the timing of its output spikes, the Tempotron can choose the spatiotemporal features that will trigger its firing for each "target" pattern, enabling it to surpass the performance of its rate-based analog, the Perceptron, and of other Perceptron-based models for learning temporal sequences (Bressloff and Taylor, 1992) that specify the desired times of the output spikes. Finally we suggest that Tempotrons with similar synaptic weights carry out very different decisions on the same set of inputs is reminiscent of the fact that responses of cortical neurons to complex stimuli (e.g., natural signals) are often poorly predicted by their receptive field structure (see, e.g., Bar-Yosef et al., 2002).

2.5 TEMPOTRON LEARNING

2.5.1 Tempotron Learning Algorithm

Given a batch of P patterns of N spike trains, with their desired classification $y^\mu = \pm 1$, $\mu = 1 \ldots P$, how should we modify the weights to yield a network that carries out the desired task? The difficulty lies in the fact that rewarding (for successful trials) or punishing (for erroneous trials) all the active synapses will in general fail to generate an appropriate set of weights. Thus, for the synaptic modification rule to be successful, it needs to identify for each pattern, which afferents contributed the most to the success or failure of the output neuron. This problem is known as *temporal credit assignment* problem. In fact, as will be shown below, this problem is resolved by a learning rule that is based on the temporal nature of the Tempotron's dynamics.

To determine the required weight modification in each trial, we build on one of the most widely used methods of deriving synaptic learning rules, namely, gradient-based learning. In this method, an appropriate cost function, which measures the performance of the system for each set of weights for a given task, is defined. The learning rule consists of changing the weights such that they carry out gradient descent in the defined cost function. In our case, the binary classification error is not an appropriate cost function for gradient-based learning. Instead, Gütig and Sompolinsky (2006) propose to use the distance between the maximal potential and the threshold as the cost function, $E = \sum_{\mu=1}^{P} E_\mu$, where the contribution to the cost from each pattern is

$$E_\mu = \left[y^\mu \left(U_{\text{thr}} - U_\mu \left(t_{\max}^\mu \right) \right) \right]_+ . \tag{2.8}$$

Here, $[x]_+ = x$ for $x > 0$ and zero otherwise. $U_\mu(t)$ is the postsynaptic potential induced by the μth input pattern (Equation 2.1) and $t_{\max}^\mu = \text{argmax}_t\, U_\mu(t)$ is the time of the global maximum of postsynaptic potential for pattern μ. Note that $E_\mu \geq 0$ and that $E_\mu = 0$ if and only if the classification error of this example is zero. Thus, given that a solution exists, minimizing E is equivalent to finding a solution to the binary classification problem.

In principle, one might compute the gradient of E and change the weight vector in small steps in the direction opposite to that of the gradient. This batch update is very costly computationally and is not biologically attractive. Instead, we use the gradient in an online mode.

Using Equations 2.1 and 2.8 for each misclassified pattern, we have

$$\frac{dE_\mu}{d\omega_i} = -y^\mu \left[\sum_{t_i^\mu < t_{\max}^\mu} u\left(t_{\max}^\mu - t_i^\mu \right) + \frac{\partial U_\mu \left(t_{\max}^\mu \right)}{\partial t} \frac{\partial t_{\max}^\mu}{\partial \omega_i} \right] . \tag{2.9}$$

The first term on the right-hand side of Equation 2.9 is simply the contribution of input from the ith afferent to the total depolarization at time $t_{\max}^\mu$. The second term vanishes since $U_\mu(t^\mu_{max})$ is an extremum point of $U_\mu(t)$.* Therefore, the Tempotron learning algorithm consists of updating the weights by $\omega_i^{new} = \omega_i^{old} + \Delta\omega_i$, with

$$\Delta\omega_i = \eta y^\mu \sum_{t_i^\mu < t_{max}^\mu} u\left(t^\mu_{max} - t_i^\mu\right), \tag{2.10}$$

where η is some positive learning rate.† Patterns are iterated until the error vanishes or until some stopping criteria is achieved. Figure 2.6 demonstrates the application of the learning rule. As seen, the Tempotron resolves the *temporal credit assignment* problem by updating only weights that were active prior and proximal to the time of maximum of the potential.‡

For simplicity, the above learning rule does not, in principle, preserve the sign of the synapse, and thus an initially excitatory synapse may become an inhibitory synapse during learning, and *vice versa*. A biological implementation of such sign reversal will require a more elaborate architecture in which each input has functionally both excitatory and inhibitory synapses onto the output neuron (e.g., by disynaptic connections) such that the net synaptic weight is their sum. More importantly, it has been shown that constraining the sign of the synapses to their initial value during learning does not qualitatively change the properties of the architecture and the learning algorithm in rate-based models (Amit et al., 1989; Brunel et al., 2004). Thus, adding a sign restriction to the Tempotron learning rule is not expected to change its qualitative behavior.

The performance of the Tempotron's learning algorithm has been tested extensively in numerical simulations (Gütig and Sompolinsky, 2006). The capacity for classification of random latency patterns agrees with the expected theoretical capacity as described in Figure 2.5b. The learning algorithm is also able to find solutions for the pairwise synchronization classification task, and other tasks described in Gütig and Sompolinsky (2006). The generalization abilities for various tasks have also been studied (Gütig and Sompolinsky, 2006, 2009).

---

* Either $U_\mu(t^\mu_{max})$ is a smooth extremum point of $U_\mu(t)$ with a vanishing slope, or t^μ_{max} corresponds to the fixed time of an inhibitory input, and therefore, is independent of the synaptic weights.

† A momentum term can be implemented by adding a fraction of the previous update to the update of synapses at each learning step (see Gütig and Sompolinsky (2006) for details). In random latency patterns, this modification accelerates learning considerably.

‡ The Tempotron learning rule as described here relies on the subthreshold potential as defined by Equation 2.1 for both "target" and "null" patterns. As noted in Urbanczik and Senn (2009), in this rule, the gradient of the cost function E_μ is not differentiable at points where there are multiple global maxima of U. This may slow learning in the case of "null" patterns. Using $U(t)$ in the case of an erroneous "target" pattern (a pattern labeled with $y = -1$) is problematic for biological implementation as this potential is not an "observable" due to reset after a spike. An alternative choice is to incorporate the effect of spike reset in an erroneous "target" pattern, by shunting all input spikes that arrive after threshold crossing. This results in a maximum, which is slightly after the threshold crossing. Note that when the maximum of the shunted potential equals the threshold, the cost function may not be continuous (Urbanczik and Senn, 2009). Nevertheless, in most of our simulations, we use the shunted potential as it is computationally more efficient and converges faster than the algorithm that uses the subthreshold potential in the problems we have studied. It should be stressed that because E is not convex (as follows from Section 2.4.3), the convergence to a global minimum is not guaranteed as is the case for other gradient descent learning rules.

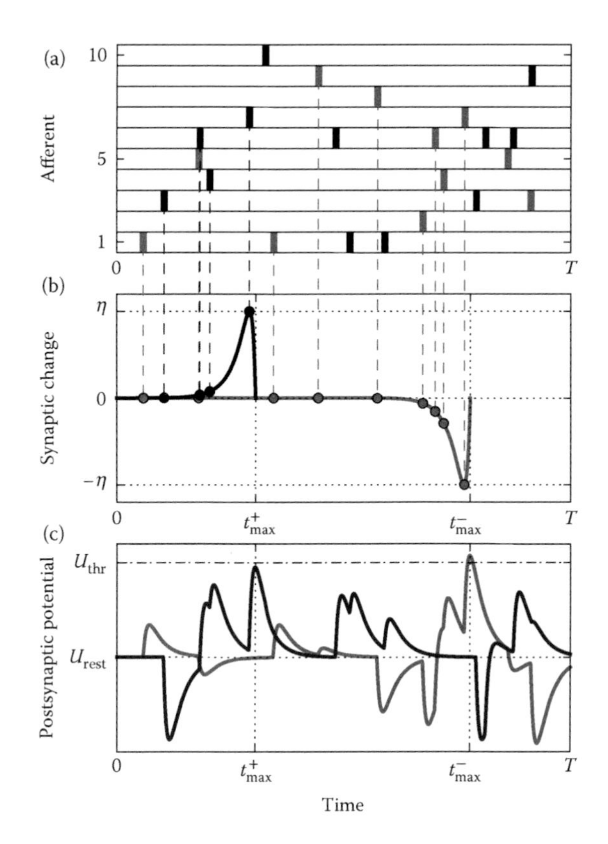

FIGURE 2.6 (a) Two input patterns (in gray and black) as described in Figure 2.2. (b) Synaptic changes in the Tempotron learning algorithm. In this case, the desired label for the black pattern is $y = +1$ and for the gray pattern, it is $y = -1$; thus, both patterns are erroneously classified (see (c)). The resulting synaptic changes (Equation 2.10) depend on presynaptic spike times (circles) relative to the corresponding timings of the postsynaptic potential maximum, t^+_{max} for the black pattern and t^-_{max} for the gray pattern. (c) The postsynaptic potential induced by each input pattern. The Tempotron integrates the incoming spikes with the same weights and postsynaptic kernel as in Figure 2.2.

2.5.2 Temporally Local Approximation of U_{MAX}

One of the challenges of implementing the Tempotron learning algorithm in a biological system is the calculation of the timing of the maximal postsynaptic potential, which is biologically implausible. One way to solve this problem is to approximate the max operation using a convolution of the postsynaptic potential with the post synaptic kernel. To derive a gradient descent rule based on this "soft max" approach, we replace the cost function of pattern μ, E^μ (see Equation 2.8), for each misclassified pattern, with the following function:

$$E^\mu_\gamma = -\frac{y^\mu}{\gamma} \ln\left[\frac{1}{T}\int_0^T e^{\gamma(U_\mu(t)-U_{thr})} dt\right]. \tag{2.11}$$

Note that for large values of γ, that is, $\gamma \gg 1$

$$E_{\gamma}^{\mu} \simeq y^{\mu}\left(U_{\text{thr}} - U_{\mu}\left(t_{\max}^{\mu}\right)\right) \tag{2.12}$$

which is the original Tempotron cost function. Thus, following an erroneous trial, the weights are updated according to

$$\Delta\omega_i = -\eta\frac{\partial E_{\gamma}^{\mu}}{\partial\omega_i} = \eta y^{\mu}\frac{\sum_{t_i}\int_0^T e^{\gamma U_{\mu}(t)}u(t-t_i)\,dt}{\int_0^T e^{\gamma U_{\mu}(t)}\,dt} \tag{2.13}$$

which converges into the original update rule in the $\gamma \rightarrow \infty$ limit. Figure 2.7a shows that the convergence of the γ rule with $\gamma \geq 10/U_{\text{thr}}$ is similar to the max rule.

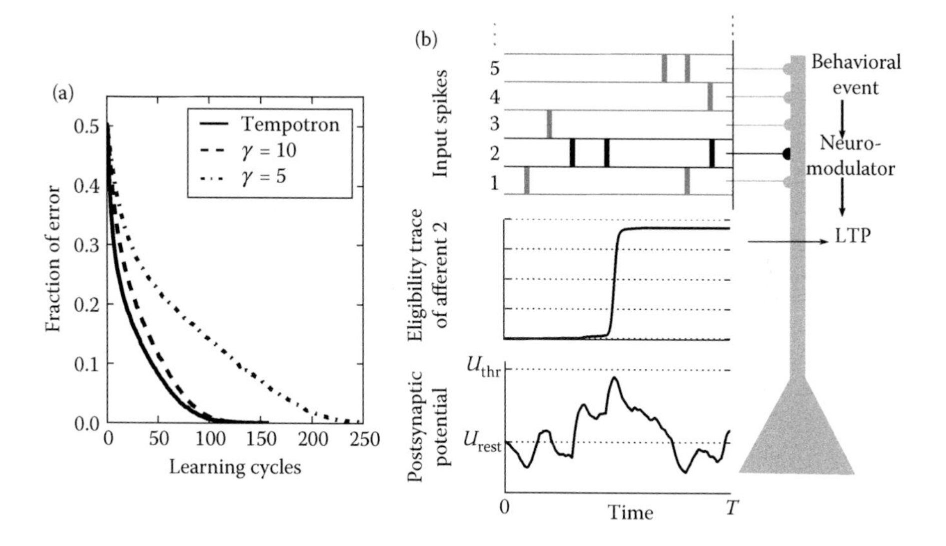

FIGURE 2.7 (a) Mean fraction of error versus number of learning cycles. Lines depict the mean error. Simulations were carried out with $N = 500$, $P = 600$, and $\eta = 10^{-3}$. Initial weights where chosen randomly from a Gaussian distribution with mean 0 and standard deviation of 10^{-4}. (b) Possible schematics of biological implementation of Tempotron learning. The top panel depicts the inputs spikes. The middle panel shows the eligibility trace as a function of time for the synapse of afferent 2 computed and maintained locally via a convolution of the postsynaptic potential, depicted in the bottom panel (in this case, we use the γ-rule). Behavioral events cause the release of neuromodulators that act as global teacher signals indicating the polarity of the required change in the synapse (LTP in this case). Plasticity is induced at each synapse according to the teacher signal and eligibility trace.

Gütig and Sompolinsky (2006) proposed a simple heuristic replacing the max Tempotron learning rule by

$$\Delta\omega_i = \eta y^\mu \sum_{t_i} \Theta\left[\int_0^T U(t)u(t-t_i)dt - \kappa\right], \tag{2.14}$$

where Θ is the Heaviside function and κ is a plasticity threshold (Gütig and Sompolinsky, 2006). This learning rule yields a reasonable performance since qualitatively, it uses similar *temporal credit assignment* as the Tempotron, namely, it updates only synapses that were active prior and proximal to large depolarization of the potential.

2.5.3 Biological Implementation of Tempotron Learning

For neuronal systems to implement a learning algorithm similar to that of the Tempotron, several requirements must be met. First, the underlying plasticity mechanism should be sensitive to the contiguity and the causal order of the timing of the presynaptic spike and the depolarization of the postsynaptic potential, as required by the rules above (Equations 2.13 and 2.14). Such sensitivity is observed in the form of long-term synaptic plasticity known as STDP. In STDP, synaptic modification occurs only if presynaptic spike occurs within a window of a few tens of milliseconds of a postsynaptic spike and potentiation requires a causal order between the pre- and postspikes. It has been suggested that STDP depends not only on the postsynaptic spike per se but also on the postsynaptic potential in a manner reminiscent of rule (2.14) above (Artola et al., 1990; Ngezahayo et al., 2000; Sjostrom et al., 2001; Lisman and Spruston, 2005; Clopath et al., 2010).

Second, the Tempotron rule is a special case of a supervised or a reward-based learning rule. In the case of the Tempotron, plasticity is activated only if error occurs and in addition, the sign of the plasticity depends on the type of error (potentiation in the case of a "target," pattern and depression for "null"). This is in contrast to the standard STDP rule, which is a form of temporal unsupervised learning where all the information required for the learning resides in the presynaptic activity and postsynaptic voltage. Indeed, several computational studies have shown that such "unsupervised" rules can extract salient statistical features in the input spike trains (Kempter et al., 1999; Song et al., 2000; Gutig et al., 2003; Guyonneau et al., 2005; Legenstein et al., 2005; Worgotter and Porr, 2005). However, the unsupervised nature of STDP does not allow for incorporating information about a desired input–output function, and therefore, is incapable of learning general input–output relations as those discussed above.

Incorporating error or reward signals into STDP such as plasticity rules can be implemented by a neuromodulatory control of the plasticity process. Indeed, it has been shown that neuromodulators can control the polarity of the synaptic plasticity (Seol et al., 2007). Plasticity can also be induced and controlled by coupling to additional synaptic inputs from feedback mechanisms, as is the case in the cerebellum (Ito and Kano, 1982).

If the supervised signal is generated by a sensory feedback signaling the outcome of the network's decision, then an additional challenge is bridging the time between the creation of the *synaptic eligibility traces*, which evaluate the contiguity of synaptic activation times and the potential depolarization, and the actual implementation of the long-term potentiation or depression following the sensory feedback (see Figure 2.7b). Such a computation can be realized by a variety of slow processes with time constants of seconds or longer which are abundant in the neuronal system. Alternatively, the supervisory signal can be in the form of instruction signal that tags the input patterns as desired "target" or "null" already at the time of their presentation. In such a case, the information necessary for implementing the plasticity rule is available at the end of the episode with no need for a delayed feedback signal.

2.6 CONDUCTANCE-BASED TEMPOTRON CLASSIFICATION

The neuron model discussed so far incorporated synaptic inputs as currents arising from the firing of the afferents. However, in real neurons, the release of neurotransmitters causes a change in the conductance of the cell's membrane. Thus, the synaptic potential is a nonlinear function of the synaptic efficacies. We review here an extension of the Tempotron model that incorporates synaptic conductances (Gütig and Sompolinsky, 2009). The extension to a conductance-based model is important to demonstrate biological plausibility, and, as will be shown, it has unexpected benefits: learning of effective integration time and invariance to dynamic time warp.

2.6.1 Conductance-Based IF Neuron

The dynamics of the postsynaptic potential of a conductance-based leaky IF neuron is given by

$$\frac{dU}{dt}(t) = -g_{\mathrm{m}}U(t) - \sum_{i=1}^{N} g_i(t)\left[U(t) - U_i^{\mathrm{rev}}\right], \qquad (2.15)$$

where g_{m} is the leak conductance of the neuron's membrane, U_i^{rev} is the reversal potential of the ith synapse, and $g_i(t)$ is its conductance at time t (all conductances are in units of 1/s). The conductances are modeled as

$$g_i(t) = \omega_i \sum_{t_i<t} \epsilon(t - t_i), \qquad (2.16)$$

where $\{t_i\}$ are the spike times of the ith afferent and $\epsilon(t)$ is a temporal kernel with unit area. For concreteness, we choose $\epsilon(t) = \tau_{\mathrm{s}}^{-1}e^{-(t/\tau_{\mathrm{s}})}$ where τ_{s} is the synaptic time constant. In this case, $\omega_i\tau_{\mathrm{s}}^{-1}$ is the maximum conductance of the ith synapse.*

* For this type of synaptic input, the resulting postsynaptic potential as well as its derivatives can be expressed analytically (Brette, 2006) enabling exact simulation of the neuron's dynamics and learning.

It is helpful to rewrite Equation 2.15 as

$$\frac{dU}{dt}(t) = -G(t)U(t) + I(t), \tag{2.17}$$

where

$$G(t) = g_{\mathrm{m}} + g_{\mathrm{syn}}(t) \tag{2.18}$$

and $g_{\mathrm{syn}}(t) = \sum_{i=1}^{N} g_i(t)$ and

$$I(t) = \sum_{i=1}^{N} g_i(t)U_i^{\mathrm{rev}}, \tag{2.19}$$

are the total conductance and total synaptic input, respectively. The total conductance determines the timescale of integration of synaptic inputs. Thus, the *effective* membrane time constant is

$$\tau_{\mathrm{eff}} = \frac{1}{G(t)}. \tag{2.20}$$

Note that while the net current flow into the neuron is determined by the balance between excitatory and inhibitory synaptic inputs, both types of inputs increase the total synaptic conductance, which in turn modulates the effective integration time of the postsynaptic cell (Bernander et al., 1991; Koch et al., 1996; Häusser and Clark, 1997). Finally, the solution of Equation 2.17 (assuming $U(0) = 0$) can be written as

$$U(t) = \int_0^t du \exp\left(-\int_u^t dz G(z)\right) I(u). \tag{2.21}$$

2.6.2 Learning Algorithm

The learning algorithm has the same error-correcting characteristics as in the current-based model. Each peak conductance is changed after each error trial according to its contribution to the maximal postsynaptic potential over the duration of the trial, leading to

$$\Delta\omega_i = \eta y^{\mu} \sum_{t_i} u_i(t_{\max}, t_i), \tag{2.22}$$

where $u_i(t, t_i)$ is the contribution of the spike at t_i to $\partial U/\partial\omega_i$ at t. Differentiating Equation 2.17 with respect to ω_i, we obtain

$$\frac{\partial u_i(t,t_i)}{\partial t} = -G(t)u_i(t,t_i) + \epsilon(t - t_i)\left(U(t) - U_i^{\mathrm{rev}}\right), \tag{2.23}$$

yielding

$$u_i(t,t_i) = \int_{t_i}^{t} du \exp\left(-\int_{u}^{t} dz G(z)\right)\epsilon(u - t_i)\left(U_i^{\text{rev}} - U(u)\right). \quad (2.24)$$

Short τ_s *approximation*: With many fast synapses impinging on the output neuron, it is often the case that $G(t)$ and $U(t)$ are smooth on the scale of τ_s. In such a case, Equation 2.24 can be approximated as

$$u_i(t,t_i) \simeq \left[\exp\left(-\frac{t - t_i}{\tau_{\text{eff}}(t,t_i)}\right) - \exp\left(-\frac{t - t_i}{\tau_s}\right)\right]\left(U_i^{\text{rev}} - U(t_i)\right) \quad (2.25)$$

where $\tau_{\text{eff}}(t,t_i) = 1/\bar{G}(t,t_i)$ and

$$\bar{G}(t,t_i) = \frac{1}{t - t_i}\int_{t_i}^{t} d\tau G(\tau) \quad (2.26)$$

Equation 2.25 has a similar structure as the current-based filter Equation 2.10. The difference is that here the effective integration time replaces τ_m and that the filter is multiplied by the *driving voltage* $U_i^{\text{rev}} - U(t_i)$. For the learning updates, this filter has to be evaluated at $t = t_{\text{max}}$. Note that as in the current-based learning, local approximations can be used as in Equations 2.13 and 2.14 with $u_i(t, t_i)$ replacing $u(t - t_i)$. We will discuss further properties of the short τ_s limit in Section 2.6.4.

2.6.3 Modifiable Effective Time Constant and Tempotron Capacity

One of the functional advantages of having synaptic inputs as conductances is that the system can tune the neuron's integration time through modulation of the strength of the synapses. As we have seen above, the current-based Tempotron works well in a range of integration windows relative to the pattern duration, quantified above by the parameter K, as shown in Figure 2.5b. In particular, for large τ_m/T (i.e., small K), the neuron loses its temporal sensitivity and the performance drops. In contrast, for the conductance-based Tempotron, the effective integration time is controllable by learning. Indeed, as Figure 2.8a shows, the conductance-based model is not sensitive to increases in the values of τ_m/T.

2.6.4 Time-Warp Invariance in a Neuron with Large Synaptic Conductance

In the above section, we have shown how the functionality of the Tempotron is enhanced by its ability to change the effective integration time through learning. Here, we show that the system can benefit from the automatic dynamic scaling of the effective time constant through the change in the input rate. In many circumstances, animals demonstrate the ability to process temporal signals even if the duration of

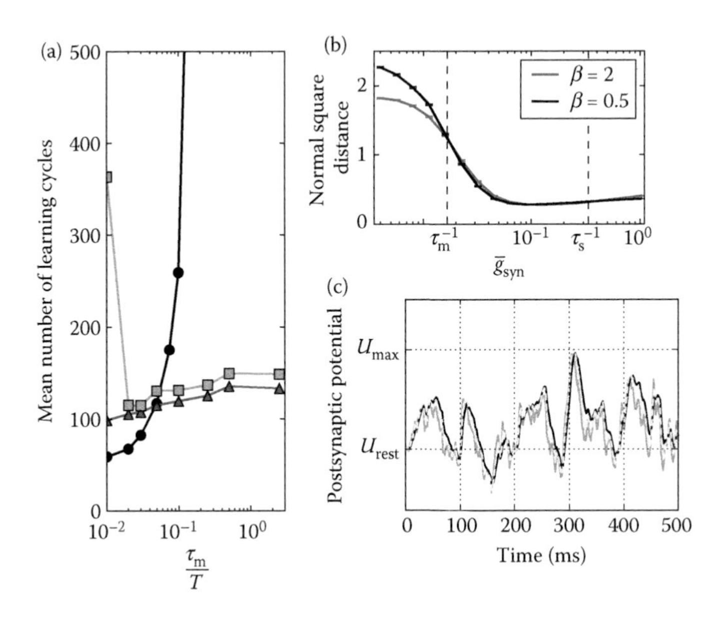

FIGURE 2.8 (a) Effect of conductance-based synapses on the capacity for learning patterns of random latency. The mean number of learning cycles required for learning versus the ratio of the membrane time constant, $\tau_m = 1/g_m$, and the pattern's duration, T, for the current-based Tempotron (circles), and the conductance-based Tempotron (triangles), and the conductance-based short τ_s approximation (squares), given by Equation 2.25. The data presented here is for $N = 500$ and $P = 1000$. Here, T was fixed at 500 ms, and τ_m varied from 6 to 2500 ms. τ_s was fixed at 5 ms. (b) The mean square distance between the voltage traces of a time-warped and an original pattern, $\left[\frac{1}{T}\int_0^T[U(\beta t,\beta) - U(t,1)]^2\,dt\right]^{1/2}\Big/\left[\frac{1}{T}\int_0^T U(t,1)^2 dt - \left(\frac{1}{T}\int_0^T U(t,1)\,dt\right)^2\right]^{1/2}$, versus the mean synaptic conductance for random input patterns and synapses. The patterns were randomly chosen as described in Section 2.4 and the synapses were drawn from Gaussian distribution with mean $1/N$ and standard deviation σ_ω (negative synapses were taken to be inhibitory synapses with efficacy $|\omega_i|$). The mean synaptic conductance given by $\bar{g}_{syn} = N\tau_s/T\sigma_\omega$ was varied by changing σ_ω. The data were taken with $N = 1000$, $\tau_s = 3$ ms, $\tau_m = 60$ ms, $T = 500$ ms. (c) The postsynaptic potential of a conductance-based Tempotron evoked by a single random pattern for the optimal value of mean conductance, $\bar{g}_{syn} = 0.08$. The plot depicts $U(\beta t, \beta)$ for three values of β: the original pattern ($\beta = 1$) in dashed gray, a time-compressed pattern ($\beta = 0.5$) in black, and a time-dilated pattern ($\beta = 2$) in solid gray.

the signals varies considerably from trial to trial. For instance, humans are able to understand speech regardless of the speed of speaking and regardless of dynamic variability in the speed of speaking over a wide range of speech velocities. Similarly, natural voluntary movements are generated with the same characteristics over a wide range of duration. Finally, recent evidence for time-warp processing has been demonstrated in the rodent olfactory system (Shusterman et al., 2011). The mechanisms of how this invariance is achieved in neuronal circuits is poorly understood. Gütig and Sompolinsky (2009) have suggested a neural mechanism for implementing such invariance based on the fact that the neuronal "internal clock" is automatically adjusted to the rate of incoming input.

The time-warp invariance induced by the synaptic conductances is valid in the regime

$$\tau_{\mathrm{m}}^{-1} \ll \bar{g}_{\mathrm{syn}}(t) \ll \tau_{\mathrm{s}}^{-1}, \tag{2.27}$$

where $\bar{g}_{\mathrm{syn}}(t)$ is the total synaptic conductance averaged over the timescale τ_{s}. To see how this invariance is accomplished in this regime, we consider the limit $\bar{g}_{\mathrm{syn}}(t)\tau_s \to 0$ and $\bar{g}_{\mathrm{syn}}(t)\tau_{\mathrm{m}} \to \infty$. In this limit

$$G(t) \approx g_{\mathrm{syn}}(t) \approx \sum_i \omega_i \lambda_i(t), \tag{2.28}$$

$$I(t) \approx \sum_i \omega_i U_i^{\mathrm{rev}} \lambda_i(t), \tag{2.29}$$

where $\lambda_i(t) = \Sigma_{t_i} \delta(t - t_i)$.

Consider now a global time warp of all input spikes by a factor β, that is, $t_i \to \beta t_i$. Changing the rate of the incoming spikes will induce a corresponding change in the timing of these pulses but not in their charge. The total charge delivered between times t and t' for a pattern with $\beta = 1$ will be approximately equal to the total charge delivered between time βt and $\beta t'$ for the same pattern with $\beta \neq 1$. Therefore, ignoring the effect of time warp on the time-scale of τ_{s}, which is short relative to the timescale of voltage modulations, synaptic current and conductance, given by Equations 2.28 and 2.29, obey the following time warp scaling relation

$$I(\beta t, \beta) \approx \beta^{-1} I(t, 1) \tag{2.30}$$

and

$$G(\beta t, \beta) \approx \beta^{-1} G(t, 1), \tag{2.31}$$

where $I(t, \beta)$ and $G(t, \beta)$ are the synaptic input and total conductance induced by a time-warped input pattern with factor β.

Substituting these relations in Equation 2.21 yields

$$U(\beta t, \beta) \approx U(t, 1).$$

This result is illustrated in Figure 2.8b, which displays the mean square error of $U(\beta t, \beta) - U(t, 1)$ induced by a random spike pattern with $\beta = 0.5$ and $\beta = 2.0$ as a function of the strength of the mean synaptic conductance. Consistent with condition (2.27) above, the error increases for small as well as large $\bar{g}_{\mathrm{syn}}$, with an optimum value between τ_{m}^{-1} and τ_{s}^{-1}. Voltage traces for the optimal mean $\bar{g}_{\mathrm{syn}}$ are presented in Figure 2.8c.

Gütig and Sompolinsky have found that using a conductance-based Tempotron learning algorithm, the IF neuron can learn time-warp invariant classification for a broader range of time-warp factors than can be achieved in the current-based Tempotron (Gütig and Sompolinsky, 2009). They have trained the Tempotron to classify a set of random patterns into two classes where in every presentation, each pattern was time warped by a random factor of up to $\beta_{\max}$ compression or dilation. The conductance-based Tempotron is able to learn these types of classification with almost zero errors up to $\beta_{max} = 2$. Further details are given in (Gütig and Sompolinsky, 2009). Furthermore, the time-warp invariant feature of the synaptic conductances helps the Tempotron learn isolated word recognition task, a problem that is challenging in part due to large global and dynamic time warp inherent in natural speech (Sakoe and Chiba, 1978).

2.7 SUMMARY AND DISCUSSION

We have focused here on a simple model of a neural decision based on information embedded in the spatiotemporal patterns of spiking events. The architecture is a single-layer network with a binary output similar to the well-known Perceptron. However, as shown here, the dynamic nature of the inputs and the output voltage yields a system, which is richer in its capabilities and is more complex in its action. In particular, the capacity of the Tempotron for classification of random patterns exceeds the capacity of the rate-based Perceptron. It scales with the number of input synapses, N, and for biologically relevant parameters, the maximal number of patterns that can be classified correctly is around $3N$ (compared to $2N$ for the Perceptron). Owing to the nonlinear decision it carries out, the Tempotron demonstrates qualitatively different behavior than the behavior of similar rate-based architectures; the set of all weights carrying out a specific classification of a given set of inputs is, in general, fractured into many small domains, allowing very different sets of weights to implement similar classifications and, in contrast, allowing similar sets of weights to implement very different classifications. This feature should be considered when analyzing neuronal data and characterizing neuronal responses. It is worth noting that this complex behavior is already present for a simple leaky IF model and is not sensitive to the details of the spike generation mechanism. We have also shown that it is possible, using the Tempotron learning algorithm, to efficiently learn to decode information carried by the relative spike times of the input afferents of the neuron, and to decode a variety of spatiotemporal neuronal codes. When applied to a conductance-based neuron model, the Tempotron learning rule enables neurons to control their effective integration time constant by balancing excitatory and inhibitory synaptic inputs. In a high-conductance regime where synaptic conductances control the integration time constant of the postsynaptic membrane, the neuronal voltage response becomes robust to temporal warping of the input spike trains.

2.7.1 Other Models for Learning Spatiotemporal Spike Patterns

The Tempotron model and its learning are based on a deterministic spiking and synaptic transmission. Other approaches to learning, spatiotemporal spike patterns have been recently proposed by Seung and coworkers, which are based on a stochastic framework. Similar to reinforcement learning algorithms (Williams, 1992), Seung's models utilize neural noise to carry out exploration in the space of the synapses during learning, a process that results in a stochastic gradient ascent in the landscape of the expected reward. In one version, the source of the stochastic response may lie in the spike generation mechanism (Xie and Seung, 2004), in which case, a neuron is considered as a Poisson spike generator with an inhomogeneous rate function defined by the instantaneous synaptic current (Xie and Seung, 2004). Alternatively, stochasticity in the action of individual synapses can act as the source of noise for the exploratory learning (Seung, 2003; Fiete and Seung, 2006). In both cases, the resulting synaptic modification rule has the form of a third-order correlator combining the input to the modified synapse, the resulting reward (or error), and the instantaneous neural noise, a more complex form than the input–output correlator in the Tempotron model.

Comparing the synaptic noise model and the Tempotron model on the random latency tasks indicated that the reinforcement algorithm is substantially slower. This is to be expected since it requires a considerable averaging to implement the gradient ascent. Furthermore, even at long time, the stochastic model does not carry out perfectly due to the inherent noise in the neural decision, unless a slow annealing procedure is adopted (Gütig and Sompolinsky, 2006).

Another probabilistic model for supervised learning in spiking networks has been proposed by Pfister et al. (2006) and Pfister (2006). Here, a neuron is modeled as a stochastic spike generator with a general inhomogeneous intensity function (which may incorporate refractoriness or other memory terms). In contrast to Seung's model, synaptic modifications are derived from an explicit calculation of the online gradients of the log-likelihood function of the desired output spike trains. Thus, learning is driven by instruction signals (i.e., the desired output spike times) and does not depend on the actual stochastic output of the neurons during the learning episodes. It is interesting that in the case of a single-layer network and the Tempotron binary decision model, the log-likelihood learning rule is similar (although not identical) to the soft max rule, Equation 2.13, where the exponential function has a stochastic interpretation as representing the instantaneous rate function for the Poisson firing of the output neuron (Pfister, 2006).

2.7.2 Open Issues

Developing a comprehensive model of neural learning and processing of spatiotemporal information in spiking networks is a major challenge for theoretical neuroscience. While the Tempotron has some attractive features, it needs to be extended in several directions to become a viable model for supervised neural learning. First, it is important to incorporate more biophysical properties of neuronal circuits, including active current at the soma and dendrites, and short-term synaptic dynamics. In fact, the Tempotron model can be extended to a more complex neuron model as long as the neuron's responses can be described by threshold crossing, as was done for conductance-based version. In such a model, the target function for gradient-based learning will still be U_{max}, which will be a complex nonlinear function of the synaptic weights. Most importantly, the error-based spike-time-dependent learning rule needs to be extended to more complex architectures, at least to feed-forward multilayer systems. Lastly, a comprehensive model of learning needs to address more general tasks, beyond binary decisions, such as multiclass decision, regressing an analog function, or generating precisely timed responses.

While the functional importance of the precise timing in the neural code is still debated, there is no question that spike times are the elementary units of the language of neurons in most of the central nervous system. Thus, understanding how spatiotemporal spike patterns encode information and how is this information decoded remains a major challenge of the theory of the brain.

ACKNOWLEDGMENTS

This work was partially supported by the Israeli Science Foundation, Israeli Ministry of Defense (MAFAT), the Chateaubriand fellowship, the Gatsby Charitable Foundation, and the William N. Skirball Fund for research in neurophysics. We thank Kamesh Krishnamurthy for help in the simulations of Figure 2.8a.

REFERENCES

Abeles, M., 1991. *Corticonics: Neural Circuits of the Cerebral Cortex*. Cambridge University Press, New York.

Amit, D., C. Campbell, and K. Wong, 1989. The interaction space of neural networks with sign-constrained synapses. *Journal of Physics A: Mathematical and General* 22, 4687.

Amit, D. J., H. Gutfreund, and H. Sompolinsky, 1985. Storing infinite numbers of patterns in a spin-glass model of neural networks. *Physical Review Letters* 55(14), 1530–1533.

Artola, A., S. Brocher, and W. Singer, 1990. Different voltage-dependent thresholds for inducing long-term depression and long-term potentiation in slices of rat visual cortex. *Nature* 347, 69–72.

Bar-Yosef, O., Y. Rotman, and I. Nelken, 2002. Responses of neurons in cat primary auditory cortex to bird chirps: Effects of temporal and spectral context. *Journal of Neuroscience* 22(19), 8619.

Bernander, O., R. Douglas, K. Martin, and C. Koch, 1991. Synaptic background activity influences spatiotemporal integration in single pyramidal cells. *Proceedings of the National Academy of Sciences* 88(24), 11569.

Bressloff, P. and J. Taylor, 1992. Perceptron-like learning in time-summating neural networks. *Journal of Physics A: Mathematical and General* 25, 4373.

Brette, R., 2006. Exact simulation of integrate-and-fire models with synaptic conductances. *Neural Computation* 18(8), 2004–2027.

Brunel, N., V. Hakim, P. Isope, J. Nadal, and B. Barbour, 2004. Optimal information storage and the distribution of synaptic weights: Perceptron versus purkinje cell. *Neuron* 43(5), 745–757.

Burak, Y., U. Rokni, M. Meister, and H. Sompolinsky, 2010. Bayesian model of dynamic image stabilization in the visual system. *Proceedings of the National Academy of Sciences* 107(45), 19525.

Carr, C. and M. Konishi, 1990. A circuit for detection of interaural time differences in the brain stem of the barn owl. *The Journal of Neuroscience* 10(10), 3227.

Clopath, C., L. Büsing, E. Vasilaki, and W. Gerstner, 2010. Connectivity reflects coding: A model of voltage-based STDP with homeostasis. *Nature Neuroscience* 13(3), 344–352.

deCharms, R. and M. Merzenich, 1996. Primary cortical representation of sounds by the coordination of action-potential timing. *Nature* 381(6583), 610–613.

Deneve, S., P. E. Latham, and A. Pouget, 1999. Reading population codes: A neural implementation of ideal observers. *Nature Neuroscience* 2(8), 740–745. 10.1038/11205.

Diesmann, M., M. Gewaltig, and A. Aertsen, 1999. Stable propagation of synchronous spiking in cortical neural networks. *Nature* 402(6761), 529–533.

Fiete, I. R. and H. S. Seung, 2006. Gradient learning in spiking neural networks by dynamic perturbation of conductances. *Physical Review Letters* 97(4), 048104.

Gardner, E., 1987. Maximum storage capacity in neural networks. *Europhysics Letters* 4(4), 481–485.

Gollisch, T. and M. Meister, 2008. Rapid neural coding in the retina with relative spike latencies. *Science* 319(5866), 1108.

Greschner, M., A. Thiel, J. Kretzberg, and J. Ammermüller, 2006. Complex spike-event pattern of transient on-off retinal ganglion cells. *Journal of Neurophysiology* 96(6), 2845.

Gutig, R., R. Aharonov, S. Rotter, and H. Sompolinsky, 2003. Learning input correlations through non-linear temporally asymmetric Hebbian plasticity. *Journal of Neuroscience* 23, 3697–3714.

Gütig, R. and H. Sompolinsky, 2006. The tempotron: A neuron that learns spike timing based decisions. *Nature Neuroscience* 9(3), 420–428.

Gütig, R. and H. Sompolinsky, 2009. Time-warp-invariant neuronal processing. *PLoS Biology* 7(7), e100141.

Guyonneau, R., R. VanRullen, and S. J. Thorpe, 2005. Neurons tune to the earliest spikes through STDP. *Neural Computation* 17, 859–879.

Häusser, M. and B. Clark, 1997. Tonic synaptic inhibition modulates neuronal output pattern and spatiotemporal synaptic integration. *Neuron* 19(3), 665–678.

Hromádka, T. and A. Zador, 2009. Representations in auditory cortex. *Current Opinion in Neurobiology* 19(4), 430–433.

Ito, M. and M. Kano, 1982. Long-lasting depression of parallel fiber-purkinje cell transmission induced by conjunctive stimulation of parallel fibers and climbing fibers in the cerebellar cortex. *Neuroscience Letters* 33(3), 253–258.

Jadhav, S., J. Wolfe, and D. Feldman, 2009. Sparse temporal coding of elementary tactile features during active whisker sensation. *Nature Neuroscience* 12, 792–800.

Johansson, R. and I. Birznieks, 2004. First spikes in ensembles of human tactile afferents code complex spatial fingertip events. *Nature Neuroscience* 7(2), 170–177.

Kempter, R., W. Gerstner, and J. L. van Hemmen, 1999. Hebbian learning and spiking neurons. *Physical Review E* 59(4), 4498–4514.

Koch, C., M. Rapp, and I. Segev, 1996. A brief history of time (constants). *Cerebral Cortex* 6(2), 93.

Leadbetter, M., G. Lindgren, and H. Rootzén, 1983. *Extremes and Related Properties of Random Sequences and Processes*. Springer, New York NY, USA.

Legenstein, R., C. Naeger, and W. Maass, 2005. What can a neuron learn with spike-timing-dependent plasticity? *Neural Computation* 17, 2337–2382.

Lisman, J. and N. Spruston, 2005. Postsynaptic depolarization requirements for LTP and LTD: A critique of spike timing-dependent plasticity. *Nature Neuroscience* 8, 839–841.

Liu, J., A. Harris, and N. Kanwisher, 2002. Stages of processing in face perception: An meg study. *Nature Neuroscience* 5(9), 910–916.

Meister, M., L. Lagnado, and D. Baylor, 1995. Concerted signaling by retinal ganglion cells. *Science* 270(5239), 1207–1210.

Minsky, M. and S. Papert, 1988. *Perceptrons: Expanded Edition*. MIT Press, Cambridge, MA, USA.

Monasson, R. and D. O'Kane, 1994. Domains of solutions and replica symmetry breaking in multilayer neural networks. *Europhysics Letters* 27(21), 85–90.

Ngezahayo, A., M. Schachner, and A. Artola, 2000. Synaptic activity modulates the induction of bidirectional synaptic changes in adult mouse hippocampus. *Journal of Neuroscience* 20, 2451–2458.

Noda, H., 1975. Sustained and transient discharges of retinal ganglion cells during spontaneous eye movements of cat. *Brain Research* 84(3), 515–529.

O'Connor, D. H., S. P. Peron, D. Huber, and K. Svoboda, 2010. Neural activity in barrel cortex underlying vibrissa-based object localization in mice. *Neuron* 67(6), 1048–1061.

Pfister, J., 2006. *Theory of Non-Linear Spike-Time-Dependent Plasticity*. PhD thesis, École Polytechnique Fédérale de Lausanne.

Pfister, J., T. Toyoizumi, D. Barber, and W. Gerstner, 2006. Optimal spike-timing-dependent plasticity for precise action potential firing in supervised learning. *Neural Computation* 18(6), 1318–1348.

Rosenblatt, F., 1962. *Principles of Neurodynamics: Perceptrons and the Theory of Brain Mechanisms*. Spartan Books, Washington DC.

Roska, B. and F. Werblin, 2003, Rapid global shifts in natural scenes block spiking in specific ganglion cell types. *Nature Neuroscience* 6(6), 600–608.

Rubin, R., R. Monasson, and H. Sompolinsky, 2010. Theory of spike timing-based neural classifiers. *Physical Review Letters* 105(21), 218102.

Sakoe, H. and S. Chiba, 1978. Dynamic programming algorithm optimization for spoken word recognition. IEEE Transaction on *Acoustics, Speech, and Signal Proceedings ASSP-26*, 43–49.

Salinas, E. and L. F. Abbott, 1994. Vector reconstruction from firing rates. *Journal of Computational Neuroscience* 1(1), 89–107. 10.1007/BF00962720.

Segev, R., E. Schneidman, J. Goodhouse, and M. Berry, 2007. Role of eye movements in the retinal code for a size discrimination task. *Journal of Neurophysiology* 98(3), 1380.

Seol, G., J. Ziburkus, S. Huang, L. Song, I. Kim, K. Takamiya, R. Huganir, H. Lee, and A. Kirkwood, 2007. Neuromodulators control the polarity of spike-timing-dependent synaptic plasticity. *Neuron* 55(6), 919–929.

Seung, H. and H. Sompolinsky, 1993. Simple models for reading neuronal population codes. *Proceedings of the National Academy of Sciences* 90(22), 10749.

Seung, H. S., 2003. Learning in spiking neural networks by reinforcement of stochastic synaptic transmission. *Neuron* 40, 1063–1073.

Seung, H. S., H. Sompolinsky, and N. Tishby, 1992. Statistical mechanics of learning from examples. *Physical Review A* 45(8), 6056–6091.

Shadlen, M., K. Britten, W. Newsome, and J. Movshon, 1996. A computational analysis of the relationship between neuronal and behavioral responses to visual motion. *The Journal of Neuroscience* 16(4), 1486–1510.

Shamir, M., 2009. The temporal winner-take-all readout. *PLoS Computational Biology* 5(2), e1000286.

Shusterman, R., C. S. Matthew, A. K. Alexei, and R. Dmitry, 2011. Precise olfactory responses tile the sniff cycle. *Nature Neuroscience* 14(8), 1039–1044. 10.1038/nn.2877.

Sjostrom, P., G. Turrigiano, and S. Nelson, 2001. Rate, timing, and cooperativity jointly determine cortical synaptic plasticity. *Neuron* 32, 1149–1164.

Song, S., K. D. Miller, and L. F. Abbott, 2000. Competitive hebbian learning through spike-timing-dependent synaptic plasticity. *Nature Neuroscience* 3, 919–926.

Thorpe, S., D. Fize, and C. Marlot, 1996. Speed of processing in the human visual system. *Nature* 381(6582), 520–522.

Thorpe, S. and J. Gautrais, 1998. Rank order coding. *Computational Neuroscience: Trends in Research* 13, 113–119.

Urbanczik, R. and W. Senn, 2009. A gradient learning rule for the Tempotron. *Neural Computation* 21(2), 340–352.

Victor, J., 2000. How the brain uses time to represent and process visual information. *Brain Research* 886(1–2), 33–46.

Wehr, M. and G. Laurent, 1996. Odour encoding by temporal sequences of firing in oscillating neural assemblies. *Nature* 384(6605), 162–166.

Williams, R. J., 1992. Simple statistical gradient-following algorithms for connectionist reinforcement learning. *Machine Learning* 8(3), 229–256. 10.1007/BF00992696.

Wolfe, J., A. Houweling, and M. Brecht, 2010. Sparse and powerful cortical spikes. *Current Opinion in Neurobiology* 20(3), 306–312.

Worgotter, F. and B. Porr, 2005. Temporal sequence learning, prediction, and control: A review of different models and their relation to biological mechanisms. *Neural Computation* 17, 245–319.

Xie, X. and H. S. Seung, 2004. Learning in neural networks by reinforcement of irregular spiking. *Physical. Review. E* 69(4), 041909.

3 Can We Predict Every Spike?

Richard Naud and Wulfram Gerstner

CONTENTS

Is it possible to predict the spike times of a neuron with millisecond precision? In the classical picture of rate coding (Adrian, 1928), single spikes do not play a role, and the question would have to be answered negatively. For rate coding in a single neuron, the relevant quantity to encode a stimulus such as pressure onto a touch sensor in the skin (Adrian, 1928) or presence of a light bar in the receptive field of a visual neuron (Hubel and Wiesel, 1959) is the number of spikes a neuron emits in a short time window of, for example, 100 ms. The timing of the spikes is considered as irrelevant. However, over the last 20 years, many researchers have shown that it is not only the temporally averaged firing rate that carries information about the stimulus, but also the exact timing of spikes. For example, spike timing has shown to be relevant to encode force amplitude and direction in touch sensors of the skin (Johansson and Birznieks, 2004) as well as the whole-field visual movements (Bialek et al., 1991) or object movement (Gollisch and Meister, 2008) in visual neurons.

If spike timing is important, a whole series of questions arises: What is the precision of spike timing if the same stimulus is repeated several times? Do spikes always appear at the same time? What would be a sensible measure of spike timing precision and reliability? Can a neuron model match the spike timing precision of a real neuron? Does it matter which neuron or what stimulus we take? If so, what would be a useful stimulus?

To answer these related questions, let us think of the following experimental protocol. An experimentalist injects a time-dependent input of, say, 20-s duration into a single neuron. The neuron responds with spikes. The experimentalist now repeats the same stimulus sequence several times. At each repetition, the neuron responds with a spike train that may or may not look similar to the previous one: some spikes appear at exactly the same time during the stimulus sequence, some are missing, some are shifted by a few milliseconds or appear at a completely different time. The

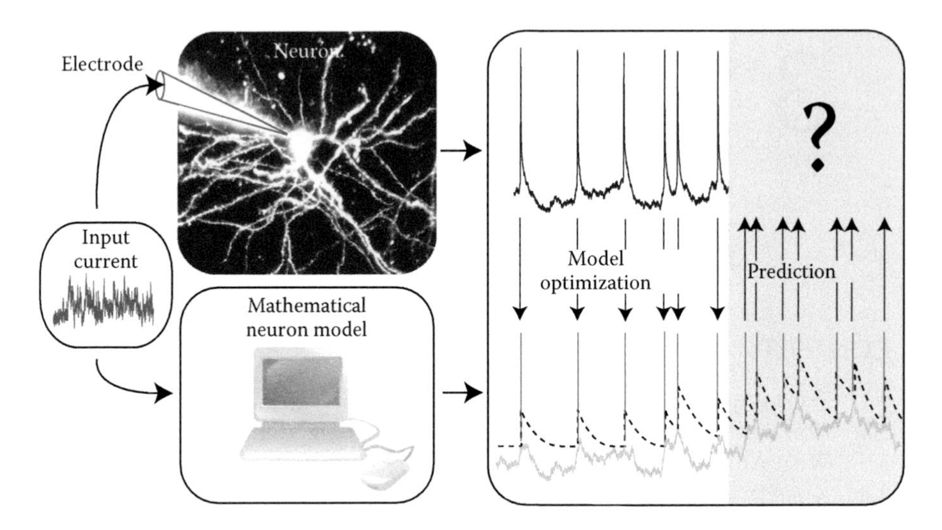

FIGURE 3.1 Schematic representation of the spike-timing prediction challenge. The same time-dependent input stimulus (left) is given to a mathematical neuron model and to a real neuron in an electrophysiological experiment. Part of the response of the real neuron is used to optimize the model parameters. The remaining part of the stimulus is injected into the model so as to predict the spike times of the real neuron. The mathematical neuron model illustrated here is made of a linear filter of the input current (bottom full trace) and a dynamic threshold (dashed black line). (Adapted from Gerstner W and Naud R. 2009. *Science* **326**, 379–380.)

information derived from this type of experiment which dates back to Bryant and Segundo (1976) and has been popularized by Mainen and Sejnowski (1995) should be sufficient to answer questions regarding precision and reliability of spike timing.

But here comes the challenge (Figure 3.1). Let us suppose that I give you the time course of the input as well as the neuron's response in each trial, but only for the first 10 s of the data. For the second half of the stimulus sequence, I give you only the time course of the input. Your task is to predict the timing of the spikes.

Will you be able to predict the timing of the spikes using an appropriate neuron model? Is your model as reliable and as precise as the real neuron? What would be the best model to choose so as to solve the task?

The above challenge has been turned into a single-neuron modeling competition that was first run by Brain Mind Institute at the Ecole Polytechnique Fédérale de Lausanne (EPFL) in Switzerland (Jolivet et al., 2008a,b) and was officially handed over to the International Neuroinformatic Coordinating Facility (INCF) in Sweden in 2009. In this chapter, we recapitulate the main questions and findings related to predicting spike times, with a special focus on the spike-time prediction competition of 2009 (Gerstner and Naud, 2009).

3.1 WHAT IS A GOOD STIMULUS TO PROBE NEURONS?

In classical electrophysiological experiments, an artificially generated input is injected in a neuron *in vitro* (Figure 3.1). In principle, the time course of the input can be chosen arbitrarily and could consist of short or long steps of different amplitudes, sequences of steps, ramps, white noise, filtered noise, or whatever comes to mind. But what is a "good" stimulus?

Since the work of Hodgkin and Huxley (1952) electrophysiologists have been using steps and ramps to characterize single-neuron responses. These types of stimulations are helpful to systematically probe the gating dynamics of ion channels under pharmacological manipulation. They can also give a qualitative classification of neuronal responses in terms of intrinsic firing patterns such as regular, fast-spiking, or bursting (Connors and Gutnick, 1990; Markram et al., 2004) but they have very little resemblance with the type of stimulus a neuron would receive in its natural environment.

Inspired by signal processing theory, the pioneering studies of Bryant and Segundo (1976) and of Marmarelis and Marmarelis (1978) used white-noise stimulation instead of step currents. However, if the aim is to drive a neuron with a stimulus that resembles as much as possible the input it would receive an *in vivo* situation, a white-noise stimulus is not sufficient. Rather, a stimulus at the soma should replace the total current flowing from the synapses to the soma while the neuron receives presynaptic input. Following a line of earlier research (Stein, 1967; Poliakov et al., 1996; Destexhe et al., 2003; Jolivet et al., 2006), the first spike timing competition in 2007 used an Ornstein–Uhlenbeck current injection with various means and variance to mimic the combined effect of a large number of synapses (Stein, 1967; Jolivet et al., 2008a). In 2008, the competition was modified (Jolivet et al., 2008a) to replace the dynamic current by dynamic inhibitory and excitatory conductances using dynamic clamp (Destexhe et al., 2003). Then in 2009, the injected current was changed to a current produced by the simulation of six populations of presynaptic neurons changing their firing rate every 200–500 ms.

3.2 HOW CAN WE MEASURE SPIKE TIMING PRECISION AND RELIABILITY?

Suppose that a single neuron is driven with multiple repetitions of the same time-dependent stimulus. The response of the neuron is recorded in each trial, so that the stimulation protocol builds up a database containing one spike train for each repetition.

If we compare the spike trains across several repetitions, different types of variability are seen depending on the system and the variance of the input (Bryant and Segundo, 1976; Mainen et al., 1995; Jolivet et al., 2006). Some spikes are seen at the same time for all repetitions, others appear at a specific time on half the repetitions and yet others do not seem to be related to a specific time. Several questions arise: First, how can we *quantitatively* compare one spike train of the neuron with another one recorded during a later repetition of the same stimulus sequence? Second, how can we quantify the reliability across the set of all spike trains recorded with the same stimulus? Finally, how can we compare the set of spike trains generated from neuronal recordings with a similar set of spike trains generated by a mathematical neuron model?

These are crucial questions which can be answered in different ways. One can compare one spike train with another one based on global features such as the intrinsic firing patterns in response to step stimuli (Connors and Gutnick, 1990; Markram et al., 2004) and ask whether a neuron model is able to reproduce the same intrinsic firing patterns (Izhikevich, 2007; Naud et al., 2008). One can focus the quantitative comparison on the shape of spikes and adaptation patterns (Druckmann et al., 2007), on the interspike interval distribution (Chacron et al., 2005) or the spike-count variability (Softky and Koch, 1993; Schaette et al., 2005).

If one focuses on spike timing, one may want to apply methods that compare spike trains in terms of a spike-train metrics (Victor and Purpura, 1996) or the coincidence rate (Kistler et al., 1997). Both measures can be used to compare a spike train from a recorded neuron in repetition n with another spike train recorded in repetition m. Both measures can also be used to compare a spike train derived from a neuron model with a spike train recorded in one of the sessions with the real neuron. Obviously, a model which achieves an optimal match in terms of spike-train metrics will automatically account for global features of the spike trains, such as interspike interval distributions.

In the INCF competition the average coincident rate was used to quantify spike-time prediction performance. The average coincidence rate can be seen as a similarity measure between pairs of spike trains that is finally averaged across all available pairs. To compute the pairwise coincidence rate, one first finds the number of spikes from the model that fall within an interval of ±4 ms around a spike from the real neuron. This is called the number of coincident events N_{nm} (at resolution $\Delta = 4$ ms). The coincidence rate is the ratio of the number of coincident events over the averaged number of events $0.5(N_n + N_m)$, where N_n is the number of spikes in the neuron spike train and N_m is the number of spikes in the model spike train (Figure 3.2). This

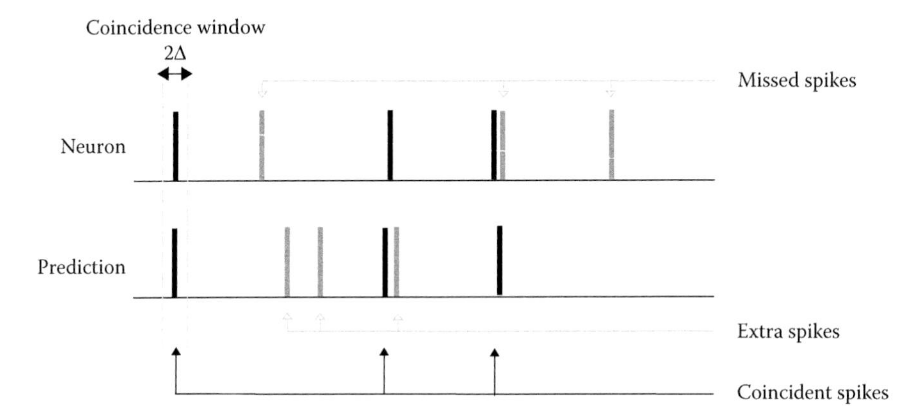

FIGURE 3.2 Counting coincident spikes for the computation of the coincidence rate Γ_{nm}. The predicted spike train (bottom) is compared to the recorded spike train (top). A predicted spike is said to be coincident (black) if it falls between $\pm\Delta$ of a recorded spike and if that recorded spike was not counted as coincident with any other predicted spike. The prediction can miss recorded spikes (gray, top) and generate extra spikes (bottom, gray). Here the total number of coincident spikes is $N_{\text{coinc}} = 3$ while there were $N_m = 6$ spikes in the predicted and $N_n = 6$ spikes in the experimental spike train. (Adapted from Figure 1.1 in Jolivet 2005. Effective minimal threshold models of neuronal activity. PhD thesis, Ecole Polytechnique Federale de Lausanne.)

ratio is then scaled by the number of coincident events, $N_{\text{Poisson}} = 2\Delta \cdot N_m N_n / T$, that are expected from a Poisson model that fires stochastically at a fixed rate N_m/T. The scaled coincidence rate is

$$\Gamma_{nm} = \frac{N_{nm} - N_{\text{Poisson}}}{\frac{1}{2}(1 - N_{\text{Poisson}}/N_n)(N_n + N_m)}. \tag{3.1}$$

Finally, the pairwise coincidence rate Γ_{nm} is then averaged across all the possible pairings of spike trains from the model with those of the neuron and this gives the averaged coincidence rate $\overline{\Gamma_{nm}}$.

The coincidence rate Γ_{nm} enables us to compare the spike timings of the model (subscript m) with that of the neuron (subscript n). If we want to know how reliable the neuron itself is, we need to measure how similar the spike trains are between two trials. To do the comparison between one neuronal spike train (index n) and another neuronal spike train (index n), we can use again the same coincidence measure that we now label Γ_{nn} so as to indicate the coincidence rate between two neuronal spike trains. The neuron-to-neuron coincidence rate, averaged over all pairs of available spike trains for the same stimulus, is a measure of the intrinsic reliability of the neuron and denoted as $\overline{\Gamma_{nn}}$. It provides an upper bound for modeling: on average, a neuron model cannot predict spikes better than the neuron itself. The averaged coincidence rate $\overline{\Gamma_{nm}}$ needs, therefore, to be compared to the upper bound provided by the intrinsic reliability $\overline{\Gamma_{nn}}$ of the neuron. Scaling $\overline{\Gamma_{nm}}$ by $\overline{\Gamma_{nn}}$ gives a number that can be interpreted as the fraction of the predictable spikes that are predicted by the model.

The coincidence rate, like the majority of spike-train metrics, has a time-scale parameter. In the above equation, the time-scale parameter Δ regulates the size of the coincidence window, and thus the level of precision of the prediction. For a very small coincidence window, the number of coincidences N_{coinc} goes to zero due to a jitter in spike timing and the finite number of spike trains. For a very large coincidence window, the coincidence rate goes to zero because it becomes insensitive to specific times of the spikes so that the difference between the prediction of a precisely tuned neuron model and that of a Poisson model with constant rate vanishes. In a large range between these two extrema, however, the model-to-neuron coincidence rate is significantly positive. Moreover, over the range roughly from 2 to 15 ms depending on the neuron and on experimental conditions (Jolivet et al., 2006, 2008b), the results measured in terms of $\overline{\Gamma_{nm}}$ do not depend on the choice of the time window Δ. In the INCF competition a window of $\Delta = 4$ ms was chosen.

It is useful to distinguish measures such as the coincidence rate (Kistler et al., 1997) or the spike train metrics (Victor and Purpura, 1996), which are both based on a comparison of a single spike train A with a second spike train B from measures that first average across all repetitions of an experiment to calculate the Peri-Stimulus Time-Histogram (PSTH) (Eggermont et al., 1983) before a comparison of the PSTH of neuron A with that of a neuron B (or of a model neuron) is performed. In the INCF challenge, rankings were based on the pairwise comparison of a model spike train with a real spike train, averaged across all repetitions of the experiment so as to determine the average coincidence rate.

Does the average coincidence rate correspond to a comparison of the PSTH? Essentially, comparing PSTHs is to compare averaged responses, whereas the average coincidence rate averages the pairwise match between spike trains. Indeed, the PSTH is calculated by averaging all the independent responses to the same input. A smoothing filter is then applied to the averaged spike trains. In many neuronal systems, the PSTH is made of a series of peaks and plateaus. The peaks correspond to spikes always coming at a precise time and the plateaus correspond to times where spikes are emitted with no specific timing. A model reproducing such a PSTH can be said to predict the spike times because such a model will emit a spike precisely at times where the neuron emits precisely timed spikes. Indeed, normalized spike-train similarity measures such as $\overline{\Gamma_{nm}}/\overline{\Gamma_{nn}}$ calculate a quantity very similar to the variance of the experimental PSTH that is explained by the model PSTH (Naud et al., 2011). The time-scale parameter of the similarity measure is equivalent to the filter time-scale applied for smoothing the PSTH. The main difference between $\overline{\Gamma_{nm}}/\overline{\Gamma_{nn}}$ and the comparison of PSTHs is that an optimization of neuron models based on the comparison of PSTH attempts to match the spike-timing variability of the model to the variability of the data. In contrast, an optimization of neuron models based on the normalized coincidence rate gives a slight advantage to deterministic neuron models, that is, those that do not correctly reproduce the intrinsic variability of the data (Naud et al., 2011).

Being aware of the similarity between PSTH comparison and coincidence rate scaled with intrinsic reliability is important to relate different studies to each other even though they use different evaluations criteria. It should be kept in mind, that there is an overestimation of the prediction performance for deterministic models using the scaled coincidence rate $\overline{\Gamma_{nm}}/\overline{\Gamma_{nn}}$ with respect to variance explained of the PSTH (Naud et al., 2011). However, there is also a small-sample bias that can overestimate the variance explained when comparing PSTHs (David and Gallant, 2005; Petersen et al., 2008; Naud et al., 2011). Values of the above 100% are possible because of the bias, but also because a model can be fitted independently on each repetition and thus allowing to take into account the experimental drifts.

3.3 WHAT ARE GOOD NEURON MODELS?

Across several years and editions of the single-neuron modeling competition, various models participated in the challenge. The models ranged from the very simple integrate-and-fire models to complete biophysical models using the Hodgkin-and-Huxley formalism and a 3D reconstruction of the morphology from another neuron of the same class. The number of state variables goes from one for the simplest models to a few hundred depending on the number of ion-channel types modeled and the number of compartments used in the discretization of space. Similarly, the number of parameters scaled with the number of state variables with five parameters for the simplest models and close to a hundred parameters for the biophysical models. Between the two extremes, various other models were used. For instance, adding nonlinear threshold, spike-triggered adaptation, and subthreshold adaptation, the Izhikevich and AdEx models* (Izhikevich, 2004; Brette and Gerstner, 2005) were used by some of the participants. Simpler versions of the complete biophysical models were also devised by reducing the number of ion channels and the number of compartments (in the same line as Pospischil et al., 2008).

A good prediction stems from the union of a good model and an efficient fitting method. The method for finding the optimal parameters should be efficient in the sense of providing a single set of optimal parameters with small computing resources. Many different methods were used by the various participants, some using the action potential shape, the subthreshold voltage dynamics, and the spike times as the observables to fit, others using only the set of spike times as observables to fit.

The participants that used biophysical models approached the problem by first constraining a significant number of parameters with published measurements of ion channel dynamics. Most of the biophysical-model participations thus reduced the parameter space to the somatic ion-channel densities only. This leaves a number of free parameters equal to the number of ion-channel species (Druckmann et al., 2007). These remaining parameters are then fit either by hand-tuning, stochastic optimization algorithms such as the genetic algorithm, or by exhaustive search when the number of ion-channel species is low.

For many of the simpler models, the optimization methods available are more efficient. Some of the most effective participations performed an exhaustive search on a small number of crucial parameters (Kobayashi et al., 2009). Convex optimization algorithms can be used to maximize the likelihood of observing the spike times (Paninski et al., 2004). This method leads to some of the top-ranking participations. When the optimization algorithm is convex, we are sure there is a single optimum solution and the optimization can run smoothly without the fear of stopping in a local minima. Another noteworthy method for fitting involves a convex, two-step procedure where in the first step the optimal passive parameters and the spike shape are determined from the voltage trace, and on the second step the parameters regulating a dynamic threshold are determined by maximizing the likelihood of the observed spike trains (Mensi et al., 2012).

The model that achieved the highest performance in 2009 was an integrate-and-fire model provided with a dynamic threshold that jumps every time there is a spike and decays back to a baseline with three different time constants (Kobayashi et al., 2009). The ratio $\overline{\Gamma_{nm}}/\overline{\Gamma_{nn}}$ for this participation was of 76.2%. The participant winning this spike-time prediction competition had extracted the membrane time constant from the voltage trace, fixed the decay time-constants of the dynamic threshold to 10, 50, and 200 ms leaving three free parameters; one parameter regulating the amount by which the threshold jumps for each of the three time scales. The optimal set of the three parameters was found by conducting an exhaustive search for the set of parameters maximizing the average coincidence. This winning model happened to be the model with the smallest number of free parameters used in the competition.

* Izhikevich and AdEx models differ only by the nonlinearity for spike initiation; quadratic in the first, exponential in the latter.

The small number of free parameters is not sufficient to explain the high performance, because for instance, a simple leaky and integrate and fire cannot predict more than 38% on the same task. The winning participant used judicious insights in deciding which model to use, which parameters could be fixed *a priori*, and which parameters required to be fitted to the specific neuron recorded.

A very small number of biophysical models participated in the competition, perhaps due to the difficulty of finding the optimal parameters for such complex models. There was a noteworthy participation using state-of-the-art optimization methods which was outside of the official competition because of being submitted after the deadline for the money prize. This submission would have ranked third if it had been submitted before the deadline. Thus, within the framework of the competition, the prediction performance of simple models is as good if not slightly better than that of the biophysical models. What this means is that when only the spike times matter we can safely replace the complex modeling of ion-channel dynamics by a simple model with formal threshold.

The relatively high prediction performance of models is not explained by the fact that the challenge is too easy. Off-the-shelf models such as hand-tuned models of pyramidal neurons or a leaky-integrate-and-fire achieved a performance around $\bar{\Gamma}_{nm}/\bar{\Gamma}_{nn} = 40\%$. Perhaps, the most important aspect for providing a good prediction in the competition was to take into account spike-frequency adaptation. The dynamic threshold of the winning submission is just one example, there are many other ways to implement adaptation in a single neuron and models that implement adaptation were systematically better than models that did not (Jolivet et al., 2008a).

So what is the best model? Accurate modeling of the refractory period is essential to predict the spike times (Kistler et al., 1997; Keat et al., 2001). More generally, if the same neuron model has to predict spike timings for stimuli with different mean firing rates, the importance of spike-frequency adaptation was recognized explicitly (Pillow et al., 2005; Jolivet et al., 2006, 2008a). State-of-the-art models now consist of models akin to the stochastic integrate-and-fire model but upgraded with an adaptation process. The adaptation makes the firing probability dependent on the timing of all recently emitted spikes. Such models are capable of predicting 75–100% of the predictable spikes (see Section 3.4). This leaves little room for improving the accuracy of encoding models. Indeed, increasing the level of detail with conductance-based adaptation or Hodgkin–Huxley ion channels does not yield substantial increases in prediction performance (Druckmann et al., 2007; Mensi et al., 2012).

How the relative refractory period decay after each spike affects directly the spike timing? Often, when a stimulus is strong, a second spike will come as soon as the relative refractory period induced by the first spike is over. Since the relative refractory period extends over a long time, a spike that occurs now has still an influence 1 s later (Mensi et al., 2012). Adaptation can be seen as cumulating the relative refractory period of multiple spikes. The multiple time scales affecting the adaptation process also become very important. The short time scales may affect the spike timing precisely as a function of the recent spiking history. The longer time scales will generally affect the number of spikes over longer periods of time.

3.4 ARE ALL NEURONS PREDICTABLE?

Every year, between 2007 and 2009, the spike prediction challenge was separated in four parts. Part A was very similar in each edition with only slight difference in experimental protocols. Parts B–D were different every year and were designed to explore different cell types, different injection sites, or the prediction of subthreshold features. In all the previous sections, the term spike timing competition referred to part A of the 2009 competition.

The different editions of the competition (part A) showed that good spike-time prediction can be achieved for different types of current injections. In 2007, the injection consisted of a current fluctuating on a single time scale and of tunable variance and mean (best performance $\bar{\Gamma}_{nm}/\bar{\Gamma}_{nn} = 82.0\%$). In 2008, the current clamp was replaced by a conductance clamp and the stimulus consisted of independent excitatory and inhibitory conductances fluctuating in time on a single time scale (best-performance $\bar{\Gamma}_{nm}/\bar{\Gamma}_{nn} = 91.4\%$). Conductance injection seems to make the neurons more predictable, but what about the statistics of the input? Changing continuously the statistics of the input on multiple timescales makes prediction more difficult, as seen in the edition of 2009 (best performance $\bar{\Gamma}_{nm}/\bar{\Gamma}_{nn} = 71.6\%$).

Can we predict the spike times in other systems than a Layer 5 pyramidal neuron in a cortical slice? This question was asked in parts B–D of the spike prediction competitions and of other independent studies. The activity of L2/3 pyramidal neurons and non-Fast spiking GABAergic neurons can be predicted with similar performances (Mensi et al., 2012). Prediction of the spike times of fast-spiking GABAergic neurons is systematically higher with $\bar{\Gamma}_{nm}/\bar{\Gamma}_{nn}$ in the range of 100% in part B of 2009 (Figure 3.3).

Real neurons receive their inputs from synapses distributed throughout their dendritic tree. The single-neuron model should model the dendritic integration of inputs. This dendritic integration is known to be highly nonlinear especially in the thick tufted L5 pyramidal cells (Larkum et al., 1999). To explore the dendritic dimension, dual electrode recordings were made with two independent injection sites: one in the soma and a second high in the dendritic tree. Spike-time prediction of $\bar{\Gamma}_{nm}/\bar{\Gamma}_{nn} = 83.8\%$ was achieved in Part C of the 2009 competition with a model similar to the one used by Larkum et al. (2004).

In the retina, spiking models of the retinal ganglion cells (RGC) can predict 91% of the variance of the PSTH (Pillow et al., 2005). Similar performances have been observed *in vivo* where 41–92% of the variance of the PSTH lateral geniculate nucleus (LGN) cells can be predicted from the activity of a single impinging RGC for spatially restricted visual stimulation (Carandini et al., 2007) as confirmed by submissions to part D in the 2009 edition of the competition.

3.5 CONCLUSION

In summary, the prediction of precise spike timing on the millisecond time scale is similar to predicting the time-dependent firing rate on the millisecond time scale. High prediction performance is possible in many neuronal systems and depends strongly on the choice of neuron model and fitting method. One important model

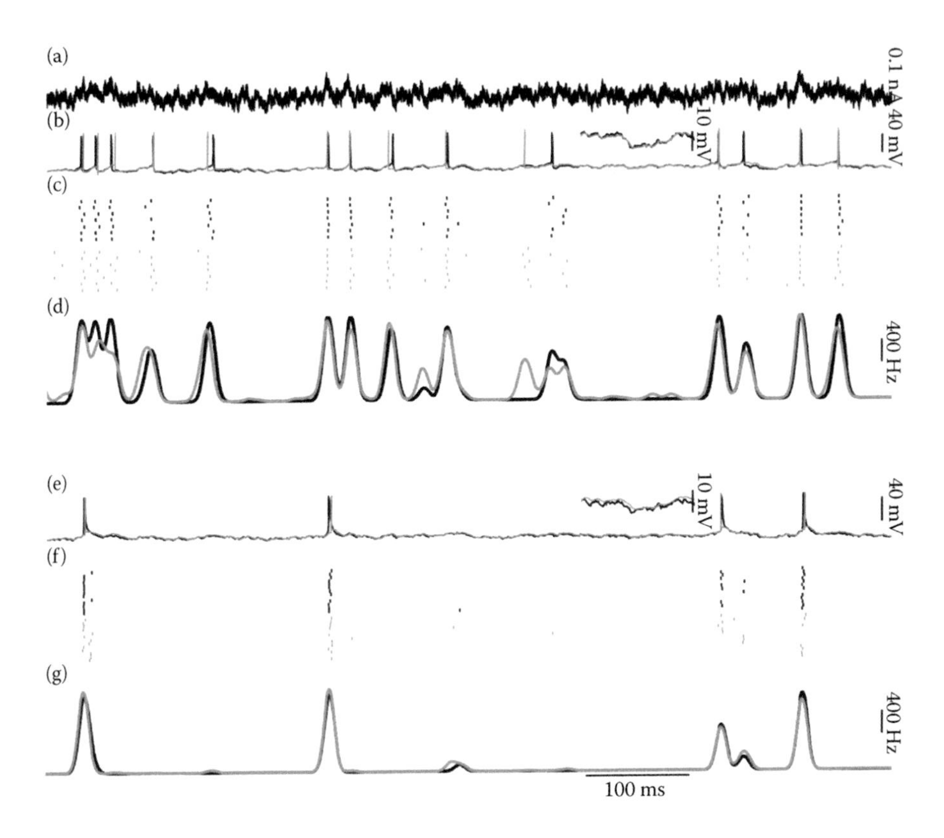

FIGURE 3.3 Predicting fast-spiking GABAergic neurons and pyramidal neurons in five from the 2009 competition. (a) The time-dependent current input models six populations of presynaptic neurons changing their firing rate every 200–500 ms. (b–d) Prediction of a GABAergic fast-spiking neuron. (b) Modeled (gray) and recorded (black) voltage traces. A zoom of 50 ms is shown in the inset. (c) Spike trains of each of the seven repetitions of the recorded (black) compared modeled (gray) spike train. (d) PSTH of the model (gray) overlaid on the PSTH of the data (black) calculated from the average of the spike trains that are then filtered with a Gaussian of 5 ms standard deviation. For the model we computed the PSTH from 1000 independent realizations while for the data we were restricted to the number of repetitions that could be recorded in the experiment. (e–g) Same as for (b–d) but predicting the activity of a pyramidal neuron from layer 5. (Courtesy of Skander Mensi.)

feature for high prediction performance is the presence of spike-frequency adaptation. The choice of the model formalism can also influence the fitting method that can be used. High-quality prediction is most of the time associated with an efficient and convex fitting method.

Can we use these results to determine the best single-neuron model? The single-neuron model of choice should be able to generalize across all the different experimental protocols and also across the possible systems and neuron types with a mere change of the model's parameters. The original competition was rewarding only the participations that could generalize across more than one of the experimental protocols. The data of the challenge will remain available in the future for benchmarking purposes, leaving the possibility for such a deed to be accomplished.*

REFERENCES

Adrian E D. 1928. *The Basis of Sensation.* W.W. Norton, New York.

Bialek W, Rieke F, de Ruyter Van Steveninck R, and Warland D. 1991. *Science* **252**(5014), 1854–1857.

Brette R and Gerstner W. 2005. *Journal of Neurophysiology* **94**, 3637–3642.

Bryant H L and Segundo J P. 1976. *Journal of Physiology* **260**(2), 279–314.

Carandini M, Horton J C, and Sincich L C. 2007. *Journal of Vision* **7**(14), 20.1–11.

Chacron M, Lindner B, Maler L, Longtin A, and Bastian J. 2005. *Proceedings of the SPIE* **5841**, 150–163.

Connors B W and Gutnick M J. 1990. *Trends in Neuroscience* **13**, 99–104.

David S V and Gallant J L. 2005. *Network* **16**(2–3), 239–60.

Destexhe A, Rudolph M, and Pare D. 2003. *Nature Reviews Neuroscience* **4**, 739–751.

Druckmann S, Banitt Y, Gidon A, and Schürmann F. 2007. *Frontiers in Neuroscience* **1**(1), 7–18.

Eggermont J J, Aertsen A M, and Johannesma P I. 1983. *Hearing Research* **10**(2), 167–90.

Gerstner W and Naud R. 2009. *Science* **326**, 379–380.

Gollisch T and Meister M. 2008. *Science* **319**, 1108–1111.

Hodgkin A L and Huxley A F. 1952. *Journal of Physiology* **117**(4), 500–544.

Hubel D H and Wiesel T N. 1959. *Journal of Physiology* **148**, 574–591.

Izhikevich E. 2004. *IEEE Transactions on Neural Networks* **15**(5), 1063–1070.

Izhikevich E M. 2007. *Dynamical Systems in Neuroscience: The Geometry of Excitability and Bursting.* MIT Press, Cambridge, MA.

Johansson R S and Birznieks I. 2004. *Nature Neuroscience* **7**(2), 170–177.

Jolivet R 2005. Effective minimal threshold models of neuronal activity. PhD thesis, Ecole Polytechnique Federale de Lausanne.

Jolivet R, Kobayashi R, Rauch A, Naud R, Shinomoto S, and Gerstner W. 2008a. *Journal of Neuroscience Methods* **169**, 417–424.

Jolivet R, Rauch A, Lüscher H, and Gerstner W. 2006. *Journal of Computational Neuroscience* **21**, 35–49.

Jolivet R, Schürmann F, Berger T, Naud R, Gerstner W, and Roth A. 2008b. *Biological Cybernetics* **99**(4), 417–426.

Keat J, Reinagel P, Reid R C, and Meister M. 2001. *Neuron* **30**, 803–817.

Kistler W, Gerstner W, and Hemmen J. 1997. *Neural Computation* **9**, 1015–1045.

Kobayashi R, Tsubo Y, and Shinomoto S. 2009. *Frontiers in Computational Neuroscience* **3**, 9.

Larkum M E, Senn W, and Luscher H R. 2004. *Cerebral Cortex* **14**(10), 1059–1070.

Larkum M, Zhu J, and Sakmann B. 1999. *Nature* **398**, 338–341.

Mainen Z F, Joerges J, Huguenard J R, and Sejnowski T J. 1995. *Neuron* **15**(6), 1427–1439.

Mainen Z F and Sejnowski T J. 1995. *Science* **268**(5216), 1503–1506.

Markram H, Toledo-Rodriguez M, Wang Y, Gupta A, Silberberg G, and Wu C. 2004. *Nature Reviews Neuroscience* **5**(10), 793–807.

Marmarelis P Z and Marmarelis V Z. 1978. *Analysis of Physiological Systems: The White-Noise Approach.* Plenum Press, New York.

Mensi S, Naud R, Avermann M, Petersen C, and Gerstner W. 2012. *Parameter Extraction and Classification of Three Neuron Types Reveals Two Different Adaptation Mechanisms* **107**(6), 1756–1775.

* http://www.incf.org/

Naud R, Gerhard F, Mensi S, and Gerstner W. 2011. *Neural Computation* **23**(12), 3016–3069.
Naud R, Marcille N, Clopath C, and Gerstner W. 2008. *Biological Cybernetics* **99**, 335–347.
Paninski L, Pillow J, and Simoncelli E. 2004. *Neural Computation* **16**, 2533–2561.
Petersen R S, Brambilla M, Bale M R, Alenda A, Panzeri S, Montemurro M A, and Maravall M. 2008. *Neuron* **60**(5), 890–903.
Pillow J, Paninski L, Uzzell V, Simoncelli E, and Chichilnisky E. 2005. *Journal of Neuroscience* **25**(47), 11003.
Poliakov A V, Powers R K, Sawczuk A, and Binder M C. 1996. *Journal of Physiology* **495**, 143–157.
Pospischil M, Toledo-Rodriguez M, Monier C, Bal T, Frégnac Y, Markram H, and Destexhe A. 2008. *Biological Cybernetics* **99**, 427–444.
Schaette R, Gollisch T, and Herz A V M. 2005. *Journal of Neurophysiology* **93**(6), 3270–3281.
Softky W and Koch C. 1993. *Journal of Neuroscience* **13**, 334–350.
Stein R B. 1967. *Proceedings of the Royal Society of London. Series B, Biological Sciences* **167**(1006), 64–86.
Victor J D and Purpura K. 1996. *Journal of Neurophysiology* **76**(2), 1310–1326.

4 Statistical Identification of Synchronous Spiking

Matthew T. Harrison, Asohan Amarasingham, and Robert E. Kass

CONTENTS

4.1 INTRODUCTION

In some parts of the nervous system, especially in the periphery, spike timing in response to a stimulus, or in production of muscle activity, is highly precise and reproducible. Elsewhere, neural spike trains may exhibit substantial variability when examined across repeated trials. There are many sources of the apparent variability in spike trains, ranging from subtle changes in experimental conditions to features of neural computation that are basic objects of scientific interest. When variation is large enough to cause potential confusion about apparent timing patterns, careful statistical analysis can be critically important. In this chapter, we discuss statistical methods for analysis and interpretation of synchrony, by which we mean the approximate temporal alignment of spikes across two or more spike trains. Other kinds of timing patterns are also of interest [2,56,73,68,23], but synchrony plays a prominent role in the literature, and the principles that arise from consideration of synchrony can be applied in other contexts as well.

The methods we describe all follow the general strategy of handling imperfect reproducibility by formalizing scientific questions in terms of statistical models, where relevant aspects of variation are described using probability. We aim not only to provide a succinct overview of useful techniques, but also to emphasize the importance of taking this fundamental first step, of connecting models with questions, which is sometimes overlooked by nonstatisticians. More specifically, we emphasize that (i) detection of synchrony presumes a model of spiking without synchrony, in statistical jargon this is a null hypothesis, and (ii) quantification of the amount of synchrony requires a richer model of spiking that explicitly allows for synchrony, and in particular, permits statistical estimation of synchrony.

Most simultaneously recorded spike trains will exhibit some *observed synchrony*. Figure 4.1 displays spike trains recorded across 120 trials from a pair of neurons in primary visual cortex, and circles indicate observed synchrony, which, in this case, is defined to be times at which both neurons fired within the same 5 ms time bin. Here, as is typical in cortex under many circumstances, the spike times are not highly reproducible across trials and are, instead, erratic. This variability introduces ambiguity in the neurophysiological interpretation of observed synchrony. On the one hand, observations of synchronous firing may reflect the existence of neuronal mechanisms, such as immediate common input or direct coupling that induce precisely timed, coordinated spiking. It is this potential relationship between observed synchrony and mechanisms of precise, coordinated spike timing that motivated the writing of this chapter. On the otherhand, the apparent variability in spike timing could be consistent with mechanisms that induce only coarsely timed spiking. For coarsely timed spikes, with no mechanisms of precise timing, there will be time bins in which both neurons fire, leading to observed synchrony. In general, we might

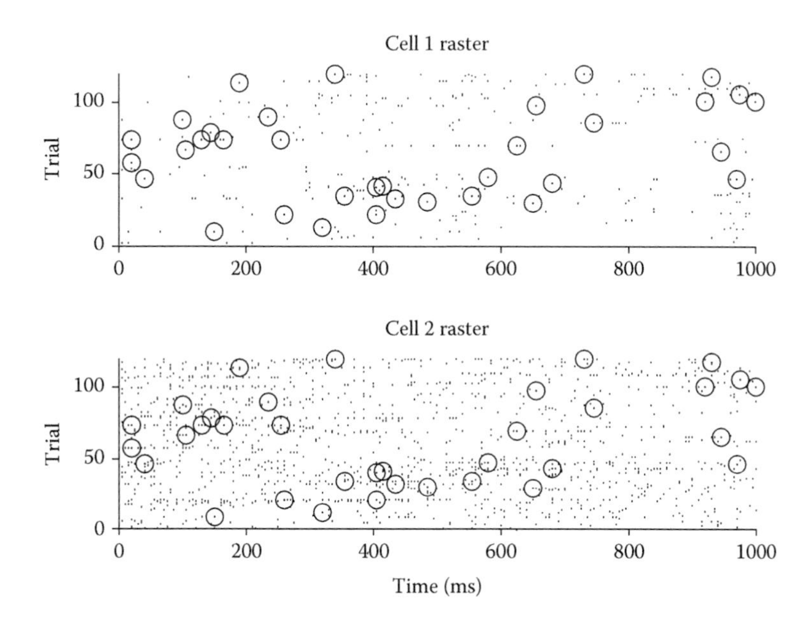

FIGURE 4.1 Raster plots of spike trains on 120 trials from two simultaneously recorded neurons, with synchronous spikes shown as circles. Here, observed synchrony is defined using time bins having 5 ms width. Spiking activity was recorded for 1 s from primary visual cortex in response to a drifting sinusoidal grating, as reported in Refs. [42] and [45].

expect that spike timing is influenced by a variety of mechanisms across multiple time scales. Many of these influences, such as signals originating in the sensory system or changes in the extracellular environment, may affect multiple neurons without necessarily inducing precise, coordinated spiking. Each of these influences will affect the amount of observed synchrony.

Observations of synchrony in the presence of apparent variability in spike timing thus raise natural questions. How much of the observed synchrony in Figure 4.1 is stimulus-related? How much comes from slow-wave oscillations? How much comes from mechanisms for creating precisely timed, coordinated spiking? How much is from chance alignment of coarsely timed spikes? Statistical approaches to such questions begin with the formulation of probabilistic models designed to separate the many influences on spike timing. Especially because of the scientific interest in mechanisms of precise spike timing, neurophysiologists often seek to separate the sources of observed synchrony into coarse temporal influences and fine temporal influences. Accordingly, many such statistical models are designed to create a distinction of time scale. In this chapter, we will discuss many of the modeling devices that have been used to create such distinctions.

We wish to acknowledge that the basic statistical question of quantifying synchrony is only relevant to a subset of the many physiologically interesting questions about synchrony. The methods described here are designed to detect and measure precise synchrony created by the nervous system and can reveal much about the temporal precision of cortical dynamics, but they do not directly address the role of synchrony in cortical information processing. Cortical dynamics may create excess synchrony, above chance occurrence, even if this synchrony plays no apparent role in downstream processing. Conversely, downstream neurons might respond to synchronous spikes as a proxy for simultaneously high firing rates even if the upstream neurons are not influenced by mechanisms that could create precise spike timing. Nevertheless, the existence of statistically detectable synchrony, beyond that explicable by otherwise apparent variation in firing patterns, not only reveals clues about cortical dynamics, but also forms a kind of *prima facie* evidence for the relevance of precise spike timing for cortical information processing.

4.2 SYNCHRONY AND TIME SCALE

In this section, we discuss what we mean by *synchrony,* which we associate with transient changes in correlation on *sub-behavioral time scales.* Our main point is that the distinction between synchrony and other types of dependence relies primarily on distinctions of *time scale.*

The cross-correlation histogram (CCH) is a popular graphical display of the way two neurons tend to fire in conjunction with each other. CCHs are described in more detail in Section 4.4.1, but, roughly speaking, the x-axis indicates the time lag (ℓ) of the synchrony and the y-axis indicates the number of observed lag-ℓ synchronous spikes. A peak (or trough) in the CCH provides evidence of correlation between the spike trains. Precisely timed synchrony is usually identified with a narrow peak in the CCH. Figure 4.2 shows a CCH taken from a pair of neurons in primary visual cortex [45]. Its appearance, with a peak near 0, is typical of many CCHs that are used to demonstrate synchronous firing. While we presume readers to be familiar with CCHs, we wish to provide a few caveats and will review throughout Section 4.4 the ways CCHs have been modified to better identify synchrony.

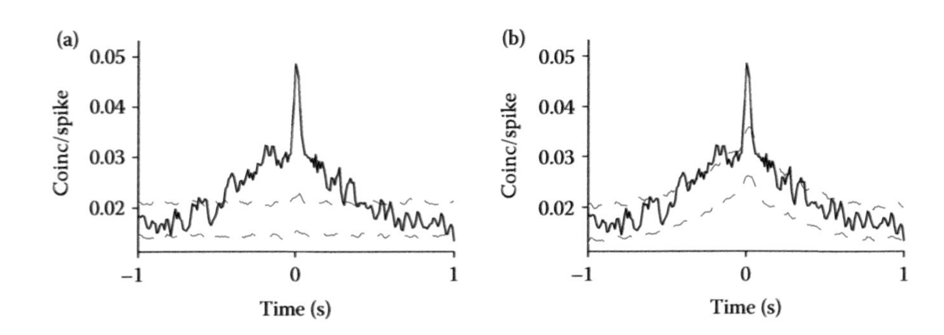

FIGURE 4.2 CCHs from a pair of neurons recorded simultaneously from primary visual cortex, as described in Ref. [45]. The x-axis displays the time lag between the two spike trains. The CCH indicates correlations only at short lags, on the scale of the x-axis. (a) Bands show acceptance region for a test of independence based on a point process model involving stimulus effects. (b) Bands show acceptance region for a test of independence based on a point process model involving both stimulus effects and slow-wave network effects, possibly due to anesthesia. (Courtesy of Ryan Kelly.)

The first thing to keep in mind when looking at a CCH is that its appearance, and therefore its biological interpretation, depends on the units of the x-axis. A 500 ms wide peak in the CCH carries a much different neurophysiological interpretation than a 5 ms wide peak, even though they would look the same if the x-axis were rescaled. A broad peak on the scale of hundreds of milliseconds can be explained by any number of processes operating on behavioral time scales, such as modulations in stimulus, response, metabolism, and so on, many of which will be difficult to distinguish among. A peak of only a few milliseconds is indicative of processes operating on sub-behavioral time scales, perhaps revealing features of the micro-architecture of the nervous system. When we use the term "synchrony" we are referring to correlated spiking activity within these much shorter time windows.

To illustrate the concept of time scale, the left column of Figure 4.3 shows several artificial CCHs that are created by mathematically identical processes. The only difference is the temporal units assigned to a statistical coupling between two spike trains. The expected shape of each CCH can be made to look like the others simply by zooming in or out appropriately. If these were real CCHs, however, the biological explanations would be quite different. Only the fourth CCH (in the left column), with a narrow 1 ms wide peak would almost universally be described as an example of synchrony in the neurophysiology literature. The third CCH with a 10 ms wide peak is more ambiguous and investigators may differ on their willingness to identify such a peak as synchrony. The other two, with widths of 100 ms or more, would typically not be identified as synchrony.

The CCHs in Figure 4.3 are unrealistic only in their simplicity. The complexity of true spike trains only further complicates issues. A real CCH will show correlations from many processes operating on a variety of time scales. Some of these correlations may be synchrony. Statistical models that distinguish synchrony from other types of correlation will necessarily involve determinations of time scale.

4.3 SPIKE TRAINS AND FIRING RATE

Statistical models of synchrony are embedded in statistical models of spike trains. In this section, we review the basic probabilistic description of a spike train as a *point process*, and we draw analogies between the conditional intensities of a point process and the conceptual notion of a neuron's instantaneous firing rate. Our main point is that there are many different ways to define the conditional intensity of a point process and there are correspondingly many ways to define firing rates. As such, when discussing firing rates it is always important to be clear about terminology and modeling assumptions. We assume that readers are familiar with the basic objects of probability and statistics, but see Appendix for a brief summary.

Since the work of Adrian [3] and Hartline [34], the most basic functional description of a neuron's activity has been its *observed firing rate*. A simple definition of observed firing rate at time t is the spike count per unit time in an interval containing t. The observed firing rate depends heavily on the length of the interval used in its definition. For long intervals, the observed firing rate is forced to be slowly varying in time. For short intervals, the observed firing rate can be quickly varying, but

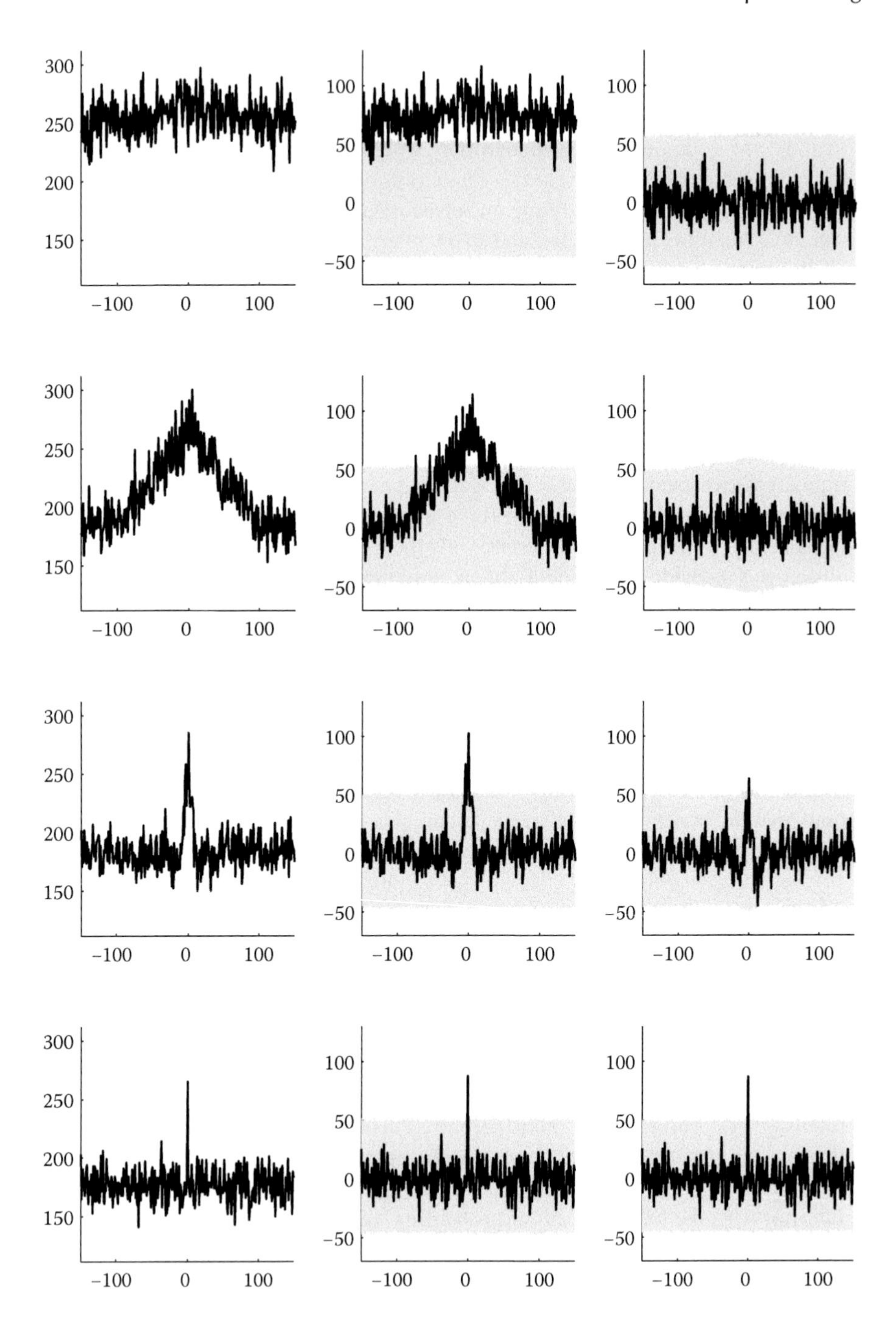

FIGURE 4.3

tends to fluctuate wildly.* This dependence on interval length complicates the use of observed firing rate, especially if one wishes to discuss quickly varying firing rates.

For statistical purposes it is convenient to conceptualize spike trains as random, with the observed rate of spiking fluctuating randomly around some theoretical time-varying *instantaneous firing rate*. This conceptualization decomposes the observed firing rate into a signal component, namely, the instantaneous firing rate, and a noise component, namely, the fluctuations in spike times. It is important to note that the statistical distinction between signal and noise is not necessarily a statement about physical randomness, but rather a useful modeling device for decomposing fluctuations in the observed firing rate. Presumably, with sufficiently detailed measurements about the various sources of neural spiking, one could deterministically predict the timing of all spikes so that the noise becomes negligible and the instantaneous firing rate begins to coincide with the observed firing rate in small intervals. Since these detailed measurements are not available, it is convenient to model their effects with some notion of randomness. As this extreme case illustrates, instantaneous firing rates do not correspond to physical reality, but reflect a particular way of separating signal and noise into a form that is, hopefully, convenient for a particular scientific or statistical investigation.

4.3.1 Point Processes, Conditional Intensities, and Firing Rates

The mathematical theory of point processes [18] is the natural formalism for the statistical modeling of spike trains. Suppose we observe n spikes (action potentials) on a single trial and we use $s_1, s_2, \ldots, s_n$ to denote the times at which the spikes occur (a spike train). Also, let us take the beginning of the observation period to be at time

FIGURE 4.3 CCHs for four simulations with correlations on different time scales. Each row corresponds to a different simulation. Each column corresponds to a different type of CCH. The x-axis is lag in milliseconds. The y-axis is observed synchrony count (1 ms bins). For each simulation, each of the two spike trains could be in either a high-firing-rate or low-firing-rate state at each instant in time. The two spike trains shared the same state and the state occasionally flipped. This shared (and changing) state creates a correlation between the two spike trains. The rate of flipping is what changes across these figures. In each case the state has a 50% chance of flipping every w ms, where $w = 1000, 100, 10, 1$ from top to bottom, respectively. (So $w = 1$ ms corresponds to "fine" temporal coupling, and $w = 1000$ ms corresponds to "coarse" temporal coupling. Of course, the distinction between "fine" and "coarse" is qualitative, not quantitative.) Left column: Raw CCHs. Middle column: Uniform-corrected CCHs (black line; see Section 4.4.3.6) and 95% simultaneous acceptance regions (gray region; see Section 4.4.3.4). Except for the units of the y-axis and negligible (in this case) edge effects, the uniform-corrected CCHs and their acceptance regions are essentially identical to using the correlation coefficient (Pearson's r between the lagged spike trains viewed as binary time series with 1 ms bins) and testing the null hypothesis of independence. Right column: 20 ms jitter-corrected CCHs (see Section 4.4.5.2) with 95% simultaneous acceptance regions.

* For tiny intervals of length δ, the observed firing rate jumps between zero (no spike in the interval) and $1/\delta$ (a single spike in the interval), depending more on the length of the interval than on any sensible notion of the rate of spiking.

0, and define $s_0 = 0$. Then $x_i = s_i - s_{i-1}$ is the ith interspike interval, for $i = 1, \ldots, n$. Given a set of positive interspike interval values $x_1, x_2, \ldots, x_n$ we may reverse the process and find $s_j = \sum_{i=1}^{j} x_i$ to be the jth spike time, for $j = 1, \ldots, n$. Similarly, in probability theory, if we have a sequence of positive random variables $X_1, X_2, \ldots$ the sequence of random variables $S_1, S_2, \ldots$ defined by $S_j = \sum_{i=1}^{j} X_i$ forms a *point process*. It is thus a natural theoretical counterpart of an observed spike train. Data are recorded with a fixed timing accuracy, often corresponding with a sampling rate such as 1 kHz. Spike trains recorded with an accuracy of 1 ms can be converted to a sequence of 0s and 1s, where a 1 is recorded whenever a spike occurs and a 0 is recorded otherwise. In other words, a spike train may be represented as a binary time series. Within the world of statistical theory, every point process may be considered, approximately, to be a binary time series.

Another way of representing both spike trains and their theoretical counterparts is in terms of counts. Let us use $N(t)$ to denote the total number of spikes occurring up to and including time t. Thus, for $s < t$, the spike count between time s and time t, up to and including time t, is $N(t) - N(s)$. The expected number of spikes per unit time is called the *intensity* of the point process. If the point process is time invariant, it is called stationary or *homogeneous*, and the intensity is constant, say $\lambda \geq 0$. Otherwise, the point process is nonstationary or *inhomogeneous*, and the intensity can be a function of time, say $\lambda(t)$. The intensity can be conceptualized as the instantaneous probability (density) of observing a spike at time t, suggestively written as

$$P(\text{spike in}(t, t + \mathrm{d}t]) = \lambda(t)\, \mathrm{d}t, \tag{4.1}$$

where $\mathrm{d}t$ denotes an infinitesimal length of time.

At first glance, the intensity of a point process feels exactly like the notion of a theoretical time-varying instantaneous firing rate that we were looking for, and it is tempting to equate the two concepts. But in many cases the basic intensity does not match our intuitive notion of an instantaneous firing rate. Consider two experimental conditions in which a neuron has a consistently larger spike count in one condition than the other. It seems natural to allow the instantaneous firing rate to depend on experimental condition, so we need a notion of an intensity that can depend on the experimental condition. Recall that biological spike trains have refractory periods so that a new spike cannot occur ("absolute refractory period") or is less likely to occur ("relative refractory period") during a brief (millisecond-scale) interval immediately following a spike. If a spike just occurred, it might be useful in some contexts to think about the instantaneous firing rate as being reduced for a short time following a spike, so we need a notion of an intensity that can depend on the history of the process. Finally, we might imagine that the baseline excitability of a neuron is modulated by some mechanism, creating a kind of gain modulation [60], that is not directly observable. If we want to model this mechanism as having an effect on the instantaneous firing rates, then we also need to consider intensities that depend on unobservable variables.

The intensity can be generalized to accommodate each of the above considerations by allowing it to depend on variables other than time. This more general notion of intensity is often called a *conditional intensity* because it is defined using conditional probabilities. It is the notion of a conditional intensity that corresponds most closely with the intuitive concept of a theoretical instantaneous firing rate. For example, if Z_t is some additional information, like experimental condition or attentional state (or both), then we could write the conditional intensity given Z_t as

$$P(\text{spike in}(t, t + dt] \mid Z_t) = \lambda(t \mid Z_t)dt. \tag{4.2}$$

Now the conditional intensity, and hence, the instantaneous firing rate, can depend on Z_t. Z_t can have both observed components, O_t, and unobserved components, U_t, so that $Z_t = (O_t, U_t)$. These components can be fixed or random, and they may or may not be time varying. When Z_t has random components, the conditional intensity given Z_t is random, too. Statistical models often draw distinctions between observed and unobserved variables and also between nonrandom and random variables. It is common to include in Z_t the internal history H_t of spike times prior to time t. Specifically, if prior to time t there are n spikes at times $s_1, s_2, \ldots, s_n$, we write $H_t = (s_1, s_2, \ldots, s_n)$. Including H_t as one of the components of Z_t allows the conditional intensity to accommodate refractory periods, bursting, and more general spike-history effects. By also including other variables, the instantaneous firing rates can further accommodate stimulus effects, network effects, and various sources of observed or unobserved gain modulation.

The conditional intensity given only the internal history, that is, $Z_t = H_t$, is special because it completely characterizes the entire distribution of spiking. There are no additional modeling assumptions once the conditional intensity is specified. Many authors reserve the term conditional intensity for the special case of conditioning exclusively on the internal history and use terms such as the Z_t-intensity or the intensity with respect to Z_t for the more general case. We will use the term conditional intensity in the more general sense, because it emphasizes our main point: conditional intensities, and hence instantaneous firing rates, necessitate the specification of exactly what information is being conditioned upon.

The simplest of all point processes is the Poisson process. It has the properties that (i) for every pair of time values s and t, with $s < t$, the random variable $N(t) - N(s)$ is Poisson distributed and (ii) for all nonoverlapping time intervals the corresponding spike count random variables are independent. Poisson processes have the special *memoryless* property that their conditional intensity given the internal history does not depend on the internal history, but is the same as the unconditional intensity: $\lambda(t|H_t) = \lambda(t)$. They cannot exhibit refractory periods, bursting, or any other spike history effects that do not come from the intensity. The intensity cannot depend on unobserved, random variables. Homogeneous Poisson processes (HPP) are time invariant in the sense that the probability distribution of a count $N(t) - N(s)$ depends only on the length of the time interval $t - s$. Its intensity is constant, say, $\lambda \geq 0$, and its entire distribution is summarized by this single parameter λ.

Observable departures from Poisson spiking behavior have been well documented (e.g., [6,62,69]). On the other hand, such departures sometimes have relatively little effect on statistical procedures, and Poisson processes play a prominent role in theoretical neuroscience and in the statistical analysis of spiking data. Certain types of non-Poisson processes can be built out of Poisson processes by using random variables and the appropriate conditional intensities. For example, if U is a random variable and we model the conditional distribution given U as a Poisson process, that is, $\lambda(t|H_t,U) = \lambda(t|U)$, then the resulting unconditional process is called a *Cox process* (or doubly stochastic Poisson process). It can be viewed as a Poisson process with a random instantaneous firing rate. Alternative analyses, not relying on Poisson processes at all, typically proceed by explicitly modeling $\lambda(t|H_t)$, thus capturing the (non-Poisson) history dependence of a spike train.

4.3.2 Models of Conditional Intensities

For statistical purposes, it is not enough to specify which type of conditional intensity (i.e., which type of conditioning) we are equating with the notion of a firing rate, but we must also appropriately restrict the functional form of the conditional intensity (i.e., the allowable "shapes" of the function λ). If the functional form is fully general, we could fit any data set perfectly without disambiguating the potential contributions to observed synchrony.* There are three common ways to restrict the functional form of conditional intensities: identical trials, temporal smoothness, and temporal covariates.

For identical trials models, we assume that our conditional intensity is identical across trials. This relates the value of the conditional intensity across multiple points in time. Another way to think about identical trials is that time, that is, the t in $\lambda(t)$, refers to trial-relative time, and that our data consist of many independent observations from a common point process. The identical trials perspective and the use of the (unconditional) intensity, or perhaps a conditional intensity given experimental condition, are what underlies the interpretation of the peri-stimulus time histogram (PSTH)† as an estimator of the firing rate. This perspective is also at the heart of the well-known shuffle-correction procedure for the CCH and for other graphical displays of dependence.

For temporal smoothness models, we assume that our conditional intensity is slowly varying in time in some specific way. Severe restrictions on temporal smoothness, such as assumptions of stationarity or homogeneity, permit the aggregation of statistical information across large intervals of time. But even mild restrictions on temporal smoothness permit the aggregation of certain types of statistical information, particularly the signatures of precise timing [7]. This perspective underlies the jitter approach to synchrony identification.

For temporal covariates models, we assume that our conditional intensity depends on time only through certain time-varying, observable covariates, such as a time-varying

* The problem is particularly easy to see for temporally discretized spike trains that can be viewed as binary time series. In this case, the intensities specify probabilities of observing a spike in each time bin and the probabilities can vary arbitrarily across time bins and across spike trains. We can choose an intensity that assigns probability one for each time bin with a spike and probability zero for each time bin without a spike. This intensity (i.e., the firing rate) perfectly accounts for the data, and we do not need additional notions, like synchrony, to explain its features. In statistical parlance, we face a problem of *nonidentifiability*: models with and without explicit notions of synchrony can lead to the same probability specification for the observed data.

† The PSTH is simply a histogram of all trial-relative spike times from a single neuron, perhaps normalized to some convenient units like spikes per unit time.

stimulus or behavior, say Z_t. Models of hippocampal place cells, directionally tuned motor–cortical cells, and orientation-specific visual–cortical cells often take this form. Symbolically, we might write $\lambda(t|Z_t) = g(Z_t)$ for some function g. When Z_t repeats, we can accumulate statistical information about λ. Or, if g has some assumed parametric form, such as cosine tuning to an angular covariate, then every interval conveys statistical information about our conditional intensity.

In general, statistical models might combine these principles for modeling conditional intensities, or perhaps, incorporate restrictions that do not fit cleanly into these three categories. The generalized linear model approach that we discuss in Section 4.4.6 allows all of these modeling assumptions to be included in a common framework. We discuss several different types of models below from the perspective of synchrony detection. Many of these models involve point processes and instantaneous firing rates, that is, conditional intensities, but they do not all use the same type of point process nor the same definition of firing rate. When interpreting a statistical model for spike trains it is crucial to understand how the model and the terminology map onto neurophysiological concepts. In particular, the term "firing rate" often refers to different things in different models.

4.4 MODELS FOR COARSE TEMPORAL DEPENDENCE

As we discussed in the Introduction, the statistical identification of synchrony requires a separation of the fine temporal influences on spike timing from the many coarse temporal influences on spike timing. In this section, we focus primarily on models that only allow coarse temporal influences on spike timing. Besides being useful building blocks for more general models, these coarse temporal spiking models are actually quite useful from the statistical perspective of *hypothesis testing*. For hypothesis testing, we begin with a model, the *null hypothesis* or the null model, that allows only coarse temporal spiking and ask if the observed data are consistent with the model. A negative conclusion, that is, a rejection of the null hypothesis, is interpreted as evidence in favor of fine temporal spiking dynamics that create excess observed synchrony, although, strictly speaking, it is really only evidence that the null model is not an exact statistical description of the data. (The Appendix contains a brief overview of the principles of hypothesis testing.) Our main point in this section is that some models are more appropriate than others for describing the coarse temporal influences on spike timing.

To simplify the discussion and to provide easy visual displays of the key concepts, we restrict our focus to the case of two spike trains and we concentrate primarily on the CCH as our way of measuring and visualizing raw dependence—for hypothesis testing, this means that *test statistics* are derived from the CCH. Section 4.4.7 discusses how to relax each of these restrictions.

4.4.1 Cross-Correlation Histogram (CCH)

We begin with a more precise description of the CCH. Suppose we have two spike trains with spike times $s_1, \ldots, s_m$ and $t_1, \ldots, t_n$, respectively. There are mn different *pairs* of spike times (s_i, t_j) with the first spike taken from spike train 1 and the second from spike train 2. Each of these pairs of spike times has a lag $\ell_{ij} = t_j - s_i$, which is just the difference in spike times between the second and the first spike. These *mn* lags are the data that are used to create a CCH. Here, a CCH is simply a histogram of all of these lags.

There are many choices to make when constructing histograms. One of the most important is the choice of bin-width, or more generally, the choice of how to smooth the histogram. Figure 4.4 shows several CCHs constructed from a single collection

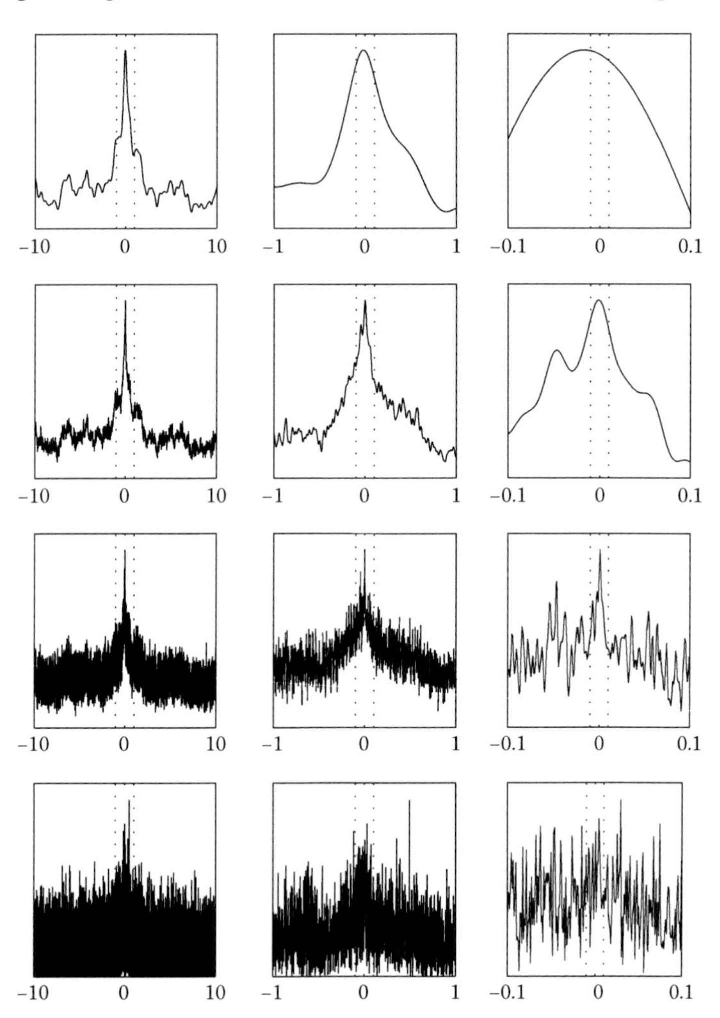

FIGURE 4.4 CCHs for two simultaneously recorded neurons in monkey motor cortex. Each column of CCHs shows a different range of lags, as indicated on the *x*-axis (in seconds). Each row of CCHs shows a different amount of smoothing. For the first two columns, the vertical dotted lines in each CCH indicate the region shown in the CCH to its immediate right. For the final column, the vertical dotted lines indicate ±10 ms. Peaks within this final region correspond to the time scales usually associated with synchrony. (Courtesy of Nicholas Hatsopoulos.)

of lags, but using different amounts of smoothing and showing different ranges of lags. By varying the degree of smoothing, one can choose to emphasize different aspects of the correlation structure. Only one of the smoothing choices reveals a narrow peak in the CCH that is suggestive of zero-lag synchrony (row 3, column 3). As there are entire subfields within statistics devoted to the art of smoothing histograms, we will not delve further into this topic, but caution that visual inspection of CCHs can be quite misleading, not in small part because of choices in smoothing.

Another practical detail when dealing with CCHs is how one deals with *edge effects*. Edge effects refer to the fact that, for a fixed and finite observation window, the amount of the observation window that can contain a spike pair varies systematically with the lag of the spike pair. For example, if a spike pair has a lag that is almost as long as the observation window, then one of the spikes must be near the beginning of the window and the other must be near the end. On the other hand, if the lag is near zero, then the spikes can occur anywhere in the observation window (as long as they occur close to each other). This creates the annoying feature that a raw CCH will typically taper to zero for very long lags, not because of any drop in correlation, but rather because of a lack of data.*

We view the CCH as a *descriptive statistic*, a simple summary of the data. Only later, when we begin to introduce various models will the CCH take on more nuanced interpretations. These modeling choices will affect such interpretations greatly. To illustrate the distinction, suppose we have finely discretized time so that the theoretical spike trains can be represented as binary time series, say $(U_k : k \in \mathbb{Z})$ and $(V_k : k \in \mathbb{Z})$, where, for example, U_k indicates spike or no spike in the first spike train in time bin k. ($\mathbb{Z}$ is the set of integers.) Suppose that we model each of the time series to be time invariant or *stationary*.† Then the expected value of the product of U_k and $V_{k+\ell}$ depends only on the lag ℓ, and we can define the expected cross-correlation function (using signal processing terminology) as

$$\gamma(\ell) = E[U_k V_{k+\ell}] = E[U_0 V_\ell]$$

It is straightforward to verify that a properly normalized CCH‡ is an unbiased estimator of the expected cross-correlation function, γ. As such, the CCH takes on additional interpretations under the model of stationarity, namely, its interpretation as a (suitable) *estimator* of a theoretical model-based quantity. Note that the terminology involving cross-correlation functions is not standardized across fields.

---

* For a contiguous observation window of length T, a common solution is to divide the CCH at lag ℓ by $T - |\ell|$, which is the amount of the observation window available to spike pairs having lag ℓ. Many authors would include this in their definition of the CCH.

† $(V_k : k \in \mathbb{Z})$ is stationary if the distribution of the vector of random variables $(U_k, U_{k+1}, \ldots, U_{k+m})$ is the same as the distribution of $(U_j, U_{j+1}, \ldots, U_{j+m})$ for all j, k, m. In other words, the process looks statistically identical in every observation interval.

‡ For data collected in a continuous observation window with T time bins and defining the CCH as a standard histogram with bin widths that match the discretization of time, we need to divide the CCH at lag ℓ by $(T - |\ell|)(2T - 1)$. This is the edge correction of footnote * (in this page) with an additional term to match the units of γ.

In statistics, for example, the cross-correlation function is usually defined as the Pearson correlation

$$\rho(\ell) = Cor(U_k, V_{k+\ell}) = \frac{\gamma(\ell) - E[U_k]E[V_{k+\ell}]}{\sqrt{Var[U_k]Var[V_{k+\ell}]}} = \frac{\gamma(\ell) - E[U_0]E[V_0]}{\sqrt{Var[U_0]Var[V_0]}}$$

which is a distribution-dependent shifting and scaling of γ (meaning that γ and ρ have the same shape as a function of ℓ). After appropriate standardization, the CCH can also be converted into a sensible estimator of ρ under the stationary model. In the following, however, we interpret the (raw) CCH only as a descriptive statistic without additional modeling assumptions, like stationarity.

4.4.2 Statistical Hypothesis Testing

The data for the CCHs in Figure 4.4 came from two simultaneously recorded neurons on different electrodes in monkey motor cortex (courtesy of Nicholas Hatsopoulos; see [37] for additional details). Like many real CCHs, these would seem to show an amalgamation of correlations on different time scales, perhaps from different processes. The broad (±500 ms) peak is easily attributable to movement-related correlations—like many neurons in motor cortex, these neurons have coarse temporal observed firing rates that modulate with movement. Smaller fluctuations on top of this broad correlation structure might be attributable to noise or perhaps finer time-scale processes. The CCH alone cannot resolve these possibilities. Careful statistical analysis is needed.

To test whether certain features of a CCH are consistent with coarse temporal spiking, we first need to create a null hypothesis, that is, a statistical model that allows for coarse temporal influences on spike timing. Our emphasis in this chapter is on this first step: the choice of an appropriate null model. Next, we need to characterize the typical behavior of the CCH under the chosen null model. This step involves a great many subtleties, because a null hypothesis rarely specifies a unique (null) distribution for the data, but rather a family of distributions. We only scratch the surface in our treatment of this second step, briefly discussing bootstrap approximation and conditional inference as possible techniques. Finally, we need to quantify how well the observed CCH reflects this typical behavior, often with a p-value. Here, we use graphical displays showing bands of typical fluctuation around the observed CCH.

4.4.3 Independent Homogeneous Poisson Process (HPP) Model

For pedagogical purposes, we begin with an overly simple model to illustrate the main ideas and introduce the terminology that we use throughout the chapter. Suppose we model two spike trains, say A and B, as independent HPPs. In this case, it does not matter whether we think about repeated trials, or a single trial, because the intensities are constant and all disjoint intervals are independent. The distribution of each spike train is specified by a single parameter, the (constant) intensity, say

λ^A and λ^B, respectively. Our model is that the spike trains only "interact" (with each other and with themselves) through the specification of these two constants. Once the constants are specified, there is no additional interaction because the spike trains are independent and Poisson.

Using the independent HPP model as a null hypothesis is perhaps the simplest way to formalize questions like, "Is there more synchrony than expected by chance?" or "Is there more synchrony than expected from the firing rates alone?" In order to be precise, these questions presume models that define "chance" and "firing rates" and "alone." The independent HPP model defines the firing rate to be the unconditional and constant intensity of an HPP. Chance and alone refer to the assumption that the spike trains are statistically *independent*: there are no other shared influences on spike timing. In particular, there are no mechanisms to create precise spike timing.

We caution the reader that concepts such as "chance" and "independence," which appear frequently throughout the chapter (and in the neurophysiology literature), can mean different things in the context of different models. For example, independence often means *conditional* independence given the instantaneous firing rates. But as we have repeatedly emphasized, there is great flexibility in how one chooses to define the instantaneous firing rates. This choice about firing rates then affects all subsequent concepts, like independence, that are defined in relation to the firing rates.*

The independent HPP model fails to account for many biologically plausible influences on coarse temporal spiking. For example, it fails to account for systematic modulations in observed firing rate following the presentation of a repeated stimulus, which would require models with time-varying firing rates. If the independent HPP model were used to detect fast-temporal synchrony, then we could not be sure that a successful detection (i.e., a rejection of the independent HPP null hypothesis) was a result of synchrony or merely a result of coarse temporal spiking dynamics that are not captured by the independent HPP model. Despite these shortcomings, the independent HPP model is pedagogically useful for understanding more complicated models.

4.4.3.1 Bootstrap Approximation

The first problem that one encounters when trying to use the independent HPP model is that the firing rates λ^A and λ^B are unknown. The independent HPP model does not specify a unique distribution for the data, but rather a parameterized family of distributions. In particular, the model does not uniquely specify the typical variation in the CCH. The typical variation will depend on the unknown firing rate parameters λ^A and λ^B. To use the model, it is convenient to first remove this ambiguity about the unknown firing rates.

One possible approach to removing ambiguity is to replace the unknown λ^A and λ^B with approximations derived from the observed data, say $\lambda^A \approx \hat{\lambda}^A$ and $\lambda^B \approx \hat{\lambda}^B$, where each $\hat{\lambda}$ is some sensible estimator of the corresponding parameter λ, such as

---

* This ambiguity of terminology is not peculiar to spike trains. Probabilities (and associated concepts, like independence and mean) are always defined with respect to some frame of reference, which must be clearly communicated for probabilistic statements to make sense. Another way to say this is that all probabilities are *conditional* probabilities, and we must understand the conditioning event to understand the probabilities [40, p. 15].

the total spike count for that spike train divided by the duration of the experiment (the maximum likelihood estimate). Replacing unknown quantities with their estimates is the first and most important approximation in a technique called *bootstrap*; see Section 4.4.3.3. Consequently, we refer to this method of approximation as a *bootstrap approximation*. (The terminology associated with bootstrap is not used consistently in the literature.)

When using a bootstrap approximation, it is important to understand the method of estimation, its accuracy in the situation of interest, and the robustness of any statistical and scientific conclusions to estimation errors. For the independent HPP model, where we usually have a lot of data per parameter, this estimation step is unlikely to introduce severe errors. For more complicated models with large numbers of parameters, the estimation errors are likely to be more pronounced and using a bootstrap approximation is much more delicate and perhaps inappropriate.

Under the null model, after we have replaced the unknown parameters with their estimated values, we are left with a single (null) distribution for the data: independent HPPs with known intensities $\hat{\boldsymbol{\lambda}}^A$ and $\hat{\boldsymbol{\lambda}}^B$, respectively.

4.4.3.2 Monte Carlo Approximation

After reducing the null hypothesis to a single null distribution, it is straightforward to describe the null variability in the CCH. Perhaps the simplest way to proceed, both conceptually and algorithmically, is to generate a sample of many independent Monte Carlo pseudo-datasets from the null distribution (over pairs of spike trains) and compute a CCH for each one. This collection of pseudo-data CCHs can be used to approximate the typical variability under the null distribution in a variety of ways. Some examples are given in Sections 4.4.3.4 and 4.4.3.7.

Monte Carlo approximation is ubiquitous in this review and in the literature. We want to emphasize, however, that Monte Carlo approximation is rarely an integral component of any method. It is simply a convenient method of approximation, and is becoming increasingly convenient as the cost of intensive computation steadily decreases. Even for prototypical resampling methods such as bootstrap, trial shuffling, permutation tests, jitter, and so on, the explicit act of resampling is merely Monte Carlo approximation. It is done for convenience and is not an important step for understanding whether the method is appropriate in a particular scientific or statistical context. In many simple cases, one can derive analytic expressions of null variability, avoiding Monte Carlo approximation entirely.

4.4.3.3 Bootstrap

A bootstrap approximation (Section 4.4.3.1) followed by Monte Carlo approximation (Section 4.4.3.2) is usually called *bootstrap* [20]. In the context here it is *bootstrap hypothesis testing*, because the bootstrap approximation respects the modeling assumptions of the null hypothesis. For the case of the independent HPP model described above, it is *parametric* bootstrap hypothesis testing, because the bootstrap approximation replaces a finite number of parameters with their estimates. It is the bootstrap approximation in Section 4.4.3.1 that is most important for understanding the behavior of bootstrap. The Appendix contains a more detailed example describing bootstrap hypothesis testing.

4.4.3.4 Acceptance Bands

We can summarize the relationship between the observed CCH and a Monte Carlo sample of pseudo-data CCHs in many ways. Eventually, one should reduce the comparison to an acceptable statistical format, such as a p-value, or perhaps a collection of p-values (maybe one for each lag in the CCH). See [7,72] for some specific suggestions. For graphical displays, one can represent the null variability as *acceptance bands* around the observed CCH. If these bands are constructed so that 95% of the pseudo-data CCHs fall completely (for all lags) within the bands, then they are 95% *simultaneous* (or global) acceptance bands. If the observed CCH falls outside of the simultaneous bands at even a single lag, then we can reject the null hypothesis (technically, the null distribution, as identified by the estimated parameters from the bootstrap approximation) at level 5%. Edge effects and smoothing choices are automatically accommodated, since the pseudo-data CCHs have all of the same artifacts and smoothness. Alternatively, if these bands are constructed so that, at each time lag, 95% of the pseudo-data CCHs are within the bands at the time lag, then they are 95% *pointwise* acceptance bands. If we fix a specific lag, say lag zero, *before collecting the data*, and if the observed CCH falls outside of the pointwise bands at that specific lag, then we can reject the null at level 5%. The consideration of many lags, however, corresponds to testing the same null hypothesis using many different test statistics, the test statistics being different bins in the CCH. This is a multiple-testing problem and a proper statistical accounting involves a recognition of the fact that, even if the null hypothesis was true, we might expect to see lags where the CCH falls outside of the pointwise bands simply because we are examining so many time lags.

4.4.3.5 Conditional Inference

For the independent HPP model, a bootstrap approximation is not the only tool available for removing the ambiguity of the unknown parameters λ^A and λ^B. If we condition on the pair of total observed spike counts for the two spike trains, say N^A and N^B, respectively, then the conditional distribution of the spike times given these spike counts no longer depends on λ^A and λ^B. In particular, the null conditional distribution is uniform, meaning that all possible arrangements of N^A spikes in spike train A and N^B spikes in spike train B are equally likely. We can use this uniquely specified null conditional distribution to describe the null variability in the CCH. This is called *conditional inference*.

As with bootstrap, once we have reduced the null hypothesis to a single distribution, Monte Carlo approximation is a particularly convenient computational tool. In particular, we can generate a Monte Carlo sample of pseudo-data (by independently and uniformly choosing the N^A and N^B spike times within their respective observation windows), construct from this a collection of pseudo-data CCHs, and then proceed exactly as before with p-values and acceptance bands. We call this *Monte Carlo conditional inference*. In this particular example, the only difference between the samples created by bootstrap and those created by Monte Carlo conditional inference is that the bootstrap samples have variable spike counts (and this variability is reflected in the resulting pseudo-data CCHs), whereas the conditional inference samples do not.

4.4.3.6 Uniform Model and Conditional Modeling

Conditional inference avoids a bootstrap approximation, which can make a critical difference in settings where the approximation is likely to be poor, usually because of an unfavorable ratio of the number of parameters to the amount of data. It is also completely insensitive to the distribution of the conditioning statistic, in this case, the pair of total spike counts. Depending on the problem, this insensitivity may or may not be desirable. For the current example, the null hypothesis specifies that the joint distribution of total spike counts (N^A, N^B) is that of two independent Poisson random variables. By conditioning on these spike counts, this particular distributional assumption is effectively removed from the null hypothesis. In essence, we have enlarged the null hypothesis. The new null hypothesis is that the conditional distribution of spike times is uniform given the total spike counts. Independent HPPs have this property, as do many other point processes. We call this enlarged model the *uniform model.* The uniform model is defined by restricting our modeling assumptions to certain conditional distributions, an approach that we call *conditional modeling.*

For the case of synchrony detection, enlarging the null hypothesis to include any distribution on total spike counts seems beneficial, because it allows for more general types of coarse temporal gain modulation in the null hypothesis without also including processes with fine temporal modulations. If one was using the independent HPP model for purposes other than synchrony detection, this enlarging of the null hypothesis might be undesirable. Nevertheless, even using the enlarged null hypothesis afforded by conditional inference, the independent HPP model is rarely appropriate for the statistical detection of synchrony, because, as we mentioned before, it does not allow for any time-varying interactions, either within or across spike trains.

4.4.3.7 Model-Based Correction

Averaging all of the pseudo-data CCHs (whether from bootstrap or Monte Carlo conditional inference) gives (a Monte Carlo approximation of) the expected null CCH. For graphical displays of the CCH it can be visually helpful to subtract the expected null CCH from the observed CCH (and from any acceptance bands).* This new CCH is frequently called a *corrected CCH.* Since the correction depends on the null model and also the procedure (such as a bootstrap approximation or conditional inference) used to reduce the null model to a single null distribution, there are many types of corrected CCHs. Corrected CCHs are useful because, under the null hypothesis, they have mean zero for all lags. (Edge effects and smoothing procedures are automatically handled, similar to acceptance bands.) Any deviations from zero of the corrected CCH are a result of either noise or a violation of the assumptions of the null distribution.

The middle column of Figure 4.3 shows *uniform-corrected CCHs,* that is, model corrected CCHs using the uniform model with conditional inference to specify a null

* In many cases, one can compute the expected null CCH without explicit generation of pseudo-datasets. This can greatly accelerate computation, although, proper hypothesis testing and the construction of acceptance bands, especially simultaneous acceptance bands, typically do require explicit generation of pseudo-datasets.

distribution, along with the associated simultaneous acceptance bands. (Using the independent HPP model with bootstrap creates almost identical figures—the only difference being slightly wider acceptance bands that account for the additional variability in spike counts.) Uniform-correction essentially subtracts a constant from the CCH,* so it cannot isolate synchrony time-scale dependencies. From Figure 4.3, we see that the CCH exceeds the acceptance bands, signaling a rejection of the null hypothesis of independence, regardless of the width of the CCH peak.

4.4.4 Identical Trials Models

Many neurophysiological experiments with repeated trials exhibit spike trains whose observed firing rates vary systematically during the trials, as evidenced by PSTHs with clearly nonconstant shapes. As we mentioned in Section 4.3.2, the PSTH shows the trial-averaged observed firing rate as a function of (trial relative) time for a single spike train. The essential idea behind identical trials models is to account for the structure of each spike train's PSTH.

When used as a null hypothesis for detecting synchrony, identical trials models attempt to decompose the influences on observed synchrony into those that result from a time-varying PSTH and those that do not. This decomposition is useful in some contexts, but it does not directly address our focus here, which is the decomposition into fine temporal influences and coarse temporal influences. In Section 4.4.4.4, we discuss trial-to-trial variability (TTV) models, which preserve much of the familiar structure and terminology of identical trials models, but attempt to better distinguish between fine and coarse temporal influences on observed synchrony.

4.4.4.1 Joint Peri-Stimulus Time Histogram (JPSTH)

We have been using the CCH as our descriptive measure of interactions between spike trains. For experiments with repeated trials, the JSPTH is another important graphical display of the joint interactions between spike trains [4]. The JPSTH is especially convenient because, unlike the CCH, it is much simpler to anticipate visually the effects of each spike train's PSTH. For two simultaneously recorded spike trains, the construction of a JPSTH begins with all pairs of trial-relative spike times (s_i, t_j) that happen to occur during the same trial, where the first spike time comes from spike train one and the second from spike train two. These pairs are similar to those that defined the CCH histogram, except that we now use trial-relative time and we only consider pairs from the same trial. The JPSTH is simply a two-dimensional histogram of these spike pairs.

The JPSTH decomposes the CCH according to spike time. If, when constructing the CCH, we only consider spike pairs that occurred during a trial and during the same trial, then this *trial-restricted CCH* can be derived directly from the JPSTH by summing (or integrating) along diagonals. The main diagonal (where $s_i = t_j$) gives

* Technically, it may not be a constant depending on edge effects and the method of smoothing, but ignoring these details this constant corresponds to an estimate of the $E[U_0]E[V_0]$ term in the definition of the Pearson cross-correlation function $\rho(\ell)$ from Section 4.4.1.

lag zero in the CCH and the off-diagonals give the other lags. In principle, then, the JPSTH shows more than the CCH. In practice, the JPSTH requires much more data to reliably distinguish signal from noise. This is especially true for synchrony-like correlations, which can easily be obscured by noise or over smoothing.

The first column of Figure 4.5 shows some PSTH, JPSTH, and trial-restricted CCH examples. Rows A and B show data corresponding to the second row of Figure 4.3, but with trials defined differently in each case. Only in the first case are the trials aligned with the shared state of the spike trains (see Figure 4.3 caption) revealing the underlying block structure of the simulation in the JPSTH. Row C corresponds to the final row of Figure 4.3. With more data and very little smoothing, a narrow ridge along the diagonal would appear in this case, but for the given data, we were unable to choose the smoothing parameters in a way that made the ridge visible. (It is visible in the trial-restricted CCH.) Rows D and E show real data recorded from anesthetized monkey primary visual cortex for a sinusoidal drifting grating stimulus (courtesy of Matt Smith; see [66] for additional details). The periodicity of the spiking response for the individual spike trains is clearly apparent in the PSTHs, and this is strongly reflected in the JPSTH.

From the hypothesis testing perspective, the JPSTH is simply another test statistic, just like the CCH. We can use any null hypothesis and proceed exactly as with the CCH. Because the JPSTH is two-dimensional, it is difficult to show acceptance bands, and we only show model-corrected JPSTHs in this chapter. In practice, of course, we strongly recommend drawing proper statistical conclusions, whether or not they are easy to visualize.

4.4.4.2 Independent Inhomogeneous Poisson Process (IPP) Model

One of the most basic identical trials models for a single spike train is the inhomogeneous Poisson process (IPP) model. The spike train on a single trial is modeled as an IPP with intensity $\lambda(t)$, where t refers to trial-relative time, and the observations across trials are modeled as independent and identically distributed (iid) observations from this same IPP.* The PSTH (normalized to the appropriate units) can be interpreted as an estimator of the intensity under this model. For two spike trains, say A and B, we will additionally assume that they are mutually independent, with respective intensities $\lambda^A(t)$ and $\lambda^B(t)$. We call this null model the independent IPP model. Note that there are two different independence assumptions: (i) independence across trials (even for a single spike train) and (ii) independence across spike trains (even for a single trial). The independent IPP model is the natural generalization of the independent HPP model to allow for time-varying firing rates. It formalizes questions like, "Is there more synchrony than expected from the firing rates alone?", with the same probabilistic concepts as in the independent HPP model, except now the intensities are allowed to be time varying.

Now that we have fixed a null hypothesis, the abstract steps of hypothesis testing proceed exactly as in Sections 4.4.3.1 through 4.4.3.7. We must find some way to

* If the experiment involves multiple conditions, then we typically restrict this model to a single condition. Another way to write this is to use the conditional intensity $\lambda(t|Z)$ given the experimental condition Z, and model the spike train as a conditionally IPP given Z.

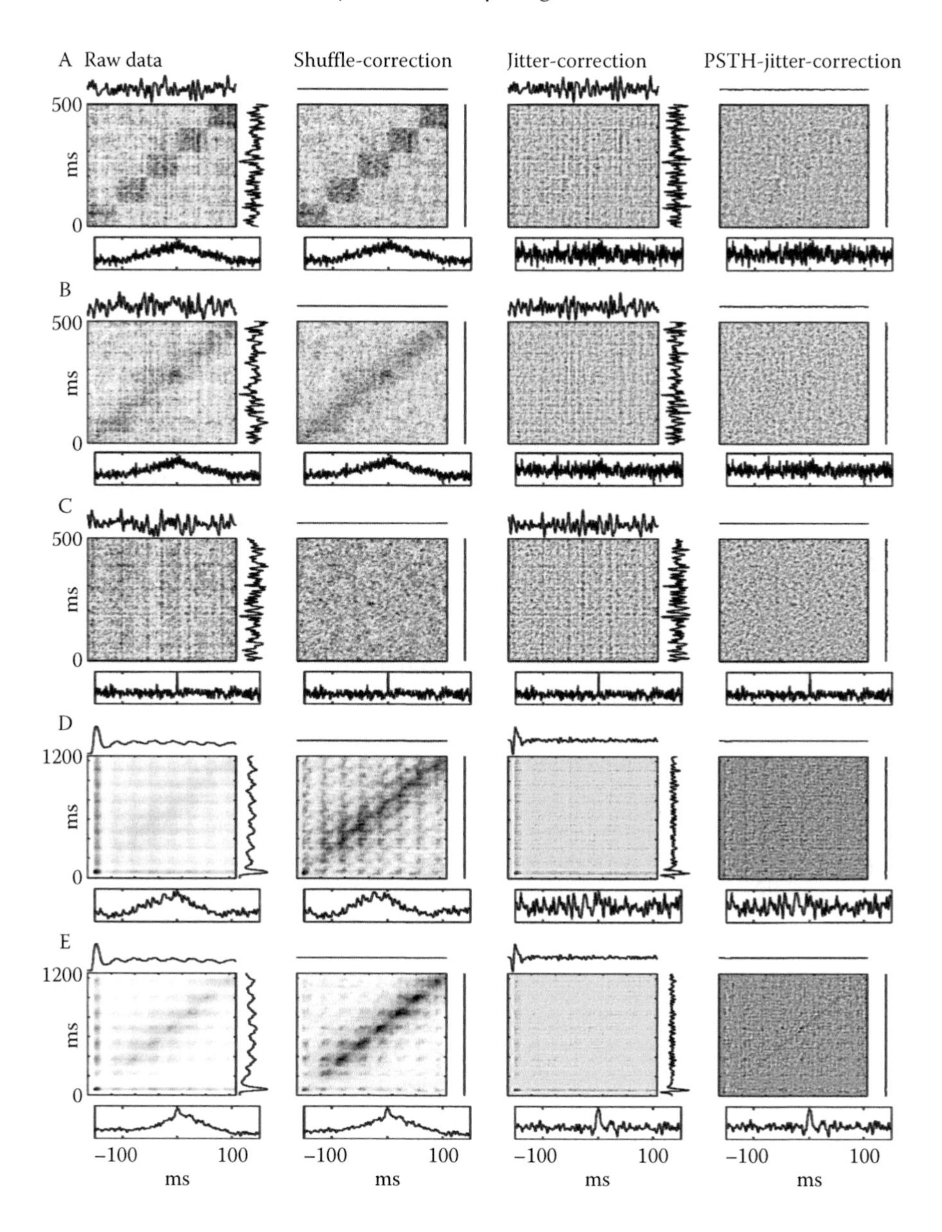

FIGURE 4.5 Examples of model correction on different datasets. Each group of plots shows a model-corrected JPSTH (central image), two model-corrected PSTHs (lines above and to the right of each JPSTH, corresponding to the two spike trains), and a model-corrected, trial-restricted CCH (plots below the JPSTH). Left-to-right, (1) raw data (no model-correction), (2) shuffle-correction, (3) 20 ms jitter-correction, (4) 20 ms PSTH-jitter-correction (Section 4.4.5.2). Row A: Data from Figure 4.3 row 2 with 500 ms trials aligned to switching of hidden coupling variable. Row B: Same as row A, but with misaligned trials. Row C: Data from Figure 4.3 row 4. Rows D and E: Real data recorded from anesthetized monkey primary visual cortex for a sinusoidal drifting grating stimulus. (Courtesy of Matt Smith.)

handle the ambiguity of the unknown intensities, often with a bootstrap approximation or with conditional inference. Then we can compute acceptance bands and/or p-values, often with Monte Carlo samples. Also we can create model-corrected graphical displays, such as CCHs or JPSTHs, to better visualize the conclusions.

There are, however, several additional subtleties as compared with Sections 4.4.3.1 through 4.4.3.7, especially if we want to use a bootstrap approximation, because the unknown intensities are now functions. Estimating a function always requires care, and typically involves, either explicitly or implicitly, additional smoothness assumptions that should be communicated along with the model. Appropriate choice and understanding of these smoothness assumptions are especially important in the context of synchrony, because synchrony requires a fine discrimination of time scale that can be strongly affected by temporal smoothing. The special case here of estimating the intensity of an IPP is closely related to the statistical topic of nonparametric density estimation, and various methods of this kind may be used to smooth the PSTH and thereby obtain good estimates of the two firing-rate functions $\lambda^A(t)$ and $\lambda^B(t)$. For example, Kass et al. [45] develop firing rate estimators using Bayesian adaptive regression splines (BARS) [21]. The bands shown in the left panel of Figure 4.2 come from this type of bootstrap. The bands indicate the middle 95% of CCH values under the null hypothesis of independence. A similar method may be applied in the non-Poisson case, using estimated conditional intensity functions based on the internal history [72].

When our primary emphasis is on testing the independence-across-spike-trains assumption in the null independent IPP model, we can avoid a bootstrap approximation by using a particular type of conditional inference procedure called the *permutation test*, or trial shuffling. The effective null hypothesis for the permutation test is much larger than the independent IPP model (but includes independent IPP as a special case), and we discuss it in the next section.

4.4.4.3 Exchangeable Model and Trial Shuffling

Consider the following nonparametric model for identical trials. For a single spike train, the trials are iid, and multiple spike trains are mutually independent. These are the same two independence assumptions as in the independent IPP model, and also the same identically distributed assumption, but we avoid any additional modeling assumptions about the spike trains. We refer to this as the *exchangeable model*, for reasons we explain below. The exchangeable model includes the independent IPP model, but it also includes all types of non-Poisson processes with refractory periods, bursting, or any other autocorrelation structure. The notion of the firing rate is associated with the conditional intensity given the internal history, namely, $\lambda^A(t|H_t^A)$ for spike train A (and similarly for other spike trains), where t refers to trial-relative time. (Recall from Section 4.3.1 that this notion of firing rate completely specifies the distribution of the spike train.)

Suppose that, for each spike train, we condition on the exact sequence of trial-relative spike times for each trial *but not the trial order*. The only remaining uncertainty in the data is how to *jointly* order the trials for each spike train. Under the exchangeable model, this conditional distribution is uniquely specified: all trial orderings are equally likely*—they are *exchangeable.* Conditioning removes all ambiguity about the firing rates $\lambda(t|H_t)$ of each spike train. For the special case of IPP spike trains, conditioning removes ambiguity about the intensities $\lambda(t)$. Furthermore, Monte Carlo sampling under this null conditional distribution is easy: uniformly and independently shuffle the trial order for each spike train to create pseudo-datasets (from which one can create pseudo-data CCHs, JPSTHs, or other test statistics). Model-based correction under the exchangeable model with conditional inference is usually called *shuffle-correction.* It is one of the most popular methods of correcting CCHs and JPSTHs so that they better reveal dependencies between spike trains that are not explained by the PSTHs. (Notice that pseudo-datasets created by trial shuffling exactly preserve the PSTH of each spike train, so shuffle-correction necessarily removes features of graphical displays that directly result from the PSTHs.) Proper hypothesis tests under this model are called *permutation tests.*

It is important to understand the scientific questions addressed by the exchangeable model (with conditional inference). By conditioning on so much of the data, the exchangeable model cannot address most questions about individual spike trains. It does, however, focus attention on the dependence between spike trains. The conditional distribution of trial orderings given the spike times precisely captures the scientific question of whether the spike trains have dependence above that which can be accounted for by identical trials, because it is the trial order of each spike train that controls which trials from spike train A are paired with which from spike train B. If there is no additional dependence, then the spike times from spike train A on a given trial, say trial i, should have no predictive power for choosing among the possible trials from spike train B to be paired with A's trial i. In this case, all trial orderings are equally likely. Alternatively, if there is additional dependence, then the spike times on A's trial i should provide some information about which of B's trials are paired with A's trial i. In this case, trial orderings are not equally likely. Hypothesis testing using the exchangeable model can, in principle, distinguish among these two cases. But they cannot make further distinctions of the time scale.

Examples of shuffle-corrected JPSTHs and CCHs can be seen in the second column of Figure 4.5. Rows D and E show how shuffle correction reduces the effects of the PSTHs on the JPSTH, more clearly revealing additional correlations. One can also see from the figures that shuffle correction and the exchangeable null model (including the independent IPP null model) are agnostic about the time scale of the interactions between spike trains. There are many plausible neuronal mechanisms, such as slow-wave oscillations or slow-temporal gain modulations, and experimental artifacts, such as imprecisely aligned trials or drifts in signal-to-noise ratio, that (i) could influence observed spike timing in multiple neurons, (ii) do not reflect

* If we have n trials, labeled $1, \ldots, n$, for each of m spike trains, then there are $n!$ permutations of the trial labels for each spike train, giving a total of $(n!)^m$ different choices for arranging the trials of all spike trains. The conditional distribution of these choices under the exchangeable model is uniform, that is, each choice has probability $(n!)^{-m}$.

processes that induce precise, coordinate spiking, and (iii) are not accounted for by the basic identical trials models because, for example, they may not be precisely aligned to the trials. Consequently, hypothesis tests based on these basic models may not isolate the fine temporal influences of observed synchrony. Different models are needed.

4.4.4.4 Trial-to-Trial Variability (TTV) Models

For investigations of the fine temporal influences on observed synchrony, the utility of identical trials models is diminished by the possible existence of shared, coarse temporal influences on observed spike timing that cannot be modeled using identical trials [9–13,17,30,53,54,73]. Examples of such influences might include slow-temporal gain modulations, slow-wave oscillations, metabolic changes, adaptation over trials, reaction time variability, imprecisely aligned trials, or drifts in the recording signal. Ideally, one would like to formulate a null hypothesis that allowed for these coarse temporal sources of correlation. The resulting hypothesis tests and model-corrected JPSTHs and CCHs would then target precise spike timing. In this section, we discuss a particular type of model formulation, often referred to as TTV models that broaden identical trials models to better account for shared, coarse temporal influences on spike timing. TTV models are not pure identical trials models, but include elements of both identical trials models and temporal smoothness models (Section 4.4.5). TTV models cause a great deal of confusion in the analysis of synchrony, and their use and interpretation requires care.

The idea behind TTV models is to preserve the basic intuitive structure of identical trials models, but to allow the firing rates to change on each trial. Coarse temporal interactions that differ across trials can now be modeled as affecting the firing rates. This allows the firing rates of different spike trains to co-modulate with each other on a trial-by-trial basis, greatly increasing the scope of possible interactions that can be modeled with the concept of firing rates. We can now test for additional interactions above those accounted for by the (trial varying) firing rates in order to better isolate additional fine temporal mechanisms of precise spike timing. Notice the distinction between time scales in this line of reasoning. Coarse temporal interactions get absorbed into the definition of firing rates, but fine temporal interactions do not. It is this distinction of time scales that make TTV models appropriate for the distinction between coarse and fine temporal influences on observed synchrony. And, crucially, it is this distinction of time scale that must be explicitly understood when interpreting TTV models in the context of synchrony or more general investigations of precise spike timing.

There are many ways to formulate TTV models mathematically. One approach is to ignore the trial structure all together; basically, to model the data as a single trial. As this abandons the concept of trials (even if the experimental protocol involves repeated trials), such models are typically not called TTV models, and we discuss them in Sections 4.4.4 and 4.4.5. A more common approach, and the one we take in this section, begins with the introduction of an unobserved process, say $U(k)$, for each trial k, that is common to all spike trains. Firing rates are defined as conditional intensities given U (and anything else of interest). For example, the firing rate of spike train A on trial k could be the conditional intensity $\lambda^A(t|H_t^A,U(k))$ and similarly for spike train B, where t is trial relative time. Since the firing rates of both A and B depend on the common value of $U(k)$, they can co-modulate across trials. In the context of IPP models, we could further assume that the spike trains are conditionally Poisson given $U(k)$, namely, that $\lambda^A(t|H_t^A,U(k)) = \lambda^A(t|U(k))$. If $U(1)$, $U(2)$, ... are iid, then this conditionally Poisson model is called a *Cox process* or a *doubly stochastic Poisson process*.* If the functional form of $\lambda(t|U)$ is fully general, meaning that changes in U across trials can completely alter all aspects of the conditional intensity, then we can fit any data set perfectly without disambiguating the potential contributions to observed synchrony (cf., footnote *, p. 88). TTV models must restrict the types of variability that are allowed across trials. For our purposes, we would expect these restrictions to involve some distinction of time scale.

With an appropriate TTV model for individual spike trains, we can formulate a null hypothesis that the spike trains are conditionally independent given the firing rates, that is, given each $\lambda(t|H_t,U(k))$. Notice that our definition of firing rate depends on the hidden variable U. This is a different definition of firing rate than in previous models we have considered. There have been a few attempts to use TTV models in this way for isolating synchrony-like correlations. For example, [13] introduces a Cox process where $\lambda^A(t|U(k)) = c^A(U(k))g^A(t) + d^A(U(k))$ for scalars $c^A(U(k))$, $d^A(U(k))$ and a function $g^A(t)$, and similarly for spike train B. In words, the time-varying shape of the intensity, $g^A(t)$, cannot vary across trials, but the total magnitude of the intensity, $c^A(U(k))$, and the baseline activity, $d^A(U(k))$, can vary across trials and can co-vary across spike trains. This allows the firing rates to model very slow gain modulations and is often called *amplitude variability* or *excitability variability*. A bootstrap approximation can be used to remove the ambiguity of the c's, d's, and λ's (with all the caveats from Section 4.4.4.2). There are many other examples of models that allow U to introduce highly specific changes in some baseline intensity $\lambda(t)$. One of the more recent is [73], which extends the amplitude variability approach by allowing the c's to also vary slowly with time according to certain specific parametric assumptions. Another recent approach to TTV is the PSTH-jitter method described in Section 4.4.5.2.

4.4.5 Temporal Smoothness Models

Pure temporal smoothness models do not make use of repeated trials, even if the experiment involves repeated trials. They are essentially single trial models and

* The Cox process is instructive for understanding the confusing terminology that surrounds TTV models. Each trial of a Cox process begins with the iid selection of a trial-specific intensity function from some distribution over possible intensities. Then the spike times (for that trial only) are generated according to an inhomogeneous Poisson process with that trial-specific intensity. For a Cox process, the trials are independent and identically distributed, and the unconditional intensity $\lambda(t)$ does not vary across trials. There would seem to be no trial-to-trial variability. In many neuroscience contexts, however, it is not the unconditional intensity that maps onto the intuitive notion of a firing rate, but rather, the trial-specific intensities, which are conditional intensities, say $\lambda(t|U)$, given the trial-specific selection, U. It is these conditional intensities that vary across trials (because U varies across trials); hence, the trial-to-trial variability terminology.

place restrictions on how quickly the firing rates (suitably defined) can change. By allowing the firing rates of multiple spike trains to co-modulate, but limiting how quickly they can co-modulate, temporal smoothness models are designed to explicitly address the issue of precise spike timing. Any coarse temporal interactions can be absorbed into the definition of the firing rates, whereas, all precise spike timing must be explained by additional mechanisms. Temporal smoothness models can often be combined with identical trials models to create TTV models. This is especially convenient for situations where the firing rates could co-modulate quickly in response to a precisely timed stimulus, but where this type of stimulus-locked, fine temporal, coordinated spiking is not reflective of the types of fine temporal mechanisms an investigator wishes to study.

4.4.5.1 Independent Inhomogeneous Slowly Varying Poisson Model

Consider the independent IPP null model (Section 4.4.4.2) *for a single trial.* As discussed in Section 4.3.2 (cf., footnote *, p. 88), this model is too general to be useful without additional constraints. For the case of two spike trains, say A and B, suppose that we further assume that the intensities $\lambda^A(t)$ and $\lambda^B(t)$ are piecewise constant over the disjoint Δ-length intervals $(k\Delta - \Delta, k\Delta]$ for integers $k \in \mathbb{Z}$. If we let λ_k^A denote the common value of $\lambda^A(t)$ for $k\Delta - \Delta < t \leq k\Delta$, and similarly for λ_k^B, then the null hypothesis is completely characterized by the λ_k^A's and λ_k^B's. We can think about this model as a sequence of small independent HPP models (Section 4.4.3), one for each Δ-length interval. For shorthand, we refer to this null model as the independent Δ-IPP model.

The piecewise constant assumption in the independent Δ-IPP model is designed to loosely capture the intuition of an intensity function that is slowly varying on time scales qualitatively larger than Δ. Certainly one could model slowly varying intensities in any number of ways. The key point is that for synchrony detection one must chose a sensible way to quantify what "slowly varying" means. Here this is quantified by Δ, which is often taken to be something on the order of 5–50 ms, but could be any value depending on the time scales of the phenomena an investigator wishes to disambiguate. The basic intuition for the independent Δ-IPP model is the same as for previous models we have considered: "Is there more synchrony than expected from the firing rates alone?" The difference is that now firing rates are only allowed to vary on time scales larger than Δ. Any other probabilistic structure in the joint spiking must be accounted for by mechanisms that we are not including in our definition of firing rates. Note that Δ, which controls time scale of the null hypothesis of no excess synchrony, is usually chosen to be somewhat larger than the time scale used in the definition of observed synchrony, the latter of which plays the role of a test statistic, and, loosely speaking, controls the time scale of the alternative hypothesis of excess synchrony.

There are two rationales for choosing the piecewise constant characterization of slowly varying intensities in the independent Δ-IPP model. First, for essentially any sensible characterization of intensities that are slowly varying on time scales qualitatively larger than Δ, we would expect these intensities to be approximately constant on Δ-length intervals. To a first approximation then, this particular null model includes all independent IPPs with appropriately slowly varying intensities. Second, the piecewise constant nature of the intensities permits a conditional inference approach to removing ambiguity about the intensities, called *jitter*, which we discuss below. This allows the model to be used successfully in practice.

To use the independent Δ-IPP model, we need a method for handling the ambiguity of the unknown λ_k's. A bootstrap approximation is one potential solution. In each Δ-length interval and for each neuron, we can approximate the constant intensity in that interval with an estimate based on the data (usually the number of spikes in the interval divided by Δ). Δ will typically be small, with each interval having only a few, often zero, spikes. Consequently, this is not a convincing approximation. Great care must be taken when using a bootstrap approximation in situations where the resulting approximation is poor, because then the sensibility of the method largely rests upon whether its eventual use, such as accessing the null variability of a CCH, is robust to large approximation errors. Assessing this robustness can be quite challenging.

A less dubious procedure, in this case, is to condition on the *sequences* of spike counts in each Δ-length interval for each spike train, that is, on the sequences $(N^A(k\Delta) - N^A(k\Delta - \Delta) : k \in \mathbb{Z})$ and $(N^B(k\Delta) - N^B(k\Delta - \Delta) : k \in \mathbb{Z})$. For the independent Δ-IPP null model, the conditional distribution of spike times given the sequence of spike counts is uniform in each Δ-length interval, regardless of the actual values of the intensities. Generating Monte Carlo pseudo-data from this null conditional distribution is trivial: independently and uniformly perturb (or "jitter") each spike in the Δ-length interval in which it occurred.

This jitter conditional inference procedure permits exact statistical testing of the independent Δ-IPP model, and, following the reasoning above, it permits approximate statistical testing of essentially any independent IPP model (whether single trial, identical trials, or TTV) with intensities that are slowly varying on time scales qualitatively larger than Δ [7]. These Poisson models, however, make precise assumptions about the joint distribution of spike counts (namely, independent Poisson random variables), whereas the conditional inference procedure is insensitive to these assumptions. Conditional inference, in this case, is effectively testing a much larger null hypothesis, and it is important to understand if this larger null model is still appropriate for the scientific task of distinguishing between fine temporal and coarse temporal influences on observed synchrony.

4.4.5.2 Δ-Uniform Model and Jitter

Here, we more carefully inspect the effective null hypothesis tested by the jitter conditional inference approach described in the previous section. Conditional inference is useful because, under the independent Δ-IPP model, the conditional distribution of spike times given the sequence of spike counts in Δ-length intervals is uniform; there is no remaining ambiguity. For convenience, we will call this the Δ-uniform conditional distribution. Distributions within the independent Δ-IPP model have the Δ-uniform conditional distribution, but there are many other distributions with this property, such as Cox processes with trial-specific intensities that are Δ-piecewise constant. Consider the class of distributions over pairs of spike trains that have the

Δ-uniform conditional distribution. We refer to this class of distributions as the *Δ-uniform model.** *The Δ-uniform model can be tested by the same jitter conditional inference procedure used for the Δ-IPP model in the previous section.*

The Δ-uniform model is a very large class of spike train models. The sequences of spike counts in Δ-length intervals, namely, $(N^A(k\Delta) - N^A(k\Delta - \Delta) : k \in \mathbb{Z})$ and $(N^B(k\Delta) - N^B(k\Delta - \Delta) : k \in \mathbb{Z})$, can be viewed as temporally coarsened versions of the original spike trains. The Δ-uniform model allows for any joint distribution over these temporally coarsened, simultaneously recorded spike trains, including any type of dependence within or across the temporally coarsened spike trains. To a first approximation, it allows for all coarse temporal interactions between spike trains, regardless of their source, where "coarse" means qualitatively longer time scales than Δ. On the other hand, since the Δ-uniform model specifies that the precise spike times are uniform subject to the temporally coarsened spike trains, it does not allow for any precise temporal interactions among spikes, including mechanisms that create precise synchrony. When used as a null hypothesis, the Δ-uniform model is designed to detect precise spike timing of any source [7].

As with the independent Δ-IPP model, conditioning on the sequences of spike counts in Δ-length intervals (i.e., conditioning on the temporally coarsened spike trains) removes all ambiguity about the unknown coarse distribution, and uniquely specifies the Δ-uniform conditional distribution. Exact quantitative inferences, such as p-values, for the Δ-uniform conditional distribution are computationally feasible in special cases [5,7,32,33], but inference usually proceeds via Monte Carlo approximation. Monte Carlo pseudo-data from the Δ-uniform conditional distribution are easily generated by a procedure called *jitter*: independently and uniformly perturb (or "jitter") each spike in the Δ-length interval in which it occurred. Hypothesis tests, acceptance bands, and model-based corrections can be based on this pseudo-data exactly as in Sections 4.4.3.2 through 4.4.3.5. The third column of Figure 4.3 shows $\Delta = 20$ ms jitter-corrected CCHs for the corresponding raw CCHs in the left column. Only the final two processes show significant synchrony, consistent with our intuition that these processes involve correlations on time scales shorter than 20 ms. 20 ms jitter-corrected JPSTHs, PSTHs, and CCHs can be seen in the third column of Figure 4.5. Rows C and E are the only rows that suggest precise spike timing.

The Δ-uniform model was originally conceived as a new modeling approach for addressing questions of precise spike timing based on the perspective that scientific questions about spike timing were largely questions about the conditional distribution of the precise placement of spikes given the coarse temporal structure of the

---

* Here is a more formal description of the Δ-uniform model using notation that also extends to more than two spike trains. Let $s = (s_1, s_2, \ldots, s_m)$ be an increasing sequence of spike times and define $C_k(s) = \#\{i: k\Delta - \Delta < s_i \le k\Delta\}$ to be the number of spikes in s that occur during the kth Δ-length jitter window. Let $C(s) = (C_1(s), C_2(s), \ldots)$ be the sequence of spike counts in jitter windows. For each $j = 1, \ldots, n$, let $\boldsymbol{S}^j = (\boldsymbol{S}_1^j, \boldsymbol{S}_2^j, \ldots, \boldsymbol{S}_{M_j}^j)$ be the increasing sequence of spikes from the jth spike train. The Δ-uniform model for discrete time is that $\boldsymbol{P}(\boldsymbol{S}^1 = \boldsymbol{s}^1, \ldots, \boldsymbol{S}^n = \boldsymbol{s}^n | \boldsymbol{C}(\boldsymbol{S}^1) = \boldsymbol{c}^1, \ldots, \boldsymbol{C}(\boldsymbol{S}^n) = \boldsymbol{c}^n)$ is constant on the set of allowable $s^1, \ldots, s^n$, namely, the set for which each s^j is an increasing sequence of spike times over the appropriate observation interval and for which each $C(s^j) = c^j$ $(j = 1, \ldots, n)$. The probability is zero elsewhere. For continuous time, the characterization of the model is the same, except we use the probability density instead of the probability. The model allows for any joint distribution on the sequences of spike counts $C(S^1), \ldots, C(S^n)$.

spike trains [1,7,19,26,27,34,36,55,56,67]. Various applications to neurophysiological questions can be found in [16,22,24,25,41,47,48,52,58–60,64–66]. It is one of the simplest and most effective models for isolating and visualizing synchrony-like correlations.

The Δ-uniform model requires an appropriately chosen test statistic, such as the observed synchrony count or the CCH, so that the hypothesis test has power for detecting the type of precise timing that is being investigated, such as synchrony. An appropriate test statistic is also important for helping to mitigate the very real possibility that the null hypothesis is rejected because of the wrong type of precise spike timing, such as refractory periods, bursting, or precise stimulus locking. A better solution is to change the null hypothesis so that it also includes those types of precise spike timing that are not being investigated. For example, [34] modify the Δ-uniform model so that individual spike trains can also have precise temporal autocorrelation structure, such as refractory periods and bursting, and they describe an algorithm, called *pattern jitter*, for exact Monte Carlo sampling from an appropriate null conditional distribution. Replacing jitter with pattern jitter ensures that statistical conclusions about precise synchrony are not artifacts of precise autocorrelation structure. For another example, [8,66] incorporate the Δ-uniform model into a TTV model by also conditioning on the observed PSTH. Monte Carlo sampling uses an algorithm called PSTH-jitter that does not jitter spikes uniformly, but rather jitters according to the trial-averaged PSTH. Replacing jitter with PSTH-jitter ensures that statistical conclusions about precise synchrony are not artifacts of stimulus-locked firing rates (at least if the trials are correctly aligned). The fourth column of Figure 4.5 shows $\Delta = 20$ ms PSTH-jitter-corrected CCHs. The utility of PSTH-jitter-correction for visualizing the JPSTH is especially apparent in rows D and E where the PSTH modulates strongly during the trial.

4.4.5.3 Heuristic Spike Resampling Methods

The jitter Monte Carlo sampling algorithm is closely related to a heuristic spike-resampling approach that we call *spike-centered jitter* (various terms are used in the literature such as jitter, dither, or teeter), which creates pseudo-datasets by jittering each observed spike in a window centered around its original location. Spike-centered jitter is motivated by the observations that (i) the resulting pseudo-data essentially preserve all coarse temporal correlations of the original spike trains, (ii) the resulting pseudo-data essentially destroy any fine temporal correlations that may have been present in the original spike trains, and consequently (iii) comparisons between the original and the pseudo-data can reveal the presence of fine temporal correlations. Most heuristic procedures are designed according to similar criteria.

Unlike the jitter procedure described in the previous section, spike-centered jitter does not correspond to a well-defined null hypothesis and cannot be used to generate meaningful p-values, hypothesis tests, or other statistical conclusions [7,34]. Like spike-centered jitter, there are many other spike resampling techniques in the literature that are not based on a true underlying model (e.g., [27,38,51,53]). Heuristic procedures are invaluable data exploration tools, but cannot be used to draw statistical conclusions and should not be discussed using statistical terminology. Proper statistical conclusions require a proper statistical model. Just as spike-centered jitter is closely related to jitter and the Δ-uniform model, we suspect that most heuristic spike resampling approaches are closely related to some model. Whenever possible, we prefer replacing heuristic procedures with more careful model-based approaches.

4.4.6 Generalized Regression Models

The point process conception described in Section 4.3 was used in Section 4.4 to provide a rigorous statistical framework for separating synchronous spiking from correlated firing that occurs at broader time scales. The framework in Sections 4.4.2 through 4.4.5 was aimed at testing specific null hypothetical (nonsynchronous) models, which progressed from the simplest null, the independent HPP model, to much more general null models that allow complicated spike patterns while assuming independent uniform conditional spiking distributions within intervals of length Δ. In this section, we describe an alternative approach, based on generalized regression models, that uses measured variables, or covariates, to describe correlation at coarse times, and provides estimates of the magnitude of excess synchronous spiking beyond that produced under independence—the null case corresponding to independence, and thus zero excess synchronous spiking. Generalized regression models parameterize the effects of synchrony, and thereby characterize the excess probability of spiking beyond that predicted by independence. To provide such information, which cannot be obtained from the hypothesis testing framework in Sections 4.4.2 through 4.4.5, the regression approach introduces additional modeling assumptions.

Let us return to the definition of instantaneous firing rate in terms of the conditional intensity function, as in Equation 4.2, and reexpress the conditional intensity in the form

$$\log \lambda(t \mid H_t) = \text{stimulus effects} + \text{history effects}, \tag{4.3}$$

where we use "stimulus effects" and "history effects" to refer to some linear combination of measured variables that represent, respectively, the response to stimulus and the spiking history prior to time t. When applied to binary data in discrete time bins, assumed to arise from a point process, Equation 4.3 defines what is often called a *generalized linear model* (GLM) [44]. In ordinary linear regression, the response values may range over a wide scale. Here, instead, the model is "generalized" to accommodate the representation of spike trains as a sequence of 0s and 1s.

One of the purposes of regression methods is to adjust effects of interest, such as stimulus effects, by removing the effects of covariates that are considered irrelevant, such as refractory effects represented as spiking history. In addition to spike history effects, generalized regression models may incorporate additional covariates that represent slow fluctuations in firing rate that occur across trials [73]. Furthermore, measured network effects related to neural inputs can be included in the generalized regression model and they, too, can be effectively removed from the analysis [42,43,46]. If, for example, each of two neurons were to have highly rhythmic spiking in conjunction with an oscillatory potential produced by some large network of cells, then the two neurons would both tend to spike near the times of the peaks of the oscillatory potential. For a high-frequency oscillation (in the mid-to-high gamma range), the peaks would be concentrated within a range of a few milliseconds, and this could be an important source of synchrony [71]. For slower oscillations, the extra spiking would occur across broader time lags (the CCH peak would be broader), but the extra spikes could mask any possible synchronous spikes that might occur—and that might carry information about the stimulus. In either case, it would be helpful to identify the oscillatory effects, and to look for additional synchronous spiking not associated with the oscillations. An illustration of the slow-wave spiking activity across an array of neural spike trains is given in Figure 4.6.

Details of regression-based synchrony identification using point process representations and generalized linear models may be found in Ref. [42]. In outline, given multiple trials of spike trains for neurons A and B, we consider them to follow point processes with conditional intensities $\lambda^A(t \mid H_t^A)$ and $\lambda^B(t \mid H_t^B)$ and we also consider the synchronous spikes to follow point processes* with conditional intensity $\lambda^{AB}(t|H_t)$. Here H_t^A and

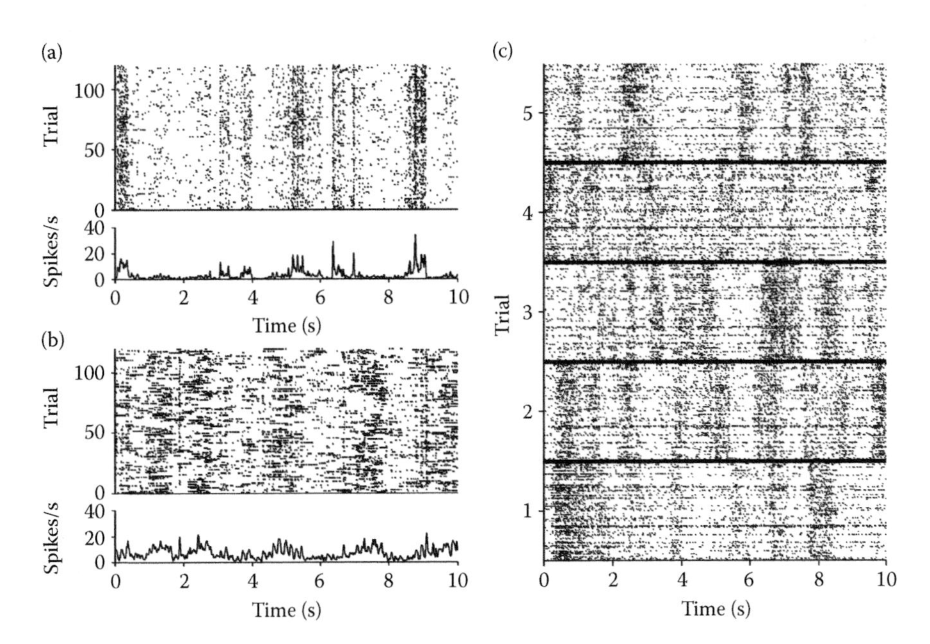

FIGURE 4.6 Neural spike train raster plots for repeated presentations of a drifting sine wave grating stimulus. (a) Single cell responses to 120 repeats of a 10 s movie. At the top is a raster corresponding to the spike times, and below is a PSTH for the same data. Portions of the stimulus eliciting firing are apparent. (b) The same plots as in (a), for a different cell. (c) Population responses to the same stimulus, for five repeats. Each block, corresponding to a single trial, is the population raster for $\nu = 128$ units. On each trial there are several dark bands, which constitute bursts of network activity sometimes called "up states." Up-state epochs vary across trials, indicating they are not locked to the stimulus. (Reprinted from RE Kass, RC Kelly, and W-L Loh, *Annals of Applied Statistics* **5**, 2011, 1262–1292.)

* Technically, we assume a family of point processes indexed by bin size δ such that as $\delta \to 0$ the conditional intensity $\lambda_\delta^{AB}(t|H_t)$ vanishes at the same rate as δ.

H_t^B are the spiking histories prior to time t for each neuron separately, within each trial, and H_t is the combined spiking history prior to time t. We define $\zeta(t)$ by

$$\zeta(t) = \frac{\lambda^{AB}(t \mid H_t)}{\lambda^A(t \mid H_t^A)\lambda^B(t \mid H_t^B)} \tag{4.4}$$

so that under independence we have the null hypothesis

$$H_0 : \zeta(t) = 1 \tag{4.5}$$

for all t, and $\zeta(t) - 1$ becomes the excess proportion of spiking above that which would occur under independence. We may also include additional variables in these conditional intensity functions. Refs. [42,43] included a variable, which we here label x, based on the joint spiking of 118 other neurons recorded from the same array. This was meant to capture slowly oscillating network effects. Thus, the individual conditional intensities would be written $\lambda^A(t \mid H_t^A, x)$ and $\lambda^B(t \mid H_t^B, x)$, the point being that neurons A and B would be more likely to fire during network bursts due to slow oscillations, as displayed in Figure 4.6. When the right-hand side of Equation 4.4 is modified to include the variable x, we change the notation on the left-hand side to become $\zeta_x(t)$ and write the independence null hypothesis as

$$H_0 : \zeta_x(t) = 1 \tag{4.6}$$

In words, Hypothesis 4.5 is the hypothesis that synchronous spiking is due entirely to the individual-neuron fluctuations in firing rate, and spike times are otherwise independent. Hypothesis 4.6 allows synchronous spiking to be due, in addition, to slow bursting activity shared by both neurons, but precise spike times are otherwise independent.

Kass et al. [43] illustrated the regression-based point process formulation by analyzing two different pairs of neurons under the assumption that $\zeta = \zeta(t)$ was time invariant. For both pairs, a bootstrap test of the null hypothesis in Equation 4.5 gave very small p-values, indicating that there was excess synchrony beyond that produced by chance from the point processes that were assumed to have generated the individual spike trains. To check whether the excess synchrony might have been due to the slow bursting activity, Kass et al. [43] estimated the parameter ζ_x for each pair. For one pair, bootstrap methods produced an estimate of excess spiking $\log_e \hat{\zeta}_x = .06$ with standard error $SE = .15$, so that the test of the null hypothesis in Equation 4.6 produced a large p-value, and excess synchrony was, apparently, produced by the slow network bursts like those visible in Figure 4.6. For the other pair, however, they obtained $\log_e \hat{\zeta}_x = .82$ with standard error $SE = .23$ giving a very small p-value for the null hypothesis in Equation 4.6 and indicating that there remained additional substantial excess synchrony, more than double the number of spikes produced under independence,* even after the bursting had been accounted for. For this pair, the excess synchrony may well have been providing stimulus-related information.

* Exponentiating .82 gives 2.27, with an approximate 95% confidence interval of (1.4, 3.6).

4.4.7 Multiple Spike Trains and More General Precise Spike Timing

We have focused on synchrony analysis, but with respect to hypothesis testing, observed synchrony merely plays the role of a *test statistic*. Our main focus has been on models for the null hypothesis of no excess synchrony, which we have primarily framed as null hypotheses for no precise spike timing of any kind. For this reason, many of the models we have been discussing are also appropriate for testing for the presence of other types of precise spike timing, such as recurring motifs of precisely timed spikes. Only the test statistic needs to change in order to increase the statistical power for detecting the phenomenon of interest. Changing the test statistic is particularly straightforward in the case of Monte Carlo approximation: simply compute the new test statistic on the original dataset and the Monte Carlo pseudo-datasets, and then proceed as before. We have focused on precise pairwise synchrony, with the implication that a relevant test statistic, such as observed synchrony or a related feature of the CCH or JPSTH, would be used. But for other types a precise spike timing, a different test statistic might be more appropriate.

For example, Grün et al. [28] define a *unitary event* (UE) as the event of spike coincidence across a constellation of neurons (in practice, consisting of more than two neurons). This is the natural generalization of the observed synchrony statistic to higher-order interactions. The test statistic is then the number of occurrences of a particular UE, or a global statistic that combines multiple UE's. These statistics can then be calibrated in the sense of model correction, or, more formally, used as the test statistic in a hypothesis test. The interpretation is exactly the same: the focal point of the analysis is the model, and unitary events play a role exactly analogous to observed synchrony or to the CCH.

UE statistics have been applied in combination with different models. In Ref. [28] the null hypothesis is that spike trains are independent HPPs* (Section 4.4.3) and a bootstrap approximation is used to remove ambiguity about the unknown intensities. That is, the Poisson rate is estimated from the spike trains, and the estimated Poisson rates are used to derive the distribution of the UE statistic under the (Poisson) model. The observed UE statistic is then compared with the derived distribution. A UE statistic is also used in Ref. [31], but using the uniform model with conditional inference (Section 4.4.3.6). The assumption of stationarity is problematic. The Δ-uniform null model (Section 4.4.5.2), tested with a UE statistic, is a natural alternative approach. As another approach to nonstationarity, Grün et al. [29] consider a slowly varying independent IPP model (Section 4.4.5.1) and propose a bootstrap approximation using a kernel-based (uniform) smoother to estimate the unknown intensities.

It is worth reemphasizing here the distinction between a statistic and the null hypothesis. In UE analysis, the statistic is the number of unitary events, which involve higher-order constellations of activity. When a UE statistic is used to reject a null hypothesis, the justifiable statistical conclusion is simply that there is evidence against that null hypothesis. While this may be regarded as a kind of *prima facie* suggestion of higher-order interaction, that conclusion requires some additional

* Depending on whether spike trains are modeled as discrete or continuous. In the discrete case (spike or no spike in time bins of finite extent), the model is stationary Bernoulli. In the continuous case, it is stationary Poisson. The two models are closely related and practically equivalent for small time bins.

assumption or analysis. For example, it may be possible to introduce a framework, such as the generalized regression framework of Section 4.4.6, according to which a large value of some point process quantity (analogous to $\zeta(t)$ in Equation 4.4) will correspond to higher-order correlation, and will make it likely that the UE statistic will reject the null hypothesis. Such a quantity could then be estimated from the data. The strict interpretation of a rejection of the Δ-uniform model is, similarly, that coarse spike timing models (quantified by Δ) are poorly fit to the spike trains. In general, the choice of statistic affects the power of the test (cf., the discussion of statistical significance in the Appendix Section "Hypothesis Tests and Monte Carlo Approximation") and, in the case of rejection of the null hypothesis, some quantification of the magnitude of the effect—via additional modeling assumptions—is likely to be helpful. Kass et al. [43] showed how the generalized regression framework can accommodate higher-order synchrony among multiple spike trains by including higher-order quantities analogous to $\zeta(t)$ defined in Equation 4.4. For example, in the case of three-way interaction among three neurons, the analogous quantity represents the excess probability of three-way synchrony above that predicted by a model involving all three pairs of two-way interactions. When the point processes are assumed to be stationary, the methods of Kass et al. [43] reduce to those of [50,62].

4.4.8 Comparisons across Conditions

Hypothesis testing is designed to detect unusual synchrony, and the related model-based correction perspective is useful for preserving the familiar formats of graphical displays, like the CCH and JPSTH, but tailoring them to reveal specific types of correlations, like synchrony. Hypothesis testing and model-based correction are much less useful, however, for drawing comparisons across conditions. Hypothesis testing focuses entirely on null models capturing the absence of the target phenomenon, like synchrony. Comparing synchrony across different conditions requires explicit models of synchrony, something that is lacking in the hypothesis testing perspective.

Consider the following hypothetical example that illustrates the ambiguities of drawing comparisons across conditions based solely upon hypothesis testing. The point of the example is to determine which of two experimental conditions (I or II) exhibits more unusual synchrony between two simultaneously recorded spike trains (A or B). The spike trains are HPP (but not independent) in each condition. Spike train A has rates of 30 and 10 spikes/s in conditions I and II, respectively. Spike train B also has rates of 30 and 10 spikes/s in conditions I and II, respectively. On average 10% of the spikes (i.e., 3 spikes/s) in condition I are forced to be precisely synchronous. On average 20% of the spikes (i.e., 2 spikes/s) in condition II are forced to be precisely synchronous. The spike times are otherwise independent. (In each case, additional precise alignments between spikes can arise by chance.)

The absolute rate of (unusual) synchronous firing is higher in condition I (3 Hz) than in condition II (2 Hz), but the percentage of spikes that are synchronous is higher in condition II (20%) than in condition I (10%). Hypothesis testing and model-based correction are useful for revealing the unusual synchrony in each condition, but they do not speak to the scientific issue of which condition has more synchrony.

As Ito and Tsuji [39] and others have noted (see also the final discussion in Staude et al. [68]), there is no uniquely compelling answer as to which condition has more synchrony in an example like this. The scientific question is too vague. The answer will depend on how one chooses to *model* synchrony. For example, in the injected synchrony model, whereby synchronous spikes are simultaneously injected into each neuron, then condition I has more synchrony than condition II, since it has more injected spikes. Alternatively, using log-linear models [49], for which synchrony is quantified by an odds ratio related to how much a spike in one spike train increases the chances of a nearby spike in the other spike train, then condition II has more synchrony than condition I, since it has a higher percentage of synchronous spikes. Section 4.4.6 provides additional details about explicit statistical models of synchrony that can be used to quantify differences across conditions within a common statistical framework.

4.5 DISCUSSION

We have focused on basic or prominent methods for the statistical identification of precisely timed synchrony. Figure 4.7 contains a summary of the main methods that we covered. More generally, any survey of the neurophysiology literature will uncover an enormous diversity of methods employed to analyze spike timing patterns, many of which we have not discussed here. (For example, we have made no mention of either spectral methods or information theory; our general remarks about the importance of proper statistical modeling apply equally to these and other approaches.) Coupled with the experimental complexity of spike data sets, this diversity can present a challenge to investigators and readers.

In reviewing the statistical thinking that underlies different approaches, our major conclusion is that a great deal of attention ought to be focused on the process by which a neuroscientific question is translated into the terms of a precise statistical model. Once a model is specified, inferential techniques can be evaluated or compared using relatively common statistical criteria. On the other hand, many fundamental concepts in neurophysiology and in the neural coding literature remain vaguely conceived in terms of many alternative precise definitions. Here, we have examined "synchrony," "chance," and "firing rate" but more generally we also have in mind "information," "population coding," and "neuronal ensembles," all of which have multiple mathematical definitions in the literature. Consideration of the meaning of these terms is tied intimately to the construction and application of statistical models, the precise specification of which is a prerequisite for quantitative data analysis. Statisticians are fond of saying that the most important part of an analysis is the mapping between the scientific problem and the statistical model (see Ref. [14] together with the published discussion and rejoinder).

Returning to the focus of this chapter, a take-home message is that precisely timed synchrony, as it is commonly understood in neurophysiology, enters into statistical models of spike trains through a consideration of time scale. This might be formalized in several ways. For example, in the Δ-uniform model (jitter) and its variants, the time scale is modeled explicitly. In many TTV approaches, a time scale is implicitly specified through the time scale at which the conditional intensity

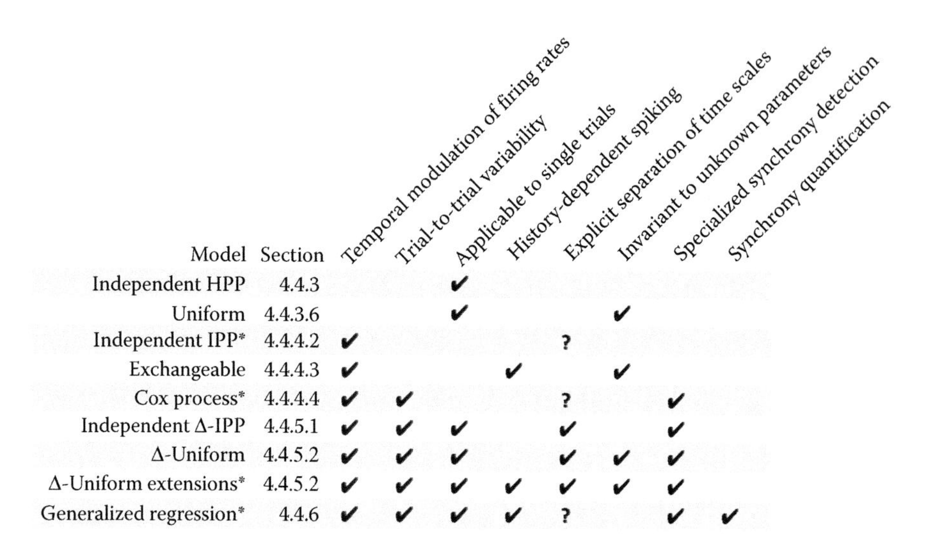

Model	Section	Temporal modulation of firing rates	Trial-to-trial variability	Applicable to single trials	History-dependent spiking	Explicit separation of time scales	Invariant to unknown parameters	Specialized synchrony detection	Synchrony quantification
Independent HPP	4.4.3			✔					
Uniform	4.4.3.6			✔			✔		
Independent IPP*	4.4.4.2	✔				?			
Exchangeable	4.4.4.3	✔			✔		✔		
Cox process*	4.4.4.4	✔	✔			?		✔	
Independent Δ-IPP	4.4.5.1	✔	✔	✔		✔		✔	
Δ-Uniform	4.4.5.2	✔	✔	✔		✔	✔	✔	
Δ-Uniform extensions*	4.4.5.2	✔	✔	✔	✔	✔	✔	✔	
Generalized regression*	4.4.6	✔	✔	✔	✔	?		✔	✔

FIGURE 4.7 Summary of the main models discussed in this chapter. Each row corresponds to a different model, which is described in the indicated section of the text. Each column corresponds to a desirable feature from the perspective of scientifically and statistically appropriate investigations of synchrony and more general precise timing. Models that allow for temporal modulation of firing rates can better account for the correlations between spike counts and temporal covariates that are observed in many experiments (cf. Sections 4.3, 4.4.4ff). Models allowing for TTV additionally account for temporal modulation on time scales spanning experimentally controlled trials (cf. Section 4.4.4.4). Some models are applicable to single trials (cf. Sections 4.4.3, 4.4.5, and 4.4.6) and some allow for history-dependent spiking, such as refractory period and bursting (cf. Sections 4.3, 4.4.4.3, 4.4.5.2, 4.4.6). Models that accommodate explicit separation of time scales allow for some degree of quantification of the temporal precision of spiking (cf. Sections 4.2, 4.4.5). The category invariant to unknown parameters signifies conditional inference for reducing a large null hypothesis to a single null distribution (as opposed to a bootstrap approximation), which increases robustness and avoids approximation errors, particularly in models with large numbers of unknown parameters (cf. Sections 4.4.3.5, 4.4.3.6, 4.4.4.3, 4.4.5.2). Some models can explicitly disambiguate dependences on fine time scales from those on coarse time scales, permitting specialized synchrony detection (cf. Sections 4.2, 4.4.5). Finally, explicit models of synchrony permit statistical synchrony quantification, as opposed to just synchrony detection (cf. Section 4.4.6). While these categories are somewhat imprecise, we hope that they are more or less identifiable with the key concepts repeated throughout the text. Our assignment of a feature to a given model is based on the observation that many simultaneously recorded spike trains appear to have correlations on a spectrum of time scales ranging from milliseconds to hours. Models with an asterisk (*) are large, flexible model classes that require careful formulation and often custom software in order to have all of the indicated features. Within these models, a question mark (?) means that the feature should be possible, but is not fully present in existing formulations.

function is assumed to vary across trials. Furthermore, it is rarely helpful to use models with no notion of time scale, such as the exchangeable-model (trial shuffling), to identify precise synchrony. A second message is that hypothesis testing and estimation are distinct statistical enterprises. As we mentioned in Sections 4.4.7 and 4.4.8 hypothesis tests are designed to reveal whether the data are inconsistent with models that do not involve precise synchrony dependence. They are not designed to quantify synchrony. Instead, by introducing further assumptions that explicitly model the synchrony time-scale dependencies (as in Section 4.4.6), we can relate various parameters associated with synchrony to other covariates, such as stimulus or behavior.

One of the current challenges in the statistical modeling of multiple spike trains is the development of scientifically appropriate and tractable models that can account for a variety of sources of dependence across a spectrum of time scales, including precise synchrony and other higher-order spatio-temporal spiking patterns. Progress is likely to be gradual, with current models informing future neurophysiological experiments, which will then lead to modeling refinements and statistical innovations.

ACKNOWLEDGMENTS

This material is based upon work supported by the U.S. National Science Foundation under Grant No. 1007593 and by the U.S. National Institutes of Health under Grant 5R01MH064537-10. The authors thank Jonathan Victor for his comments that improved the text.

APPENDIX

PROBABILITY AND RANDOM VARIABLES

Random variables are used as theoretical counterparts to data. A useful mathematical convention is to represent data with lowercase letters like x and to identify the corresponding random variables with capital letters like X. Thus, for example, x could be a measured spike count and X the corresponding random variable, which our statistical model might assume follows a Poisson distribution. We write the probability of X taking a value between a and b as $P(a < X < b)$. Two random variables X and Y are said to be independent if for all numbers a, b, c, d we have $P(a < X < b \text{ and } c < Y < d) = P(a < X < b)P(c < Y < d)$. Equivalently,* X and Y are independent if, having observed that the value of Y is y, the conditional probability of X taking a value between a and b, which we write as $P(a < X < b|Y = y)$, satisfies $P(a < X < b) = P(a < X < b|Y = y)$ for every a, b, y, that is, X and Y are independent if knowing the value of Y does not change any of the probabilities associated with the outcomes of X (and, conversely, knowing X does not change the probabilities associated with Y). When data values occur sequentially across time, the value at time t is often written as x_t and the corresponding random variable would be written as X_t. Such sequential data and random variables are usually called *time series*. For theoretical purposes, the time index is allowed to take all integer values, both negative and positive. To emphasize this, we write the theoretical time series as $(X_t : t \in \mathbb{Z})$, where $\mathbb{Z}$ represents all the integers.

---

* We are ignoring mathematical subtleties treated by measure theory.

Statistical Models and Scientific Questions

To assess variation in spiking that could lead to synchronous events "by chance" we must begin with a statistical model. Statistical models describe variation in terms of probability. For example, the simplest probabilistic description of the variation in the spike counts in a particular time interval, across trials, would be to assume they follow a Poisson distribution. A more general specification would be to assume that, on every trial, the timing of spike trains follows a Poisson process. Even more generally, we might assume that on every trial the timing of spike trains follows a point process. The main idea is that the statistical model replaces data with hypothetical, random quantities; the real-world variation in the data is modeled using the mathematics of probability theory. Then, using the statistical model, we can state precisely what we mean when we say that synchrony could arise by chance.

Here, we are using "model" in the widest statistical sense. A model could be restricted to involve a small number of free parameters (such as a time-invariant firing rate) or it could be allowed to be very flexible, with a possibly infinite number of parameters (the firing rate could be allowed to vary in mildly constrained ways across both time and experimental conditions). We stress this point because certain statistical methods are occasionally called "model free." Such jargon is misleading. Rigorous statistical assessments always rely on statistical models.

It might be nice to be able to validate a statistical model compellingly, so that it could be considered a "correct" description of measured variation. This is essentially never possible. An important aspect of data analysis involves assessment of how well a model fits the data (e.g., [15]). In applying such goodness-of-fit methods, however, the question, "Does the model fit the data?" is really a short-hand for the more accurate question, "How well does the model fit for our specified scientific purpose?" For some purposes it may be entirely reasonable to assume a Poisson model for spike counts, while for other purposes—even with the same data—it could be grossly misleading.

In particular, one of the key ideas behind the introduction of statistical models is to incorporate within them some notion of regularity, or signal, which is distinguished from noise. Thus, a data-based "sample" mean is distinguished from a theoretical or "population" mean; a data-based firing rate is distinguished from a theoretical firing rate; and a data-based cross-correlation function is distinguished from a theoretical cross-correlation function. Each of these theoretical constructs has a definition, which is embedded in the statistical model.

Hypothesis Tests and Monte Carlo Approximation

To evaluate whether synchronous firing, indicated by a peak in the CCH, could be due to chance, we introduce a baseline statistical model according to which (1) the spiking activity of each of the two neurons is represented by a point process, and (2) the two point processes are statistically (conditionally) independent (under a specified conditioning). The part of the baseline model that specifies independence is called the *null hypothesis* and is labeled H_0. A statistical test of H_0 is based on some *test statistic*, which becomes a random variable Z, and when applied to a particular set of data the test statistic takes a value z_{obs}. The p-value is then the probability $p = P(Z > z_{\text{obs}}|H_0)$ (or, sometimes, $p = P(|Z| > |z_{\text{obs}}||H_0))$ where the notation indicates that the probability is based on the assumption that H_0 is correct. The p-value is, therefore, an indication of the likelihood of a large value of the test statistic "due to chance alone," in the absence of genuine statistical dependence between the neuronal spike trains. When the p-value is very small (e.g., much less than .01) we would conclude that if H_0 is correct, a very rare event would have occurred and, rare events being rare, therefore, H_0 is probably not correct: there is evidence against independence, and in favor of some mechanism that produces extra synchrony.

As a simple pedagogical example, suppose we model two neurons, A and B, as 5 Hz HPPs, and we observe their spike trains repeatedly over 100 1-s intervals. As a test statistic, we may compute the number of instances of a spike from neuron A and spike from neuron B occurring within 5 ms of each other. We could take the null hypothesis to be that the two processes are independent, and we can then compute the probabilities associated with each possible number z_{obs} of spike co-occurrences, within a 5 ms window, for 100 sets of 1-s observation times. This would allow us to find the p-value. If the p-value were small, then we would say there is evidence of synchrony *under the assumption of homogeneous Poisson firing at 5 Hz.*

This logic of hypothesis testing is very widely accepted. Its key features are that a null hypothesis specifies a statistical model, which is used to compute the p-value for a particular test statistic. The statistical subtleties and concerns center especially on the choice of model. In the example we just mentioned, the last italicized phrase is worrisome: What if the spiking activity of either neuron is not well-modeled as a Poisson process at 5 Hz? In addition, in some contexts the test statistic may be problematic, potentially, if it is sensitive to departures from the null hypothesis other than those of central scientific interest (for then the p-value might be small for some reason other than the presence of excess synchrony). Finally, a further issue is that many procedures do not produce an exact p-value, but only an approximate one, though in many cases the approximation may be sufficiently accurate for practical purposes.

From a statistical perspective, when an investigator proposes a method of testing for synchrony, we want to know what statistical model is used for the null hypothesis, and whether the p-value is valid in the sense of equalling, at least approximately, the putative probability under H_0.

We have not yet specified how the p-value would be computed, but a very general approach is to use computer simulation to generate sets of pseudo-data spike trains repeatedly, under the assumptions specified by H_0, and then to compute the value of the test statistic for each set of pseudo-data. For instance, one might generate 10,000 sets of pseudo-data, compute the test statistic for each, and then examine the proportion of sets of pseudo-data for which the test statistic exceeds z_{obs}. This proportion would be very nearly equal to the p-value (it is typically accurate to at least 2 digits). In practice, even if the Poisson assumption is applied, one does not know the firing rate and instead "plugs in" the value of firing rate observed in the data. This kind of plug-in estimation is known as a bootstrap approximation and the entire procedure is known as *bootstrap* hypothesis testing. Plug-in estimation produces accurate p-values when there is enough data to ensure that, so long as the null hypothesis is true, the estimate is reliable.

REFERENCES

1. M Abeles and I Gat, Detecting precise firing sequences in experimental data, *Journal of Neuroscience Methods* **107**, 2001, 141–154.
2. M Abeles and G L Gerstein, Detecting spatiotemporal firing patterns among simultaneously recorded single neurons, *Journal of Neurophysiology* **60**(3), 1988, 909–924.
3. ED Adrian, The impulses produced by sensory nerve endings: Part i, *J. Physiol.* (London) **61**, 1926, 49–72.
4. AM Aertsen, GL Gerstein, MK Habib, and G. Palm, Dynamics of neuronal firing correlation: Modulation of "effective connectivity", *Journal of Neurophysiology* **61**(5), 1989, 900–917.
5. A Amarasingham, Statistical methods for the assessment of temporal structure in the activity of the nervous system, PhD thesis, Division of Applied Mathematics, Brown University, 2004.
6. A Amarasingham, T-L Chen, S Geman, MT Harrison, and D Sheinberg, Spike count reliability and the Poisson hypothesis, *Journal of Neuroscience* **26**, 2006, 801–809.
7. A Amarasingham, MT Harrison, N Hatsopolous, and S Geman, Conditional modeling and the jitter method of spike re-sampling, *Journal of Neurophysiology* **107**, 2012, 517–531.
8. A Amarasingham, MT Harrison, and S Geman, Jitter methods for investigating spike train dependencies, *Computational and Systems Neuroscience* (COSYNE), 2007.
9. W Bair, E Zohary, and WT Newsome, Correlated firing in macaque visual area mt: Time scales and relationship to behavior, *Journal of Neuroscience* **21**(5), 2001, 1676–1697.
10. SN Baker and GL Gerstein, Determination of response latency and its application to normalization of cross-correlation measures, *Neural Computation* **13**, 2001, 1351–1377.
11. Y Ben-Shaul, H Bergman, Y Ritov, and M Abeles, Trial to trial variability in either stimulus or action causes apparent correlation and synchrony in neuronal activity, *Journal of Neuroscience Methods* **111**(2), 2001, 99–110.
12. CD Brody, Slow variations in neuronal resting potentials can lead to artefactually fast cross-correlations in the spike trains, *Journal of Neurophysiology* **80**, 1998, 3345–3351.
13. CD Brody, Disambiguating different covariation types, *Neural Computation* **11**, 1999, 1527–1535.
14. EN Brown and RE Kass, What is statistics? (with discussion), *American Statistician* **63**(2), 2009, 105–123.
15. EN Brown, R Barbieri, V Ventura, RE Kass, and LM Frank, The time-rescaling theorem and its application to neural spike train data analysis, *Neural Computation* **14**(2), 2002, 325–346.
16. DA Butts, C Weng, J Jin, C-I Yeh, NA Lesica, J-M Alonso, and GB Stanley, Temporal precision in the neural code and the timescales of natural vision, *Nature* **449**, 2007, 92–96.
17. G Czanner, U Eden, S Wirth, M Yanike, W Suzuki, and E Brown, Analysis of between-trial and within-trial neural spiking dynamics, *J Neurophysiol* **99**, 2008, 2672–2693.
18. DJ Daley and DD Vere-Jones, *An Introduction to the Theory of Point Processes*, Vol. 2, New York: Springer, 2007.
19. A Date, E Bienenstock, and S Geman, *On the temporal resolution of neural activity*, Tech. Report, Division of Applied Mathematics, Brown University, 1998.
20. AC Davison and DV Hinkley, *Bootstrap Methods and their Application*, Cambridge: Cambridge University Press, 1997.
21. I Dimatteo, CR Genovese, and RE Kass, Bayesian curve-fitting with free-knot splines, *Biometrika* **88**(4), 2001, 1055–1071.
22. G Dragoi and G Buzsáki, Temporal encoding of place cells by hippocampal cell assemblies, *Neuron* **50**, 2006, 145–157.

23. J-M Fellous, PHE Tiesinga, PJ Thomas, and TJ Sejnowski, Discovering spike patterns in neuronal responses', *Journal of Neuroscience* **24**(12), 2004, 2989–3001.
24. S Fujisawa, A Amarasingham, MT Harrison, and G Buzsáki, Behavior-dependent short-term assembly dynamics in the medial prefrontal cortex, *Nature Neuroscience* **11**, 2008, 823–833.
25. S Furukawa and JC Middlebrooks, Cortical representation of auditory space: Information-bearing features of spike patterns, *Journal of Neurophysiology* **87**, 2002, 1749–1762.
26. GL Gerstein, Searching for significance in spatio-temporal firing patterns, *Acta Neurobiologiae Experimentalis* **64**, 2004, 203–207.
27. S Grün, Data-driven significance estimation for precise spike correlation, *Journal of Neurophysiology* **101**, 2009, 1126–1140.
28. S Grün, M Diesmann, and A Aertsen, Unitary events in multiple single-neuron spiking activity: I. Detection & significance, *Neural Computation* **14**(1), 2002, 43–80.
29. S Grün, M Diesmann, and A Aertsen, Unitary events in multiple single-neuron spiking activity: II. Nonstationary data, *Neural Computation* **14**(1), 2002, 81–119.
30. S Grün, A Riehle, and M Diesmann, Effects of cross-trial nonstationarity on joint-spike events, *Biological Cybernetics* **88**(5), 2003, 335–351.
31. R Gütig, A Aertsen, and S Rotter, Statistical significance of coincident spikes: Count-based versus rate-based statistics, *Neural Computation* **14**(1), 2002, 121–153.
32. MT Harrison, *Discovering compositional structures*, PhD thesis, Division of Applied Mathematics, Brown University, 2004.
33. MT Harrison, Accelerated spike resampling for accurate multiple testing controls, *Neural Computation* **25**, 2013, 418–449.
34. MT Harrison and S Geman, A rate and history-preserving algorithm for neural spike trains, *Neural Computation* **21**(5), 2009, 1244–1258.
35. HK Hartline, The receptive fields of optic nerve fibers, *American Journal of Physiology* **130**, 1940, 690–699.
36. N Hatsopoulos, S Geman, A Amarasingham, and E Bienenstock, At what time scale does the nervous system operate?, *Neurocomputing* **52**, 2003, 25–29.
37. NG Hatsopoulos, Q Xu, and Y Amit, Encoding of movement fragments in the motor cortex, *Journal of Neuroscience* **27**(19), 2007, 5105–5114.
38. Y Ikegaya, G Aaron, R Cossart, D Aronov, I Lamp, D Ferster, and R Yuste, Synfire chains and cortical songs: Temporal modules of cortical activity, *Science* **304**(5670), 2004, 559–564.
39. H Ito and S Tsuji, Model dependence in quantification of spike interdependence by joint peri-stimulus time histogram, *Neural Computation* **12**, 2000, 195–217.
40. H Jeffreys, *Scientific Inference*, Cambridge: Cambridge University Press, 1931.
41. LM Jones, DA Depireux, DJ Simons, and A Keller, Robust temporal coding in the trigeminal system, *Science* **304**, 2004, 1986–1989.
42. RE Kass and RC Kelly, A framework for evaluating pairwise and multiway synchrony among stimulus-driven neurons, *Neural Computation* **24**, 2012, 2007–2032.
43. RE Kass, RC Kelly, and W-L Loh, Assessment of synchrony in multiple neural spike trains using loglinear point process models, *Annals of Applied Statistics* **5**, 2011, 1262–1292.
44. RE Kass, V Ventura, and EN Brown, Statistical issues in the analysis of neural data, *Journal of Neurophysiology* **94**, 2004, 8–25.
45. RE Kass, V. Ventura, and C. Cai, Statistical smoothing of neuronal data, *Network-Computation in Neural Systems* **14**(1), 2003, 5–16.
46. RC Kelly, MA Smith, RE Kass, and T-S Lee, Local field potentials indicate network state and account for neuronal response variability, *Journal of Computational Neuroscience* **29**, 2010, 567–579.

47. T Lu and X Wang, Information content of auditory cortical responses to time-varying acoustic stimuli, *Journal of Neurophysiology* **91**, 2003, 301–313.
48. P Maldonado, C Babul, W Singer, E Rodriguez, D Berger, and S Grün, Synchronization of neuronal responses in primary visual cortex of monkeys viewing natural images, *Journal of Neurophysiology* **100**, 2008, 1523–1532.
49. L Martignon, G Deco, K Laskey, M Diamond, W Freiwald, and E Vaadia, Neural coding: Higher-order temporal patterns in the neurostatistics of cell assemblies, *Neural Computation* **12**(11), 2000, 2621–2653.
50. L Martignon, H von Hasseln, S Grün, A Aertsen, and G Palm, Detecting higher-order interactions among the spiking events in a group of neurons, *Biological Cybernetics* **73**, 1995, 69–81.
51. A Moekeichev, M Okun, O Barak, Y Katz, O Ben-Shahar, and I Lampl, Stochastic emergence of repeating cortical motifs in spontaneous membrane potential fluctuations in vivo, *Neuron* **53**(3), 2007, 413–425.
52. Z Nádasdy, H Hirase, A Czurkó, J Csiscvari, and G Buzsáki, Information content of auditory cortical responses to time-varying acoustic stimuli, *Journal of Neurophysiology* **91**, 2003, 301–313.
53. MW Oram, MC Wiener, R Lestienne, and BJ Richmond, Stochastic nature of precisely timed spike patterns in visual system neural responses, *Journal of Neurophysiology* **81**, 1999, 3021–3033.
54. Q Pauluis and SN Baker, An accurate measure of the instantaneous discharge probability, with application to unitary joint-event analysis, *Neural Computation* **12**(3), 2000, 647–669.
55. A Pazienti, M Diesmann, and S Grün, Bounds on the ability to destroy precise coincidences by spike dithering, *Lecture Notes in Computer Science: Advances in Brain, Vision, and Artificial Intelligence*, pp. 428–437, Berlin: Springer, 2007.
56. A Pazienti, PE Maldonado, M Diesmann, and S Grün, Effectiveness of systematic spike dithering depends on the precision of cortical synchronization, *Brain Research* **1225**, 2008, 39–46.
57. Y Prut, E Vaadia, H Bergman, I Haalman, H Slovin, and M Abeles, Spatiotemporal structure of cortical activity: Properties and behavioral relevance, *Journal of Neurophysiology* **79**, 1998, 2857–2874.
58. A Renart, J de la Rocha, P Bartho, L Hollander, N Parga, A Reyes, and KD Harris, The asynchronous state in cortical circuits, *Science* **327**, 2010, 587–590.
59. F Rieke, D Warland, R van Steveninck, and W Bialek, *Spikes: Exploring the Neural Code*, Cambridge, Ma: MIT Press, 1997.
60. A Rokem, S Watzl, T Gollisch, M Stemmler, AVM Herz, and I Samengo, Spike-timing precision underlies the coding efficiency of auditory receptor neurons, *Journal of Neurophysiology* **95**, 2006, 2541–2552.
61. E Salinas and TJ Sejnowski, Gain modulation in the central nervous system: Where behavior, neurophysiology, and computation meet, *The Neuroscientist* **7**(5), 2001, 430.
62. E Schneidman, MJ Berry, R Segev, and W Bialek, Weak pairwise correlations imply strongly correlated network states in a neural population, *Nature* **440**(7087), 2006, 1007–1012.
63. S Shinomoto, H Kim, T Shimokawa, N Matsuno, S Funahashi, K Shima, I Fujita, H Tamura, T Doi, K Kawano et al., Relating neuronal firing patterns to functional differentiation of cerebral cortex, *PLoS Computational Biology* **5**(7), 2009, e1000433.
64. T Shmiel, R Drori, O Shmiel, Y Ben-Shaul, Z Nadasdy, M Shemesh, M Teicher, and M Abeles, Neurons of the cerebral cortex exhibit precise interspike timing in correspondence to behavior, *Proceedings of the National Academy of Sciences* **102**(51), 2005, 18655–18657.

65. T Shmiel, R Drori, Oshmiel, Y Ben-Shaul, Z Nadasdy, M Shemesh, M Teicher, and M Abeles, Temporally precise cortical firing patterns are associated with distinct action segments, *Journal of Neurophysiology* **96**, 2006, 2645–2652.
66. MA Smith and A Kohn, Spatial and temporal scales of correlation in primary visual cortex, *Journal of Neuroscience* **28**(48), 2008, 12591–12603.
67. E Stark and M Abeles, Unbiased estimation of precise temporal correlations between spike trains, *Journal of Neuroscience Methods* **179**, 2009, 90–100.
68. B Staude, S Grün, and S Rotter, Higher-order correlations in non-stationary parallel spike trains: Statistical modeling and inference, *Frontiers in Computational Neuroscience* **4**, 2010, 16, 1–17.
69. P Tiesinga, JM Dellous, and TJ Sejnowski, Regulation of spike timing in visual cortical circuits, *Nature Reviews Neuroscience* **9**, 2008, 97–107.
70. DJ Tolhurst, JA Movshon, and AF Dean, The statistical reliability of signals in single neurons in cat and monkey visual cortex, *Vision Research* **23**, 1983, 775–785.
71. PJ Uhlhaas, G Pipa, B Lima, L Melloni, S Neuenschwander, D Nikolić, and W Singer, Neural synchrony in cortical networks: History, concept and current status, *Frontiers in Integrative Neuroscience* **3**, 2009. Available at: http://www.frontiersin.org/integrative_neuroscience/10.3389/neuro.07/017.2009/full
72. V Ventura, C Cai, and RE Kass, Statistical assessment of time-varying dependence between two neurons, *Journal of Neurophysiology* **94**, 2005, 2940–2947.
73. V Ventura, C Cai, and RE Kass, Trial-to-trial variability and its effect on time-varying dependence between two neurons, *Journal of Neurophysiology* **94**, 2005, 2928–2939.
74. AEP Villa, IV Tetko, B Hyland, and A Najem, Spatiotemporal activity patterns of rat cortical neurons predict responses in a conditioned task, *Proc Natl Acad Sci USA* **96**, 1999, 1106–1111.

5 Binless Estimation of Mutual Information in Metric Spaces

Ayelet-Hashahar Shapira and Israel Nelken

CONTENTS

5.1 INTRODUCTION

The quantification of the association between stimuli and the neural responses they evoke is one of the major concerns of sensory neuroscience. Currently, it is widely accepted that the temporal structure of a spike train is an essential aspect of the neuronal responses. For example, studies in the cat have shown that this temporal structure carries substantial information about the sensory environment (Middlebrooks et al. 1994; Furukawa et al. 2000). As the temporal structure of auditory stimuli is uniquely complex, the auditory system is an excellent model system to study the contribution of spike timing to the neural code (Nelken et al. 2005). In consequence, the development and validation of methods of measuring the amount of information carried by spike timing are highly relevant in the context of the auditory system.

The use of mutual information (MI) as an analytical tool for this purpose has become increasingly common in neuroscience in general and in auditory research, in particular. The MI is a measure of the strength of the association between two random variables. The MI, $I(S;R)$, between the stimuli S and the neural responses R

is defined in terms of their joint distribution $p(S,R)$. When this distribution is known exactly, the MI can be calculated as

$$I(S;R) = \sum_{s\in S, r\in R} p(s,r)\log_2\left(\frac{p(s,r)}{p(s)p(r)}\right)$$

where $p(s) = \sum_{r\in R} p(s,r)$ and $p(r) = \sum_{s\in S} p(s,r)$ are the marginal distributions over the stimuli and responses, respectively.

However, the joint distribution of stimuli and experimentally measured responses is not known exactly. Instead, it has to be estimated from the experimental data. With common experimental settings, the underlying distributions of neuronal responses are too severely undersampled for adequate estimation of the MI using naïve approaches. The bias of the MI estimator due to finite sampling is considerable, and may be substantially larger than the actual information. Inevitably, it becomes necessary to coarsen the representation of the neural responses. Such coarsening is implicit, for example, in the use of spike counts for quantifying responses—the temporal aspects of the responses, which may be informative about the stimuli, are ignored.

Calculating the MI using the "plug-in" MI estimator of the joint distribution histograms with bias correction requires on average more than one count per bin of the joint distribution (Treves and Panzeri 1995; Golomb et al. 1997), greatly reducing the resolution with which the responses can be represented. In consequence, the MI between the stimuli and the underlying distribution of the coarsened representation of the responses will be in general smaller than the "True MI." For example, in our experimental design, we have responses that are 100 ms long to 11 different stimuli. If we bin the responses in 10 ms bins and take spike trains that contain up to six spikes, we already get 2310 possible bins in the joint distribution $\left(11\binom{10}{6}\right)$. In our experiment, we have only 10 repetitions for each kind of stimulus, for a total of 110 trials, not nearly enough for applying Golomb's rule of thumb.

While the estimation of lower bounds based on coarse representations of the responses may be sufficient for some applications, they may not be sufficient when MI values calculated under different conditions have to be compared with each other. For example, we measured the responses of neurons in cat auditory cortex in response to virtual acoustic space stimuli (VAS, broad band stimuli that mimic all the relevant acoustic information available when listening in free field) and in response to modified stimuli, in which some of the physical cues for space have been removed (see Las et al. 2008 for some of these data). To study the effect of physical cues on the sensory responses, we want to compare the MI between the full VAS and the responses they evoke on the one hand, and the MI between the modified VAS and the responses they evoke on the other hand. Such comparisons cannot be rigorously made with lower bounds only.

A number of suggestions have been made for computing MI without explicit calculation of the joint distribution as an intermediate step. One common approach is to use a decoding algorithm. A decoding algorithm assigns a stimulus to each response, thereby generating a binning scheme on the response space, which is in fact a coarser version of the full stimulus–response joint distribution. A confusion matrix is generated by applying the decoder to the data and the MI is estimated from the confusion matrix. Owing to the information-processing inequality, decoding results are a lower bound on the MI. Decoding methods have been extensively used for studying neural codes in the past (Bialek et al. 1991; Rolls et al. 1997; Treves 2003). In auditory research, Middlebrooks and his coworkers studied the coding of space in auditory cortex of cats (Furukawa and Middlebrooks 2002) using decoding algorithms based on artificial neural networks.

A specific decoding scheme was suggested by Victor and Purpura (1996), who suggested to perform the decoding through the use of a metric that captures physiological-based intuitions regarding the similarity and dissimilarity between firing patterns. Their edit distance (ED) algorithm quantifies the dissimilarity between a pair of spike trains based on the cost needed to transform one spike train into the other using only certain elementary steps. They then decoded the responses by comparing the distances from each response to all other responses in the data set, and assigning each response to the stimulus with the smallest average distance to that response.

A fundamentally different approach for MI estimation has been suggested by Victor (2002). He introduced a binless approach to MI calculations (BINL) for continuous data. This approach avoids the need for *a priori* binning of the neural responses. In this approach, estimates of the density $p(x,y)$ are derived from the "distance" between a sample point and its "nearest neighbors"—small distances correspond to high-probability density and large distances to low-probability density. Thus, the probability ratio that is necessary to calculate the MI is replaced by a distance ratio, together with correction factors that depend on the dimensionality of the response space (see below). Victor applied this algorithm for estimating MI in response ensembles that had identical spike count, embedding them in Euclidian spaces with the same number of dimensions as the number of spikes in the response. This is also the limitation of the technique—layering responses by spike counts will typically result in small sets of responses with any given spike count, leading to large bias in the estimates of the MI in each layer.

Here, we explore the use of binless methods in metric spaces that avoid the layering by spike counts as suggested by Victor (2002). We first show that the methodology is sane by proving a (not very strong) convergence theorem. We then apply one specific instance of this methodology to spike trains generated from models that are derived from actual responses to VAS in cat auditory cortex. In this case, we can show that the estimated MI approaches nicely the theoretical MI at the limit of a lot of data, and may still usefully approximate it with experimentally reasonable amount of data.

5.2 METHODS

5.2.1 Neural Data

The experiments were approved by the animal use and care committee of the Hebrew University Hadassah Medical School. Extracellular activity was recorded from primary auditory cortex (A1) and the anterior ectosylvian sulcus (AES) of halothane-anesthetized cats.

The stimuli consisted of short noise bursts (100 ms) filtered through head-related transfer functions to mimic sounds arriving from azimuths between −75° and 75°. The sounds were presented to the animal through sealed earphones. Each stimulus was repeated 10 times, with an interstimulus interval of 1000 ms (onset to onset). All directions were intermixed in a pseudorandom order. For more details, see Las et al. (2008).

5.2.2 Algorithm for Calculating MI in Metric Spaces

5.2.2.1 Overview

We begin by describing the theory behind this approach, and then consider the technical and computational details. The information that the neuronal responses contains about the sensory stimulus measures the extent to which its responses can allow an observer to tell which stimulus was presented. Thus, information measures the extent to which the distributions of the responses to different stimuli are distinct. Since we have a sample of each distribution, we can hope to estimate these separations from the distances between the responses, in some abstract "response space." If the response space is an ordinary (Euclidean) one, information can be estimated from the pairwise distances between nearest responses to the same stimulus, and nearest responses to different stimuli (Kozachenko and Leonenko 1987). This estimate is highly efficient, but it critically relies on the Euclidean nature of the space, and in particular, on the fact that it has a definite dimension. However, the geometry of spike trains is not Euclidean. Instead, it has multiple Euclidean components, one for each dimension (specifically, the spike trains with n spikes form an n-dimensional space, parameterized by the time of the n spikes; see Chapter 1, this book). The nearest-neighbor method works but it has to be applied separately to each partition (Victor 2002) since the relationship between nearest neighbors and information depends on the dimension. This partitioning leads to practical difficulties since there may not be enough responses in each partition to form a useful estimate.

To avoid this partitioning, we take a different approach here. We define the geometry of the space of spike trains via the ED metric of Victor and Purpura (1996). This puts all the responses in the same space, and hence avoids the partitioning. However, a new problem arises: the need to define the effective dimension of this space. Below, we show how this can be done by using a simple heuristic based on the number of similar responses within spheres of given radii. We then show how this effective dimension can be incorporated into the nearest-neighbor approach, resulting in accurate estimates of information for very limited samples. We refer to that combination of the metric space method with binless estimator as the "BINL-ED method."

5.2.2.2 Computational Details

The binless approach requires all distances between pairs of trials to be nonzero. Therefore, trials that do not fit this assumption need to be treated separately. This is accomplished in two stages. First, trials with no spikes (if such trials were present) were pooled to form the *null* group of trials. All the other trials were grouped together into the *spikes* group. The joint distribution of groups and stimuli was estimated by the corresponding histogram. The information contributed by this partition was calculated from the histogram using the successive reduction method of Nelken et al. (2005). This contribution will be denoted as $I_{initial}$. Next, since responses could repeat exactly more than once (so that the distance between them is 0), the trials in the spikes group were divided into subgroups, one for each response that repeated at least twice, and one group of trials consisting of responses that did not repeat more than once. This last group is termed the *nonzero* group of trials. Each group could be distributed differently across stimuli. The information contributed by this partition was calculated as above and denoted as $I_{partition}$.

The distance between any two trials in the *nonzero* group was positive, and in consequence the binless approach could be used in this group. The BINL-ED algorithm estimates the ratio between the overall probability for observing a response and the conditional probability of observing this response given a stimulus by using the ratio of the distance between that response and the nearest one among all trials with the same stimulus (λ) and among all trials (λ^*). For this purpose, we first calculated for each stimulus the following distances by the metric space method: the distance to the nearest neighbor within the same stimulus (λ), the distance to the nearest neighbor over all stimuli (λ^*), and the distance of the second nearest neighbor (λ_2). We defined h_j as the distance ratio λ_2/λ of trial j.

$I_{nonzero}$ is estimated by

$$I_{nonzero} \approx \frac{1}{N_{nonzero} \ln 2} \sum_{j=1,\ldots,N} r_j \ln\left(\frac{\lambda_j}{\lambda_j^*}\right) - \frac{1}{\ln 2} \sum_{k=1,\ldots,K} \frac{N_k}{N} \ln \frac{N_k - 1}{N - 1}$$

where N is the total number of responses, K is the number of stimuli, N_k is the number of responses to stimulus k in the *nonzero* group, and r_j is the *local dimension* of the response j.

$I_{nonzero}$ turned to be highly biased. To estimate this bias, we repeated the same calculations for surrogate data sets in which the trials were randomized across stimuli. The mean $I_{nonzero}$ value of 10 such randomizations was used as an estimate of the bias and was subtracted from the $I_{nonzero}$ estimated from the data.

In the original application of the binless method, r_j was fixed and equal to the common number of spikes in each response (since the responses were layered by spike counts). In the metric space defined by the ED, it is unclear what this dimensionality should be. The choice here is to use the average r conditioned on the observed distance ratio h_j. The local dimension r_j of the metric space in the neighborhood of trial j was estimated by

$$r_j = \frac{\sum_{r=1}^{\infty} r p(h_j|r)P(r)}{\sum_{r=1}^{\infty} p(h_j|r)P(r)}$$

Here, $P(r)$ is an *a priori* distribution of dimension, common to all trials, and h_j is the distance ratio as defined above. In a Bayesian spirit, $p(h_j|r)$ is used to reweigh $P(r)$ based on the observation h_j. Finally, r_j is the mean of that reweighted distribution, where r runs over all possible spike numbers in the responses. As the *a priori* distribution, $P(r)$, we used the spike count distribution calculated from all the responses, smoothed by a hamming window of width three. The expression of the conditional probability we used is $p(h_j|r) = rh_j^{-r-1}$, as can be established following Victor (2002).

The three contributions for the MI are now combined as

$$I_{total} = I_{initial} + \frac{N_{spikes}}{N_{null} + N_{spikes}}\left[I_{partition} + \frac{N_{nonzero}}{N_{spikes}} I_{nonzero}\right]$$

where N_{spikes}, N_{null}, and $N_{nonzero}$ are the total number of trials in the *spikes*, *null*, and *nonzero* groups, respectively (Victor 2002).

5.2.3 Validating MI Estimates

We used the neural responses recorded in cat auditory cortex to generate inhomogeneous Poisson models for spike trains. The models covered 128 ms, from 10 ms after stimulus onset to 38 ms past stimulus offset. To generate the rate function for each stimulus, we started with the peri-stimulus time histograms (PSTHs) for each of the stimuli calculated with a bin width of 1 ms. The 2D matrix containing the 11 responses × 128 time bins was smoothed using a 10 ms Hamming window along the time dimension and a three-bin triangular window along the stimulus dimension. To facilitate the estimation of the MI between stimuli and responses for these inhomogeneous Poisson models, firing rates were averaged across 4 ms windows (with no overlap) and the average value was assigned to each of the 1 ms bin inside this window (see Figure 5.1, middle panel).

To calculate the MI for the models, we had to tally the probability of all possible spike trains and all stimuli. Owing to computational limitations, we were limited to spike trains 32 bins long and with no more than six spikes. We, therefore, collapsed each 4 ms block into one time bin by adding the firing rates in the 1 ms bins. This resulted in a 11 × 32 matrix for which we were able to calculate the MI directly from the joint probability of all possible response patterns in 32 bins and each virtual direction (as in Nelken et al. 2005). This value was termed True MI. We reasoned that the MI between stimuli and responses for the expanded model of 128 bins and the True MI that we calculated from the 32 bin matrix would be essentially the same, with deviations only due to the rare events in which more than one spike would occur in a 4 ms window.

The results are based on models derived from the responses of 20 neurons that spanned the whole range of firing rates that appeared in the data and the range of MIs as calculated with spike counts. For all these neurons, no more than 10% of trials had six spikes or more, to fit the assumptions for the calculation of the True MI. Also, we confirmed that the contribution of the trials with more than six spikes to the MI, estimated using spike counts was smaller than 0.01 bit.

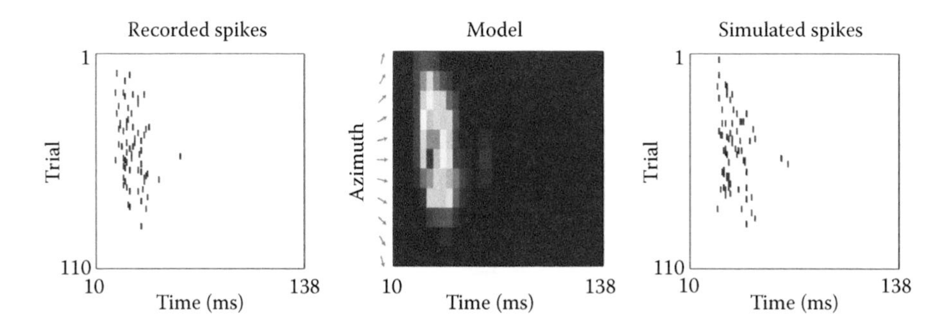

FIGURE 5.1 Simulated responses. Recorded responses ordered by 11 different azimuths (–75°:15°:+75°) of a unit from field AES (left), the model constructed for the same unit (middle), and 110 responses (right) simulated from the model. The general response pattern, as well as the onset nature of the original responses, are preserved in the simulated spike trains. On the other hand, the simulated spike trains seem more variable than the actual responses, partially due to the smoothing inherent in the generation of the model, and possibly also due to the use of inhomogeneous Poisson processes for generating the simulated data.

The simulations were based on the full 128 ms models. Spike trains were simulated in 1 ms resolution with these models for each virtual direction. Spikes were selected with probabilities set by the firing rates of the model, and the choices at different time bins were independent of each other. We simulated $N = 10$, 20, 40, 80, and 160 repetitions of each stimulus. Then, we estimated the MI using the BINL-ED method with the simulated spike trains. Each simulation was repeated 10 times. By comparing these estimates with the True MI for the model, we could compute the bias and variance of the estimators. The True MI was subtracted from the average MI across all 10 repeats and the difference squared to produce the bias. The variance was estimated by the average of the squared deviations of the individual MI values from their mean. The mean squared error (MSE) is the sum of these two terms.

Figure 5.1 displays one such model for a unit recorded in AES (Las et al. 2008). The recorded spikes are shown in the left panel. The model fitted with these spikes is shown in the middle panel. The right panel shows spikes simulated using that model. Note the larger dispersion of the spikes generated by the model relative to the recorded ones. This is due to the smoothing used in generating the models, but may also have to do with the use of inhomogeneous Poisson models for generating the spikes. The seemingly larger temporal precision of the original data may actually make the estimation of the MI easier with real data.

5.3 RESULTS

5.3.1 Theory

We present here a calculation that clarifies the relationships between the underlying joint probability distribution of stimuli and responses on the one hand and the metric used for calculating the distance between responses on the other hand. It suggests a natural condition that ensures convergence to the correct MI in the limit of large amounts of data.

Assume the data is given as N stimulus–response pairs (x_i, y_i). Here, i is an index running over all trials, x_i is the response recorded at trial i, and y_i is the stimulus played at trial i. For large N, the MI will be approximately

$$\mathrm{MI} = \frac{1}{N}\sum_i \log_2 \frac{p(x_i|y_i)}{p(x_i)}$$

where $p(x,y)$ is the joint distribution of stimuli and responses, and $p(x)$ is the marginal distribution of the responses.

Fix a small ε. A natural way to use the metric to estimate the probabilities in the expression above is to try

$$\log_2 \frac{p(x_i|y_i)}{p(x_i)} \approx \log_2 \frac{\left|\{(x,y)|d(x,x_i) < \varepsilon, y = y_i\}\right|/N_i}{\left|\{(x,y)|d(x,x_i) < \varepsilon\}\right|/N}$$

where $|\ldots|$ stands for the number of elements of the set and N_i is the total number of trials of stimulus y_i. Denote $U_\varepsilon(x) = \{\bar{x}|d(\bar{x},x) < \varepsilon\}$. Then, we have approximately

$$\left|\{(x,y)|d(x,x_i) < \varepsilon\}\right|/N \approx \int_{U_\varepsilon(x_i)} p(\bar{x})d\bar{x}$$

$$\left|\{(x,y)|d(x,x_i) < \varepsilon, y = y_i\}\right|/N_i \approx \int_{U_\varepsilon(x_i)} p(\bar{x}|y_i)d\bar{x}$$

$$\tilde{p}(x) = \frac{\int_{U_\varepsilon(x_i)} p(\bar{x})d\bar{x}}{\int_{U_\varepsilon(x_i)} d\bar{x}}, \quad \tilde{p}(x|y) = \frac{\int_{U_\varepsilon(x_i)} p(\bar{x}|y)d\bar{x}}{\int_{U_\varepsilon(x_i)} d\bar{x}}$$

Both $\tilde{p}(x)$ and $\tilde{p}(x|y)$ are probability distributions, and it can be easily seen that they are the conditional distribution and margin, respectively, of $\tilde{p}(x,y) = \tilde{p}(x|y)p(y)$, where $p(y)$ is the margin of the original distribution $p(x,y)$. In fact, $\tilde{p}(x), \tilde{p}(x|y)$ are smoothed versions of $p(x)$, $p(x|y)$, where the smoothing is performed by integrating over metric neighborhoods of radius ε. We hope now that

$$\log_2 \frac{p(x_i|y_i)}{p(x_i)} \approx \log_2 \frac{\tilde{p}(x_i|y_i)}{\tilde{p}(x_i)}$$

and therefore, replace the true probability ratio by the approximation based on the metric neighborhood, so that our approximation to the MI based on estimation in metric neighborhoods boils down to

$$\mathrm{MI}_{\varepsilon} \approx \frac{1}{N}\sum_{i}\log_2\frac{\tilde{p}(x_i|y_i)}{\tilde{p}(x_i)} \xrightarrow[N\to\infty]{} \sum_{x,y}p(x,y)\log_2\frac{\tilde{p}(x|y)}{\tilde{p}(x)}$$

Thus, the difference between the True MI and its approximation by local neighborhoods has to do with the mismatch between the two distributions $p(x,y)$ and $\tilde{p}(x,y)$. The difference between the two expressions is

$$\begin{aligned}\mathrm{MI}-\mathrm{MI}_{\varepsilon} &= \sum_{x,y}p(x,y)\left[\log_2\frac{p(x|y)}{p(x)}-\log_2\frac{\tilde{p}(x|y)}{\tilde{p}(x)}\right]\\ &= \sum_{x,y}p(x,y)\left[\log_2\frac{p(x|y)}{\tilde{p}(x|y)}-\log_2\frac{p(x)}{\tilde{p}(x)}\right]\\ &= \sum_{y}p(y)D_{KL}(p(x|y)\|\tilde{p}(x|y))-D_{KL}(p(x)\|\tilde{p}(x))\end{aligned}$$

where D_{KL} is the Kullback–Leibler (KL) divergence between the two probability distributions.

We conclude that a natural condition for the convergence of the MI estimate based on metric neighborhoods is the convergence of the smoothed joint distribution $\tilde{p}(x,y)$ to the original $p(x,y)$ in the sense of the KL divergence as ε goes to 0. This ensures that the two KL divergences in the expression above will converge to 0, and therefore, that the MI estimate based on the metric neighborhood will converge to the true one. Thus, convergence requires a consistency between the KL divergence and the metric d—smoothing over small metric convergence should result in small KL divergence between the original and the smoothed distributions. In summary, the attempt to use general metric neighborhoods for estimating the MI is sane under a natural continuity condition on the joint distribution of stimuli and responses.

5.3.2 Simulations

While the above argument provides a rigorous criterion for convergence, it does not establish how rapidly convergence occurs in practice. To do this, here we use simulations using response models that are derived from actual data to check under what conditions the performance of the method is good enough for actual calculations.

Figure 5.2 shows the models derived from the responses of the 20 units, sorted by their True MI. These are 2D plots of the firing rate as a function of time (on the abscissa) and stimulus direction (on the ordinate). Firing rates are depicted by colors. At the top of each panel, the two numbers represent maximal firing rate

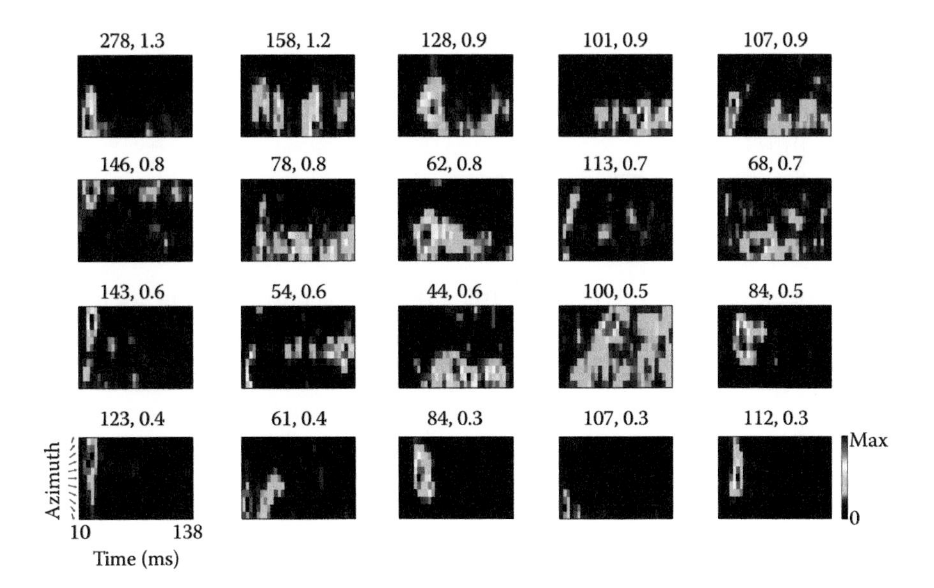

FIGURE 5.2 **(See color insert.)** All models. Models were derived from the responses of 20 units sorted by their MI. Firing rates are depicted by colors from blue, which always represent 0, to red, which varies between panels. The numbers at the top of each panel are the maximal firing rate (sp/s, left) and the MI of the model (True MI in bits, right). Unit 18 (row 4, column 3) is displayed in Figure 5.1.

(corresponding to the red of the color scale, with blue always corresponding to 0) and the MI between stimuli and responses for that model.

Almost all the models included a well-timed onset responses ($n = 16/20$, 80%) at least to some of the virtual directions. Later response components are present in half of models ($n = 10/20$, 50%). It is clear from Figure 5.2 that the rate functions show temporal modulations that vary with stimulus in many cases ($n = 12/20$, 60%). As these temporal aspects of the neural activity may play a central role in the read-out of the responses, these observations support our efforts to develop a method that takes more than just spike counts into account.

5.3.3 Estimating MI

Distances have been calculated using the Victor–Purpura editing distance metric (Victor and Purpura 1996). This metric has a parameter (q), serving as a measure of its temporal resolution. We used temporal resolutions (in ms) of $1/q = 2, 3, 5, 10, 20, 50, 100, \infty$. The parameter $1/q$ plays a role somewhat similar to that of ε in the theoretical argument above—it determines the size of the metric neighborhoods.

Figure 5.3 illustrates the effect of the different values of N and q on the estimated MI. The estimated MI values are plotted against the MI values calculated directly from the models (True MI, see Section 5.2 for details). Figure 5.3a–c shows the MIs calculated with $N = 10$, 20, and 160, and $1/q = 20$ ms. The scatter of the estimated MI values decreased with the increase in the number of trials per stimulus. The

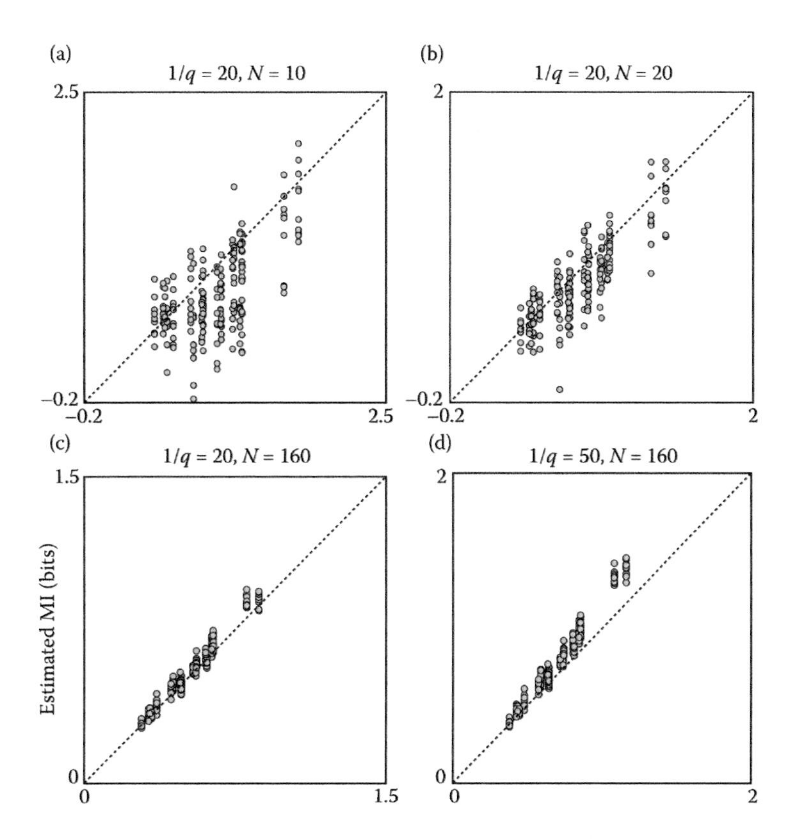

FIGURE 5.3 Estimated MI and True MI. MI estimates, using the BINL-ED method, are plotted against the True MI values, for all 20 models. Each model was simulated 10 times. The estimates were produced with four different combinations of number of trials/stimulus and temporal resolutions. (a) $N = 10$ trials/stimulus, $1/q = 20$ ms; (b) $N = 20$ trials/stimulus, $1/q = 20$ ms; (c) $N = 160$ trials/stimulus, $1/q = 20$ ms; (d) $N = 160$ trials/stimulus, $1/q = 50$ ms.

variance decreased substantially but some bias was left even with $N = 160$ trials. Figure 5.3c and d illustrates the effect of changing $1/q$ from 20 to 50 ms, for $N = 160$ trials. In this example, the bias increased at the lower temporal resolution, without much effect on the variance, presumably because with $1/q = 50$ ms, the effective size of the metric neighborhoods is too large.

We analyzed those properties of models that influenced the estimation error for $N = 20$ and $1/q = 20$. Figure 5.4a shows the square root MSE (RMSE) for each unit (ordinate) against the firing rate, averaged for the whole duration of the model (abscissa). The error increased with firing rates, so that, generally, the error in the estimate was smallest for units with lower firing rates (less than 5 sp/s averaged over the 128 ms of the model). The correlation coefficient between RMSE and firing rate was 0.71. Figure 5.4b plots the RMSE against the number of trials used to calculate the binless contribution $I_{nonzero}$ (see Section 5.2 for details). Here too the error and $N_{nonzero}$ are correlated ($r = 0.79$), suggesting that the estimation and debiasing of $I_{nonzero}$ are a significant contribution to the imprecision of these estimators.

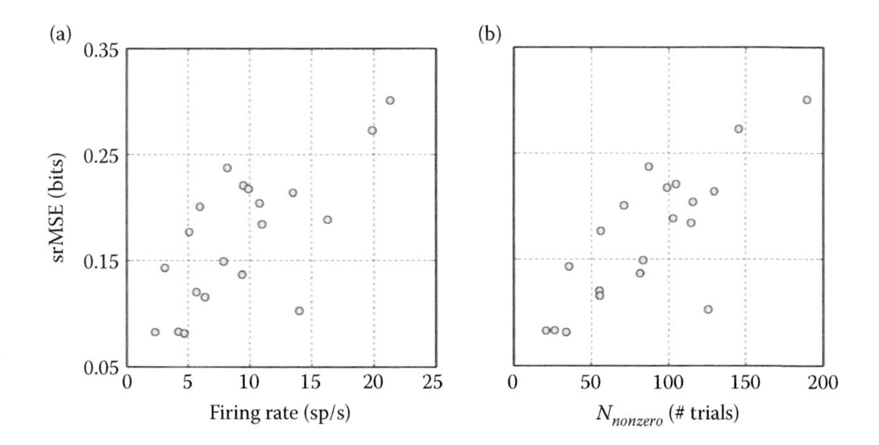

FIGURE 5.4 Neuronal response pattern affects estimation error. Square root of the MSE (RMSE) plotted against the cell's average firing rate attained from the simulation (a) and against the number of trials left in the $N_{nonzero}$ group for each model (b).

5.3.4 Validation

Figure 5.5 summarizes the performance of the estimators by plotting the square root of the bias, the standard deviation (STD) and the RMSE against the number of trials per stimulus (N) at each temporal resolution ($1/q$). For unbiased estimators, these are expected to decrease like $1/\sqrt{N}$ (the dotted line in Figure 5.5). As a rule of thumb, estimators that achieve RMSE values under 0.15 bits can be considered reliable for our data. This value is marked with a gray dashed line in Figure 5.5.

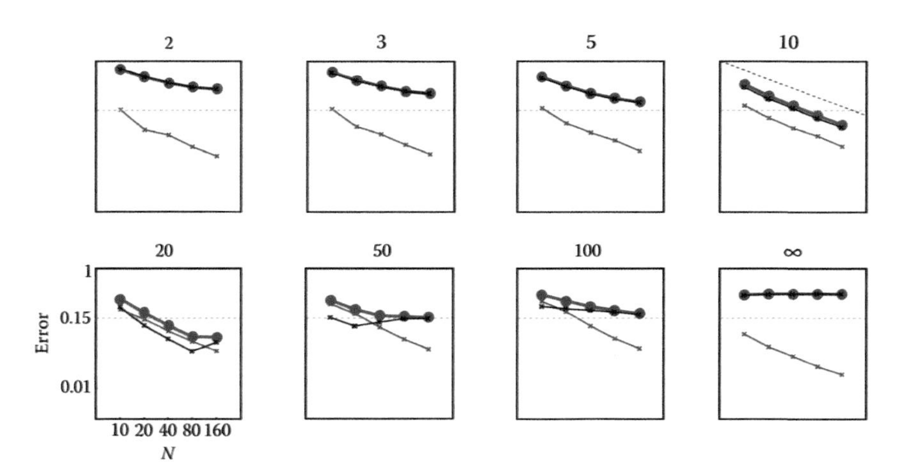

FIGURE 5.5 **(See color insert.)** Validation of the BINL-ED method. The bias (blue x), standard deviation (green x), and RMSE (red circles) are plotted as a function of the number of trials/stimulus (10, 20, 40, 80, 160) on a logarithmic scale. The gray dashed lines mark the reliability value of the error. The red dashed line shows the expected rate of decrease of the bias and variance, with a slope of −0.5 on this log–log plot.

The STD (green) decreases with sample size for all q. Moreover, it also has the expected linear trend with a slope of -0.5 for all q values, as expected. The square root of the bias (blue) shows a monotonic decrease only for the higher temporal resolutions values ($1/q$: 2, 3, 5, 10 ms). The square root of the bias does not decrease with N for low temporal resolution ($N = 160$ and $1/q = 20$ ms; $N \geq 40$ and $1/q = 50$ ms; $N \geq 20$ and $1/q \geq 100$ ms). This suggests that at the lower temporal resolutions, the size of the metric neighborhood that is effectively used to calculate the MI is too large. At temporal resolutions of 10 ms, the square root of the bias decreased with the expected slope for an asymptotically unbiased estimator. This temporal resolution corresponded to the amount of smoothing that we imposed on the rate functions, and therefore, may be due to the way we derived the models rather than from cortical mechanisms.

For most combination of parameters, the bias is greater than the variance, contributing most of the MSE. This is common in MI estimators (Paninski 2003). Nevertheless, in a few cases (e.g., $1/q = 20$ ms, $N = 20$, 40, 80 and $1/q = 50$ ms, $N = 10$, 20), the variance was greater than the bias.

Figure 5.6 replots the same data as a function of temporal resolution, with the number of trials N as a parameter. For $N = 20$ trials, the most reliable estimation requires a temporal resolution of $1/q = 20$ ms (RMSE = 0.18 bits). Although this RMSE is somewhat larger than the boundary at 0.15 bits, it is nevertheless already smaller than the RMSE of the MI between spike counts and stimuli.

For $N \geq 40$ repetitions, our estimation process produces reliable results with $1/q = 20$ ms (with root MSE of 0.1, 0.07, and 0.07 bits for $N = 40$, 80, and 160,

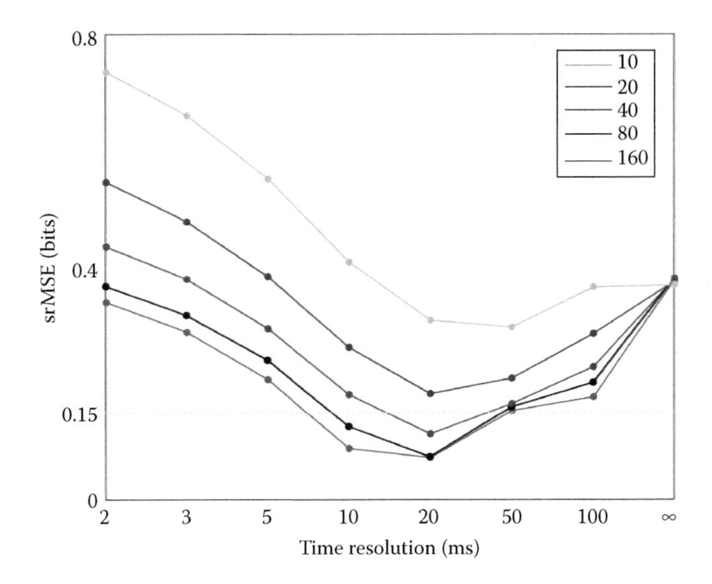

FIGURE 5.6 **(See color insert.)** Estimation quality dependence on temporal resolution. The RMSE of the estimation is plotted against the temporal resolution ($1/q$) used for the number of trials/stimulus (blue: 160; green: 80; red: 40; cyan: 20; purple: 10). The gray dashed line marks the level below which we considered the estimates reliable.

respectively). For $N = 80$ and 160, we achieved reliable estimation also for $1/q = 10$, and the results with $1/q = 20$ were comparable to both of them. These results are again consistent with the effective temporal resolution of the data, at about 10 ms. The fact that we achieved better estimation of the True MI while taking into account temporal information suggests that the time variation of the rate functions indeed carries significant amount of information about the stimuli.

5.4 DISCUSSION

The estimation of MI between stimuli and neural responses is becoming an important analytical tool in neurobiological research. In principle, the MI is a single number summary of a joint distribution. As such, it seems that it would not be more difficult to estimate than, for example, a mean and a variance (Nemenman et al. 2004). Nevertheless, achieving a precise estimate with existing methods is hard. Theoretical aspects of MI estimation have been most rigorously treated by Paninski (2003), who highlighted the severe problems encountered when trying to estimate MI from data.

Here, we present a new method that achieves reasonable estimates of MI for simulated data generated from time-dependent rate functions derived from experimental data. Our method requires at least 20 presentations of each stimulus for achieving stable estimates. The method combines two previously published methods, decoding in metric spaces using editing distance (Victor and Purpura 1996) and binless estimation of the MI (Victor 2002). The metric space approach as originally suggested (Victor and Purpura 1996) explicitly considers the topology of spike trains and, therefore, takes into account the temporal aspect of the response. However, in the original paper, it used decoding to generate confusion matrices from which the MI was derived. Since decoding is a data reduction step, there is no guarantee that it will extract the full information between stimuli and responses. The binless method, because it requires layering the data by spike counts, uses the available data points suboptimally. In the new method, we combine the advantages of the metric space and binless methods to achieve useful MI estimates based on a surprisingly limited amount of experimental data. For example, Golomb et al. (1997) established that for plug-in estimators based on matrices of joint counts, it is necessary to have at least one measurement in each cell of the matrix to achieve reasonable estimates of the MI. In our data, the number of possible response values may be very high. With spike patterns of at most six spikes within 128 ms, assuming an effective temporal resolution of about 10 ms, there are ~13 effectively different temporal positions for each spike, and therefore, a few thousands of possible response values. However, stable results can be achieved with 40 repeats of each stimulus, and useable results with only 20.

The proposed method is only one way to perform MI estimation in metric spaces. As shown theoretically, the method is rather general, requiring only a natural consistency condition between the metric and the KL divergence. Our results suggest that much of the bias is due to the particular method of calculating the binless estimate, the local dimension measurement, for example. We believe that future work may uncover better ways of doing this.

REFERENCES

Bialek, W., F. Rieke, R. R. de Ruyter van Steveninck, and D. Warland. 1991. Reading a neural code. *Science* 252(5014):1854–1857.

Furukawa, S., and J. C. Middlebrooks. 2002. Cortical representation of auditory space: Information-bearing features of spike patterns. *Journal of Neurophysiology* 87(4):1749–1762.

Furukawa, S., L. Xu, and J. C. Middlebrooks. 2000. Coding of sound-source location by ensembles of cortical neurons. *Journal of Neuroscience* 20(3):1216–1228.

Golomb, D., J. Hertz, S. Panzeri, A. Treves, and B. Richmond. 1997. How well can we estimate the information carried in neuronal responses from limited samples? *Neural Computation* 9(3):649–665.

Kozachenko, L. F., and N. N. Leonenko. 1987. Sample estimate of the entropy of a random vector. *Problemy Peredachi Informatsii* 23(2):9–16.

Las, L., A. H. Shapira, and I. Nelken. 2008. Functional gradients of auditory sensitivity along the anterior ectosylvian sulcus of the cat. *Journal of Neuroscience* 28(14):3657–3667.

Middlebrooks, J. C., A. E. Clock, L. Xu, and D. M. Green. 1994. A panoramic code for sound location by cortical neurons. *Science* 264(5160):842–844.

Nelken, I., G. Chechik, T. D. Mrsic-Flogel, A. J. King, and J. W. Schnupp. 2005. Encoding stimulus information by spike numbers and mean response time in primary auditory cortex. *Journal of Computational Neuroscience* 19(2):199–221.

Nemenman, I., W. Bialek, and R. de Ruyter van Steveninck. 2004. Entropy and information in neural spike trains: Progress on the sampling problem. *Physical Review E* 69(5):056111.

Paninski, L. 2003. Estimation of entropy and mutual information. *Neural Computation* 15(6):1191–1253.

Rolls, E. T., A. Treves, M. J. Tovee, and S. Panzeri. 1997. Information in the neuronal representation of individual stimuli in the primate temporal visual cortex. *Journal of Computational Neuroscience* 4(4):309–333.

Treves, A. 2003. Computational constraints that may have favoured the lamination of sensory cortex. *Journal of Computational Neuroscience* 14(3):271–282.

Treves, A., and S. Panzeri. 1995. The upward bias in measures of information derived from limited data samples. *Neural Computation* 7(2):399–407.

Victor, J.D. 2002. Binless strategies for estimation of information from neural data. *Physical Review E, Statistical, Nonlinear, and Soft Matter Physics* 66(5 Pt 1): 051903.

Victor, J. D., and K. P. Purpura. 1996. Nature and precision of temporal coding in visual cortex: A metric-space analysis. *Journal of Neurophysiology* 76(2):1310–1326.

6 Measuring Information in Spike Trains about Intrinsic Brain Signals

Gautam Agarwal and Friedrich T. Sommer

CONTENTS

6.1 BACKGROUND

Periodic activity is ubiquitous among neural populations of the brain. Although its constituent rhythms often do not directly reflect the temporal structure of stimuli or behavioral tasks, it is modulated by both, as observed in measures of mass activity such as the local field potential (LFP) and electroencephalogram (EEG) [Buz06]. In a variety of systems, these rhythms are found to coincide with the periodic spiking of single neurons. Here, we will consider the situation where the activity of a neuron is influenced at the same time by a stimulus and an intrinsic oscillation, a special case, but a surprisingly common and important one (see Chapter 11 and the examples below). So far, the complex and uncontrollable nature of ongoing activity has made it impossible to reveal its exact role in the neural code. In fact, traditional approaches for analyzing the activity in sensory neurons treat the variability due to ongoing activity as noise. In this chapter, we present a new analytic approach that combines autoregressive (AR) modeling and information theory to characterize the information that spikes carry about intrinsic periodic signals. Combined with one of the established methods to estimate the information in spike trains about external stimuli [RWvSB99], our method allows a direct comparison of the different influences on a neuron's activity in terms of encoded information. This method can be applied to recorded data, and can help test functional theories about ongoing neural activity, of which there are several promising, although so far unproven, ones.

One functional theory proposes that the frequency mixing of sensory signals with intrinsic neural oscillations could increase the available bandwidth in neuronal signal transmission. In engineering, a scheme called frequency division multiplexing enables multiple signals to be transmitted simultaneously, where each is assigned to a distinct frequency band. Such a scheme may be used to encode both odor category and identity in the zebrafish olfactory bulb [FHL04]. In the retina, the oscillatory structure of spike timing at the gamma frequency may convey information to the lateral geniculate nucleus (LGN) in addition to the transmission of visual features via the traditionally described rate code [KWV+09]. Further, it has been proposed that the vibrissa system uses internal periodic reference signals that employ frequency mixing for stabilization [AHZ97] and for computing the position of vibrissal contact in head-centered coordinates [ALSK02, KM06].

Another functional theory hypothesizes that periodic patterns in spike trains serve as "labels" for routing stimulus information dynamically in the cortical network [FRRD01, FNS07, TFS08] (see also Chapter 7). Specifically, the theory predicts that a downstream region becomes sensitive to a group of neurons upon matching the phase and frequency of the upstream region's activity. It has been shown that pairs of brain areas, in fact, do exhibit attention-dependent changes in coherence at gamma frequencies, in support of this theory [FRRD01]. Although this downstream propagation of oscillatory activity has been shown to correlate with visual performance [WFMD06], it is still a current subject of investigation whether there is a causal involvement of this mechanism in cortico-cortical communication.

A third theory suggests that the relative phase between a spike and a periodic reference signal could encode stimulus information. The most prominent example of such a "phase code" is found in the hippocampus. Specifically, as an animal moves through a place cell's receptive field, the cell spikes at earlier phases of the theta rhythm (~8 Hz) in the hippocampal LFP, an observation known as phase precession [OR93]. Thus, a place cell's firing profile relates jointly to ongoing hippocampal activity and physical location.

A common assumption in all three theories is that neurons share their information capacity between representing the features of a stimulus and the properties of an intrinsic reference signal. Current measures of phase-locking between spike trains and reference signals, such as spike-field coherence [JM01], do not provide a direct comparison of the amounts of information a spike train carries about these two sources. Our new method builds on approaches that have been successfully employed to characterize the relationship between spiking and stimuli, such as computing spike-triggered ensembles (STEs) and estimating mutual information [RMvSB99]. However, the components of intrinsic activity that affect spiking are oscillatory, making it important to account for temporal autocorrelation in the information estimate. We tackle this problem in four steps: extract a periodic component of the reference signal by demodulation; fit this component with an AR model to capture its periodic dynamics; use the AR model to estimate how the information provided by past spikes degrades over time; and compute the mutual information between spikes and periodic component of the reference signal conditioned on the information provided by the past spikes. The resulting method allows the quantification, in terms of information rate, of how much of its information capacity a neuron devotes to representing the ongoing rhythmic activity.

In what follows, we consider a common situation in which spikes are recorded from a neuron, and in parallel, other brain-intrinsic reference signals are registered. For example, such reference signals could be local field potentials, rhythmic spiking of neurons in another brain region, or the synaptic input of a neuron measured by whole-cell recordings.

6.2 PREREQUISITES

To set up a method for estimating the information in spike trains about intrinsic brain signals, we can build on a rich literature about analyzing the relationship between stimuli and spikes.

6.2.1 Lessons from Analyzing Stimulus Encoding

How a stimulus $s(t)$ affects the firing of neurons is a classical question of systems neuroscience, for example, [PGM67]. In typical experimental designs, neural recordings are performed during repeats of the same stimulus or other repeatable behavioral conditions. One can then average the neural response across stimulus repeats; the number of spikes occurring at different time points during stimulus presentation is called the *peristimulus time histogram (PSTH)*.* Optimization techniques, such as adaptive kernel estimation [ROS90], have been proposed to estimate the stimulus-induced *instantaneous firing rate* $r(t)$ from the PSTH. For a given number of repeats, $r(t)$ and the intertrial variability can be estimated [SL03]. Note that all reproducible stimulus-locked influences on the firing of a neuron, both periodic and aperiodic, are captured by the PSTH.

Another important concept is the *STE*, that is, the empirical distribution of stimuli that coincide with the occurrence of a spike. If spiking in a short time bin is represented by a binary indicator variable $X(t)$, then the STE is described by the conditional distribution $p(S|X = 1)$. Various methods have been developed for estimating the information that neural responses convey about a stimulus from the instantaneous firing rate or the STE [EP75, RWvSB99, BT99]. The common goal of these methods is to estimate the mutual information between a time-varying stimulus $s(t)$ and the neural firing $x(t)$:

$$\begin{aligned} I(X;S) &:= D[p(X,S) \mid p(X)p(S)] \\ &= \sum_{x=0,1} \int_s ds\, p(X=x,S=s) \log_2 \frac{p(X=x,S=s)}{p(X=x)p(S=s)} \\ &= \sum_{x=0,1} p(X=x) \int_s ds \frac{p(X=x,S=s)}{p(X=x)} \log_2 \frac{p(X=x,S=s)}{p(X=x)p(S=s)} \\ &=: \sum_{x=0,1} p(X=x) I(X=x;S) \end{aligned} \quad (6.1)$$

* To characterize the full structure of the neural response, one has to generate histograms of all occurring spike patterns, that is, single spikes, spike pairs with different intervals, spike triplets, and so on. However, often, the analysis is restricted to single spikes.

Equation 6.1 is the definition of mutual information, and $D[p|q]$ is the Kulback–Leibler (KL) divergence, or relative entropy, between two distributions p and q; see [CT91]. The expression $I(X = x;S)$ denotes the average mutual information between a spike $X = 1$ or an empty bin $X = 0$ and the stimulus. If the spike rate is sufficiently low compared to the sampling rate (i.e., most time bins are empty), one can assume that $p(S|X = 0) \approx p(S)$ and one can neglect the $X = 0$ term [RWvSB99].

There are two approaches to estimate the average mutual information between a spike and the stimulus $I(X = 1;S)$. The direct method [RWvSB99, BSK+00] uses distributions over the neuron's response:

$$I(X = 1;S) = \int_s ds\, p(S = s) \frac{p(X = 1|S = s)}{p(X = 1)} \log_2 \frac{p(X = 1|S = s)}{p(X = 1)}$$

$$= \frac{1}{I_S} \int_0^{I_S} dt \frac{r(t)}{\bar{r}} \log_2 \left(\frac{r(t)}{\bar{r}} \right) \text{bit/spike} \tag{6.2}$$

with $\bar{r}$ as the mean spike rate (see [BSK+00] for details of the derivation). In Equation 6.2, the integral over stimuli is replaced with an integral over time, an approximation that works well if the stimulus duration I_S is long enough and the stimulus distribution is sufficiently rich. This approach relies on the availability of recordings performed during exact repeats of the reference signal, which is the stimulus in this case. However, intrinsic neural activity is a reference signal without direct experimental control, and thus this approach is not applicable.

An alternative approach is to express the average mutual information in a spike by distributions over the stimulus [RWvSB99, BT99]:

$$I(X = 1;S) = \int_s ds\, p(S = s|X = 1) \log_2 \frac{p(S = s|X = 1)}{p(S = s)}$$

$$= D[p(S|X = 1)|p(S)] \tag{6.3}$$

In this formulation, one has to estimate the raw stimulus ensemble ($p(S)$) and the stimulus ensemble conditioned on the response ($p(S|X = 1)$). The KL divergence between the posterior after an observation and the prior denoted in Equation 6.3 is also called Bayesian surprise [IB05]. The full distributions in Equation 6.3 are often approximated as Gaussian [RWvSB99], allowing one to compute Equation 6.3 in closed form from the first and second moments (i.e., mean and covariance) of the data.

6.2.2 Extracting Oscillatory Structure from Brain Signals and Estimating Information Rates

We now describe how information theory can be applied to assess how much information a spike train carries about a simultaneously recorded, periodic brain signal. The first important step in this endeavor is to expose the oscillating structure in the brain-intrinsic signal. One natural way to achieve this is to adopt a signal representation of the form $Z(t) = A(t)e^{i\phi(t)}$, where $A(t)$ is a slowly changing amplitude that modulates a faster changing phase $\phi(t)$. To capture the signal structure, the representation is adapted to the recorded signal; for example, if the signal's power spectrum is peaked, the center frequency is matched by the average increase of $\phi(t)$ in time. There are several advantages of a signal representation in complex space. One is that in many systems of the brain, the phase corresponds to a feature that seems to modulate neural activity (a phenomenon described as "phase-locking") [KWV+09, OR93]. Additionally, complex representations lend themselves to the construction of simple models for oscillatory dynamics: a waveform evolves cyclically simply by advancing its phase by a fixed amount over time.

For signals that have power concentrated in a distinct narrow frequency band, demodulation can be achieved by applying the Hilbert transform (see our example in the next section). However, in general, intrinsic brain signals contain a broad spectrum of frequencies, requiring additional constraints to arrive at a unique demodulation of amplitude/phase structure [SS10], perhaps even in a nested scheme involving different frequency bands [CED+06, CK10, LSK+05].

To estimate the information between spikes and intrinsic oscillatory neural activity, a naive approach would be to simply substitute the stimulus S in Equation 6.3 with the oscillatory reference signal (henceforth termed Z). However, the problem with this approach is that the plain mutual information $I(X;Z)$ does not account for the information about Z that persists from previous spikes by virtue of Z's autocorrelated temporal structure: for a highly regular reference signal, there is less potential for "surprise" when a new spike arrives. To capture this history effect, we develop a method for computing the conditional mutual information $I(Z;X|\text{history of } X)$.

6.3 INFORMATION IN LGN SPIKES ABOUT RETINAL OSCILLATIONS

We demonstrate methods for estimating the information that a spike conveys about a periodic reference signal in neurons of the lateral geniculate nucleus of cats during visual stimulation with natural movies [WWV+07, KWV+09, WHS10]. Using whole-cell recordings in the LGN, it is possible to detect not only the spikes a cell produces but also the individual excitatory postsynaptic potentials (EPSPs) that constitute its input from the retina. We focus on an example from a fraction of cells in the LGN in which the retinal EPSPs are periodic. These retinal oscillations are not phase-locked to the stimulus [KWV+09] and thus can serve as a case study.

6.3.1 DEMODULATION OF RETINAL OSCILLATIONS

To obtain a continuous reference signal that captures periodic activity in the retina, we demodulate the EPSP time series by convolving it with a complex Morlet wavelet

$$w(t,f) = Ce^{2\pi ift}e^{-t^2/2\sigma_t^2} \tag{6.4}$$

centered at frequency f, with temporal width σ_t and normalization factor C, resulting in a complex waveform $Z(t)$. The center frequency and the bandwidth of the wavelet

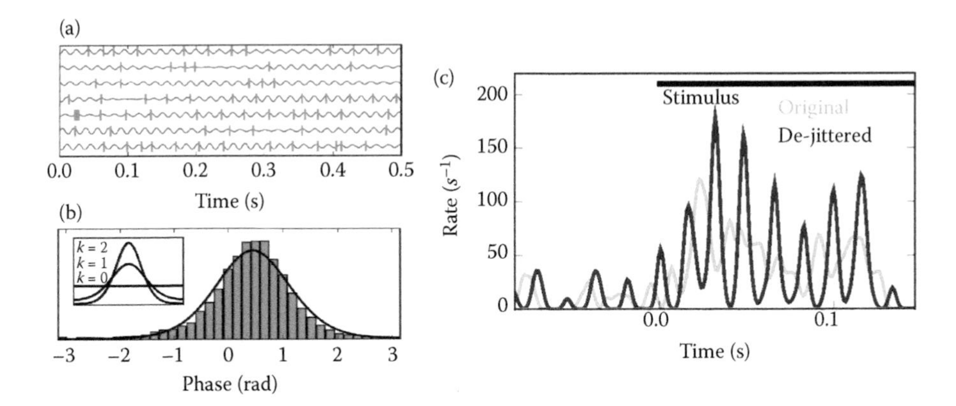

FIGURE 6.1 Modulation of spiking in a thalamic relay cell by a reference signal extracted from the retinal inputs. (a) Raster plot of thalamic spikes (vertical bars) overlaid on the oscillatory reference signal (i.e., filtered retinal EPSPs, continuous trace). (b) Histogram of preferred phase of spiking approximates a von Mises distribution. Inset: von Mises distributions for different concentration parameters. (c) Time course of spiking upon stimulus presentation (gray), and after de-jittering the oscillation phase (black). (Adapted from K. Koepsell et al. *Frontiers in Systems Neuroscience*, 3, 2009.)

were fitted to the peak in the power spectrum of the EPSP spike train; see [KWV+09]. Figure 6.1a depicts the real component of the wavelet-filtered train of EPSPs for an example cell, along with its spiking output. The distribution of phases of the analytic signal measured during spikes in the thalamic relay cell is shown in Figure 6.1b. It is clearly peaked, suggesting that spikes are preferably elicited at a certain phase of the reference signal. The PSTH of the first 150 ms following stimulus onset is shown by the gray trace in Figure 6.1c. To demonstrate the correlation between the phase of the retinal input and thalamic spikes, we used the analytic signal to de-jitter the phase variability across stimulus repeats [KWV+09]. After de-jittering, the strong modulation of spiking by the retinal oscillations becomes apparent (black trace in Figure 6.1c). It follows that applying Equation 6.2 to the de-jittered PSTH results in a higher information rate, indicating that the spike train carries information about the reference signal [KWV+09]. Biological gamma oscillations overlap in frequency with signals from the environment, such as power line noise and monitor refresh rates. Therefore, careful controls were performed in [KWV+09] to exclude artifacts induced by such signals. Specifically, the event histograms were verified to be flat with respect to the line and monitor cycles. Further, the small amounts of line power detectable in the raw signal of some recordings were found not to affect the timing of detected synaptic events and spikes: filtering out the line signal left the event times unchanged.

6.3.2 Spike-Triggered Distributions of the Reference Signal

It has been common in the literature to assess the spike triggered distributions of the phases of reference signals [KWV+09, OR93]. Analogous to the spike-triggered phase histogram in Figure 6.1b, one can also form the spike-triggered histogram of

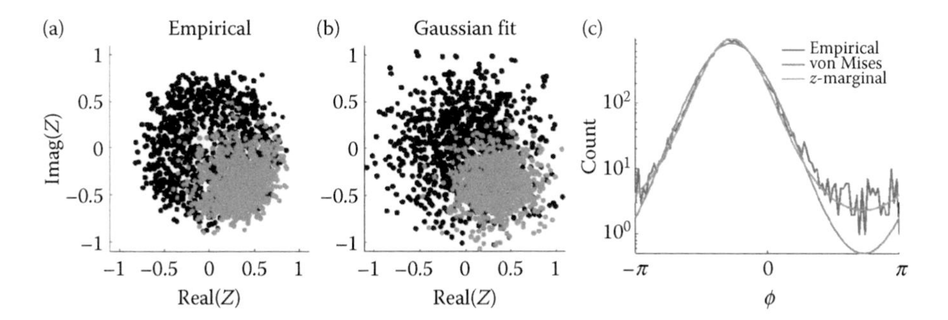

FIGURE 6.2 **(See color insert.)** Fitting the correlation between the analytic reference signal and spikes for the one-dimensional case. (a) Distributions of raw ensemble ($p(Z)$, black) and STE ($p(Z|X = 1)$, red) of the analytic signal in the complex plane. (b) Complex circular-symmetric Gaussian fits of raw ensemble and STE. Restricting the fit to be circular-symmetric results in a distribution that is isotropic in the complex plane. (c) The spike-triggered phase distribution (blue), fit with von Mises distribution (green) and with marginal of complex Gaussian (red) from Equation 6.9.

the complex analytic signal (Figure 6.2a). We use a complex Gaussian to fit the spike-conditional distribution

$$p(Z = z|X = 1) = \frac{1}{\pi \det(P_0)} e^{-(\bar{z}-\bar{w}_0)^T P_0^{-1}(z-w_0)} = \mathcal{CG}_{Z=z}(w_0, P_0) \tag{6.5}$$

Note that the exponent in Equation 6.5 is real if P_0 is Hermitian, that is, $P_{0,ij} = \overline{P_{0,ji}}$. We restrict the Gaussian to be circular-symmetric, that is, with zero pseudocovariance, which provides satisfactory fits (see Figure 6.2a and b for the one-dimensional case) and simplifies the ensuing analysis [NM93].

The Gaussian fits of the raw and the spike-triggered distributions (Figure 6.2b) are able to capture the structure of the empirical distributions (Figure 6.2a) quite well. Further, Figure 6.2c demonstrates that the empirical spike-triggered phase distribution (blue line) is well fit by the marginal of the complex Gaussian, in fact, more accurately than by a von Mises distribution (i.e., circular Gaussian*):

$$\mathcal{M}(\phi;\kappa,\mu) = e^{\kappa\cos(\phi-\mu)}/I_0(\kappa) \tag{6.6}$$

with I_0 the modified Bessel function of zeroth order, which has been commonly used in the past to fit spike-triggered phase distributions, for example, [KWV+09].

6.3.3 Autoregressive Models of the Reference Signal

A periodic reference signal contains autocorrelations and thus a series of spikes phase-coupled to it will convey redundant information. This redundancy must be accounted for when estimating the information content of a single spike. For the example cell in

* Not to be confused with circular-symmetric Gaussian, which are defined in complex space; circular Gaussians are distributions in phase, over the range $[-\pi, \pi]$.

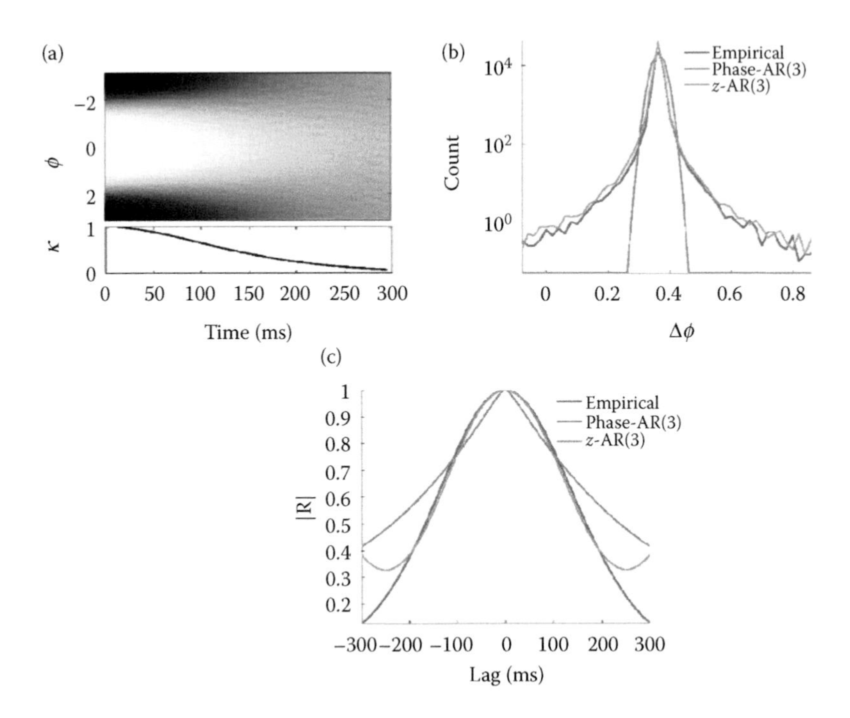

FIGURE 6.3 (**See color insert.**) Autoregressive modeling of reference signal dynamics. (a) Following a spike, the phase distribution gradually broadens (upper panel), seen in the decrease of the concentration of the von Mises fit (lower panel). (b) Distribution of the phase difference in consecutive time steps (instantaneous frequency). The empirical distribution (blue) is well fitted by the AR model of the analytic signal (red) but only poorly fitted by the AR model of phase (green). (c) The autocorrelation of the analytic signal is well approximated by the z-AR model for 200 ms; here, the magnitude of the autocorrelation is shown for clarity.

Figure 6.1, the empirical phase distribution stays peaked following a spike, only gradually broadening (Figure 6.3a). In consequence, subsequent spikes convey less information about the state of the reference signal than they would if this distribution quickly dissipated. To take this effect into account, one has to model the oscillatory autocorrelation of the reference signal, which we choose to do using an AR model [Jon80].

The reference signal, represented as a discrete-time analytic signal, is fitted by an AR(p) process

$$\mathbf{z}_t = A\mathbf{z}_{t-1} + \mathbf{u}_t \tag{6.7}$$

with $\mathbf{z}_t := (z_t, z_{t-1}, \ldots, z_{t-p})^T$, A is a companion matrix with the fitted AR coefficients, and $\mathbf{u}$ is a circular-symmetric Gaussian variable with covariance matrix Q (i.e., sampled from the distribution described in Equation 6.5). AR coefficients and Q were estimated using the Burg method [Bur67]. To calculate the p-dimensional spike-triggered average and spike-triggered covariance of the analytic signal for a cell, the p values leading up to each spike were tabulated, and their mean and covariance computed. AR models of different orders p were compared to find one that provides a good fit of the data, while resulting in stable estimates of signal evolution ($p = 3$ for our example cell (Figure 6.4b)). As an alternative option to Equation 6.7, we also evaluate an AR(3) model fit to the instantaneous phase, $\Delta\phi(t)$.*

We compare how the two models capture the empirical distribution of the reference signal during a spike (Figure 6.2c), and as it evolves (Figure 6.3b). For both cases, the phase-AR model mandates a circular Gaussian distribution that fails to replicate the empirical distribution (green curves). In contrast, the complex-AR model is able to fit both distributions (red curves) as well as the autocorrelation of the reference signal (Figure 6.3c). On first glance, the disparity in the phase statistics of the two models may seem surprising. To understand this result, it is important to recall that the phase distribution of a complex Gaussian is not Gaussian. Consider a univariate complex Gaussian (6.5)

$$\mathcal{CG}_Z(\mu,\sigma^2) \propto e^{-1/\sigma^2 \|z-\mu\|^2} \propto e^{-1/\sigma^2 (r^2+s^2-2rs\cos\phi)} \propto p(\phi|r)p(r) \tag{6.8}$$

with $r = \|z\|$, $s = \|\mu\|$, and ϕ the angle between the two vectors. The conditional distribution of the phase given a fixed amplitude r is a circular Gaussian (6.6): $P(\phi|r) \propto \mathcal{M}(\phi;\kappa = sr,0)$. However, the marginal distribution of the angle resulting from integrating r in Equation 6.8 out [UA80] is

$$p(\phi) = \frac{1}{2\pi}\left[1 + \sqrt{\pi}Ae^{A^2}\{1 + \mathrm{erf}(A^2)\}\right]e^{\frac{A^2}{\cos^2\phi}} \tag{6.9}$$

with $A = (s\cos\phi/\sqrt{2\sigma})$ and erf(x) the Gauss error function. The marginal phase distribution (6.9) is periodic and peaked at ϕ but it is heavy-tailed (i.e., not a circular Gaussian). We conclude that the AR model in complex space is the only adequate option of the two to describe the heavy-tailed phase distributions in the reference signal.

6.3.4 Model of the Signal Distribution Conditioned on the Spike History

We have explained in Section 6.3.2 how the STE of the reference signal can be modeled. Another key component for computing the information rate a spike conveys about the reference signal is a model of the distribution of the analytic signal given the history of spikes $p(z|\text{history})$. Here, we make the simplifying assumption that the spike history can be sufficiently characterized by the time T since the last spike, that is, $p(z|\text{history}) \approx p(z|T)$. This distribution can be described by the AR model (6.7) fitted to the reference signal. The mean and the covariance of $p(z|T = \tau)$ in τ time bins after the last spike are

$$w_\tau = A^\tau w_0 \tag{6.10}$$

* AR models make the assumption that the modeled quantities are Gaussian. Thus, an important criterion for determining which of the modeling approaches is more adequate is to check whether the modeled quantities can be approximated by Gaussian distributions.

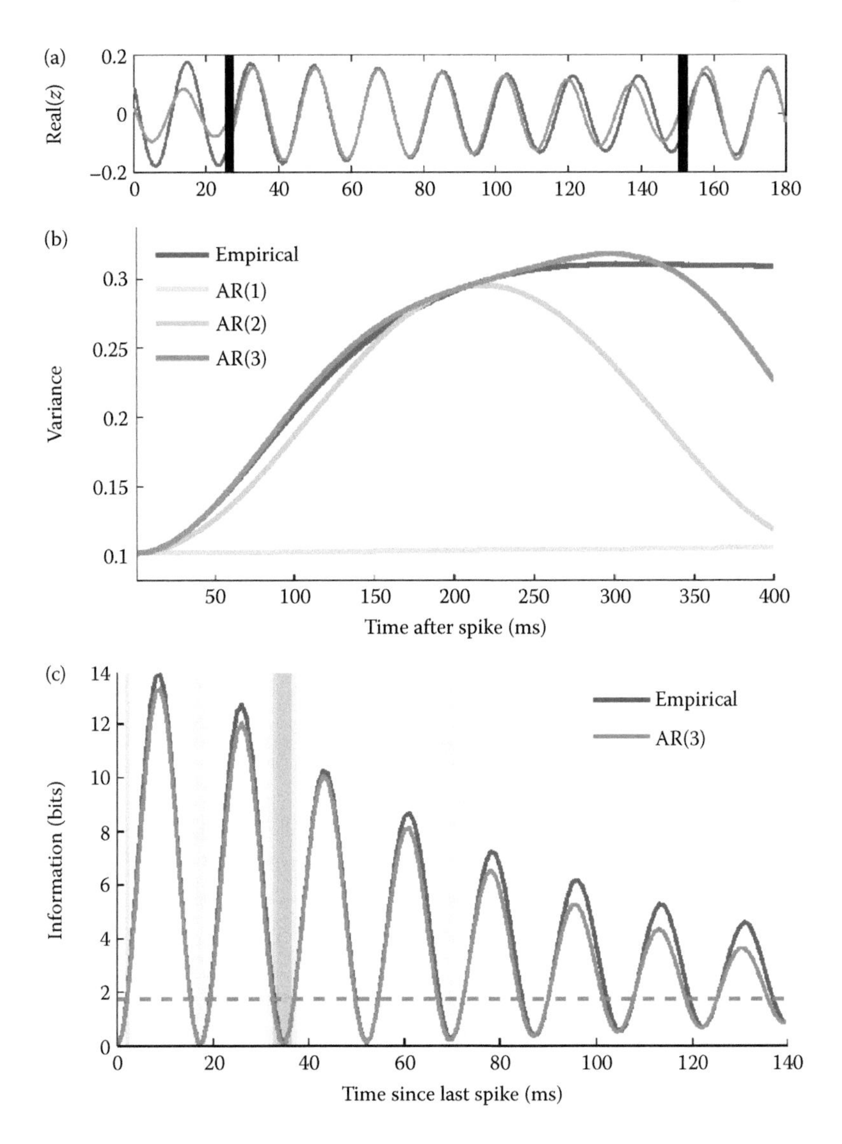

FIGURE 6.4 **(See color insert.)** Estimating the variance of the analytic signal after a spike as a function of the interval *ISI* since the spike. (a) When a spike is observed (black bars), the AR estimate of the reference signal is updated by the spike-triggered average. This estimate is propagated through the AR model but gradually diverges from the actual signal. (b) The variance of the analytic signal increases following a spike, and is best approximated by of the AR(3) model. (c) The information (Equation 6.12) provided by a spike following different interspike intervals (ISIs), as measured by the KL divergence of the empirical distribution (blue), or the AR(3) model (red), from the STE. Dashed line indicates average information per spike. Darker green strips in the background mark ISIs where more spikes fall.

$$P_\tau = A^\tau \Sigma (A')^\tau + \sum_{i=0}^{\tau-1} A^i P_0 (A')^i \tag{6.11}$$

with w_0 and P_0 set to spike-triggered mean and covariance, respectively [Jon80].

Figure 6.4b depicts how the covariance, and thus the uncertainty, of the Z-distribution increases with time in the period following a spike. Increasing the order of the AR model to 3 improves its ability to capture the evolution of the covariance; further increases in model order lead to unstable estimates that diverge to infinity (data not shown).

Armed with parametrized estimates of the Z-distribution during and following a spike, we can now calculate the information rate of a spike train and its corresponding reference signal.

6.3.5 Information That a Spike Carries about a Periodic Intrinsic Brain Signal

At time $T = \tau$ following the last spike, the reference signal is approximated by the distribution $p(Z|T)$. If a new spike is observed, this estimate is updated to the spike-triggered distribution $p(Z|X = 1)$ (see Section 6.3.2). The difference between these two distributions determines the additional information that the new spike conveys about the reference signal Z. The average information conveyed by the spike train is calculated using the conditional mutual information [CT91]:

$$\begin{aligned}
I(Z;X\mid T) &:= \int_\tau d\tau\, p(T=\tau) D[p(Z,X\mid T=\tau) \mid p(Z\mid T=\tau)p(X\mid T=\tau)] \\
&= \int_\tau d\tau\, p(T=\tau) \sum_{x=0,1} \int_z dz\, p(Z=z, X=x\mid T=\tau) \\
&\quad \times \log_2 \frac{p(Z=z, X=x\mid T=\tau)}{p(Z=z\mid T=\tau)p(X=x\mid T=\tau)} \\
&= \int_\tau d\tau\, p(T=\tau) \sum_{x=0,1} p(X=x\mid T=\tau) \int_z dz\, p(Z=z\mid X=x, T=\tau) \\
&\quad \times \log_2 \frac{p(Z=z\mid X=x, T=\tau)}{p(Z=z\mid T=\tau)} \\
&= \int_\tau d\tau \sum_{x=0,1} p(X=x, T=\tau) D[p(Z\mid X=x, T=\tau) \mid p(Z\mid T=\tau)] \\
&= \sum_{x=0,1} p(X=x) \int_\tau d\tau\, p(T=\tau\mid X=x) D[p(Z\mid X=x, T=\tau) \mid p(Z\mid T=\tau)] \\
&=: \sum_{x=0,1} p(X=x) I(Z; X=x|T)
\end{aligned} \tag{6.12}$$

In Equation 6.12, the KL divergence is between distributions over complex variables. For two complex univariate Gaussian distributions*

$$D[\mathcal{CG}_Z(\mu_1,\sigma_1^2) \mid \mathcal{CG}_Z(\mu_0,\sigma_0^2)] = \frac{1}{\ln 2}\left(\frac{\sigma_1^2}{\sigma_0^2} + \frac{\|\mu_1 - \mu_0\|^2}{\sigma_0^2} - \ln\frac{\sigma_1^2}{\sigma_0^2} - 1\right) \quad (6.13)$$

Analogous to Equation 6.3, one can again assume for low spike rates that the absence of a spike in a time bin is uninformative about z, that is, $p(Z|X = 0, T = \tau) \approx p(Z|T = \tau)$, and therefore, only the conditional mutual information of spikes $I(Z;X = 1|T)$ contributes to the information rate. The conditional distribution of silent periods before spikes in $I(Z;X = 1|T)$ can be estimated by the empirical distribution of ISIs:

$$p(T = \tau | X = 1) = p(ISI) = \frac{1}{M}\sum_{i=1}^{M} \delta(\tau - ISI_i) \quad (6.14)$$

where ISI_i is the interspike interval preceding spike i. Inserting Equation 6.14 into Equation 6.12, and observing that $p(Z|X = 1,T) \approx p(Z|X = 1)$, we arrive at the average information a spike carries about Z:

$$I(Z;X = 1|T) \approx \frac{1}{M}\sum_{i=1}^{M} D[\, p(Z|X = 1) | p(Z|ISI_i)] \quad (6.15)$$

Equation 6.15 has the following interpretation. A new spike provides an updated estimate of Z, $p(Z|X = 1)$, replacing the previous estimate based on the preceding spike $p(Z|T = \tau)$. The information of a new spike is given by the KL divergence between the update and the previous estimate, the so-called Bayesian surprise, similar to Equation 6.3. Thus, the mean information of a spike about Z is given by the surprise averaged over the interspike intervals observed in the data. Consequently, if a spike occurs at an opposite phase from the previous spike, it conveys a larger amount of information than if it occurs at the same phase because the former elicits a larger surprise. In practice, few spikes occur at such ISIs (Figure 6.4c), and so the average information content of the spike train is closer to the minima of Equation 6.12.

In the case of our example cell, the average information per spike according to Equation 6.15 is 1.76 bits/spike. By comparison, the previous estimate of the information about the phase of the intrinsic signal was 0.83 bits/spike, and the information about the visual stimulus was estimated as 0.14 bits/spike [KWV+09].

6.4 DISCUSSION

We have presented a method to estimate the mutual information between a spike train and intrinsic brain signals, which may be derived, for example, from whole-cell

* We consider here the Z distribution for the current time point. Alternatively, one may consider the p-dimensional distribution associated with the AR(p) model that captures Z's full dynamical state.

signals, LFPs, or other spike trains that were recorded in parallel. The new approach provides an estimate that accounts for redundancy in the information conveyed by subsequent spikes. In contrast, a previous method [KWV+09] does not account for this redundancy. The core idea of the current approach is to estimate *conditional mutual information* rather than mutual information as has become the standard for estimating information rates about stimuli [RWvSB99, BT99]. The new approach relies on a combination of AR modeling of the reference signal with the estimation of conditional information. Specifically, the AR model generates an evolving estimate of the signal's prior distribution that extends through the periods where no spike is observed. The conditional mutual information is then computed as the averaged Bayesian surprise [IB05] between the spike-triggered distribution and the prior distribution, over all ISIs.

As an example, we use the new method to investigate spikes and synaptic inputs detected from whole-cell recordings in the LGN. The timing of synaptic inputs is influenced by retinal gamma oscillations. Previous work had shown that spikes carry significant information about the phase of the synaptic input's analytic signal. We find that an AR model of the oscillation must be fit to the complex analytic signal, rather than its phase representation. Using the AR model, we estimate the mutual information between the analytic signal and a spike, subtracting the remaining information from the previous spike. The resulting information estimates differ from those previously reported [KWV+09]. It must be emphasized, however, that it is not clear to what extent a neuron or a local network can maintain an estimate of the reference signal by projecting the information provided by a spike into the future. In this sense, the new approach provides a conservative estimate of mutual information.

Our example offers a first foray into combining information theory and AR modeling, which may merit several modifications. First, this method is restricted to circular-symmetric complex Gaussians, which might not be a good approximation for the spike-triggered average, in general. Second, standard AR models cannot describe nonstationary signals, which are common in brain activity. This may be addressed by utilizing a Kalman filter-based approach, proposed for describing nonstationary features in the hippocampal LFP [NWBB09]. Third, the estimate of the reference signal $p(z|\text{history})$ may be refined by accumulating the estimates provided by several spikes, rather than being reset by the last spike. This may be achieved by utilizing a Kalman filter, which provides a mechanism to accumulate information from a series of measurements. Fourth, AR models cannot capture nonlinear features of the reference signal. Some of these may be captured by utilizing nonlinear extensions such as the extended Kalman filter [GR99] or the particle filter [AMGC02].

The present work shows that spike-triggered methods, developed for analyzing the responses of sensory neurons to stimuli, can be extended to describe the relationship between oscillatory intrinsic brain signals and spike trains. In the context of neural responses to stimuli, many interesting approaches have been proposed, some of which might be potentially applicable for intrinsic brain signals as well [SPPS04, PS06]. Another interesting extension of this work is to explore alternative ways of signal demodulation. Here, we use a standard demodulation technique, the analytic signal; however, other approaches to demodulation [SS10] may reveal signal components that are more informative with respect to the spike train.

To conclude, we return to the utility of this approach in evaluating the functional models of neural oscillations presented at the beginning of this chapter. In each scenario, spikes are found to occur preferentially at certain phases of the reference signal. Currently, such phase-coupling is quantified by the spike-field coherence (SFC) between spike train and reference signal [JM01, WFMD06]. While the measure of SFC has great merit in identifying phase-coupling, it usually is not assessed in terms of information rate (but see [BT99]). By providing such a measure, the proposed method allows a direct comparison of the information that the spike train carries about the reference signal and about the stimulus. This allows one to estimate the fraction of its information capacity a neuron devotes to an intrinsic oscillation in different functional contexts, that is, the routing of additional sensory information at multiple frequencies (theory 1) or of top-down attentional modulation (theory 2). The third theory we reviewed posits that the stimulus itself governs phase-coupling. In such a case, the respective influences of the stimulus and intrinsic activity on the spike train can no longer be considered independently, requiring that our approach be extended to evaluate such a joint dependency.

In closing, we hope that the presented method outlines a framework that may serve to disentangle the rich interplay of intrinsic and extrinsic factors that shape neural activity.

ACKNOWLEDGMENTS

We thank Kilian Koepsell for contributing valuable ideas and providing some of the software. J. Hirsch, X. Wang, and V. Vaingankar, University of Southern California, permitted the use of their experimental data and provided many helpful discussions. Jascha Sohl-Dickstein provided feedback on the application of conditional mutual information. Joe Goldbeck read an earlier version of the manuscript and gave substantial input for improving it. This work has been supported by NSF grant IIS-0713657 (FS) and an NIH NRSA postdoctoral fellowship (GA). Some of the data analysis was done using IPython [PG07], NumPy/SciPy [Oli07], and MATLAB® (MathWorks, Natick, MA). All figures were produced using MATLAB or Matplotlib [BHM+05].

REFERENCES

[AHZ97] E. Ahissar, S. Haidarliu, and M. Zacksenhouse. Decoding temporally encoded sensory input by cortical oscillations and thalamic phase comparators. *Proceedings of the National Academy of Sciences of the United States of America*, 94(21):11633, 1997.

[ALSK02] K.F. Ahrens, H. Levine, H. Suhl, and D. Kleinfeld. Spectral mixing of rhythmic neuronal signals in sensory cortex. *Proceedings of the National Academy of Sciences of the United States of America*, 99(23):15176, 2002.

[AMGC02] M.S. Arulampalam, S. Maskell, N. Gordon, and T. Clapp. A tutorial on particle filters for online nonlinear/non-Gaussian Bayesian tracking. *IEEE Transactions on Signal Processing*, 50(2):174–188, 2002.

[BHM+05] P. Barrett, J. Hunter, J.T. Miller, J.C. Hsu, and P. Greenfield. Matplotlib—A portable Python plotting package. *Astronomical Data Analysis Software and Systems XIV ASP Conference Series*, 347:91–95, 2005.

[BSK+00] N. Brenner, S.P. Strong, R. Koberle, W. Bialek, and R.R.R. Steveninck. Synergy in a neural code. *Neural Computation*, 12(7):1531–52, 2000.

[BT99] A. Borst and F.E. Theunissen. Information theory and neural coding. *Nature Neuroscience*, 2:947–57, 1999.

[Bur67] J.P. Burg. Maximum entropy spectral analysis. *Proceedings of the 37th Meeting of the Society of Exploration Geophysics*, 1967.

[Buz06] G. Buzsáki. *Rhythms of the Brain*. Oxford University Press, USA, 2006.

[CED+06] R.T. Canolty, E. Edwards, S.S. Dalal, M. Soltani, S.S. Nagarajan, H.E. Kirsch, M.S. Berger, N.M. Barbaro, and R.T. Knight. High gamma power is phase-locked to theta oscillations in human neocortex. *Science*, 313(5793):1626–1628, 2006.

[CK10] R.T. Canolty and R.T. Knight. The functional role of cross-frequency coupling. *Trends in Cognitive Sciences*, 14(11):506–515, 2010.

[CT91] T.M. Cover and J.A. Thomas. *Elements of Information Theory*. Wiley, New York, 1991.

[EP75] R. Eckhorn and B. Pöpeel. Rigorous and extended application of information theory to the afferent visual system of the cat. II. Experimental results. *Biological Cybernetics*, 17(1):71–7, 1975.

[FHL04] R.W. Friedrich, C.J. Habermann, and G. Laurent. Multiplexing using synchrony in the zebrafish olfactory bulb. *Nature Neuroscience*, 7(8):862–71, 2004.

[FNS07] P. Fries, D. Nikolić, and W. Singer. The gamma cycle. *Trends in Neurosciences*, 30(7):309–16, 2007.

[FRRD01] P. Fries, J.H. Reynolds, A.E. Rorie, and R. Desimone. Modulation of oscillatory neuronal synchronization by selective visual attention. *Science*, 291(5508):1560, 2001.

[GR99] Z. Ghahramani and S.T. Roweis. Learning nonlinear dynamical systems using an em algorithm. *Advances in Neural Information Processing Systems 11: Proceedings of the 1998 Conference*, pp. 431–437, 1999.

[IB05] L. Itti and P. Baldi. Bayesian surprise attracts human attention. In J. Platt Y. Weiss, B. Scholkopf, editors, *Advances in Neural Information Processing Systems 18, NIPS 2005*, pp. 547–554, MIT Press, Cambridge, MA, 2005.

[JM01] M.R. Jarvis and P.P. Mitra. Sampling properties of the spectrum and coherency of sequences of action potentials. *Neural Computation*, 13(4):717–749, 2001.

[Jon80] R.H. Jones. Maximum likelihood fitting of arma models to time series with missing observations. *Technometrics*, 22(3):389–395, 1980.

[KM06] D. Kleinfeld and S.B. Mehta. Spectral mixing in nervous systems: Experimental evidence and biologically plausible circuits. *Progress of Theoretical Physics Supplement*, 161:86, 2006.

[KWV+09] K. Koepsell, X. Wang, V. Vaingankar, Y. Wei, Q. Wang, D.L. Rathbun, W.M. Usrey, J.A. Hirsch, and F.T. Sommer. Retinal oscillations carry visual information to cortex. *Frontiers in Systems Neuroscience*, 3, 2009.

[LSK+05] P. Lakatos, A.S. Shah, K.H. Knuth, I. Ulbert, G. Karmos, and C.E. Schroeder. An oscillatory hierarchy controlling neuronal excitability and stimulus processing in the auditory cortex. *Journal of Neurophysiology*, 94(3):1904–1911, 2005.

[NM93] F.D. Neeser and J.L. Massey. Proper complex random processes with applications to information theory. *IEEE Transactions on Information Theory*, 39(4):1293–1302, 1993.

[NWBB09] D.P. Nguyen, M.A. Wilson, E.N. Brown, and R. Barbieri. Measuring instantaneous frequency of local field potential oscillations using the Kalman smoother. *Journal of Neuroscience Methods*, 184(2):365–374, 2009.

[Oli07] T.E. Oliphant. Python for scientific computing. *Computing in Science & Engineering*, 9(3):10–20, 2007.

[OR93] J. OKeefe and M.L. Recce. Phase relationship between hippocampal place units and the EEG theta rhythm. *Hippocampus*, 3(3):317–330, 1993.

[PG07] F. Pérez and B.E. Granger. IPython: A system for interactive scientific computing. *Computing in Science & Engineering*, 9(3):21–29, 2007.

[PGM67] D.H. Perkel, G.L. Gerstein, and G.P. Moore. Neuronal spike trains and stochastic point processes: I. the single spike train. *Biophysical Journal*, 7(4):391–418, 1967.

[PS06] J.W. Pillow and E.P. Simoncelli. Dimensionality reduction in neural models: An information-theoretic generalization of spike-triggered average and covariance analysis. *Journal of Vision*, 6(4), 2006.

[ROS90] B.J. Richmond, L.M. Optican, and H. Spitzer. Temporal encoding of two-dimensional patterns by single units in primate primary visual cortex. I. Stimulus-response relations. *Journal of Neurophysiology*, 64(2):351–69, 1990.

[RWvSB99] F. Rieke, D. Warland, R.R. van Steveninck, and W. Bialek. *Spikes: Exploring the Neural Code*. MIT Press, Cambridge, 1999.

[SL03] M. Sahani and J.F. Linden. Evidence optimization techniques for estimating stimulus-response functions. In K. Obermayer S. Becker, S. Thrun, editors, *Advances in Neural Information Processing Systems 15: Proceedings of the 2002 Conference*, pp. 109–116. MIT Press, 2003.

[SPPS04] E. Simoncelli, L. Paninski, J. Pillow, and O. Schwartz. Characterization of neural responses with stochastic stimuli. *The Cognitive Neurosciences*, pp. 327–338, 2004.

[SS10] G. Sell and M. Slaney. Solving demodulation as an optimization problem. *IEEE Transactions on Audio, Speech, and Language Processing*, 18(8):2051–2066, 2010.

[TFS08] P. Tiesinga, J.M. Fellous, and T.J. Sejnowski. Regulation of spike timing in visual cortical circuits. *Nature Reviews Neuroscience*, 9(2):97–107, 2008.

[UA80] J. Uozumi and T. Asakura. First-order probability density function of the laser speckle phase. *Optical and Quantum Electronics*, 12(6):477–494, 1980.

[WFMD06] T. Womelsdorf, P. Fries, P.P. Mitra, and R. Desimone. Gamma-band synchronization in visual cortex predicts speed of change detection. *Nature*, 439(7077):733–736, 2006.

[WHS10] X. Wang, J.A. Hirsch, and F.T. Sommer. Recoding of sensory information across the retinothalamic synapse. *The Journal of Neuroscience*, 30(41):13567, 2010.

[WWV+07] X. Wang, Y. Wei, V. Vaingankar, Q. Wang, K. Koepsell, F.T. Sommer, and J.A. Hirsch. Feedforward excitation and inhibition evoke dual modes of firing in the cat's visual thalamus during naturalistic viewing. *Neuron*, 55(3):465–478, 2007.

7 Role of Oscillation-Enhanced Neural Precision in Information Transmission between Brain Areas

Paul H. Tiesinga, Saša Koželj, and Francesco P. Battaglia

CONTENTS

7.1 INTRODUCTION

A fundamental question in neuroscience is how is information encoded in neural activity (Ermentrout et al., 2008; Tiesinga et al., 2008). The visual system is often used as a model to study neural encoding because both the anatomy and the response properties of the cells are well characterized. When a visual stimulus is presented, a large number of cells in the visual cortex respond (Kenet et al., 2003; Tsodyks et al., 1999). Within this activated population, information may be represented in the average firing rate across a group of neurons during a short-time interval (Shadlen and Newsome, 1998), in the precise spike times produced during that interval (Bair, 1999; Bair and Koch,1996), in the correlations between a group of neurons (Averbeck and Lee, 2006), or in the correlation of spikes times with an oscillation in the local field potential (LFP) (Kayser et al., 2009). A large number of experiments have been conducted in different neural systems to determine the reliability of these features and their information content.

We will use computational approaches to illustrate how oscillations can increase spike time precision, thereby defining cell assemblies of neurons that fire at similar times. Models further show that this improved spike time precision increases the postsynaptic impact, leading to better transmission of information from one cell assembly to another. Because there is convergence in cortical circuits, a single neuron receives inputs from multiple cell assemblies, each representing a different stimulus. Depending on the current goal of the organism, some cell assemblies represent more relevant information than others; hence, it is necessary to select those cell assemblies from amongst the inputs. We show using models that phase shifting of oscillations can help one input cell assembly dominate the output of a receiving cell assembly.

The computational examples highlight the possible role of spike times in neural coding and information transmission and suggest how neural oscillations can improve both the neural coding and the transmission thereof between brain areas. We discuss these results within the context of a number of recent experiments that have explored the communication between brain areas such as the hippocampal projection to prefrontal cortex (PFC) (Benchenane et al., 2010, 2011) and between early visual areas (Gregoriou et al., 2009; Womelsdorf et al., 2007). Our approach is to study the similarity of spike trains generated by a pool of uncoupled neurons and their impact on postsynaptic cells (Reyes, 2003; Tiesinga, 2004; Tiesinga and Toups, 2005).

7.2 RESULTS

In the following sections, we will use simulations to illustrate the main points of the chapter. In the Appendix, we provide the overall setup of the model simulations, references to papers where details can be found regarding the model and its implementation, as well as the list of the key parameters for each figure.

When information is represented in the precise spike times of neurons, these spike times should be reproducible across trials, which are multiple repeated presentations of the same stimulus, where the state of the subject or slice is kept as constant as possible. In Figure 7.1, we show a hypothetical rastergram, in which each spike time is represented by a tick, whose abscissa is the spike time and whose ordinate is the trial index. In the rastergram, there are vertical streaks of spike times due to spike alignments, which we refer to as events, and which are also visible as peaks in the histogram. In the following, we are interested in quantifying the reliability, the fraction of trials during which a spike is present during the event, and quantifying the precision,

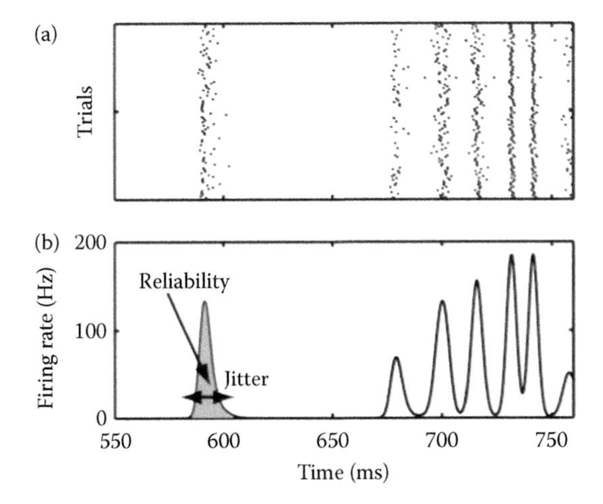

FIGURE 7.1 Model neurons driven by injection of a fluctuating current produce reliable and precise spike trains. A model neuron with Hodgkin–Huxley-type sodium and potassium channels and a leak current was driven by an aperiodic current waveform obtained by filtering white noise (Tiesinga and Toups, 2005). The same current was injected on multiple trials. (a) In the rastergram, each spike is represented as a dot with the spike time as the abscissa and the trial index as the ordinate. Events are vertical streaks in the rastergram that represent spike alignments. (b) In the histogram, the instantaneous firing rate is plotted, in which each event is represented by a peak. The reliability is the fraction of trials during which there is a spike during the event, indicated as the area under the peak (arrow). The spike time jitter is the standard deviation of the spikes that are part of the event, as indicated by the double-headed arrow. Precision is the numerical inverse of the jitter.

which is the inverse of the standard deviation (jitter) of the spike times belonging to an event. There are, broadly speaking, two approaches to quantifying precision and reliability. First, one can assign each spike time to an event and calculate the event precision and reliability and then average them across events. Effective automatic methods exist to accomplish this goal, which are characterized in terms of a few heuristic parameters (Toups et al., 2011). The second method is to use composite measures that produce a number that reflects the degree of spike alignment between pairs of spike trains, and that corresponds to a combination of reliability and precision, the relative contribution of which is characterized by a single parameter. In one measure, here referred to as the R-reliability, all spikes are replaced by a Gaussian with width σ to generate a vector for each spike train. The overlap between spike trains is the inner product between the corresponding pair of vectors, normalized by their lengths. The R-reliability measure is obtained by averaging the overlaps across all pairs (Schreiber et al., 2003, 2004). In the Victor–Purpura (VP) measure, the distance between spike trains is the cost of transforming one into the other through two elementary operations: inserting a spike at a cost of 1 and moving a spike time at a cost of q per unit time (Victor and Purpura, 1996). A bounded similarity measure, such as R-reliability, can always be transformed into a distance measure, such as the VP distance, and vice versa. For instance, a small mean VP distance corresponds to an "R-similarity" value close to one.

7.2.1 Building Cell Assemblies

Our general setup is illustrated in Figure 7.2. The model neuron (Tiesinga and Toups, 2005) is driven by a current drive, which is the sum of three components: an aperiodic stimulus waveform that contains the information to be encoded, a periodic current that represents the modulatory effects of synchronous oscillations, and a noise current that represents stochastic effects. The resulting spike trains can be interpreted in two ways. First, they could be the result of multiple trials on which the same stimulus was presented; in that case, the noise current represents the trial-to-trial variability of inputs and intrinsic neural activity. This interpretation covers the result of intracellular recordings *in vitro* during which an aperiodic plus periodic current was injected or where the aperiodic current was injected in combination with a pharmacologically induced oscillation (Fellous et al., 2001; Hunter et al., 1998; Hunter and Milton, 2003; Mainen and Sejnowski, 1995; Markowitz et al., 2008). It also covers *in vivo* experiments, where a sensory stimulus was presented; in this case, the aperiodic current represents the synaptic inputs generated by the stimulus (Buracas et al., 1998; Butts et al., 2007; Elhilali et al., 2004; Fellous et al., 2004;

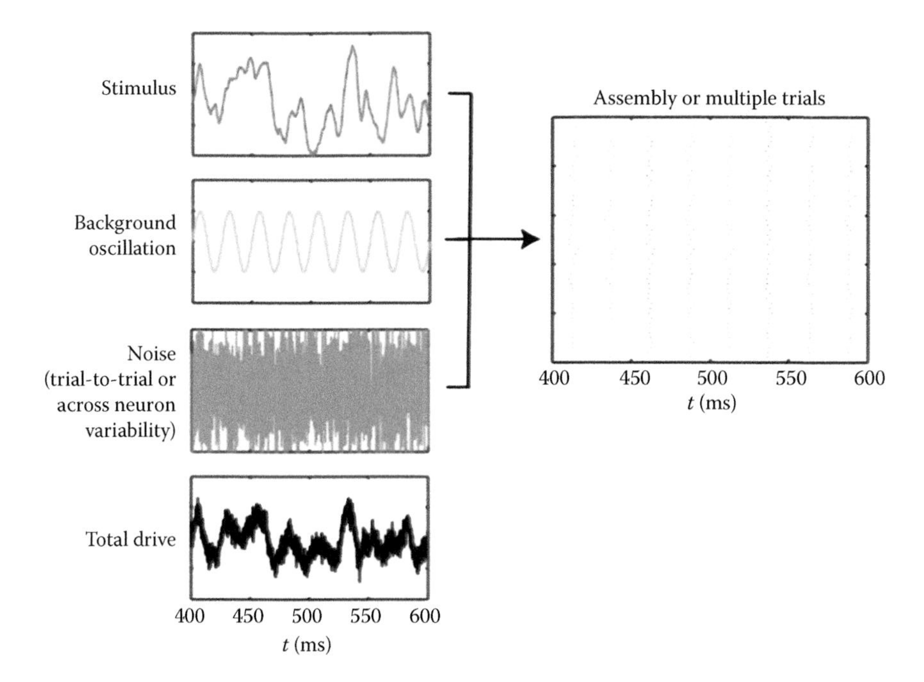

FIGURE 7.2 The synaptic inputs a neuron receives *in vivo* can be approximated as the sum of three components. A cell assembly is simulated *in vitro* or *in silico* by injecting into a neuron a current consisting of these three components. (Left) From top to bottom: the time-varying stimulus, assumed to be aperiodic, and the background oscillation, both of which are the same across neurons in the assembly (or across trials) and the noise, which refers to the component that varies between neurons (across trials). The sum of these three components forms the total drive shown in the bottom panel. (Right) Rastergram: when the total drive is injected in multiple neurons (across multiple trials), the spike times vary between neurons (across trials).

Reinagel and Reid, 2002), and the periodic current represents the ongoing oscillations in a particular frequency band (Kayser et al., 2009). In general, the phase and amplitude of the model oscillation would need to vary from trial to trial to mimic the experimental situation. In the second interpretation, the spike trains are the response of similar behaving cells on a single trial. We will refer to these as a pool of neurons or, alternatively, a cell assembly, which is driven by a common input, the aperiodic and periodic current, and a private input, which differs from cell to cell (Deweese and Zador, 2004). This interpretation assumes that the effect of recurrent connections can either be neglected or be incorporated in an effective way as a modulation of intrinsic cell properties. We consider this interpretation to be the starting point for further investigations that explicitly incorporate the network structure (Tiesinga and Sejnowski, 2010). Nevertheless, important insights can be gained from this simple approach. When a rastergram is shown in the following, the index on the y-axis could thus be either the trail or the cell index.

7.2.2 Noise Sensitivity and Reliability Resonances

Neurons are reliable when they are driven by a fluctuating input (Mainen and Sejnowski, 1995). This is perhaps best explained using as an example a leaky integrate-and-fire neuron driven by current injection. This drive then results in a stable voltage trajectory in time, to which a neuron returns when it is perturbed by a brief and weak current pulse (see Figure 3 in Tiesinga and Toups, 2005). This implies that noise will make the membrane potential fluctuate around this time-varying mean, but it will never deviate much from it (Tiesinga and Toups, 2005). A similar picture holds when there are action potentials: the neuron will reach threshold at a particular time, which will fluctuate due to noise, but both the membrane potential and the sequence of spike times will never deviate much from the sequence obtained in the absence of noise (see Figure 3 in Tiesinga and Toups, 2005). This means that each spike will happen with perfect reliability but the spike time jitter in these simple cases will be proportional to the noise standard deviation (Cecchi et al., 2000). The same happens for a neuron that is driven by the injection of a constant current, except in that case the jitter accumulates from spike to spike. Hence, after a few spikes, the events overlap to such an extent that they are not distinguishable anymore (see Figure 4 in Tiesinga et al., 2008).

This picture will hold only up to a certain level of noise because for medium noise levels, there will be spike failures, which reduce the reliability, although the precision will in general still be high (Tiesinga and Toups, 2005). For a high level of noise, there will be noise-induced spikes and both reliability and precision will be low. The firing rate will follow the time course of the drive waveform, for periodic drives it will do so at a particular phase lag (Brunel et al., 2001). The precise quantification of what constitutes low, medium, and high noise levels for one specific neuron model is found in Tiesinga and Toups (2005). The qualitative characterization presented here is more broadly applicable.

The key aspect is that not every state will be equally stable against noise, in the sense that the noise strength for which the above-mentioned transitions occur will vary (Hunter et al., 1998; Hunter and Milton, 2003). Instead of determining the noise strength

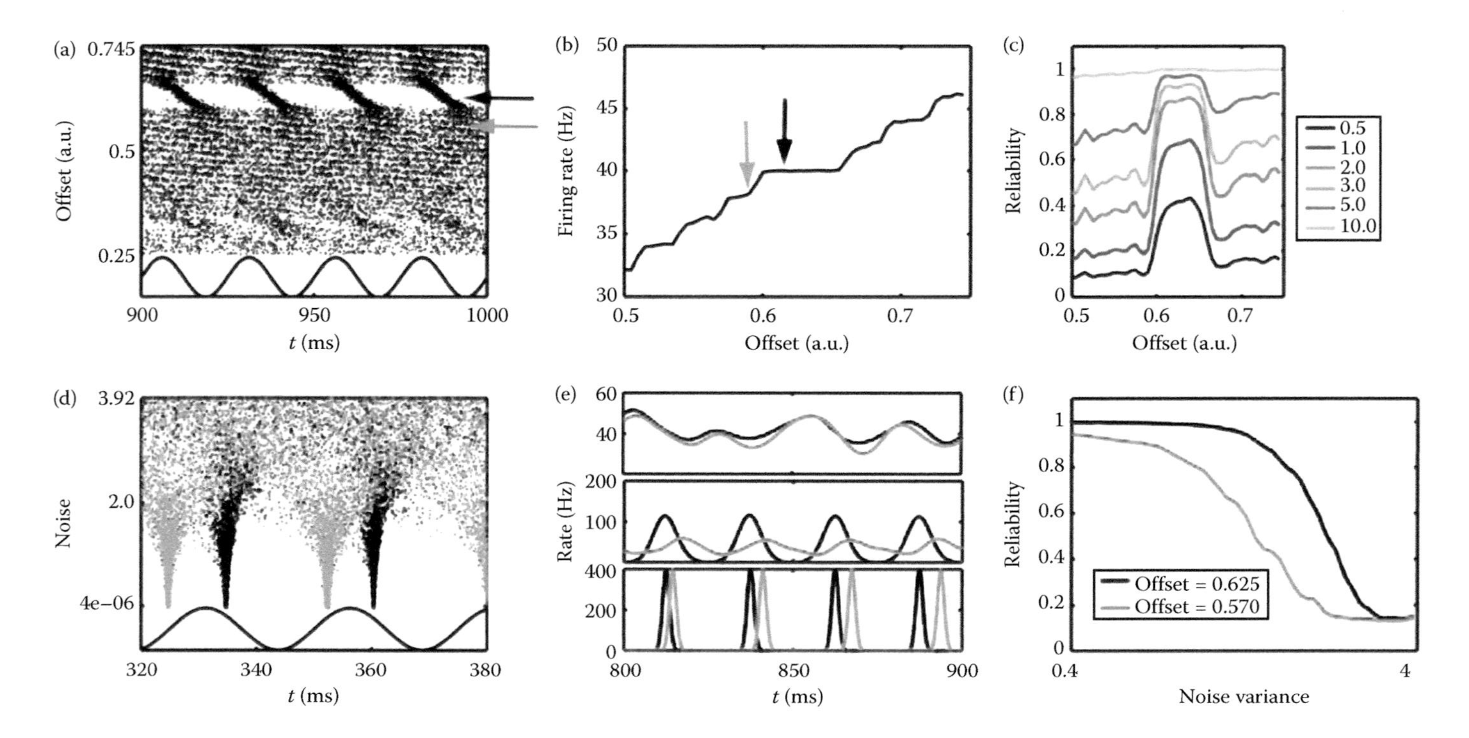
(a)
0.745
0.5
0.25
Offset (a.u.)
900
950
1000
t (ms)
(b)
50
45
40
35
30
Firing rate (Hz)
0.5
0.6
0.7
Offset (a.u.)
(c)
1
0.8
0.6
0.4
0.2
0
Reliability
0.5
0.6
0.7
Offset (a.u.)
0.5
1.0
2.0
3.0
5.0
10.0
(d)
3.92
2.0
4e−06
Noise
320
340
360
380
t (ms)
(e)
60
40
200
100
400
200
0
Rate (Hz)
800
850
900
t (ms)
(f)
1
0.8
0.6
0.4
0.2
0
Reliability
0.4
4
Noise variance
Offset = 0.625
Offset = 0.570

FIGURE 7.3

at which these qualitative transitions occur, we instead fix the noise strength and use the R-reliability to assess the overall reliability and precision of the response.

In Figure 7.3a–c, we drive a model neuron with a sinusoidal drive with a frequency of 40 Hz and vary the overall level of depolarizing current (also referred to as offset). The results are not specific to gamma band frequencies, and generalize to other frequencies. In Figure 7.3a, the compound rastergram is shown, in which we show for each current value 50 trials. When the neuron is precise, the spike trains will form streaks (e.g., Figure 7.3a, red arrow), whereas if it is not precise, the ticks representing the spikes will be smeared. The appearance of the streaks will be different from the single parameter case shown in Figure 7.1: they will be at an angle with the vertical because the spike times vary with the level of depolarizing current. The red arrow in Figure 7.3a indicates where the neuron locks one to one with the sinusoidal current: on each cycle of the drive, the neuron produces one spike, with a phase that shifts to earlier times in the cycle as the level of depolarization increases. The locking region corresponds to a step in the firing rate versus offset curve (Figure 7.3b). On this step, the firing rate is constant and only the spike phase varies. These steps were observed experimentally in PFC (Tiesinga et al., 2002) and hippocampus (McLelland and Paulsen, 2009).

In Figure 7.3c, we plot multiple R curves, each for a different σ value. For low σ values, the measure is sensitive to spike timing, and a clear peak emerges, representing the high precision at resonance. As σ increases, the peak height relative to the shoulder (baseline) decreases until it disappears between $\sigma = 5$ and 10 ms.

This model data illustrates an important point. The level of depolarization increases the firing rate of the neuron until it is close to the oscillation frequency at which point the neuron phase locks and both the precision and the reliability are maximal (Schreiber et al., 2004). When the frequency of the periodic drive is varied

FIGURE 7.3 **(See color insert.)** Phase locking to a sinusoidal drive leads to a resonance in the precision; resonant states are the most robust against noise. (a) The compound rastergram is used to show the effect of varying the level of the depolarizing current (offset). For each value, we plot 50 trials, with the offset increasing from bottom to top. At the bottom, we show the sinusoidal driving current. The blue and red arrows in panels (a) and (b) indicate an offset of $I = 0.57$ μA/cm^2 and $I = 0.625$ μA/cm^2, respectively. (b) The firing rate generally increases with the value of the offset, but the curve also shows plateaus where the firing rate does not change with the offset. (c) The plateau at which the firing rate is equal to the frequency of the driving current (40 Hz) also corresponds to a peak in the R-value. The R-value is a compound measure that increases both with precision and reliability and takes a temporal resolution σ as parameter. When σ increases, the measure becomes less sensitive to time differences between the spike trains. For $\sigma = 10$ ms, the R curve does not display a resonance anymore. (d) A compound rastergram with spike trains on multiple trials for different values of the noise variance. For clarity, only five trials are shown per noise value, but the reliability calculations were performed using all 50 trials. The spike trains with offset $I = 0.57$ μA/cm^2 (blue dots) are more noise sensitive than those for 0.625 μA/cm^2 (red dots), which is on the resonance step. (e) We show smoothed histograms representing the instantaneous firing rate for noise variances $D = 4 \times 10^{-6}$, 2.24, and 3.84 mV^2/ms. The colors represent the two offset values as in the preceding panels. (f) This difference in sensitivity is visible as a shift in the R-reliability versus noise curve shown with the same color code.

instead of the depolarizing current, a similar resonance is obtained (Fellous et al., 2001; Haas and White, 2002; Hunter et al., 1998; Hunter and Milton, 2003; Lawrence, 2008; Pike et al., 2000).

The specific value of the resonance frequency for each of the cell types has important functional consequences because it determines which circuit elements can engage in what oscillation (Klausberger and Somogyi, 2008). In the PFC, pyramidal cells displayed a resonance at a lower frequency (5–20 Hz, roughly theta or alpha band) than fast-spiking interneurons, which resonated for frequencies starting in the theta band and reaching the gamma band (5–50 Hz) (Fellous et al., 2001). Similar results were obtained for pyramidal cells and fast-spiking neurons in rodent hippocampus (Pike et al., 2000). Modern genetic techniques allow for the expression of green fluorescent proteins in specific cell types (Luo et al., 2008). This helps to record *in vitro* from specific cells, including those that are only present in small numbers. This strategy was used to determine the sensitivity to gamma frequency inputs of neurons in the rodent cortex. The increase in firing rate with the amplitude of a 40 Hz sinusoidal current was compared to the influence on the firing rate of increasing the level of depolarizing current (Otte et al., 2010). G42 fast-spiking interneurons were most sensitive to gamma drives, whereas layer 5 intrinsically bursting pyramidal cells were the least sensitive, with the G30 and GIN interneurons showing an intermediate sensitivity. The functional consequences of these sensitivity values on cortical information processing remain to be assessed.

7.2.3 Information Coding

A neuron is coding information when different inputs are mapped onto different spike trains, which can be reliably distinguished by an ideal observer, or more importantly, postsynaptic neurons. For the purpose of this review, we are not interested in how the precise features of the input are mapped onto spikes, just that it is done reliably and that stimuli can be distinguished. The Strong method is appropriate for this purpose (Strong et al., 1998). In essence, the mutual information is estimated by subtracting the trial-to-trial variability from the total variability. Specifically, the spike train response across multiple representations of the same stimulus is turned into a binary vector, where a one indicates the presence of a spike in the time bin, whereas a zero signals the absence of a spike. The mutual information is estimated as the entropy of the distribution of binary words across time (total variability) minus the entropy across trials at one time point averaged across all time points (trial-to-trial variability). We use a similar strategy, but without binning the spike trains, as illustrated in Figure 7.4. We construct a distance matrix between a set of spike trains obtained in response to stimulus 1 combined with those obtained in response to stimulus 2. We apply a clustering method to this distance matrix to find two clusters. Each spike train now has two labels: the index of the stimulus that generated it, and the cluster label. When these two labels are the same, apart from a possible permutation in index values, the two stimuli are perfectly distinguishable using these spike trains. The distinguishability is quantified as the mutual information between the two labels, which is equal to one for perfect distinguishability and approaches zero for label pairs that are uncorrelated. This procedure has been detailed in Toups et al.

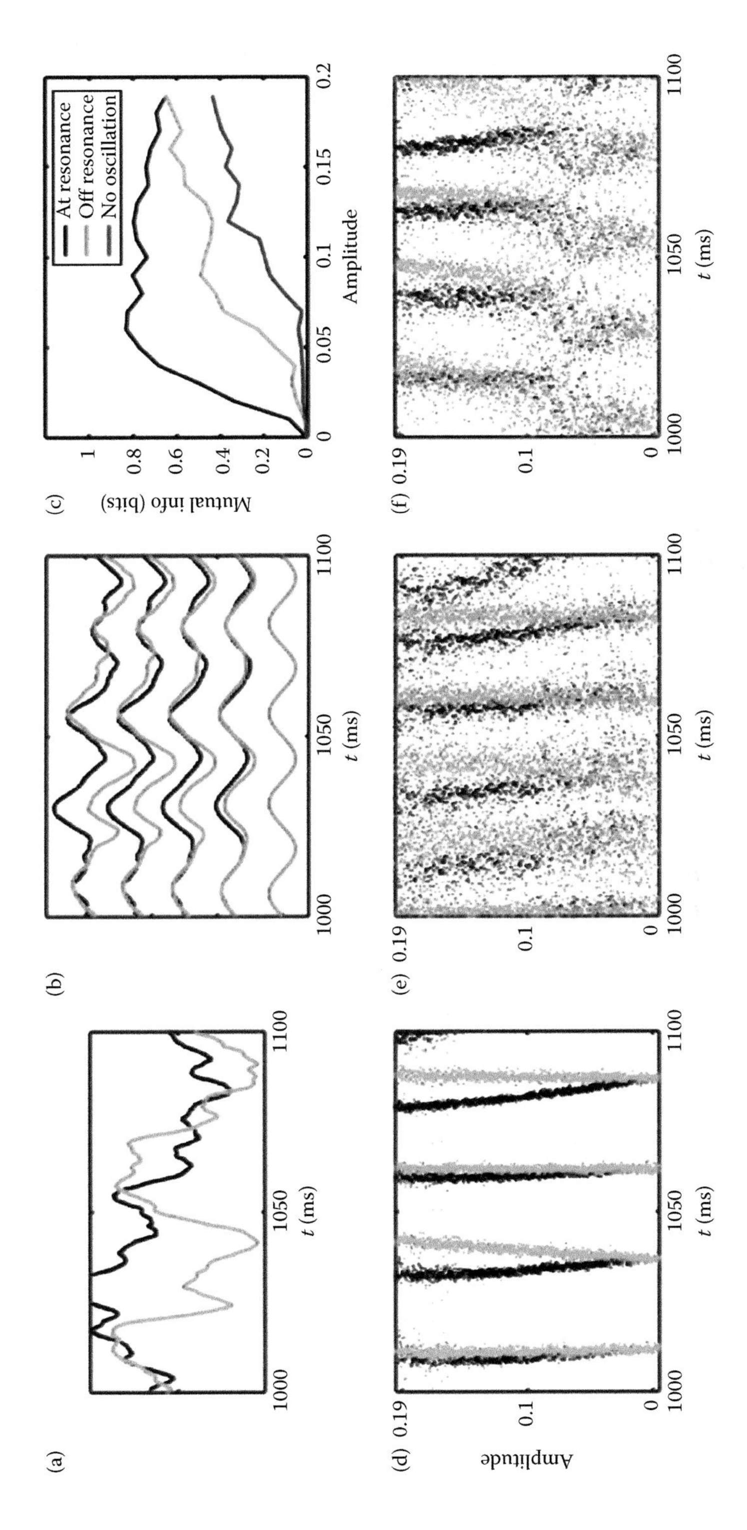

(a)
1000
1050
1100
t (ms)
(b)
1000
1050
1100
t (ms)
(c)
At resonance
Off resonance
No oscillation
Mutual info (bits)
1
0.8
0.6
0.4
0.2
0
0
0.05
0.1
0.15
0.2
Amplitude
(d) 0.19
0.1
0
Amplitude
1000
1050
1100
t (ms)
(e) 0.19
0.1
0
1000
1050
1100
t (ms)
(f) 0.19
0.1
0
1000
1050
1100
t (ms)

FIGURE 7.4

(2011). We quantify the distance between the responses to two different aperiodic stimuli (Figure 7.4a) as a function of their strength (Figure 7.4b and c). The resulting difference in spike trains is plotted in a composite rastergram. For each parameter value, the amplitude in this case, we plot a number of trials, but use as ordinate the parameter value plus a smaller offset. The response to a single stimulus is a jittered spike train. The response to two stimuli is only distinguishable when the difference in spike times is bigger than the trial-to-trial variability due to jitter. By plotting the mutual information versus stimulus amplitude, one can find the onset of this point and determine for which strength distinguishability becomes perfect (Figure 7.4c). We compare the response to spike trains that were generated when the neuron was on the resonance step (Figure 7.4d), to when it received a periodic drive, but was not on the resonance step (Figure 7.4e), and to when there was no periodic drive at all (Figure 7.4f). The best results were obtained during resonance because the precision was higher than in the two other cases, with the distinguishability in the absence of a periodic drive being the worst (Figure 7.4c).

The two spike trains are distinguishable because locking to a periodic drive means that neurons fire at a fixed (constant) phase when there is no aperiodic stimulus waveform injected. When the stimulus is injected, it causes small phase shifts, forward or backward, depending on its sign. In essence, this is due to a rate (depolarizing current) to phase transformation that takes place on the phase locking step (McLelland and Paulsen, 2009; Tiesinga et al., 2002), as discussed in the context of Figure 7.3. For this mechanism to work, phase locking should be stable against noise and the neuron should forget the stimulus-induced phase shift on the previous cycle and let the phase shift be determined only by the value of the stimulus during the current cycle. The noise stability was investigated using *in vitro* studies that show

FIGURE 7.4 **(See color insert.)** Neurons modulated by an oscillatory current encode stimulus information more efficiently than those without an oscillatory current, with the maximum benefit being obtained during resonance conditions. (a) Two different aperiodic stimulus waveforms, red (stimulus 1) and blue (stimulus 2), respectively, were injected on 50 trials into the neuron on top of either a constant depolarizing current or a sinusoidal current. (b) The amplitude of the stimulus was varied. We show the case with a periodic current present; for clarity, only a small segment of the 5 s long stimulus is shown and, for each amplitude, the curve is offset by a fixed amount. From bottom to top, the amplitude is increased and the difference between red and blue curves becomes more pronounced. (c) We used a clustering algorithm (Toups et al., 2011) to find two clusters of spike trains and determined the match between the clustering-determined labeling and the stimulus number (1 or 2) from which the spike train was obtained, in terms of the mutual information. We compare three cases: with periodic current injection and with a level of depolarizing current corresponding to either (black) a resonance or (blue) outside the resonance range, or (red) without a periodic current. The mutual information is plotted as a function of the amplitude of the stimulus. (d–f) The behavior of the mutual information is visualized by the rastergram. The two sets of trials are plotted on top of each other, with the stimulus number represented by the color of the dots as before. (d) For an on-resonance neuron, the precision is high and the difference in spike times between the two drives becomes immediately visible for nonzero amplitude. (e) Off-resonance, the precision is reduced and only for the highest amplitudes is a systematic difference between spike times visible. (f) When there is no periodic current, the response is imprecise for low amplitudes and the difference between spike times due to the two different stimuli is not visible.

that interneurons can sustain phase locking in the gamma frequency range (Tiesinga et al., 2002), whereas hippocampal pyramidal cells *in vitro* cannot (McLelland and Paulsen, 2009). However, the latter are able to sustain theta-frequency phase locking. A theoretical investigation utilizing the so-called phase return map found that at the center of the resonance step the cycle-to-cycle correlations were the weakest. Therefore, during resonance, the neuron can support efficient information transmission because of its stable phase locking and low cycle-to-cycle correlations (Tiesinga et al., 2002). Another study demonstrated by indirect means that the oscillation-induced precision increase improved the distinguishability of spike trains produced by different input stimuli (Schaefer et al., 2006).

7.2.4 Information Transmission and Precision

Synchronous inputs drive a cell better than asynchronous inputs (Marsalek et al., 1997). Hence, the impact of a cell assembly is higher when it is responding with precise and reliable spike trains. Therefore, the precision and reliability of neurons characterized *in vitro* will tell us about the possible impact of *in vivo* assemblies. This type of coding has to be compared with the case when the mean firing rate of an assembly is transmitted to another neuron in the absence of synchronous volleys (Litvak et al., 2003; van Rossum et al., 2002). The signal-to-noise ratio then depends on the degree of correlations among the assembly members, the number of neurons participating, and the length of the interval over which the firing rate is calculated (Shadlen and Newsome, 1998).

We performed simulations to show the impact of synchrony and to illustrate that the computation made possible by synchrony of inhibitory inputs is different from that made possible by synchronous excitatory inputs (Figure 7.5a). We generated the synaptic inputs as a sequence of volleys, with the mean number of spikes in a volley following a Poisson distribution and with the precision characterized by the standard deviation σ of a Gaussian distribution (Tiesinga et al., 2004). The interval between the volleys was chosen to minimize interference between consecutive volleys. For the fast excitatory currents, mediated by amino-3-hydroxy-5-methyl-4-isoxazolepropionic acid (AMPA) receptors, this interval was taken to be 200 ms. When the volley was precise (Figure 7.5b, $\sigma = 1$ ms), the volley elicited a postsynaptic spike with a high probability, whereas when it was not precise (Figure 7.5c, $\sigma = 15$ ms), it produced a depolarization, but in most cases no spike was produced. The mean probability of eliciting a spike in response to a volley was plotted as a function of the precision (Figure 7.5a, black curve). It increased rapidly with precision, with a mid point at approximately 0.2 ms^{-1}, after which it saturated because further improvement in precision did not lead to additional spikes. Note that, for different parameter settings, for instance, when the input volley contained more spikes, a given volley could elicit two spikes in short succession. This parameter range is not considered here.

An inhibitory volley is very effective at preventing a spike just after its arrival time, but that effect falls off rapidly in time. The effect of synchronous inhibition is often compared to its effect when it is spread out uniformly across time. That is, to what extent the degree of inhibitory synchrony is able to modulate the mean firing rate. For this type of comparison, rhythmic inhibition produced by an oscillation is the most appropriate. When presented as precise volleys with a 25 ms interval

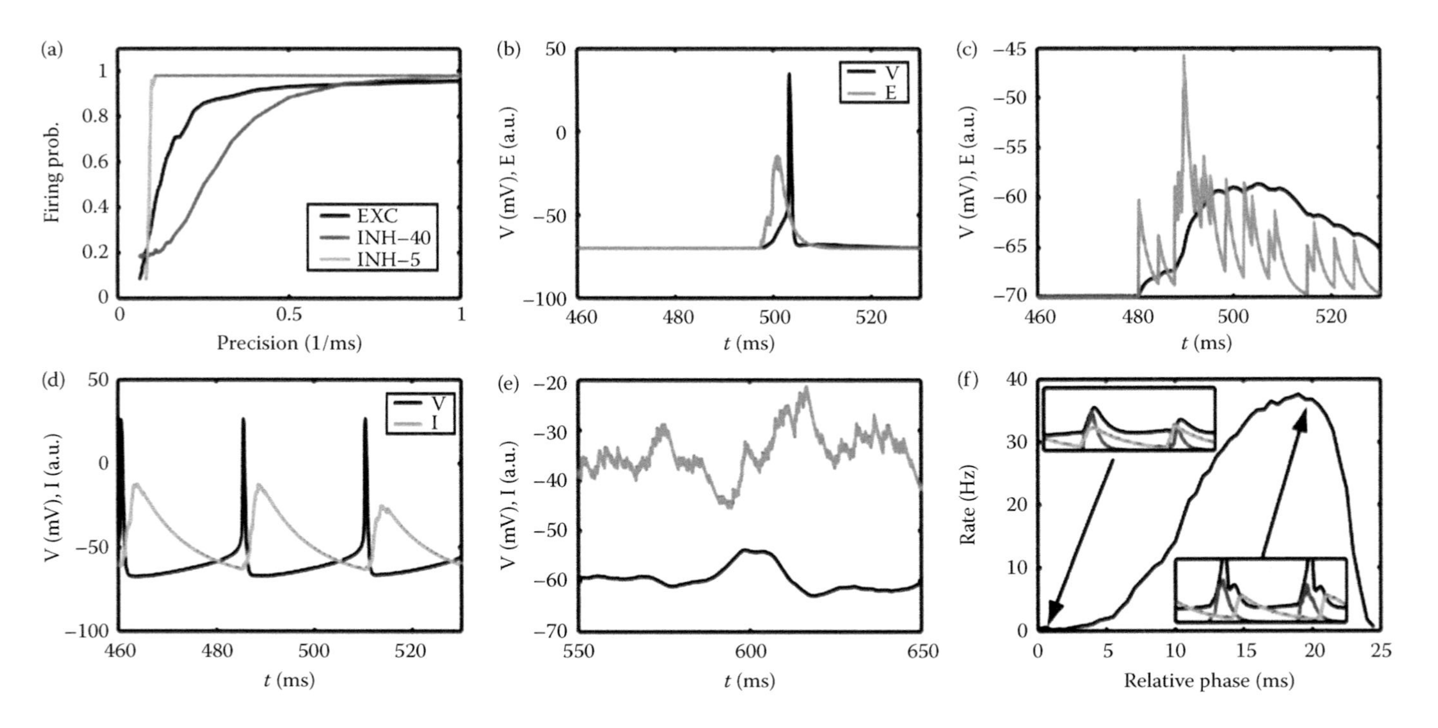
(a)
Firing prob.
Precision (1/ms)
EXC
INH−40
INH−5
(b)
V (mV), E (a.u.)
t (ms)
V
E
(c)
V (mV), E (a.u.)
t (ms)
(d)
V (mV), I (a.u.)
t (ms)
V
I
(e)
V (mV), I (a.u.)
t (ms)
(f)
Rate (Hz)
Relative phase (ms)

FIGURE 7.5

between them, corresponding to a 40 Hz gamma oscillation, the model neuron spiked with high probability once every cycle (Figure 7.5d), close to the time when the inhibitory conductance was at its lowest value, that is, just before the arrival of the next inhibitory volley. The level of lowest conductance depends not only on the interval between volleys, but also on the temporal spread of the constituent spikes (Tiesinga et al., 2004). When the volleys were imprecise, the neuron did not fire (Figure 7.5e). This modulatory effect of synchrony was reduced when the interval was made longer (Tiesinga et al., 2004), in which case the neuron fired for virtually all precisions (Figure 7.5a, light gray curve labeled INH-5). Hence, for inhibitory modulation, oscillations in the gamma frequency band are the most effective.

In general, a neuron receives both excitatory and the inhibitory inputs, each of which could have a component consisting of synchronous volleys. Under the parameter setting where the excitatory volleys are reliable and precise enough to generate a postsynaptic spike, periodic inhibitory volleys generate transmission windows, in which the excitatory inputs can drive the cell (Tiesinga and Sejnowski, 2010), and outside of which they do not lead to spikes. This mechanism generates a modulation of the firing rate, or equivalently of the spike probability during a cycle, as a function of the relative phase between excitation and inhibition (Figure 7.5f). This offers the possibility to affect stimulus selection using phase shifting.

In summary, increased synchrony of excitatory inputs will always increase their impact, whereas the synchrony of inhibitory inputs can modulate the neuron's firing rate only when they are rhythmic with a frequency commensurate with the decay constant of inhibition. For $GABA_A$ receptor-mediated currents, the decay constant corresponds to the gamma frequency range (Bartos et al., 2007). The combination of synchronous excitation and inhibition creates transmission windows, which form a mechanism for stimulus selection through phase shifting.

FIGURE 7.5 Synchronous excitatory volleys are effective in driving postsynaptic neurons, whereas inhibitory synchrony modulates the firing rate when the oscillation frequency is in the gamma frequency range. The model neuron was driven by a periodic sequence of volleys, each consisting of synchronous inputs. The volleys were characterized by their precision, the inverse of the standard deviation of the constituent spike times, that is, their degree of synchrony; their average number of spikes (proportional to the reliability); and their period. In the simulations, the precision was varied to determine the impact of the degree of synchrony of the spikes. (a) The firing rate as a function of precision for (black) excitatory volleys and (light gray) inhibitory volleys with an interval of 200 ms and (gray) inhibitory volleys with an interval of 25 ms. (b) A precise excitatory volley recruits a spike at a short latency, (c) whereas an imprecise volley with the same mean number of spikes does not elicit a spike. We plot (light gray) the excitatory conductance and (black) the membrane potential as a function of time. The conductances are rescaled to fit in the graph and are thus expressed in arbitrary units. (d) Precise inhibition allowed for one spike per cycle, (e) whereas imprecise inhibitory volleys blocked spiking in the neuron. The (light gray) inhibitory conductance and (black) membrane potential is plotted as a function of time. (f) When there is both synchronous inhibition and excitation, excitation can only elicit an action potential when the excitatory volley arrives when the inhibition has decayed to a sufficient extent. The insets show the voltage (black), excitatory (gray), and inhibitory (light gray) conductance for two values of the relative phase (expressed as a delay in ms) as indicated by the arrows.

7.2.5 Modulation of Information Processing through Oscillations

Oscillations can improve the ability to distinguish different stimuli by making the response more precise. After this processing step with oscillations, the stimulus would be represented by a sequence of synchronous volleys. The question then becomes, how the sequence of volleys is transmitted to downstream areas, and in the case of converging inputs, how the response to one stimulus is selected relative to the other. In this section, we address the first issue in a setup that is representative of a cell assembly in one cortical area projecting to a cell assembly in a downstream, cortical area. For instance, within the context of the visual system, this could correspond to a group of cells in V2 projecting to a group of cells in V4. We implement this setup by driving a neuron with excitatory (E) synaptic inputs derived from the phase-locked cell assemblies (e.g., Figure 7.4d) together with synchronous inhibitory volleys with a period of 25 ms (oscillation frequency 40 Hz). These inhibitory inputs generate a periodic sequence of windows during which the neuron can respond with a spike to its excitatory inputs. We study how the transmission depends on the phase of the oscillations that generated the E inputs in the area of origin relative to the phase of the local inhibition. As mentioned before, when the aperiodic stimulus waveform has a low amplitude, spike volleys will occur at a specific phase relative to the underlying periodic current that is determined by the level of depolarization. By varying the phase of the periodic inhibition, the excitatory inputs go from arriving at an optimal phase (relative to the inhibition), for which the output spikes are phase locked to the 40 Hz rhythm, to arriving at the worst phase, for which the output firing rate is low. This leads to a strong modulation, between 0 and 40 Hz, of the output firing rate by phase (Figure 7.6a). The overall chain of events is further illustrated in Figure 7.6f: when the excitatory input volleys arrive at the same time as the inhibitory volleys, there are very few output spikes (Figure 7.6f, top), whereas when the E inputs arrive later, they generate a spike before the next inhibitory volley arrives (Figure 7.6f, bottom). This mimics the results shown in Figure 7.5f. A phase relative to inhibition for which the E input makes the neuron fire is said to be in the transmission window.

When the amplitude of the aperiodic current is increased, the phase of the spike volley varies from cycle to cycle and thus deviates more from the constant phase determined by the overall level of depolarization of the input assembly. Hence, if the constant phase was in the transmission window, some E volleys will fall outside the transmission window, thereby reducing the firing rate. As a consequence, the firing rate will decrease as a function of stimulus amplitude (Figure 7.6b, curve labeled "2"). By contrast, if this fixed phase was outside the transmission window, some spike volleys will be moved into the transmission window by the aperiodic stimulus, thereby increasing the firing rate. Hence, for that case, the firing rate increases with stimulus amplitude (Figure 7.6b, curve labeled "1"). As a consequence, the modulation of firing rate with phase is less deep for larger stimulus amplitudes (Figure 7.6a). Taken together, this implies that modulation of signal transmission by phase can only operate in a tight parameter regime wherein the stimulus only causes small shifts in the spike times.

The R-reliability of the output spike trains versus stimulus amplitude resembles the behavior of the firing rate. Responses to phases in the transmission window have

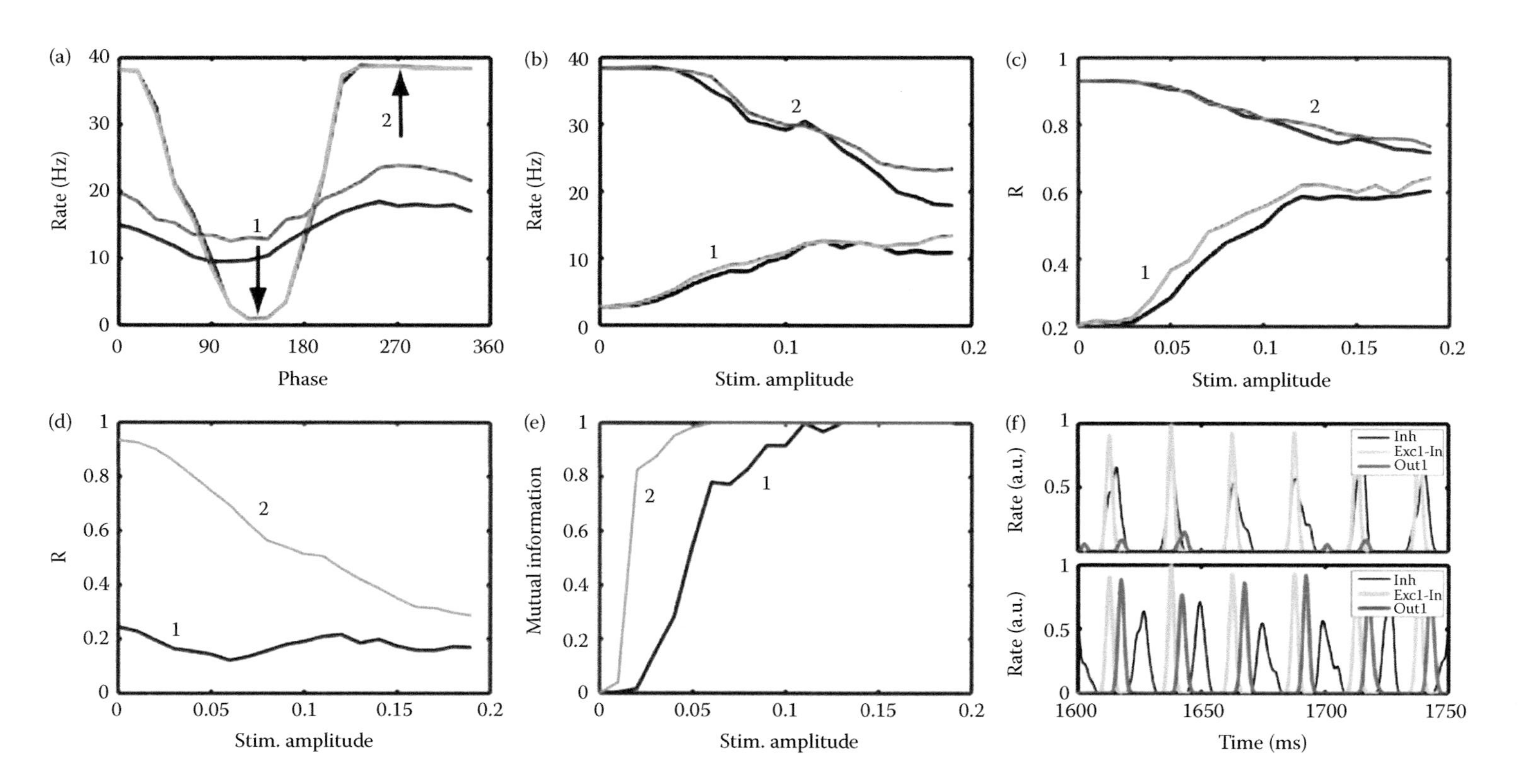
(a)
Rate (Hz)
Phase
1
2
(b)
Rate (Hz)
Stim. amplitude
1
2
(c)
R
Stim. amplitude
1
2
(d)
R
Stim. amplitude
1
2
(e)
Mutual information
Stim. amplitude
1
2
(f)
Rate (a.u.)
Inh
Exc1-In
Out1
Rate (a.u.)
Inh
Exc1-In
Out1
Time (ms)

FIGURE 7.6

a high reliability, which decreases with stimulus strength (Figure 7.6c, curve labeled "2"), whereas phases outside the transmission window lead to a low reliability (Figure 7.6c, curve labeled "1").

We also determined whether an observer, such as a postsynaptic neuron, would be able to determine the identity of the aperiodic stimulus that gave rise to the particular set of spike trains. We first characterized the similarity of the spike trains generated using each of the two stimuli by determining the R-reliability as the average across all pairs with one spike train in response to stimulus 1 and the other in response to stimulus 2. For E volleys with a phase in the transmission window, the reliability was high but decreased with stimulus amplitude (Figure 7.6d, curve labeled "2"). By contrast, for the ones with a phase outside the transmission window, the R-reliability was low, and stayed low when the stimulus amplitude was increased (Figure 7.6d, curve labeled "1"). In principle, these low values for the reliability could be due to systematic differences between the responses to the two stimuli, supporting distinguishability, or due to trial-to-trial variability within each set, which would not contribute to distinguishability. To address this issue, we used the clustering procedure to assign one of two labels to each spike train and compare this label to the label of the stimulus that gave rise to this spike train, quantified in terms of the mutual information between the two labels. The responses to inputs with a phase in the transmission window could be distinguished for stimulus amplitudes as low as 0.02, whereas when the phase was outside this window, a higher stimulus strength was required (Figure 7.6e). This implies that the reduced similarity was due to the high overall variability of the spike trains obtained in response to the same stimulus.

7.2.6 Stimulus Selection through Phase Shifting

In the preceding section, we examined how distinguishable to a postsynaptic neuron the responses of one cell assembly are, when it is driven by one of two possible stimulus waveforms. Now, we turn to a situation where the postsynaptic neuron

FIGURE 7.6 **(See color insert.)** The transmission of E volleys is modulated by the phase of local inhibition. (a) A neuron received phase-locked E volleys as shown in Figure 7.4 together with periodic inhibitory volleys of which the phase was varied. The resulting firing rate as a function of inhibitory phase is shown for each of the two stimulus waveforms, in red and blue, respectively. Label 2 indicates the phase for which the E inputs arrive in the transmission window, whereas label 1 is for a phase outside the transmission window. We plot (b) the firing rate and (c) the R-reliability as a function of the stimulus amplitude for two phases, one inside the transmission window ("2") and one outside ("1"). (d) The similarity between the responses generated by stimulus 1 and stimulus 2 and (e) the distinguishability quantified as the mutual information as a function of stimulus amplitude for the two phases as labeled. (f) We plot (black) inhibitory volleys and (green) excitatory inputs and (red) the output spikes they generate. The histogram of each set of spike trains was convolved with a Gaussian with standard deviation of 1 ms and normalized by the maximum value across all stimulus conditions. For (top) a phase outside the transmission window, only a few output spikes result, whereas when the phase is inside the transmission window, there is a robust output just before the inhibitory volleys arrive.

receives inputs from multiple cell assemblies, each representing a different stimulus. This setup could represent a neuron in a downstream area receiving inputs from different groups of neurons in the same cortical area or from groups in different cortical areas. We aim to determine whether it is possible to represent only one of the multiple input cell assemblies in the neuron's output, thereby achieving stimulus selection. We thus examined the impact of a mixture of inputs from two cell assemblies on the postsynaptic neuron. We used 50 input spike trains from cell assembly 1 and also 50 from cell assembly 2 (*mix condition*); these were compared to two control conditions for which no stimulus selection had to take place, with all 50 input spike trains coming from cell assembly 1 (*stim 1 condition*) or all 50 coming from cell assembly 2 (*stim 2 condition*). Within the context of the visual system, the mix condition would correspond to the case where there are two stimuli in the receptive field of the output cell in V4, which is driven by two groups of neurons in upstream area V2 responding selectively to only one of the two stimuli. The control cases would correspond to only a single stimulus in the receptive field. We kept the phase of the inhibitory input constant, and instead varied the phase of the periodic input to the cell assembly that generated the input spike trains. The phase of cell assembly 2 was in the transmission window, and that of cell assembly 1 was varied across the entire range.

Selection of an input could be said to have been achieved, either when the output firing rate matches that in response to one of the two stimuli or when the occurrence times of the output volleys match those in response to one of the two stimuli. To investigate the latter type of selection, we constructed the spike time histograms of the output spike trains, smoothed them by a Gaussian and determined the cross-correlation between the histograms obtained in the *mix* condition and those obtained for either the *stim 1* or the *stim 2* condition. Because the cross-correlation is normalized, differences in firing rate between conditions are ignored and only the similarity in spike volley times is considered. By contrast, the clustering-based mutual information will always lead to full distinguishability when there is a difference in firing rates. For this reason, we do not use it for this situation. The cross-correlation never reaches one, but there are peaks at values close to 1; however, for most phases, the similarity is much lower (Figure 7.7). At the peak of the cross-correlation, the volley times of the *mix* condition match that of one of the single-stimulus results. This was due to the fact that one E volley, when arriving in the transmission window, was enough to make the neuron spike and that a later volley could not make the neuron spike because of the afterhyperpolarization caused by the first spike. This means that the volley that arrives first will make the neuron spike at a short latency, but with some variability so that it is not necessarily true that the stimulus-induced phase shift can still be used to determine the time course of the aperiodic stimulus. The order of the volleys for each stimulus is determined by their fixed gamma phase and allows for stimulus selection by modulating this relative to the phase of the inhibitory volleys.

This chain of events is broken when the stimulus amplitude is so high that the resulting phase shifts lead to a different order of the volleys of the respective cell assemblies that provide the input spike trains. In mathematical terms, let ϕ_1 and ϕ_2 be the phase of cell assembly 1 and 2, respectively, $\Delta\phi_1(k)$ and $\Delta\phi_2(k)$ the phase shift

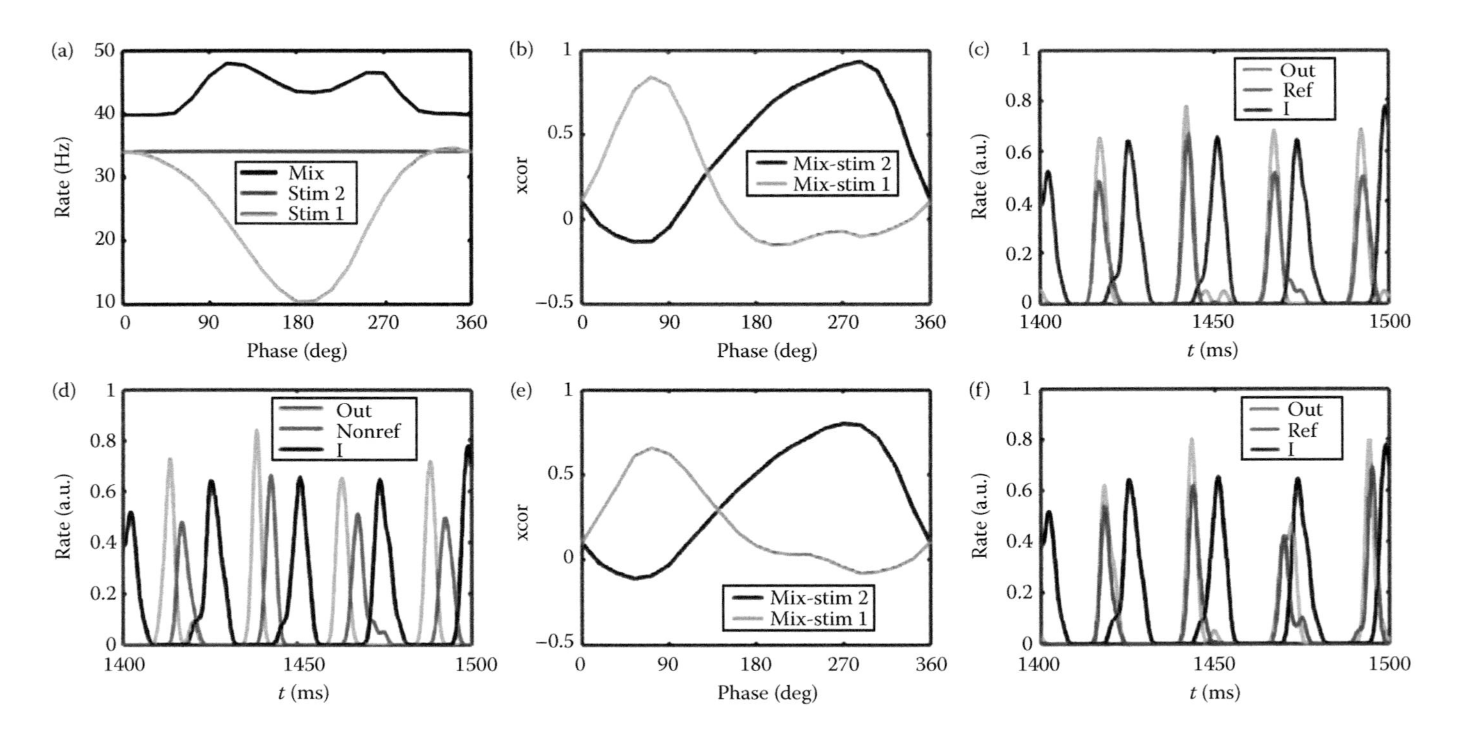
(a)
Rate (Hz)
Mix
Stim 2
Stim 1
Phase (deg)
(b)
xcor
Mix-stim 2
Mix-stim 1
Phase (deg)
(c)
Rate (a.u.)
Out
Ref
I
t (ms)
(d)
Rate (a.u.)
Out
Nonref
I
t (ms)
(e)
xcor
Mix-stim 2
Mix-stim 1
Phase (deg)
(f)
Rate (a.u.)
Out
Ref
I
t (ms)

FIGURE 7.7

induced by the aperiodic stimulus on cycle k, respectively, and ϕ_{inh} the inhibitory phase. Then, stimulus selection can be achieved when $\phi_1 + \Delta\phi_1(k) < \phi_2 + \Delta\phi_2(k)$ and $\phi_1 + \Delta\phi_1(k) < \phi_{inh}$ for all cycles k. This is possible when $\phi_2 - \phi_1 > \max(\Delta\phi_1(k) - \Delta\phi_2(k))$ and $\phi_{inh} - \phi_1 > \max(\Delta\phi_1(k))$, where the maxima are taken with respect to the cycle index k. If these conditions are not met, then the maximum in the cross-correlation is reduced (Figure 7.7e) as a result of the overall reduced overlap (Figure 7.7f).

7.3 DISCUSSION

We have shown that in the context of cell assemblies, precision and reliability of individual neurons driven by an appropriate fluctuating stimulus current (Figure 7.2) are relevant for cortical information processing because postsynaptic impact depends on the degree of synchrony, which is a combination of reliability and precision (Figure 7.5). Achieving precision and reliability, and the modulation thereof, is facilitated by the presence of oscillations because a phase locking step is reached by matching the neuron's firing rate to the oscillation frequency, at which point the precision attains its highest value (Figure 7.3). An aperiodic stimulus waveform can be encoded effectively in terms of small phase shifts when the neuron is phase locked (Figure 7.4). Inhibitory synchrony in the gamma frequency range can gate the propagation of volleys generated by the phase-locked cell assemblies by creating transmission windows, which can be accessed through phase shifting (Figures 7.6 and 7.7). In the following, we discuss these findings in the context of *in vitro* and *in vivo* experiments and cognitive performance.

We have chosen to keep the model as simple as possible to most clearly demonstrate the role of spike time precision in the representation and transmission of information.

FIGURE 7.7 Stimulus selection can be achieved by modulating inhibitory phase when stimuli are phase locked but arrive at different phases. (a) The model neuron received excitatory synaptic inputs that were generated by cell assemblies driven by a periodic current and one of two aperiodic currents representing stimulus 1 or 2, together with synchronous inhibitory volleys arriving at a 25 ms interval. We consider three cases. First, the excitatory volleys derive from only one stimulus, either (light gray) stimulus 1 or (gray) stimulus 2, or (black) are the sum of volleys from two cell assemblies; this is referred to as the *mix* condition. The input from cell assembly 2 is always at a fixed phase, in the transmission window, whereas the phase of cell assembly 1 inputs is varied and displayed as abscissa. We plot the output-firing rate as a function of this phase. (b) We determined for each set of output spike trains an instantaneous firing rate by convolving the spike times with a Gaussian with unit standard deviation. The cross-correlation of the mix condition with the *stim 1* or *2* condition is plotted as a function of the cell assembly 1 phase. (c,d) The instantaneous firing rate as a function of time for (black) the inhibitory volley and (light gray) the output in the *mix* condition together with (gray) (c) the single stimulus condition with the highest cross-correlation ("ref") and (d) the one with the lowest cross-correlation ("nonref"). In the first case, the light gray and gray curves overlap in time but do not have the same height or width, whereas in the second case, they do not overlap. These data were obtained for a stimulus amplitude of 0.01. (e,f) When the stimulus amplitude was 0.10, the cross-correlation (panel e, conventions as in b) was reduced and the overlap between the *mix* condition and the selected single stimulus condition (panel f, conventions as in c) was reduced.

The results are nevertheless representative for more realistic neural systems. The model utilized here can be extended to incorporate a number of other features of neural circuits, which could possibly improve the stimulus selection results presented here or make them more robust. For instance, there are recurrent connections that can generate synchrony in the absence of a fluctuating drive. This leads to a competition between synchrony generated by common input and that generated by recurrent connections. Stimulus selection, such as presented in Figure 7.7, could also be due to network effects where the first input to make the network spike recruits inhibition that prevents the other input from being effective (Borgers and Kopell, 2008), which means that an intrinsically generated gamma rhythm may not be necessary for stimulus selection. Furthermore, we used point neurons, but real neurons have extended dendritic trees with nonlinear voltage-gated channels. The combination of nonlinear dendritic integration and spatial preference on the dendritic tree of one projection compared to the other could also mediate stimulus selection, without the need for oscillations (Gasparini and Magee, 2006; Mel et al., 1998). Future studies of stimulus selection will, therefore, need to incorporate these features of neural circuits to assess their contribution.

7.3.1 Computational Role of Oscillations in Other Models

The model simulations presented here suggest that the rate-to-phase transformation implemented by the phase shift with depolarizing current on the resonance step can be used to encode a time-varying stimulus waveform. An application of this idea is the many are equal (MAE) calculation. When neurons are on the resonance step and receive the same level of input (i.e., depolarizing current in the model setting), they fire at the same rate and at the same phase, thereby producing synchronous volleys, which make a postsynaptic cell spike. When the neurons do not receive the same level of input, they are either not on the resonance step or at different phases on the resonance step. In either of these two circumstances, the cells will not elicit spikes from the postsynaptic cell. This mechanism was suggested for the concentration-independent detection of odors (Brody and Hopfield, 2003). Hopfield also showed how a small-amplitude gamma oscillation on top of a stimulus waveform could improve coding because it integrated stimulus information across 150 ms (Hopfield, 2004). Note that in that study, the gamma oscillation is a perturbation, in contrast to the situation studied in this review, in which the stimulus is the perturbation (Fiser et al., 2004).

7.3.2 Spike Phase Can Encode Stimulus Features

Two recent analyses of *in vivo* recordings have shown that the gamma phase of spikes can contain information about stimulus features such as orientation. Specifically, the spike phase of individual neurons in macaque visual cortex varies with the level of activation as expressed by the instantaneous firing rate (Vinck et al., 2010), which is comparable to the level of depolarization in the models here. Furthermore, the relative phase difference between multiple neurons in cat visual cortex also contains information about orientation (Havenith et al., 2011). The LFP in the auditory cortex locks to the envelope of the auditory stimulus waveform. When the measured responses are aligned on trials onset, the LFP phase is reproducible and contains additional information compared with the spike rate (Kayser et al., 2009). Another example consistent with phase coding has been found in the hippocampus: there, theta (around 8 Hz) shapes cell assembly activation into orderly sequences that get repeated at each theta cycle. This gives rise to the phenomenon of theta phase precession (Hafting et al., 2008; O'Keefe and Recce, 1993) by which hippocampal place cells and entorhinal grid cells (neurons that activate as the rat transits a well-defined region of space, the place field) fire at a late theta phase upon the animal's entrance into the place fields, and shift to earlier and earlier firing phases as the animal progresses into the place field. From the population point of view, this means that within each theta cycle, cell groups with place fields further and further along the animal's path will fire in distinctive sequences (Skaggs et al., 1996). Therefore, the phase at which the spikes are fired is informative about animal position (Jensen and Lisman, 2000). The actual mechanism of phase precession is not known in detail; however, in one proposed theory, the input level (which encodes animal position) would be translated into a firing phase by a mechanism similar to what is shown in Figure 7.3 (Harris et al., 2002; Mehta et al., 2002). An alternative hypothesis is that phase precession is generated by interference between the global theta oscillations and subthreshold oscillations in the membrane potential at a slightly higher frequency (about 1 Hz higher) (Lengyel et al., 2003; O'Keefe and Recce, 1993). Such oscillations have indeed been observed in entorhinal cortex grid cells (see, e.g., Giocomo et al., 2007). Note that the optimal "locking region" of Figure 7.3 corresponds to an input with a rate similar to the underlying oscillation.

7.3.3 Phase Locking, Spike Timing Precision, and Significance for Cognitive Functions

The timing precision increase induced by oscillations has a beneficial effect on information coding and transmission. One would expect that this increased computational efficiency is reflected in cognitive abilities, which indeed has been shown in experiments that combine electrophysiology with behavior in a number of contexts. Many of these studies focus on gamma oscillations, a hallmark of cortical activity during the awake, active state. In monkey visual cortex, the degree by which action potentials maintain a consistent phase relationship with the underlying LFP in the gamma frequency range (as it would be in the "locking region" of Figure 7.3b) is predictive of subject performance as measured by reaction times (Womelsdorf et al., 2006). Interestingly, this "spike-to-field" coherence (SFC) may differ in different cortical columns: when it is higher in the columns with a receptive field encompassing the stimulus to be attended, the monkey responds faster. Conversely, high SFC in columns whose receptive field includes a distracter stimulus is correlated with a reduced performance level. These results may be interpreted in the light of the MAE hypothesis described above: as many neurons in one cortical column fire synchronously, they have a better chance to make the postsynaptic cell spike, and therefore to exert an influence on downstream cortical areas (in this case, facilitating information transfer from visual cortices to the areas involved with the generation of a behavioral response). In fact, gamma coherence is related to ongoing computations for functions other than visual perception and attention. For example, both LFP power and SFC are increased in the monkey parietal (LIP) cortex in the delay period of a working memory task (Pesaran et al., 2002). By contrast, in transgenic mice lacking N-methyl-D-aspartate (NMDA) receptors on hippocampal parvalbumin-positive interneurons, gamma oscillations were increased (Korotkova et al., 2010), but this led to a decreased precision of the place field of CA1 pyramidal neurons, as well as with deficits in spatial short- and long-term memory. Hence, excessive fast synchronization (in this case, together with disrupted theta synchronization) may not always be helpful for cognitive function.

7.3.4 Gamma Rhythms and Communication through Coherence

When communication between brain areas is mediated through cell assemblies producing synchronous volleys locked to gamma rhythms, modulation of phase of these volleys relative to the local inhibitory rhythm can achieve signal rerouting and stimulus selection. This is the idea behind the principle of communication through coherence (CTC) (Fries, 2005), which was studied in macaque and cat visual cortex using simultaneous recordings on multiple electrodes. When two sites have coherent gamma oscillations, with a specific phase offset between them, the oscillatory power at the two sites fluctuates together, whereas for other values of the phase offset, the correlations are less (Womelsdorf et al., 2007). Thus, oscillations may turn on and off communication between two cortical sites: in each brain region, the oscillatory inhibitory drive from the interneurons (Figures 7.5f and 7.6a) will permit neuronal firing only in a certain time window within each gamma cycle. When the windows at the two sites overlap, neurons at the two locations have a chance to determine each other's firing. On the other hand, when the windows do not intersect, communication between areas is much less efficient. Similar coherence phenomena have been demonstrated between cortical areas further apart in the brain, such as between frontal eye fields (in the frontal lobe) and V4 (Gregoriou et al., 2009), or between the cortex and the spinal cord (Schoffelen et al., 2005), hinting at a general role of neuronal oscillations in communication throughout the nervous system. CTC implies that the relative phase between different brain areas is relevant and is modulated by cognitive demands.

7.3.5 Neural Oscillations and Cell Assembly Formation

As we hypothesized above, one consequence of the sculpting of spike timing by oscillations is that cell populations in a given structure may fire in a synchronous fashion. This may give these populations an advantage when it comes to their effect on downstream neural areas. This idea was first proposed by Hebb (1949). Hebb's cell assemblies were tightly connected groups of cells that tended to fire synchronously. Spike patterns that suggest the emergence of cell assemblies have been observed in the hippocampus (Harris et al., 2003). In that study, the activity of one cell could be better predicted when the activity of all remaining recorded cells was included as a predictor together with the strongest correlates of hippocampal firing, such as rat position, theta phase. This predictive power was substantiated by the appearance of synchronized activation of neuronal groups, taking place on a typical time scale of ~20 ms, that is, the typical duration of a gamma cycle, highlighting the importance of this type of oscillation in shaping neural activity. In the rat PFC, similar synchronized activity patterns appeared in a so called Y-maze experiment (Benchenane et al., 2010); at the fork and after achievement of the performance criterion, that is, under the same circumstances, cortico-hippocampal coherence is observed. This increased synchronization may favor the formation of persistent cell assemblies by means of spike timing-dependent plasticity (Cassenaer et al., 2007; Benchenane et al., 2011).

7.3.6 Properties of Gamma Oscillations in Cortical Area V1

In the preceding text, we have discussed the role of oscillations in multiple frequency bands, including theta, but have mostly emphasized the possible functional role of gamma oscillations, which depended on synchronous inhibition generated by tightly synchronized inhibitory interneurons. Experimental studies have found and characterized the role of gamma oscillations in different parts of the visual cortex, hippocampus, and PFC, as reviewed. The tightly synchronized gamma oscillations at a constant frequency are a model idealization and may not be representative of real neural systems. For instance, the LFP signals measured during visual stimulation in area V1 of anesthetized macaque monkeys were statistically not significantly different from white noise filtered by gamma-frequency-range band-pass filter and could not be modeled as resulting from a burst of precisely spaced volleys (Burns et al., 2011). Furthermore, the frequency of gamma oscillations is contrast-dependent (Ray and Maunsell, 2010, 2011), which is consistent with our model simulations (Buia and Tiesinga, 2006; Tiesinga and Sejnowski, 2010), but has been interpreted as implying that gamma synchronized discharges cannot be used for communication between brain areas (Ray and Maunsell, 2010), a claim that requires further substantiation. These studies are important because they give insights into what features of models are realistic and which are not, and furthermore, they provide the constraints under which communication through gamma rhythms must operate.

7.4 CONCLUSION

Taken together, the computational studies and the reviewed experimental studies point to an important role for oscillation-induced increases in spike time precision in boosting impact on postsynaptic neurons and modulating information transmission between different cortical areas. These effects occur in various frequency bands of which we highlighted gamma. The role of oscillations in setting spike times and improving spike time precision is difficult to assess *in vivo* if the phase and amplitude of the underlying oscillation varies from trial to trial. To address this issue, Agarwal and Sommer report in Chapter 6 on an information-theoretical method to shift spike times to compensate for the variability of the underlying oscillation, thereby making it possible to better quantify spike timing precision *in vivo*. While oscillations play a role in shaping spike time precision in a number of situations in different brain areas, it is in some cases possible to achieve efficient temporal coding in the absence of oscillations. For instance, Gawne reviews in Chapter 15 how the relative timing of spikes, even in the absence of oscillations, can form the basis for a neural code representing stimulus properties in cortical area V4.

APPENDIX

The models used, and subsequent analysis presented here, have been described in earlier publications. We highlight the relevant parameters and their value and refer to the original publications for details. The neuron was modeled using the Hodgkin–Huxley formalism. The model had voltage-gated potassium (I_K) and sodium channels (I_{Na}), as well as a leak current (I_L). It was driven by a time-varying fluctuating current $I_{inj}(t)$, a noise current $C_m\xi(t)$, excitatory synaptic inputs of the AMPA type (I_{AMPA}), and inhibitory inputs of the $GABA_A$ type ($I_{GABA,A}$), which led to the following differential equation for the membrane potential V (in mV):

$$C_m \frac{dV}{dt} = -I_{Na} - I_K - I_L - I_{AMPA} - I_{GABA,A} + I_{inj} + C_m\xi.$$

The details and the other differential equations for the kinetic variables are given in Tiesinga and Jose (2000). For the simulations presented here, the synaptic inputs $I_{AMPA} = g_e s_e(V - E_{AMPA})$, $I_{GABA,A} = g_i s_i(V - E_{GABA})$, and the injected current $I_{inj}(t) = I_0 + A_p \cos 2\pi ft + A_s I_{stim}(t)$ are the most important. Currents I_0 (representing the level of constant depolarizing current), A_p (the amplitude of the sinusoidal current), and A_s (the amplitude of the aperiodic current) are expressed in μA/cm², the unitary conductances g_e and g_i are expressed in mS/cm², and the noise strength D (variance of ξ) is expressed in mV²/ms. The stimulus current I_{stim} had a zero mean and a standard deviation of 1 and was generated as described in Toups et al. (2011), except that for each of the two stimuli, the procedure was started with a different seed for the random number generator. The synaptic inputs are generated using an input spike train, where each spike results in either (Figure 7.5) an exponentially decaying conductance pulse with amplitude 1 ($\exp(-t/\tau)$, the decay constant τ is 2 and 10 ms for excitatory and inhibitory inputs, respectively) or an alpha function with maximum 1 ($(t/\tau)\exp(1 - t/\tau)$, the time constant τ is 1 and 4 ms for excitatory and inhibitory inputs, respectively). The detailed settings of either the synaptic or the driving current is given for Figures 7.1 through 7.4 in Table 7.1, for Figure 7.5 in Table 7.2, and for Figures 7.6 and 7.7 in Table 7.3.

TABLE 7.1
Parameter Values Used to Construct the Injected Current, $I_{inj}(t)$, for Figures 7.1 through 7.4

Figure	I_0 (μA/cm²)	A_p (μA/cm²)	A_s (μA/cm²)	D (mV²/ms)
7.1	0.26	0.0	1.0	0.04
7.2	0.625	0.1	0.0	0.004
7.3a–c	0.25–0.745	0.1	0.0	0.004
7.3d–f	0.625 and 0.57	0.1	0.0	4×10^{-6} to 4.0 (50 steps)
7.4c	7.4d–f combined	See below		
7.4d	0.625	0.2	0–0.19 (20 steps)	0.004
7.4e	0.75	0.2	0–0.19 (20 steps)	0.004
7.4f	0.625	0	0–0.19 (20 steps)	0.004

TABLE 7.2
Parameter Values of the Periodic Synchronous Volleys Used in the Simulations for Figure 7.5

Figure	I_0 (μA/cm²)	N_E	σ_E (ms)	g_E (mS/cm²)	N_I	σ_I (ms)	g_I (mS/cm²)	f v (Hz)
7.5a (black)	−0.5	20	1–15.5 (30 steps)	0.011	–	–	–	5
7.5a (gray)	2.0	–	–	–	160	1–15.5 (30 steps)	0.0075	40
7.5a (light gray)	0.2	–	–	–	160	1–15.5 (30 steps)	0.0075	5
7.5b	−0.5	20	1	0.011	–	–	–	5
7.5c	0.5	20	15	0.011	–	–	–	5
7.5d	2.0	–	–	–	160	1	0.0075	40
7.5e	2.0	–	–	–	160	15	0.0075	40
7.5f	0.0	150	1	0.02	150	1	0.05	40

Note: The noise current had $D = 0$ mV²/ms, g_E and g_I denote the unitary conductance of excitatory and inhibitory synaptic conductance, respectively, that had an exponential form with decay constants of 2 and 10 ms, respectively. N_E (N_I) is the number of spikes in each volley, σ_E (σ_I) is their standard deviation, and f_v is the (oscillation) frequency corresponding to the interval between consecutive volleys.

TABLE 7.3
Parameter Values for the Construction of Excitatory and Inhibitory Synaptic Inputs for Figures 7.6 and 7.7

Figure	I_0 (μA/cm²)	D (mV²/ms)	N_E	A_s (in)	g_E (μS/cm²)	N_I	σ_I (ms)	g_I (μS/cm²)	f (Hz)
7.6a	0	0.004	50	0.01–0.19	1.9	90	2	2	40
7.6b–e	0	0.04	50	0.00–0.19 (20 steps)	1.9	90	2	2	40
7.6f	0	0.04	50	0.01	1.9	90	2	2	40
7.7a–d	0	0.04	50	0.01	1.9	90	2	2	40
7.7e–f	0	0.04	50	0.09	1.9	90	2	2	40

Note: Each unitary synaptic input was represented by an alpha function, $(t/\tau)\exp(1 - t/\tau)$, which is normalized such that the maximum is 1 and $\tau = 1$ ms for excitatory inputs and 4 ms for inhibitory inputs, respectively. The excitatory inputs are generated by sampling 50 spike trains from a cell assembly generated as in Figure 7.4. The parameters were $I_0 = 0.625$ μA/cm², $A_p = 0.2$ μA/cm², $D = 0.004$ mV²/ms, and A_s as listed in the table; for Figure 7.6, $A_p = 0.2$ μA/cm² and for Figure 7.7, $A_p = 0.1$ μA/cm²; in both cases, the frequency f of the sinusoidal current was 40 Hz.

REFERENCES

Averbeck, B.B. and Lee, D. 2006. Effects of noise correlations on information encoding and decoding. *Journal of Neurophysiology 95*, 3633–3644.

Bair, W. 1999. Spike timing in the mammalian visual system. *Current Opinion in Neurobiology 9*, 447–453.

Bair, W. and Koch, C. 1996. Temporal precision of spike trains in extrastriate cortex of the behaving macaque monkey. *Neural Computation 8*, 1185–1202.

Bartos, M., Vida, I., and Jonas, P. 2007. Synaptic mechanisms of synchronized gamma oscillations in inhibitory interneuron networks. *Nature Reviews 8*, 45–56.

Benchenane, K., Peyrache, A., Khamassi, M., Tierney, P.L., Gioanni, Y., Battaglia, F.P., and Wiener, S.I. 2010. Coherent theta oscillations and reorganization of spike timing in the hippocampal-prefrontal network upon learning. *Neuron 66*, 921–936.

Benchenane, K., Tiesinga, P.H., and Battaglia, F.P. 2011. Oscillations in the prefrontal cortex: A gateway to memory and attention. *Current Opinion in Neurobiology 21*, 475–485.

Borgers, C. and Kopell, N.J. 2008. Gamma oscillations and stimulus selection. *Neural Computation 20*, 383–414.

Brody, C.D. and Hopfield, J.J. 2003. Simple networks for spike-timing-based computation, with application to olfactory processing. *Neuron 37*, 843–852.

Brunel, N., Chance, F.S., Fourcaud, N., and Abbott, L.F. 2001. Effects of synaptic noise and filtering on the frequency response of spiking neurons. *Physical Review Letters 86*, 2186–2189.

Buia, C. and Tiesinga, P. 2006. Attentional modulation of firing rate and synchrony in a model cortical network. *Journal of Computational Neuroscience 20*, 247–264.

Buracas, G.T., Zador, A.M., DeWeese, M.R., and Albright, T.D. 1998. Efficient discrimination of temporal patterns by motion-sensitive neurons in primate visual cortex. *Neuron 20*, 959–969.

Burns, S.P., Xing, D., and Shapley, R.M. 2011. Is gamma-band activity in the local field potential of V1 cortex a "clock" or filtered noise. *The Journal of Neuroscience 31*, 9658–9664.

Butts, D.A., Weng, C., Jin, J., Yeh, C.I., Lesica, N.A., Alonso, J.M., and Stanley, G.B. 2007. Temporal precision in the neural code and the timescales of natural vision. *Nature 449*, 92–95.

Cassenaer, S. and Laurent, G. 2007. Hebbian STDP in mushroom bodies facilitates the synchronous flow of olfactory information in locusts. *Nature 448*, 709–713.

Cecchi, G.A., Sigman, M., Alonso, J.M., Martinez, L., Chialvo, D.R., and Magnasco, M.O. 2000. Noise in neurons is message dependent. *Proceedings of the National Academy of Sciences of the United States of America 97*, 5557–5561.

Deweese, M.R. and Zador, A.M. 2004. Shared and private variability in the auditory cortex. *Journal of Neurophysiology 92*, 1840–1855.

Elhilali, M., Fritz, J.B., Klein, D.J., Simon, J.Z., and Shamma, S.A. 2004. Dynamics of precise spike timing in primary auditory cortex. *The Journal of Neuroscience 24*, 1159–1172.

Ermentrout, G.B., Galan, R.F., and Urban, N.N. 2008. Reliability, synchrony and noise. *Trends in Neurosciences 31*, 428–434.

Fellous, J.M., Houweling, A.R., Modi, R.H., Rao, R.P., Tiesinga, P.H., and Sejnowski, T.J. 2001. Frequency dependence of spike timing reliability in cortical pyramidal cells and interneurons. *Journal of Neurophysiology 85*, 1782–1787.

Fellous, J.M., Tiesinga, P.H., Thomas, P.J., and Sejnowski, T.J. 2004. Discovering spike patterns in neuronal responses. *The Journal of Neuroscience 24*, 2989–3001.

Fiser, J., Chiu, C., and Weliky, M. 2004. Small modulation of ongoing cortical dynamics by sensory input during natural vision. *Nature 431*, 573–578.

Fries, P. 2005. A mechanism for cognitive dynamics: Neuronal communication through neuronal coherence. *Trends in Cognitive Sciences 9*, 474–480.

Gasparini, S. and Magee, J.C. 2006. State-dependent dendritic computation in hippocampal CA1 pyramidal neurons. *The Journal of Neuroscience 26*, 2088–2100.

Giocomo, L.M., Zilli, E.A., Fransen, E., and Hasselmo, M.E. 2007. Temporal frequency of subthreshold oscillations scales with entorhinal grid cell field spacing. *Science (New York, NY) 315*, 1719–1722.

Gregoriou, G.G., Gotts, S.J., Zhou, H., and Desimone, R. 2009. High-frequency, long-range coupling between prefrontal and visual cortex during attention. *Science (New York, NY) 324*, 1207–1210.

Haas, J.S. and White, J.A. 2002. Frequency selectivity of layer II stellate cells in the medial entorhinal cortex. *Journal of Neurophysiology 88*, 2422–2429.

Hafting, T., Fyhn, M., Bonnevie, T., Moser, M.B., and Moser, E.I. 2008. Hippocampus-independent phase precession in entorhinal grid cells. *Nature 453*, 1248–1252.

Harris, K.D., Csicsvari, J., Hirase, H., Dragoi, G., and Buzsaki, G. 2003. Organization of cell assemblies in the hippocampus. *Nature 424*, 552–556.

Harris, K.D., Henze, D.A., Hirase, H., Leinekugel, X., Dragoi, G., Czurko, A., and Buzsaki, G. 2002. Spike train dynamics predicts theta-related phase precession in hippocampal pyramidal cells. *Nature 417*, 738–741.

Havenith, M.N., Yu, S., Biederlack, J., Chen, N.H., Singer, W., and Nikolic, D. 2011. Synchrony makes neurons fire in sequence and stimulus properties determine who is ahead. *The Journal of Neuroscience 31*, 8570–8584.

Hebb, D.O. 1949. *The Organization of Behavior* (New York, Wiley & Sons).

Hopfield, J.J. 2004. Encoding for computation: recognizing brief dynamical patterns by exploiting effects of weak rhythms on action-potential timing. *Proceedings of the National Academy of Sciences of the United States of America 101*, 6255–6260.

Hunter, J., Milton, J., Thomas, P., and Cowan, J. 1998. Resonance effect for neural spike time reliability. *Journal of Neurophysiology 80*, 1427–1438.

Hunter, J.D. and Milton, J.G. 2003. Amplitude and frequency dependence of spike timing: Implications for dynamic regulation. *Journal of Neurophysiology 90*, 387–394.

Jensen, O. and Lisman, J.E. 2000. Position reconstruction from an ensemble of hippocampal place cells: Contribution of theta phase coding. *Journal of Neurophysiology 83*, 2602–2609.

Kayser, C., Montemurro, M.A., Logothetis, N.K., and Panzeri, S. 2009. Spike-phase coding boosts and stabilizes information carried by spatial and temporal spike patterns. *Neuron 61*, 597–608.

Kenet, T., Bibitchkov, D., Tsodyks, M., Grinvald, A., and Arieli, A. 2003. Spontaneously emerging cortical representations of visual attributes. *Nature 425*, 954–956.

Klausberger, T. and Somogyi, P. 2008. Neuronal diversity and temporal dynamics: The unity of hippocampal circuit operations. *Science (New York, NY) 321*, 53–57.

Korotkova, T., Fuchs, E.C., Ponomarenko, A., von Engelhardt, J., and Monyer, H. 2010. NMDA receptor ablation on parvalbumin-positive interneurons impairs hippocampal synchrony, spatial representations, and working memory. *Neuron 68*, 557–569.

Lawrence, J.J. 2008. Cholinergic control of GABA release: Emerging parallels between neocortex and hippocampus. *Trends in Neurosciences 31*, 317–327.

Lengyel, M., Szatmary, Z., and Erdi, P. 2003. Dynamically detuned oscillations account for the coupled rate and temporal code of place cell firing. *Hippocampus 13*, 700–714.

Litvak, V., Sompolinsky, H., Segev, I., and Abeles, M. 2003. On the transmission of rate code in long feedforward networks with excitatory-inhibitory balance. *The Journal of Neuroscience 23*, 3006–3015.

Luo, L., Callaway, E.M., and Svoboda, K. 2008. Genetic dissection of neural circuits. *Neuron 57*, 634–660.

Mainen, Z.F. and Sejnowski, T.J. 1995. Reliability of spike timing in neocortical neurons. *Science (New York, NY) 268*, 1503–1506.

Markowitz, D.A., Collman, F., Brody, C.D., Hopfield, J.J., and Tank, D.W. 2008. Rate-specific synchrony: Using noisy oscillations to detect equally active neurons. *Proceedings of the National Academy of Sciences of the United States of America 105*, 8422–8427.

Marsalek, P., Koch, C., and Maunsell, J. 1997. On the relationship between synaptic input and spike output jitter in individual neurons. *Proceedings of the National Academy of Sciences of the United States of America 94*, 735–740.

McLelland, D. and Paulsen, O. 2009. Neuronal oscillations and the rate-to-phase transform: mechanism, model and mutual information. *The Journal of Physiology 587*, 769–785.

Mehta, M.R., Lee, A.K., and Wilson, M.A. 2002. Role of experience and oscillations in transforming a rate code into a temporal code. *Nature 417*, 741–746.

Mel, B.W., Ruderman, D.L., and Archie, K.A. 1998. Translation-invariant orientation tuning in visual "complex" cells could derive from intradendritic computations. *The Journal of Neuroscience 18*, 4325–4334.

O'Keefe, J. and Recce, M.L. 1993. Phase relationship between hippocampal place units and the EEG theta rhythm. *Hippocampus 3*, 317–330.

Otte, S., Hasenstaub, A., and Callaway, E.M. 2010. Cell type-specific control of neuronal responsiveness by gamma-band oscillatory inhibition. *The Journal of Neuroscience 30*, 2150–2159.

Pesaran, B., Pezaris, J.S., Sahani, M., Mitra, P.P., and Andersen, R.A. 2002. Temporal structure in neuronal activity during working memory in macaque parietal cortex. *Nature Neuroscience 5*, 805–811.

Pike, F.G., Goddard, R.S., Suckling, J.M., Ganter, P., Kasthuri, N., and Paulsen, O. 2000. Distinct frequency preferences of different types of rat hippocampal neurones in response to oscillatory input currents. *The Journal of Physiology 529 Pt 1*, 205–213.

Ray, S. and Maunsell, J. 2010. Differences in gamma frequencies across visual cortex restrict their possible use in computation. *Neuron 67*, 885–896.

Ray, S. and Maunsell, J. 2011. Different origins of gamma rhythms and high-gamma activity in macaque visual cortex. *PLoS Biology 9*, e1000610.

Reinagel, P. and Reid, R.C. 2002. Precise firing events are conserved across neurons. *The Journal of Neuroscience 22*, 6837–6841.

Reyes, A.D. 2003. Synchrony-dependent propagation of firing rate in iteratively constructed networks *in vitro*. *Nature Neuroscience 6*, 593–599.

Schaefer, A.T., Angelo, K., Spors, H., and Margrie, T.W. 2006. Neuronal oscillations enhance stimulus discrimination by ensuring action potential precision. *PLoS Biology 4*, e163.

Schoffelen, J.M., Oostenveld, R., and Fries, P. 2005. Neuronal coherence as a mechanism of effective corticospinal interaction. *Science (New York, NY) 308*, 111–113.

Schreiber, S., Fellous, J.M., Tiesinga, P., and Sejnowski, T.J. 2004. Influence of ionic conductances on spike timing reliability of cortical neurons for suprathreshold rhythmic inputs. *Journal of Neurophysiology 91*, 194–205.

Schreiber, S., Whitmer, D., Fellous, J.M., Tiesinga, P., and Sejnowski, T.J. 2003. A new correlation-based measure of spike timing reliability. *Neurocomputing 52–54*, 925–931.

Shadlen, M. and Newsome, W. 1998. The variable discharge of cortical neurons: Implications for connectivity, computation, and information coding. *The Journal of Neuroscience 18*, 3870–3896.

Skaggs, W.E., McNaughton, B.L., Wilson, M.A., and Barnes, C.A. 1996. Theta phase precession in hippocampal neuronal populations and the compression of temporal sequences. *Hippocampus 6*, 149–172.

Strong, S., Koberle, R., de Ruyter van Stevenick, R., and Bialek, W. 1998. Entropy and information in neural spike trains. *Physical Review Letters 80*, 197–2000.

Tiesinga, P., Fellous, J.M., and Sejnowski, T.J. 2008. Regulation of spike timing in visual cortical circuits. *Nature Reviews 9*, 97–107.

Tiesinga, P.H., Fellous, J.M., Jose, J.V., and Sejnowski, T.J. 2002. Information transfer in entrained cortical neurons. *Network 13*, 41–66.

Tiesinga, P.H., Fellous, J.M., Salinas, E., Jose, J.V., and Sejnowski, T.J. 2004. Inhibitory synchrony as a mechanism for attentional gain modulation. *Journal of Physiology Paris 98*, 296–314.

Tiesinga, P.H. and Sejnowski, T.J. 2010. Mechanisms for phase shifting in cortical networks and their role in communication through coherence. *Frontiers in Human Neuroscience 4*, 196.

Tiesinga, P.H.E. 2004. Chaos-induced modulation of reliability boosts output firing rate in downstream cortical areas. *Physical Review E 69*, 031912.

Tiesinga, P.H.E. and Jose, J.V. 2000. Robust gamma oscillations in networks of inhibitory hippocampal interneurons. *Network-Comp Neural 11*, 1–23.

Tiesinga, P.H.E. and Toups, J.V. 2005. The possible role of spike patterns in cortical information processing. *Journal of Computational Neuroscience 18*, 275–286.

Toups, J.V., Fellous, J.M., Thomas, P.J., Sejnowski, T.J., and Tiesinga, P.H. 2011. Finding the event structure of neuronal spike trains. *Neural Computation 23*, 2169–2208.

Tsodyks, M., Kenet, T., Grinvald, A., and Arieli, A. 1999. Linking spontaneous activity of single cortical neurons and the underlying functional architecture. *Science (New York, NY) 286*, 1943–1946.

van Rossum, M.C., Turrigiano, G.G., and Nelson, S.B. 2002. Fast propagation of firing rates through layered networks of noisy neurons. *The Journal of Neuroscience 22*, 1956–1966.

Victor, J.D. and Purpura, K.P. 1996. Nature and precision of temporal coding in visual cortex: A metric-space analysis. *Journal of Neurophysiology 76*, 1310–1326.

Vinck, M., Lima, B., Womelsdorf, T., Oostenveld, R., Singer, W., Neuenschwander, S., and Fries, P. 2010. Gamma-phase shifting in awake monkey visual cortex. *The Journal of Neuroscience 30*, 1250–1257.

Womelsdorf, T., Fries, P., Mitra, P.P., and Desimone, R. 2006. Gamma-band synchronization in visual cortex predicts speed of change detection. *Nature 439*, 733–736.

Womelsdorf, T., Schoffelen, J.M., Oostenveld, R., Singer, W., Desimone, R., Engel, A.K., and Fries, P. 2007. Modulation of neuronal interactions through neuronal synchronization. *Science (New York, NY) 316*, 1609–1612.

Section II

Spike Timing

Coding, Decoding, and Sensation

8 Timing Information in Insect Mechanosensory Systems

Alexander G. Dimitrov and Zane N. Aldworth

CONTENTS

8.1 INTRODUCTION

What role does timing play in communicating information throughout nervous systems? This question is usually coupled with other questions about the biological coding of sensory information. Here, we discuss temporal dynamics as both a signal to be communicated, and a method to communicate information. When information about stimulus identity is explicitly decoupled from information about other stimulus parameters (e.g., timing), the analysis and our understanding of neural function improve. In this chapter, the core question we ask is whether detailed spike patterns are used to code sensory information. To formalize this question, we proceed as follows. First, we characterize the precision of individual spikes. Next, we determine the information that spike patterns carry about the stimulus, above and beyond the contribution of individual spikes. We do this by assessing the responses in two ways: by assuming that the spike pattern is nothing more than the collection of individual spikes that it contains, and by considering simple spike patterns as potentially unique symbols. As we will show, the two approaches differ in their predictions. That difference, which can be fairly large, is direct evidence that spike patterns carry information, and an indication about the type of information they carry.

The questions related to the biological implementation of temporal coding by individual neurons have generated a large amount of controversy over the last three decades [62,86,95]. While it has been shown that neurons can respond to temporally varying stimuli with high precision [4,10,21,65], it is not yet known to what (if any) extent this precision is used to represent aspects of the stimulus beyond its own temporal modulation. This has led to the suggestion that temporal coding, as defined in for example, Ref. [99], is not implemented in individual neurons, and rather that the temporal precision of neuronal responses to stimuli is used only to represent high-frequency components of the stimulus. This is considered a form of rate coding [8,11]. Linear stimulus reconstruction is a popular form of analysis that implicitly assumes the occurrence of such rate coding [7,87,100]. This technique has been successfully employed in several sensory systems to gain insight into how information about stimuli is contained in neural responses. However, comparisons of information rates obtained through linear reconstruction techniques with information rates obtained with consideration of temporal patterns of spikes have shown that the linear reconstruction technique consistently underestimates the information available in the response by up to 67%. These results have been demonstrated in different sensory modalities and in different phyla, including salamander retina [68,105], cat retina [78], cat thalamus [83,84,103], primate medical temporal area (MT) [11], fish electric lateral line (ELL) [15], fly visual cells [7,21], and cricket cercal interneurons [3]. While this suggests that temporal patterns of spikes are in fact an important component of the neural code, it remains an open question how these responses represent stimuli in a manner different than that predicted by linear reconstruction.

In the studies reported here, we examined the extent to which temporal encoding is implemented in two insect systems: in filiform hair afferents and interneurons of the cercal system of the house cricket, *Acheta domesticus,* and the campaniform sensilla on the wing surfaces and hinges of the hawkmoth, *Manduca sexta.* The cercal sensory system in the house cricket *A. domesticus* [59,60,90,96] mediates the detection and analysis of low-velocity air currents in the cricket's immediate environment. This sensory system is capable of detecting the direction and dynamic properties of air currents with great accuracy and precision, and can be thought of as a near-field, low-frequency extension of the animal's auditory system. The afferent neurons innervate the filiform sensilla synapse with a group of approximately 30 local interneurons and approximately 20 identified projecting interneurons that send their axons to motor centers in the thorax and integrative centers in the cricket's brain. Here, we discuss interneurons 10-2a and 10-3a, a subset of these projection interneurons. Similar to the afferents, these interneurons are also sensitive to the direction and dynamics of air current stimuli [70,96,100]. The campaniform sensilla on the wing surface and hinge of the hawkmoth *M. sexta* [28,37] is common to all winged insects and their associated neurons provide information about deformation and bending of wings to aid in flight control, among other behaviors [23,31,40,92,93,98]. Both the filiform sensilla of the cricket and the campaniform sensilla of the moth are formed through modifications to the cuticle, and are each innervated by a single afferent neuron [55]. Primary sensory afferent neurons in insects are generally very tightly locked to the stimulus [35,96], and information-theoretic measures have shown that such neurons are capable of representing stimuli with very high fidelity [8,19,34,90,104].

The previous studies in invertebrate sensory systems, including the cricket cercal system, indicated that linear coding schemes have difficulty describing the stimuli preceding short-interval, high-temporal frequency doublets [20,24,27,89]. Therefore, we narrow our investigation to study only short-interval doublets, analogous to the study of bursts in other sensory systems [36,66,81,83]. Here, we assume that any pair of spikes may potentially form a doublet; the subsequent analysis makes clear which pairs need to be analyzed beyond their constituent spikes. To clarify the role that timing and temporal precision play in sensory processing, we addressed several related questions:

- Are temporal patterns of spikes reliably elicited by specific stimuli?
- Does spike timing reliability lead to increased capacity to transmit information?
- Do temporal patterns represent novel stimulus features?
- What aspects of the stimulus are being encoded and communicated to later stages of the nervous system?
- Can apparent temporal encoding by precise spike sequences be explained by simple modification to models of spikes as independent information carriers?

The general approach we took was to determine if temporal spike patterns elicited in response to sensory stimuli contain more, or different, information about the stimulus waveform than would be predicted from a simple linear analysis of individual spikes. To that effect, we examined linear stimulus reconstruction, a form of analysis that implicitly assumes the implementation of a linear rate-coding scheme [7,87,99,100]. To obtain an estimate of the rate at which information about the stimulus is encoded in the neural response (the mutual information rate), the stimulus reconstruction method makes explicit assumptions about the aspects of the stimulus that are encoded in the neural response (the reconstruction filter) and how they are encoded by the neural response (by independent single spikes). By contrast, "direct" methodologies [21,76,97], which attempt to empirically estimate the stimulus-conditioned response probability, allow exact estimates of the mutual information transmission rates of neurons with few assumptions, but provide no estimates of the stimulus quantities encoded or the coding scheme implemented by the neurons. Consequently, calculations of mutual information using the direct method can include contributions due to temporal patterns of spikes, as well as the spike rate assumption from the stimulus reconstruction methodology. The comparisons of information rates calculated using the two methods show that linear methods routinely underestimate the true amount of information contained in neural activity [8]. An open question in neural coding is whether this discrepancy arises because neurons use temporal encoding to represent the stimulus space (a possibility explicitly rejected by linear reconstruction), or whether the information gap is caused by other nonlinearities [99].

Our first step was to determine if stimulus-elicited, short-interval spike doublets occurred with greater precision than would be expected, based on the observed statistics of single spikes. Specifically, we determined if the timing of spikes in short-interval doublets had a higher covariance than would be predicted from an analysis of the jitter in the stimulus–response timing of isolated spikes [3,6]. Next, we developed models to examine the extent to which such differences in temporal precision might affect the ability of neurons to transmit information about the sensory environment. With these models, we determined if the stimuli associated with temporal patterns of spikes were significantly different than what was predicted by linear reconstruction. For this analysis, we developed linear and nonlinear models for decoding spike doublets, and compared the capabilities of these two types of decoding schemes for representing the stimuli that elicited such patterns of spikes. We demonstrate that short-interval spike doublets convey information at higher rates than predicted by the assumptions of linear coding, and that the stimuli associated with such patterns are predicted better by second-order models than by linear models. This indicates that these neurons employ a temporal encoding scheme.

8.2 RESULTS

8.2.1 Variability in Stimulus–Response Latency

The biophysical processes underlying sensory transduction, synaptic integration, spike initiation, and synaptic transmission are not perfectly deterministic, and some significant degree of "jitter" in stimulus-to-spike latency is always observable after repeated presentation of identical stimuli [6,10,64,67,101,106]. A broader ensemble of nonrepeated stimuli allows us to capture not only biophysical uncertainty, but also latency variance caused by the fact that multiple stimuli are represented by the same response (response invariance) [3,16,17,25]. This is an important distinction in the context of comparing linear reconstruction techniques with other measures of information rates in neural systems, since both the biophysical and functional variability in each spike is implicitly included in the construction of the linear kernels.

We employ two methods to characterize these distinct (but related) aspects of this spike timing variability. The first method purely assesses biophysical uncertainty by estimating spike onset jitter in response to repeated presentation of white-noise waveform (frozen noise) [4,6,10,65]. By conditioning on repeated stimuli, the raster method attempts to measure response variability solely due to biophysical sources. Figure 8.1 demonstrates the effects of such variability. It shows one typical segment from an experiment in which a 10 s stimulus was repeated 100×. Figure 8.1a3 shows a 500 ms segment of this white-noise stimulus waveform, a1 shows the corresponding raster plot of the spike trains elicited by the 100 repeated presentations of that waveform segment. Figure 8.1a2 is the histogram of these spike occurrence times

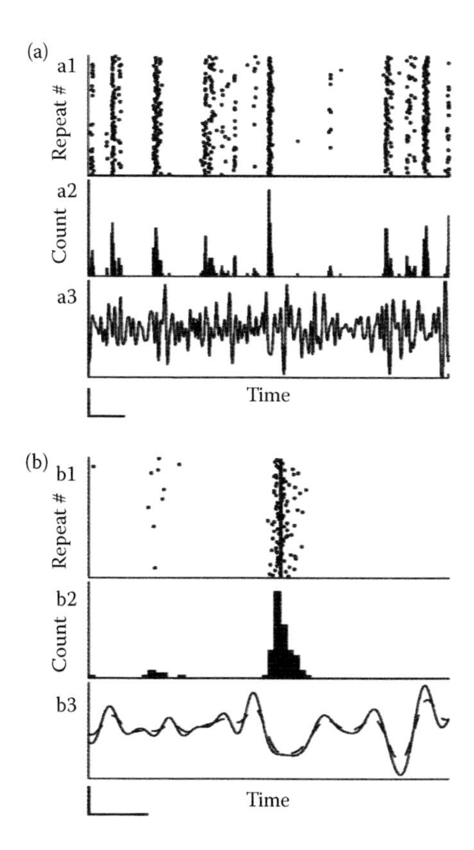

FIGURE 8.1 An experiment to illustrate the jitter problem, using a mechanosensory interneuron. a1, Raster of the spike times elicited by 100 repetitions of this waveform. a2, Histogram of this raster plot using 1 ms time bins. a3, A 0.5 s segment of the Gaussian noise stimulus waveform. (b) The central 60 ms segment of the data shown in (a) at an expanded timescale. b1, Spike-time raster plot after realignment of the spikes in the distribution shown in (b) (solid line). b2, Histogram as in a2. b3, Solid line, The actual stimulus waveform presented to the system. Dashed line: Mean of all stimulus segments realigned to the evoked spikes (i.e., the STA).

(peristimulus time histogram, PSTH). This segment of data demonstrates the wide range of stimulus-to-spike latency distributions that are typically elicited by white-noise stimuli. The three panels in Figure 8.1b show the central 60 ms segment of the data shown in Figure 8.1a at an expanded timescale, which captures the sharpest distribution of spike times from that segment.

Figure 8.1b3 illustrates the result of calculating the spike-triggered average (STA) of the stimulus waveform segments from the subset of data shown in Figure 8.1b. The solid line in Figure 8.1b1 is the raster plot of spikes after their realignment into precise registration (spike triggering). The histogram of these spikes would appear as a single solid line in b2 (i.e., all spikes fall into a single bin). The dashed trace in b3 is the average of all 100 stimulus segments after their realignment (the STA for these spikes). For comparison, the solid line in that bottom panel is the actual stimulus waveform that had been presented repeatedly to the animal, and was therefore the real mean waveform leading up to any spikes. It is clear that the STA yields a distorted estimate of the true waveform. The causes for this effect are also obvious—the STA technique implicitly assumes that the spike times are precise within the resolution selected by the researcher, whereas in reality, the biological system exhibits some intrinsic temporal variability. The latter is explicitly captured by the method of dejittering discussed next.

The second method is the dejittering technique that assesses temporal uncertainty with nonrepeated white-noise stimuli [3,16,17,25,43]. The effect that this approach is addressing can be illustrated in Figure 8.1a as well. In addition to the biophysical jitter revealed by repeated stimuli, we can also talk about functional jitter, in which the same response (e.g., the same single spike appearing near 8150 and 8300 ms) is associated with different stimuli. Hence, the noise of that otherwise unique response has additional components: different variance (larger near 8300 ms), as well as different latency with respect to any feature that may be driving the response.

To characterize these noise sources as well, we dejitter the responses [3,25]. To achieve that, we explicitly model the combined temporal jitter of neural responses. We model an observed waveform $z(\tau)$ as a waveform x, drawn from a distribution of stimuli, and then shifted at random relative to the response time, that is, $z(\tau) = x(\tau - t)$. Assuming that stimulus and time shift are drawn independently, the distribution of an observed waveform z is then

$$p(z) \equiv p(x)p(t) = \mathcal{N}(x;\bar{x},C_x)\mathcal{N}(t;0,\sigma_t^2). \tag{8.1}$$

Here, x is the stimulus drawn from a multivariate normal distribution $\mathcal{N}$ with mean $\bar{x}$ and covariance C_x, and t is the time shift drawn from a univariate normal distribution with mean 0 and variance σ_t^2. For each observed response-triggered waveform, we infer the optimal pair of shift and stimulus, (t^*, $x^* = z(\tau + t^*)$) as the solution to

$$t^* = \arg\max_t \mathcal{N}(z(\tau + t);\bar{x},C_x)\mathcal{N}(t;0,\sigma_t^2). \tag{8.2}$$

Note that here, we are assuming (and enforcing) the mean of the t distribution to be $\bar{t} = 0$, around the time of occurrence of a neural response. For computational purposes, it is more efficient to write Equation 8.2 in terms of the negative log likelihood of the transformed observation. This monotonic transformation does not change the position of any extremum, but dramatically increases the numerical precision. The nonconstant portion of the log likelihood is a quadratic form of the variables, and hence a distance

$$d((z,t),(\bar{x},0)) = (z(\tau + t) - \bar{x})C_x^{-1}(z(\tau + t) - \bar{x})^T + t^2 / \sigma_t^2. \tag{8.3}$$

A minimal distance here implies maximal likelihood in Equation 8.2. To dejitter a dataset, we initialize $p(x)$ with the estimates of the raw mean and covariance, $\bar{z}$, C_z, and $p(t)$ with a physiologically relevant σ_t. We iterate the procedure, re-estimating the parameters $\bar{x}$, C_x, and σ_t after each iteration until a predetermined level of convergence is reached.

The results of a complete analysis of a stimulus–response dataset from cricket cercal interneuron IN10–3 are shown in Figure 8.2. All instances of isolated single spikes (preceded and followed by spike-free intervals of ≈30 ms) were extracted from a 33 min recording, yielding a sample size of ≈13,600 stimulus segments. Figure 8.2a shows raster plots of a random representative subset of 100 of the sample segments of stimulus–response data. Each horizontal raster line is color coded to indicate the stimulus velocity versus time, with colors toward red indicating positive velocity and colors toward blue representing negative velocity. The black dot at time 0 indicates the time of occurrence of the spike elicited by the waveform within that particular sample. All the stimulus waveforms in Figure 8.2a are locked in their alignment to the spike occurrence times. Figure 8.2b shows the result of dejittering these waveforms. The redistribution of the stimulus–response raster lines resulting from this dejittering procedure results in a visibly sharpened registration of the stimulus waveforms across the raster set and a visibly scattered deregistration of the spike occurrence times. The distribution of shift times to achieve the optimal dejittered mean had a σ_t value of 2.2 ms (Figure 8.2d). Figure 8.2c compares the uncorrected spike-conditioned mean (i.e., the STA; red trace) with the dejittered mean (blue trace) across all 13,600 samples in the dataset. The dejittered mean stimulus had a peak amplitude 2.6× higher than that of the STA. The dejittered mean also had a greater relative proportion of its power extending into the higher-frequency range than did the STA, as can be seen from a comparison of the temporal waveforms and power spectral densities of the STA and dejittered mean waveforms (Figure 8.2e,f).

Figure 8.3 shows the comparison of both analyses in the systems of interest. The values along the x axis indicate the standard deviation (SD) of the variability in stimulus–spike latency assessed using the dejittering method (mean across the population denoted by the vertical dotted line), whereas the value on the y axis indicates the SD of the variability in stimulus–spike latency assessed using the raster method (mean across the population is shown with the horizontal dotted line).

The left panel shows the results of the dejittering- and raster-based estimates of jitter for both filiform and campaniform sensilla. Here, both estimates of jitter uncertainty are much smaller than for the interneurons. The mean raster-based estimate of jitter for the six filiform-afferent neurons was 0.3 ms, and 0.4 ms for all five of the campaniform neurons, compared with 1.3 ms discussed earlier for the cercal interneurons. The mean dejitter-based estimate of the uncertainty for the six filiform-afferent neurons was 0.7 ms, and 0.9 ms for the five campaniform neurons, compared with 2.1 ms for the cercal interneurons. In contrast to the case with the interneurons, the two uncertainty estimates show significant correlation across the population of 11 afferent neurons ($R = 0.76$, 95% CI = [0.29, 0.93]), though as in the case with the interneurons, the estimate from dejittering is always larger than estimates from rasters. In the right panel, we plot the results for cricket cercal interneurons. Here, again the dejittering method consistently gives a larger value for the variability (mean jitter value of 2.1 ms, compared with 1.3 ms for the raster method). However, in this case, there is no strong correlation between the two measures ($R = 0.06$ across the 40 cells, 95% CI = [–0.26, 0.36]). We discuss the functional significance of this observation in Section 8.2.4.

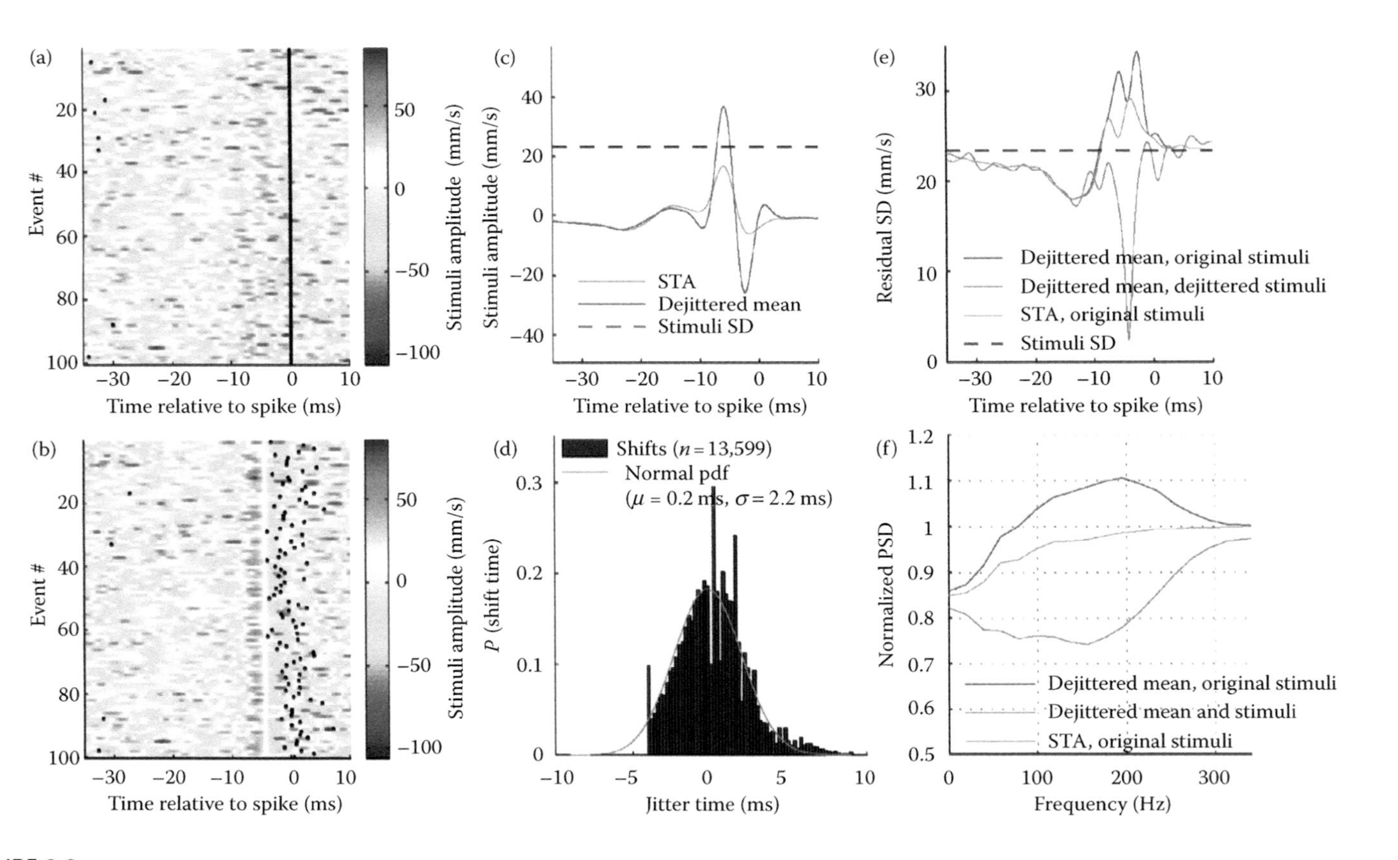
(a)
Event #
Time relative to spike (ms)
Stimuli amplitude (mm/s)
(b)
Event #
Time relative to spike (ms)
Stimuli amplitude (mm/s)
(c)
Stimuli amplitude (mm/s)
STA
Dejittered mean
Stimuli SD
Time relative to spike (ms)
(d)
P (shift time)
Shifts (n = 13,599)
Normal pdf
(μ = 0.2 ms, σ = 2.2 ms)
Jitter time (ms)
(e)
Residual SD (mm/s)
Dejittered mean, original stimuli
Dejittered mean, dejittered stimuli
STA, original stimuli
Stimuli SD
Time relative to spike (ms)
(f)
Normalized PSD
Dejittered mean, original stimuli
Dejittered mean and stimuli
STA, original stimuli
Frequency (Hz)

FIGURE 8.2

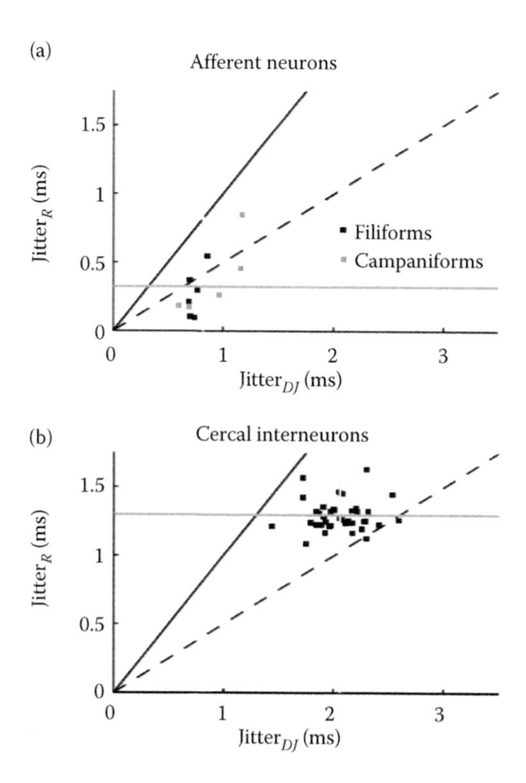

FIGURE 8.3 Temporal precision of isolated single spikes for afferents and cercal interneurons. The value along the abscissa shows single-spike precision assessed by the dejittering algorithm (population mean shown as a vertical dotted line). The value along the ordinate shows single-spike precision assessed by a raster-based analysis for the same cells (population mean shown as a horizontal dotted line). Each cell is represented by a single point. The solid black line denotes where the two methods should give equal results, whereas the dashed black line shows a factor of 2×, for reference.

FIGURE 8.2 **(See color insert.)** Characteristics of a dejittered mean stimulus. (a) Random representative subset of 100 sample segments of stimulus–response data aligned to the time of spike occurrence. The vertical black line at $t \approx 0$ is the raster plot of spikes superimposed on the color-coded traces of air-current velocity versus time. (b) The same random subset of 100 segments of stimulus–response data from (a) but now dejittered so that their alignment is based on minimal variance of the stimuli. (c) STA (red trace) and dejittered mean (blue trace). The standard deviation (STD) of the stimulus is indicated with the dashed black line. (d) Histogram of shift times (blue) compared with a Gaussian having an STD of 2.2 ms (red line). (e) STD of the residual under different dejittering conditions. The fully dejittered mean (green), which we assume is how the system interpreting the activity has the smallest residual. (f) Power spectra of the residuals, normalized by power spectra of stimulus segments. The colors and abbreviations are the same as in (e).

8.2.2 Measurement of Pattern Variability

The variability in spike latency of a single spike plays an important role in determining how much information can be encoded in a neuron's activity. However, it is not yet completely clear whether all spikes experience equal variability regardless of prior activity, or whether the immediate spiking history within a cell can affect the variability of subsequent spikes. To address this question, we measured the variability of doublet spiking in our population of cells to repeated presentations of a white-noise stimulus. If variability of spike latency was truly independent of spiking history, we would expect the average variability of spike timing to be approximately 1.3 ms, as in the case for isolated single spikes (see Figure 8.3). In addition, we would expect that the variability of interspike intervals (ISIs) would be even larger, since in that case, an ISI would be the sum (more properly, the difference) of two independent random variables. In this case, the variance of an ISI would be equal to the sum of the variances of the component spikes' jitter.

Doublet events that were consistently elicited by repeated presentations of the stimulus were identified with a modified version of the event-identification protocol of Berry and colleagues [6]. To avoid confounding results due to changing levels of adaptation, we excluded initial repetitions of the stimulus where the average firing rate was greater than 120% of the average firing rate across all trials. The adapted responses to repeated trials of the stimulus were then binned into histograms at 1 ms resolution and thresheld to define firing boundaries of events. The doublets were extracted from the collections of all events, taking care that not more than 20% of the trials contained contaminating spikes. Varying this exclusion threshold between 10% and 90% of the trials did not greatly affect the results of the correlation analysis. For each doublet event, the timing of the first and second spikes of the doublet on each trial was extracted and pooled across all events and all cells by ISI. The jitter (SD of the first spike time across trials) and the correlation coefficients between the first and second spike in the doublet were then calculated. Simple exponential models of the form

$$jitt(\text{ISI}) = ae^{-\text{ISI}/\tau} + b \tag{8.4}$$

and

$$R(\text{ISI}) = a_1 e^{-(\text{ISI}/\tau_1)} + a_2 e^{-(\text{ISI}/\tau_2)} \tag{8.5}$$

were fit to the jitter and correlation data, respectively, where ISI represents the mean interspike interval and a, b, and τ represent the parameters fit in the optimization. For all exponential equations, the number of parameters used to fit the data was determined by selecting the model with the lowest value of the *Akaike Information Criterion* (AIC) [12]. Fits for the coefficients, 95% confidence intervals on the coefficients, and 95% confidence intervals on predictions from the models were obtained with least-squares fitting using the routines *nlinfit*, *nlparci*, and *nlpredci* from the MATLAB® Statistics toolbox.

Figure 8.4 summarizes the results of the analysis for 40 neurons. Figure 8.4a shows 25 of the responses from a single IN 10-2a to 85 presentations of a stimulus that on the average elicited a doublet of 2.6 ms (same cell as in Figure 8.5a and b). The upper and lower plots show the raster and PSTH of the spiking activity, respectively. The temporal precision of the first and second spikes, as measured by the SD of the distributions were 0.3 and 0.5 ms, respectively. Figure 8.4b shows spiking

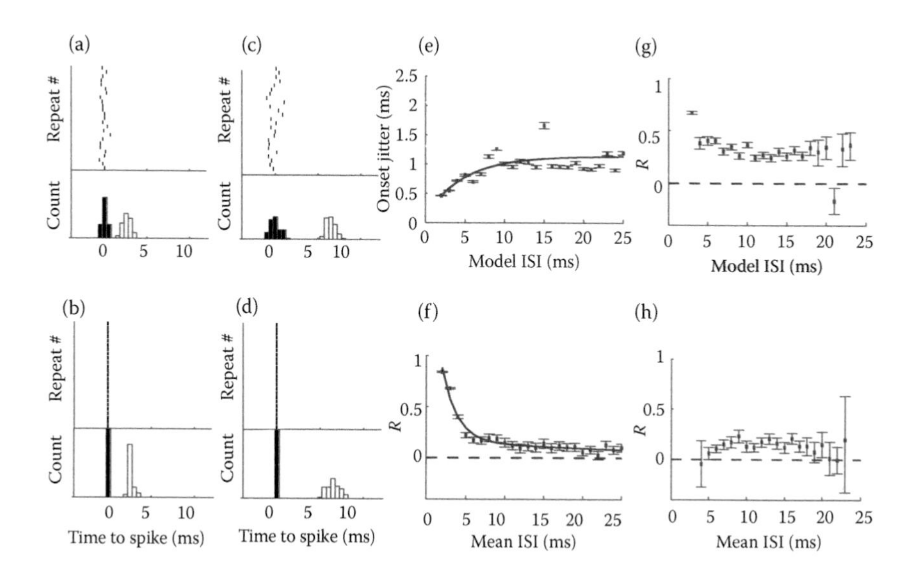

FIGURE 8.4 Spike–spike interactions in doublet patterns recorded in cricket interneurons. (a) Upper trace: A raster plot showing 25 of 85 responses to repeated presentations of a white noise stimulus, recording from the same cell as shown in Figure 8.5. The cell consistently responded to the stimulus by firing a doublet (first spike shown in dark, second spike in gray) with average ISI of 2.6 ms. (a) Lower trace: PSTH of all 85 responses from the raster, with the shading convention conserved. (b) Upper and lower traces: Raster plot and PSTH showing the same data from (a) aligned here relative to the time of the first spike in the doublet ($t = 0$) rather than to the timing of the stimulus. This shows the variability in ISI across presentations of a single stimulus. (c) and (d) Data from a second doublet event (mean ISI = 6.5 ms) from the same interneuron, data presentation is conserved. (e) Jitter of arrival time of first spike in repeatable doublets recorded from 40 different interneurons in 32 animals, as a function of ISI. The black curve shows the model fit to data (Equation 8.4), with shaded area representing 95% confidence envelope around predictions from the model. The top horizontal line shows population mean of single-spike jitter from frozen-noise method. (f) Estimate of correlation coefficient between the first and second spikes in repeatable doublets (from the same data set as in e). The error bars represent 95% confidence limits on estimation of correlation coefficient. The solid black line shows correlation coefficient as a function of ISI modeled as a double exponential (Equation 8.5), with ±95% confidence interval on predictions from the model shown by the shaded gray region. (g) Pooled population results of correlation for six filiform afferents from *A. domesticus.* (h) Pooled population results of correlation from five campaniform afferents from *M. sexta.* As in the single-cell results in (f), we see a significant correlation in the timing of nearby spikes, although the population correlation is essentially constant over the observed time interval.

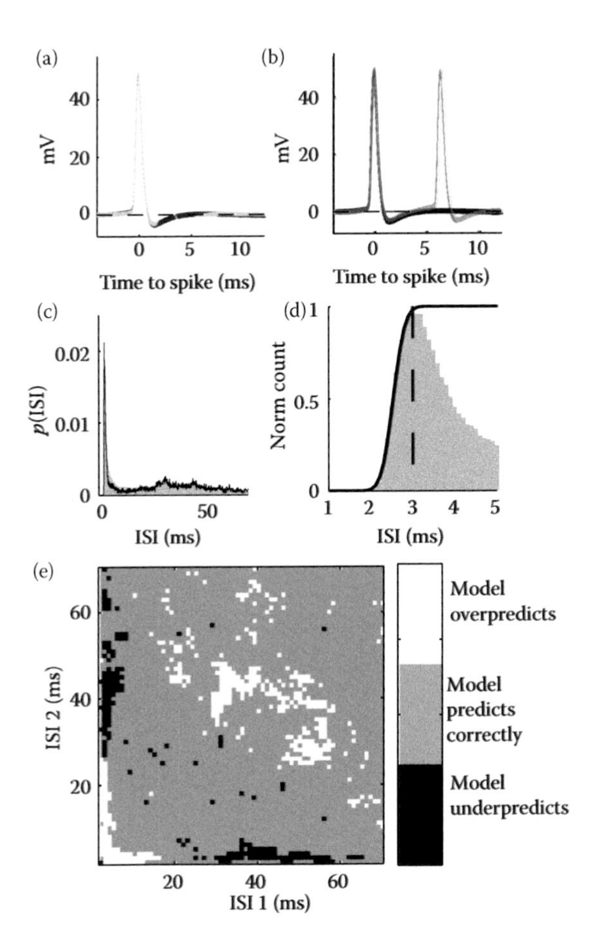

FIGURE 8.5 Statistics of doublet spiking. (a) ±1 SD envelope showing intracellular voltage waveform relative to the resting membrane potential of isolated single spikes (dark) and isolated short doublets of ISI 2.6 ms (gray) from a single recording in the interneuron of class 10-2a. The dashed black line denotes mean-resting membrane potential (0 mV). (b) ±1 SD of intracellular waveform from the same recording as in (a), this time with a doublet of ISI 6.5 ms (dark, $n = 26$). (c) ISI histogram of data from recordings in (a) and (b) at 0.1 ms resolution (black line), as well as compilation data from 40 cells of classes 10-2a and 10-3a (gray-shaded area). (d) Normalized ISI histogram of population data from panel (c), with timescale reduced to 1–5 ms. The dark line shows the recovery function, with the black dashed line showing limits of fit to recovery function. (e) Difference between independent model and measurements from data of joint probability of consecutive ISIs. The positive (white) values represent overestimation by the independent model, whereas negative (dark) values represent underpredictions by the independent model.

from the same event, but now conditioned on the first spike of the event rather than the time of the stimulus. The precision of the ISI, as measured by the SD of the difference between the second and first spike times was <0.3 ms, with a correlation coefficient R of 0.8 between the timing of the first and second spikes. Note that the overall ISI response to the stimulus was more precise than the onset latency of either of the individual spikes times. In this case, the *a priori* assumption that temporal precision of response is independent of recent spike history can clearly be rejected.

Figure 8.4c and d shows raster data and a PSTH for a second event from the same recording as in Figure 8.3a and b. The mean ISI of this second event was 6.5 ms compared to 2.6 ms in the previous case, whereas the precision of both spikes within the doublets were similar to the previous case (0.6 and 0.5 ms for the first and second spikes of the doublet, respectively). Here, however, the distribution of the ISI is slightly larger relative to the two spikes that compose it (precision = 0.7 ms, $R = 0.23$), although still slightly smaller than expected if the two spikes were independent (0.8 ms, found by taking the square root of the sum, of the squared SDs for each spike).

From the data presented in Figure 8.4a–d, we see that there is clearly a correlation between the previous spike history and stimulus–response precision, at least for these two sample-firing events in a single cell. We also see that there seems to be a decrease in this correlation with increasing time since the last spike. To increase the statistical power of our examination of the temporal precision of ISI events, we pooled the data from 7753 doublet-firing events occurring in recordings from 40 interneurons. We first use this larger data set to see if there is a systematic variation in the onset precision of a pattern of spikes dependent, on the subsequent ISI. The results of this analysis are shown in Figure 8.4e. Here, we see that very short doublets are tightly locked to the timing of the stimulus, with an SD across trials (jitter) of less than 0.5 ms for ISIs of 2 and 3 ms. Longer-duration ISIs have relatively larger values of jitter, reaching a plateau of >1.1 ms for ISIs of 25 ms or more. The onset jitter as a function of the following ISI was modeled using a simple exponential (Equation 8.4) with best-fit coefficients and 95% confidence intervals: $\tau_1 = -1.0 \pm 1.0$ ms, $a = 4.8 \pm 5.2$ ms, and $b = 1.1 \pm 0.1$ ms. The asymptotic value of the onset jitter (b) was similar to the mean stimulus–response jitter of single spikes measured during repeated presentations of frozen-noise stimuli (1.3 ms, Figure 8.3b). The resulting model is shown in Figure 8.4e as the solid black line, with ±95% confidence intervals of the fit shown with the shaded gray regions.

In Figure 8.4f the same pooled data is used to calculate the correlation between the first and second spikes in the doublets as a function of the average ISI of the doublets. What we see in the pooled data confirms what we saw in our earlier example from the single cell. ISIs had correlations significantly different from 0 to approximately 35 ms, and spikes in doublets with short ISIs (<5 ms) have correlations of 0.3 or higher. This means that stimulus events that, on average, elicit short-doublet ISIs almost always produced the same response pattern, whereas stimuli that on average produced ISIs of 10 ms or longer produced sets of doublets with more variable ISIs, as well as the more variable onset demonstrated in Figure 8.4e. The change in correlation coefficient as a function of ISI was modeled as a double exponential using Equation 8.5 with the following best-fit parameters and 95% confidence intervals: $a_1 = 2.3 \pm 0.8$, $\tau_1 = 1.7 \pm 0.5$ ms, $a_2 = 0.2 \pm 0.1$, and $\tau_2 = 28.9 \pm 10.7$ ms.

We examined the covariance between the timing of the first and second spikes in a doublet across repeated presentations of a stimulus as a function of the mean ISI between the two spikes in other systems as well. Figure 8.4h shows the results for the pooled population of six filiform-afferent neurons, part of the cercal transduction mechanism. As in the results for the postsynaptic interneurons, we see that there is a significant correlation in the timing of nearby spikes, leading to greater consistency in the ISI interval than would be expected if the timing of consecutive spikes were independent of each other. However, unlike the interneuron case, here, we did not see the extremely high (>0.8) correlation for the very short intervals, or the clear exponential decay in the ISI correlation as the mean ISI got progressively larger. Note, due to the high firing rate of these cells (sustained firing rates of 84.8 ± 12.8 spikes/s, mean ± SD), we did not observe enough samples to calculate the correlation for ISIs greater than 23 ms. Figure 8.4g shows the results of the same analysis for the pooled data from the five campaniform afferents. Here, the difference with the cercal interneurons is even more striking, as there is very little correlation in the timing of nearby spikes for any of the observed ISI values.

8.2.3 Statistics of Doublet Activity

Our working hypothesis is that sensory systems can use short-interval spike doublets to represent stimulus waveforms that are significantly different from the sum of two (STA). The latter, our null hypothesis, is know as the linear superposition hypothesis [87].

In Figure 8.5, we show the statistics associated with temporal patterns of spikes recorded under these conditions. Figure 8.5a shows the mean ± 1 SD of the membrane potential during single-spike firing events (dark, n = 10,701 events) as well as during short doublets of ISI = 2.6 ms (gray, n = 464 events) from a single recording of giant interneuron 10-2a. We see that for these short-doublet events, the second spike occurs while the membrane is still hyperpolarized from the first spike. In contrast, Figure 8.5b shows the single-spike events superimposed with a doublet event with ISI = 6.5 ms (gray, n = 26 events) from the same recording. In this case, we see that the voltage across the cell membrane has returned to the resting membrane potential (denoted with the broken black line) before the second spike occurs.

Figure 8.5c shows the probability of occurrence of all ISIs of less than 70 ms (the ISI histogram, binned at 0.1 ms resolution) from the same recording as in Figure 8.5a and b (black line). In addition, the combined ISI histogram from 40 different cells of class 10-2a and class 10-3a, recorded under the same stimulus conditions, is shown with the gray shade. In the case of the data from the single cell (black line), >85% of all ISIs were of 70 ms or less, whereas in the data pooled across all cells (gray shade), >90% of the ISIs occurred in this interval. The histogram from the single cell is well within the range of the population data. The ISI histogram contains three clear peaks; one at 44 ms, one at 31 ms, and the tallest peak at 3 ms, which lies just at the edge of the observed hard refractory period for this cell (2 ms). Note that the peaks at 44 and 31 ms correspond to firing rates of 23 and 32 Hz, respectively, which in turn corresponds to the region of peak stimulus–response coherence from analyses associated with stimulus reconstruction [18,49,63,100]. This means that from the perspective of linear rate encoding implicit in stimulus reconstruction, spikes with ISIs in the range of 31–44 ms would carry the most information about the stimulus.

Figure 8.5d shows an expanded view of the ISIs from 2 to 5 ms in the population histogram, with the y axis normalized to 1 at the most often-occurring ISI (3 ms).

At this time base, it becomes clear that the ISIs from the minimum observed (2 ms) to the modal value (3 ms) follow a sigmoidal curve. Berry and Meister [5] showed that the relative refractory period of a neuron can be well described by modeling this sigmoidal curve as a cumulative density function of the ISI probability in this range. In this spirit, we fit our data with a normal cumulative density function (CDF mean = 2.5 ms, SD = 0.2 ms) for modeling—see Figures 8.3b and 8.7.

To determine whether or not correlations between spikes could be explained simply by doublet spike patterns, we looked at patterns of two consecutive ISIs (i.e., triplet patterns). If each doublet event was independent of the preceding and following spiking activity, then the joint probability $p(ISI_1 = x, ISI_2 = y)$ could be determined by taking the product of the two marginal probabilities, $p(ISI_1 = x)p(ISI_2 = y)$, which we label as the independent joint distribution, $p_{ind}(x, y)$. We tested this hypothesis for our pooled ISI data by comparing $p(x, y)$ to $p_{ind}(x, y)$. The regions where the two probability distributions are not significantly different from each other indicate where consecutive ISIs are independent of each other. Figure 8.5e shows regions where the two models are different at the 95% significance level (after Bonferroni correction for multiple comparisons [30]). The independent model overpredicts the probability in two separate regions lying along the diagonal: the first for consecutive ISIs of approximately 5 ms or less, the second for consecutive ISIs of approximately 30–50 ms (white regions). These correspond with the peak regions from the ISI histogram in Figure 8.5c. The independent model simultaneously underpredicts the probability of a short ISI being either preceded or followed by a silent period of 30–40 ms (black regions). We note that the relatively enhanced probability of a long silent period preceding short-ISI doublet events could be explained by the presence of a slow voltage-dependent conductance [58]. Voltage-dependent Ca conductances are known to exist in these cells [57,74].

8.2.4 Contribution of Pattern Variability to Information Transmission

To rigorously determine the effects the observed precision in ISIs had on a cell's ability to transmit information about the stimuli, we measured information-theoretic quantities such as entropy and mutual information in our data and models. In direct method calculations, the mutual information rate is calculated as the difference between two entropies: the total response entropy and the entropy of the response conditioned on a stimulus. The total response entropy determines how much bandwidth a cell has available for representing stimuli, whereas the conditional entropy reflects how much of that response pattern bandwidth is used to represent the same stimulus. In this context, a cell could use ISIs with relatively small conditional entropy to transmit more information about a stimulus than ISIs that exhibit relatively large conditional entropies.

8.2.4.1 Modeling of ISI Timing Precision

Three models of ISI variability were constructed to elucidate the mechanisms of ISI precision seen in real cells (Figure 8.5). For the first two models, the onset jitter (defined as the SD in the first spike timing across repeated presentations of identical stimulus waveforms) was fixed at 1.3 ms, which is the observed across-trial jitter of isolated single spikes in recordings from 40 recorded cells (Figure 8.3b). In model 1, the timing of the two spikes in each trial were drawn from two independent normal distributions with SD of 1.3 ms (independent ISI model, iISI). In model 2, spike times were drawn independently from normal distributions as in the iISI model; however, the second spikes that occurred within a refractory period were moved by a Gaussian random variable with SD determined from the recovery function fit to the ISI histogram (Figure 8.5d) [5] (refractory ISI model, rISI). This approximated the presence of a refractory period. For the third model, both onset jitter of the doublet and variability within the doublet were matched to values observed from the data (Equations 8.4 and 8.5) (data-matched ISI model, dISI). Correlations between the spikes were imposed by multiplying the time of the first spike (mean time of occurrence = 0) by R and adding a sample from a normally distributed random variable with variance

$$\sigma^2 = \sigma_0^2(1 - R^2) \tag{8.6}$$

where σ_0 is the SD of the first spike. This preserved the unconditioned variance of the second spike time, but constrained the variance of the ISI between the first and second spike (the covariance) to be less than the sum of the variances of the two separate distributions.

The results of the simulations are shown in Figure 8.6. Figure 8.6a–f uses the same plotting convention as Figure 8.4a–d, with each of the three models being displayed in its own vertical column. Although all three models produced variable spike timing raster plots and PSTHs (Figure 8.6a,c, and e, upper and lower plots, respectively), the distributions of both spikes relative to the time of the first spike (effectively the ISI—Figure 8.6b,d, and f) are distinct between the different models, with correspondingly variable amounts of correlation between the first and second spike times. Figure 8.6g shows how the correlation coefficient between spike times evolves for each model for average ISIs varied between 2 and 65 ms. Note that the correlation for the third model explicitly matches the correlation coefficients calculated from the data in Figure 8.4f. The correlations found both in actual data (Figure 8.4f) and in the dISI model significantly exceed those for independent and refractory models for ISIs less than 30 ms. The second and third models represent the precision of doublet spiking according to biophysically plausible mechanisms, whereas the first model shows doublet spiking as predicted by strict interpretation of the assumptions of linear reconstruction analysis, that is, independence between spikes. Although the first model has first and second spikes that nominally occur independently of each other, small amounts of correlation are induced by the fact that the earliest spike was always attributed to the first spike distribution, even if it was actually generated from the second spike distribution.

Figure 8.6h shows the contribution to the conditional entropy due to variability in both ISI and timing of pattern onset for each of the three models discussed above, all as a function of ISI. The conditional entropy curve for the dISI model is lower than the curves for the other two models over the entire range tested here, and substantially so for short ISIs. Since dISI matches data from real cells, whereas models 1–2 represent decreasingly strict interpretation of linear reconstruction, this indicates

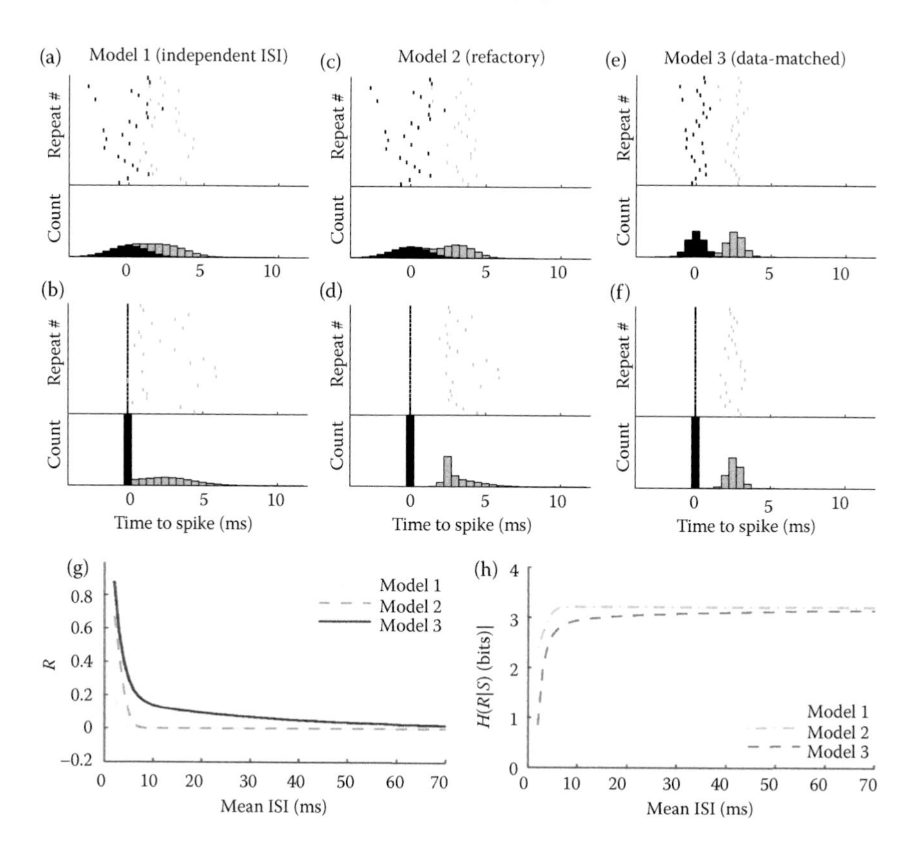

FIGURE 8.6 Three models of spike–spike interactions in doublet patterns. The plotting conventions are conserved as in Figure 8.4. (a) Upper trace: Raster plot of response from independent ISI (iISI) cell model to repeated presentations of a stimulus that reliably elicits a doublet with mean ISI of 2.6 ms. Both the first (dark) and second (gray) spikes in the doublet are drawn independently from normal distributions with means of 0 and 2.6 ms, respectively, and SD of 1.3 ms. (a) Lower trace: Standard PSTH of raster from upper trace. (b) Upper and lower traces: Raster plot and PSTH showing the same data from (a) with each row aligned to the time of occurrence of the first spike in the response. (c and d) Refractory ISI (rISI) model of doublet behavior enforcing a relative refractory period between nearby spikes, using recovery function from Figure 8.5c and jitter SD of 1.3 ms. (e and f) Data-matched (dISI) model of doublet behavior, where the relative timing of spikes is determined by Equations 8.4 and 8.5. (g) Correlation coefficient between the timing of first and second spikes of doublets drawn from the three models as a function of ISI. Note that the correlation of Model 3 matches the exponential model from Figure 8.4f by design. (h) Conditional entropy (Equation A.1) of response pattern as a function of mean ISI for all three models.

that the assumptions of linear reconstruction overestimate the conditional variability of spike patterns. In the information-theoretic framework shown here, this means that a given doublet pattern is capable of transmitting more information about the stimulus than predicted from linear reconstruction assumptions. Specifically, if a cell on average gives a 4 ms doublet response to repeated presentations of a stimulus, the iISI model predicts that the conditional entropy of the response would be 3.09

bits, whereas the dISI model predicts that it would only be 2.43 bits. This means that from this specific response event, the relative reduction in the stimulus discrimination ability of dISI model due to noise entropy would be $2^{(3.09-2.43)}$ or approximately two-thirds as large as for the iISI model.

To determine how much more information could be transmitted overall in neurons using the ISI-correlated precision seen in our cells, we calculated mutual information rates on each of our models using the "direct" methodology [21]. To do this, we first calculated the total response entropy using only doublet patterns (i.e., the ISI histogram) for each of our cells. We compared these values to the actual response entropy for each cell, estimated using the context-tree-weighting algorithm (CTW). The results are shown in Figure 8.7a. Here, each point represents data from a single cell, with the x axis indicating the estimation of the total response entropy from that cell using the CTW method, and the y axis indicating the model estimate of the response entropy described above. In the plots throughout Figure 8.7, the various models fit to the exemplar cell shown in Figure 8.4a–d appear in the middle of the point clouds. The points lie along the diagonal, indicating that reducing the response dimensionality to consider ISIs independently does not significantly reduce the calculated value of the response entropy, in spite of the fact that there are correlations in neighboring ISIs (e.g., Figure 8.5e).

We fit the free parameters for each of our three models in Figure 8.6 (jitter values, recovery function, and ISI correlation) to each of our 40 interneurons. The resulting conditional entropy rates for each of the three models and for each of the 40 cells are shown in Figure 8.7b. Here, the x axis shows the conditional entropy rate estimated using the CTW method [97]. The points that lie along the diagonal match the stimulus-conditioned variability seen in real neurons, whereas points above or below the diagonal represent overpredictions and underpredictions of conditional variability, respectively. Since dISI matches the temporal precision parameters from the real data, we expect that it should also be predictive of the conditional entropy of the real cells. We note that this could potentially provide a simple way of estimating information-theoretic quantities from relatively few parameters. Figure 8.7b shows that this is indeed the case—the dISI model tends to match the actual conditional entropy calculated in the cell most closely, with results from the other models tending to lie above the diagonal. This means that a strict interpretation of the assumptions of linear reconstruction (iISI) overpredicts the amount of conditional entropy present in the neural activity, and refractory dynamics (rISI) are not sufficient to describe the low variability seen in these neurons.

The information rate for each model was calculated by taking the difference between the total entropy rate and the conditional entropy rate. These values are plotted in Figure 8.7c versus the amount of mutual information calculated using the CTW method. As in the case of the conditional entropy, the dISI model tended to give the closest match to data, with the iISI and rISI models yielding progressively lower estimates of information rate due to their larger relative-conditional entropies.

We were concerned with determining how the precise spiking patterns seen in our data affect the ability of these cells to transmit information, and specifically how the assumptions of linear reconstruction might lead to reduced estimates of information rates. Since our models reflect varying degrees of the assumptions implicit in linear

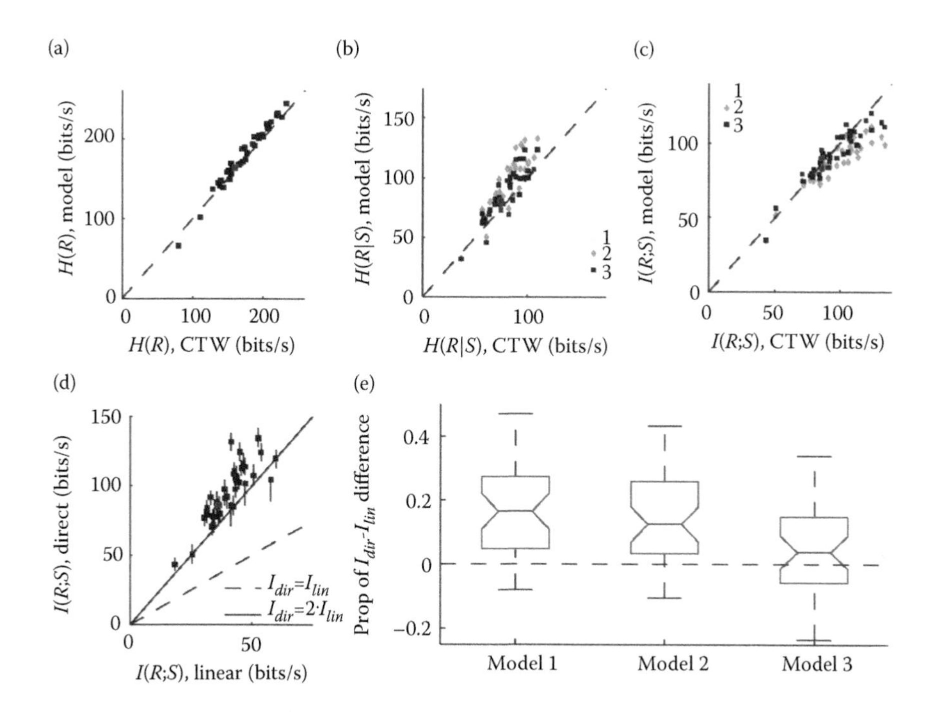

FIGURE 8.7 Comparison of information-theoretic quantities. (a) Total response entropy rate for 40 neurons as measured using the context-tree-weighting (CTW) technique (x axis) versus the modeled total response entropy (y axis). In panels (a) through (d), the gray points indicate values from the cell in Figure 8.4, dashed black lines indicate unity between the x and y axes. (b) Response entropy rate conditioned on a stimulus event as measured by CTW methods (x axis) versus models of the conditional entropy. (c) Mutual information about the stimulus contained in the response patterns, calculated as the difference between total and conditional entropies of the response. The x axis shows the result of CTW estimation for each cell, y axis shows information calculation based on each of the three models. (d) Comparison of mutual information measure using linear stimulus reconstruction approach (x axis) with estimation from CTW method. The solid black line indicates $I_{dir} = 2 \cdot I_{lin}$. (e) Boxplot showing how much of the proportional difference of information between methodologies ($I_{dir} - I_{lin}$) can be explained by varying temporal assumptions built in our models. For each of the three models, the boxplot shows the fraction of the information explained by the difference between that model and the direct method estimate from panel (d), that is, $prop(x) = (I_{dir} - I_{modx})/(I_{dir} - I_{lin})$.

reconstruction methods, we compared the modeled information rates with the rates obtained using linear reconstruction for each cell in our sample. Figure 8.7d shows a comparison between linear reconstruction information rates and rates obtained using the CTW direct method. In almost all cases, the linear method misses more than half of the information available in the spike train. To compare this with the models, we assessed what proportion of the difference between the linear and direct method calculations could be explained by the difference between information from the direct method and our models. The results of this comparison are presented in the boxplot of Figure 8.7e, which shows the lower and upper quartiles and the median value for the proportion of information difference explained across cells. The median values

TABLE 8.1
Slope and *Y*-Intercept Coefficients and Their Respective 95% CIs from Linear Regression between Five Different Parameters of Models and the Proportional Difference $(I_D - I_L)/I_D$ between Direct and Linear Reconstruction Methods of Information Calculation

Predictor Variable	Slope	±95% CI	*Y*-Intercept	±95% CI	*R*	±95% CI
$Jitt_{DJ}$	0.11	[0.05 0.16]	0.35	[0.24 0.46]	0.54	[0.28 0.73]
$Jitt_{RB}$	0.10	–[0.03 0.23]	0.44	[0.27 0.62]	0.25	[–0.07 0.52]
Firing rate	0.00	–[0.00 0.00]	0.54	[0.47 0.60]	0.16	[–0.16 0.45]
Burstiness	0.06	–[0.13 0.24]	0.56	[0.51 0.60]	0.10	[–0.22 0.40]
$\sigma_{recovery}$	–0.00	–[0.00 0.00]	0.57	[0.56 0.59]	–0.01	[–0.32 0.30]

Note: Also shown is the correlation coefficient *R* and its 95% CIs. Variables: $Jitt_{DJ}$, temporal precision of isolated single spikes from the dejittering method; $Jitt_{RB}$, temporal precision of isolated single spikes from the raster-based method; *firing rate*, sustained firing rate of the cell during stimulation; *burstiness*, proportion of all doublets in recordings that have ISIs of 8 ms or less; $\sigma_{recovery}$, SD of normcdf fit for recovery function from refractory period [5].

between different models are significantly different at the 95% level if they do not fall within the range of the notch on the respective boxplot. In the data presented here, models 1 and 2 described significant, though statistically indistinguishable proportions of the information difference (16.5% and 12.5%, respectively).

To further investigate the relationship between several measured quantities and the difference in information measures (Figure 8.7d), we used linear regression to determine how well each measurement could predict the information, with the results shown in Table 8.1. The value of the variance obtained using the "dejittering" technique was the only measured value significantly correlated to the information difference, whereas the precision value from the raster-based method was not significantly correlated. This observation combined with the previously observed lack of correlation between the two variables implies that the discrepancies in information are best explained by accounting for response invariances. This result follows from our previous observation of the lack of correlation between the two variables. The timing variability due to repeated stimulus presentations (the only component of the jitter captured in the raster-based method) affects information calculations using both direct method and stimulus reconstruction approaches. However, the variability in stimulus–response latency for different response events (the component is captured by dejittering) affects stimulus reconstruction-based information estimates; hence, the reason that it is a good predictor of the gap between the two methods of estimating information.

8.2.5 What Do Patterns Represent?

We wished to determine whether the temporally precise spike doublets represented stimuli that were significantly different from those that preceded single spikes, or from those that would be predicted by linear reconstruction analysis. To do this, we used the following model-based test. For a set of doublets with a specific ISI, a model of the doublet-conditioned stimulus ensemble was generated by taking the mean and covariance across that sample set (schematically depicted for 2 ms ISI in Figure 8.8a and d). This model is referred to as the doublet-triggered stimulus model (DTSM). To build a model of the same ISI that was consistent with the

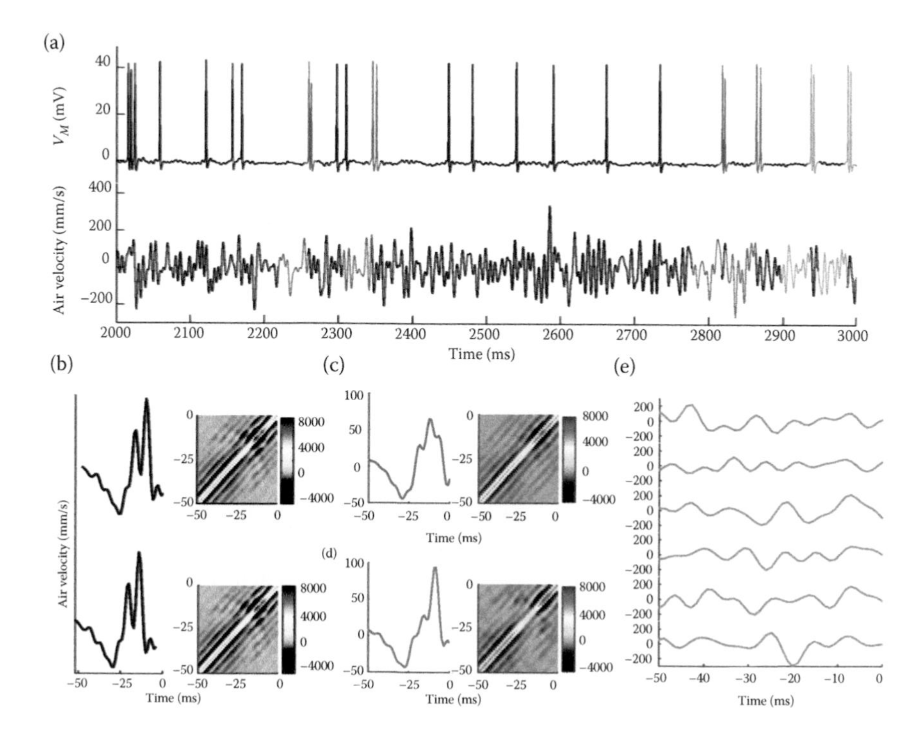

FIGURE 8.8 **(See color insert.)** Schematic of modeling event-conditioned stimuli. (a) Simultaneous recording of 1 s of white noise wind stimulus (bottom trace) and intracellular membrane potential (upper trace) from the same interneuron as in Figure 8.6. Well-isolated response patterns are divided into isolated single-spike responses (blue) and ~2 ms doublets (red and cyan). The response patterns that are not sufficiently isolated are not considered in subsequent analysis (black). The 50 ms of the stimulus preceding the second spike of the response pattern is highlighted in matching colors (bottom trace). (b) Upper panel: Gaussian model of 50 ms of stimulus preceding an isolated single-spike response, consisting of a mean (blue, left panel) and covariance (right panel, color scale in mm^2/s^2) of the entire single-spike-conditioned stimulus ensemble. (b) Lower panel: Same Gaussian model as in upper panel, offset by 2 ms. (c) Synthetic Gaussian model of stimulus preceding 2 ms doublets, obtained by summing the means from panel (b) (cyan, left panel), and summing and then constraining the covariances (Equation A.3). (d) Gaussian model (mean, red, and covariance) of 50 ms of stimulus preceding isolated doublet response patterns with 2 ms ISIs, based on 90% of the doublet-conditioned stimulus ensemble. (e) Selection of six of the stimulus samples that elicited a 2 ms doublet response and that were not used to build the Gaussian model in panel (d), to be used later for likelihood testing.

stimulus-reconstruction methodology, we followed the following procedure. First, we collected stimulus segments associated with isolated single-spike events, and took the Gaussian approximation of the ensemble as above (Figure 8.8a and b). We call this the singlet-triggered stimulus model (STSM). We then took two copies of the same "singlet" model, offset them by the specified ISI, and summed the models (Appendix, Equations A.2 through A.4). This produced a model of the stimuli associated with a doublet that was an extension of the assumptions of linear reconstruction. We denote this model as the synthetic doublet-triggered stimulus model (sDTSM, Figure 8.8c). This procedure provided us with two successively stronger testable hypotheses: (1) The stimuli preceding doublet-spiking events were not different from stimuli preceding single spikes, and (2) that the stimuli preceding such doublet patterns could be predicted by an appropriately combined pattern of the stimuli preceding a single spike. Under this second null hypothesis, there are potential infinite pairs of stimulus–response codewords, limited only by the temporal precision of the stimulus–response relationship: a 2.0 ms ISI could represent a different stimulus pattern than a 2.1 ms pattern, and so on. To properly test these two hypotheses within the constraints of the available data, we examined doublet patterns with at most 1 ms precision. The models were validated by 10× cross-validation (see Appendix). Finally, to reduce artifacts associated with the structure of the band-limited stimulus, we projected all models and test data into a reduced-dimensional space. Six of the 294 examples of the test data excluded during the 10× validation for a 2 ms doublet pattern are shown in Figure 8.8e. Note the variability in individual waveforms relative to the model means shown in Figure 8.8c and d.

We examined how tuned neurons were to a specific stimulus "feature," by using the feature selectivity index (FSI) recently developed by Escabi and colleagues [32]. Briefly, the FSI is determined by comparing how similar each stimulus sample in a spike-triggered stimulus ensemble is to the stimulus mean. It gives a value on a zero-to-one scale that describes how similar each element in the spike-triggered ensemble is to the spike-triggered mean. We calculated the FSI for all afferent interneurons in our sample, in each case using both the regular spike-triggered ensemble and mean as well as the dejittered ensemble and mean [2,19].

The results of the feature-selectivity analysis are shown in Table 8.2, which summarizes the mean ±95% CI values for the FSI pooled across the relevant populations for all interneurons and afferents, both before and after dejittering. Across the population of interneurons, there is a large increase in the FSI after dejittering, whereas neither afferent population shows a significant change.

TABLE 8.2
FSI Values for Neuron Types

Neuron Type	*n*	Mean FSI w/o Dejittering	±95% CI	Mean FSI w/ Dejittering	±95% CI
Cercal interneurons	40	0.34	[0.32 0.35]	0.57	[0.56 0.58]
Filiform afferents	6	0.56	[0.52 0.61]	0.52	[0.46 0.58]
Campaniform afferents	5	0.50	[0.42 0.58]	0.53	[0.48 0.57]

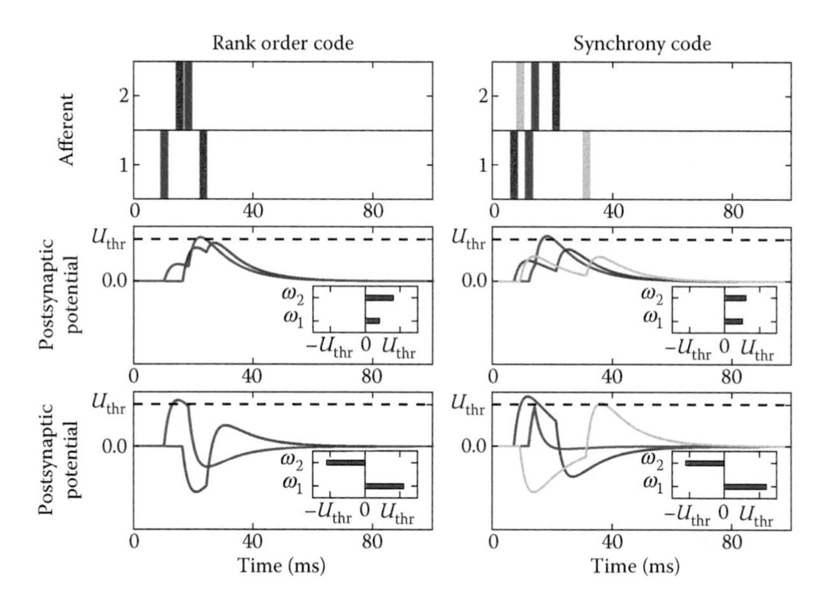

FIGURE 2.3 Implementation of rank order (left column) and synchrony (right column) codes by the Tempotron. The input patterns are presented in different colors in the top row and consist of a single spike arriving from each of the two input afferents with a certain time difference. The bottom two rows depict the postsynaptic potential traces evoked by the different input patterns. Insets depict the synaptic efficacies of the two input afferents.

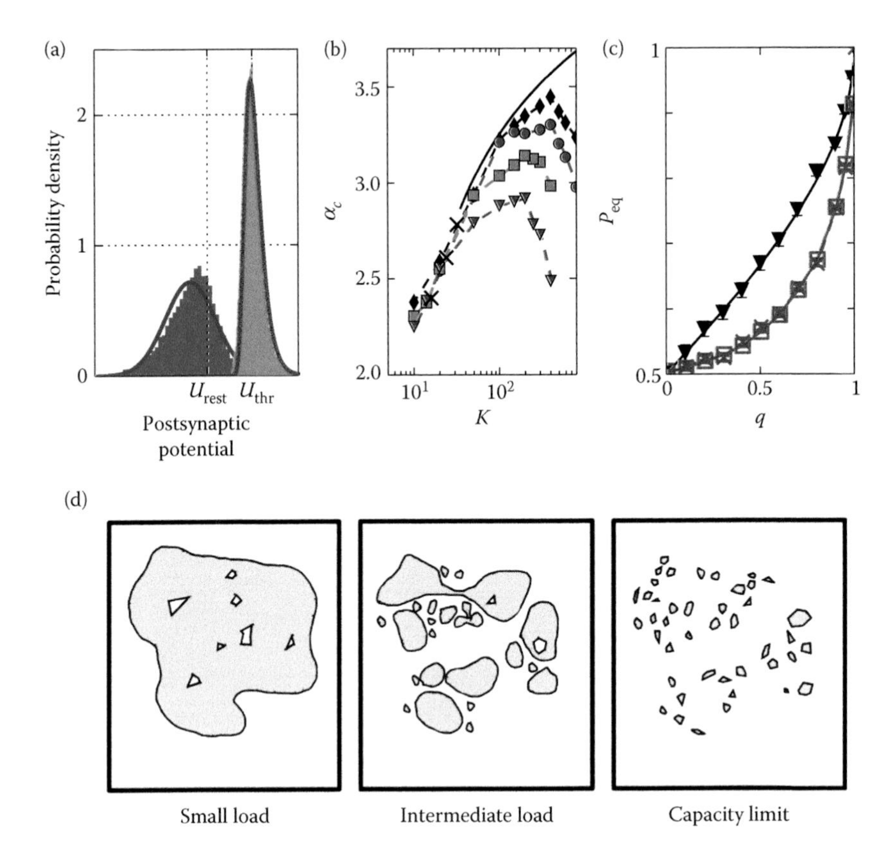

FIGURE 2.5 (a) Probability density of $U(t)$ (blue) and probability density of U_{max} (red). The lines correspond to a Gaussian with the mean and variance of the postsynaptic potential, $U(t)$, and a Gumbel law with median at U_{thr} and a scale parameter that was fitted to the data. The data were measured with $K = 400$, $\alpha = 1.68$, $N = 500$, and 34 samples. (b) Capacity α_c of the Tempotron versus K. Lines with symbols show results of the Tempotron learning algorithm for networks of different size ($N = 250$, triangles; $N = 500$, squares; $N = 1000$, circles; $N = 2000$, diamonds). The solid line and × symbols depict fitted theoretical predictions with one free parameter. (c) Probability that two neurons will classify a random pattern in the same manner, P_{equal}, versus the correlation coefficient between their weight vectors, q, for the Perceptron (theory and simulations in black), the Tempotron (blue × symbols, $K = 100$), and the Hodgkin–Huxley (red squares, input pattern duration of $T = 1.5$ s) models. See Rubin et al. (2010) for details of the Hodgkin–Huxley model. (d) The evolution of solution space with increasing load. Each square represents the space of all weight vectors and every point represents one N-dimensional vector. Gray regions represent solutions to the classification task. The solution space is connected only for a small number of patterns, and breaks to many clusters as the load increases. At the capacity limit, the solution space consists of many small clusters spread evenly throughout the weight space.

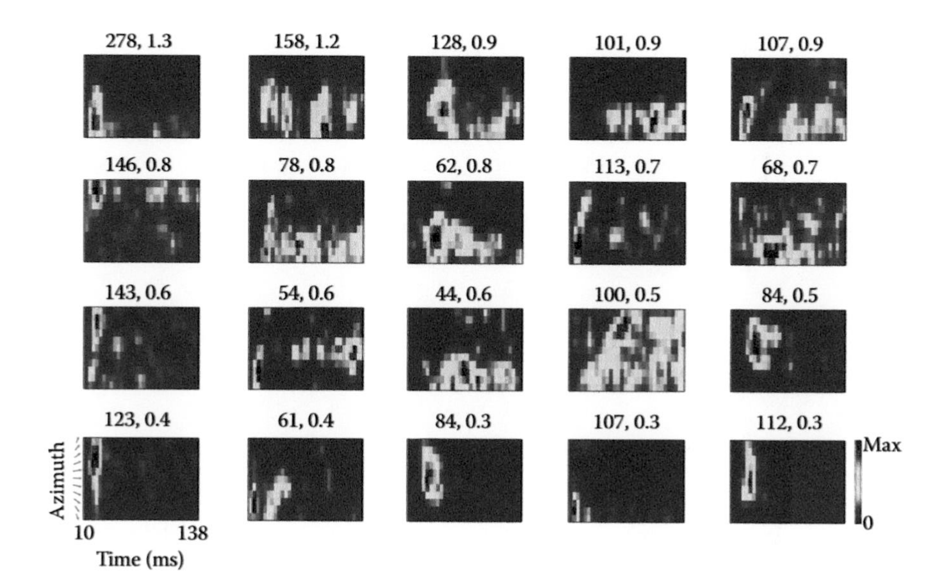

FIGURE 5.2 All models. Models were derived from the responses of 20 units sorted by their MI. Firing rates are depicted by colors from blue, which always represent 0, to red, which varies between panels. The numbers at the top of each panel the maximal firing rate (sp/s, left) and the MI of the model (True MI in bits, right). Unit 18 (row 4, column 3) is displayed in Figure 5.1.

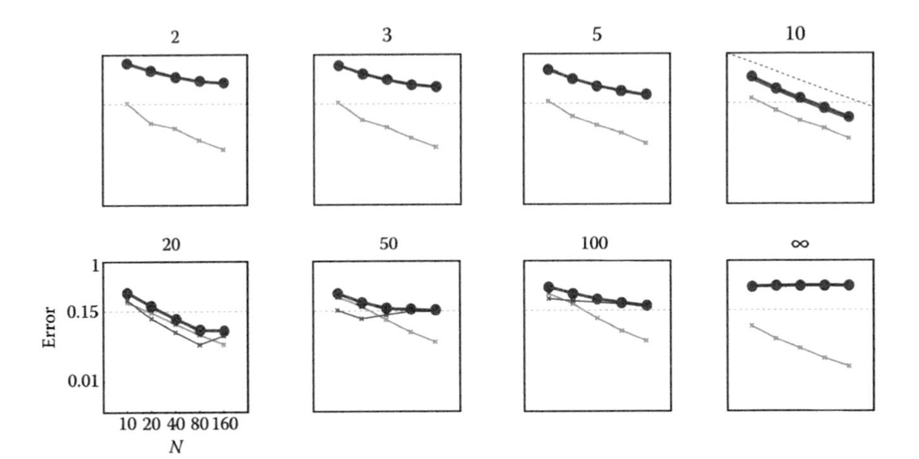

FIGURE 5.5 Validation of the BINL-ED method. The bias (blue x), standard deviation (green x), and RMSE (red circles) are plotted as a function of the number of trials/stimulus (10, 20, 40, 80, 160) on a logarithmic scale. The gray dashed lines mark the reliability value of the error. The red dashed line shows the expected rate of decrease of the bias and variance, with a slope of −0.5 on this log–log plot.

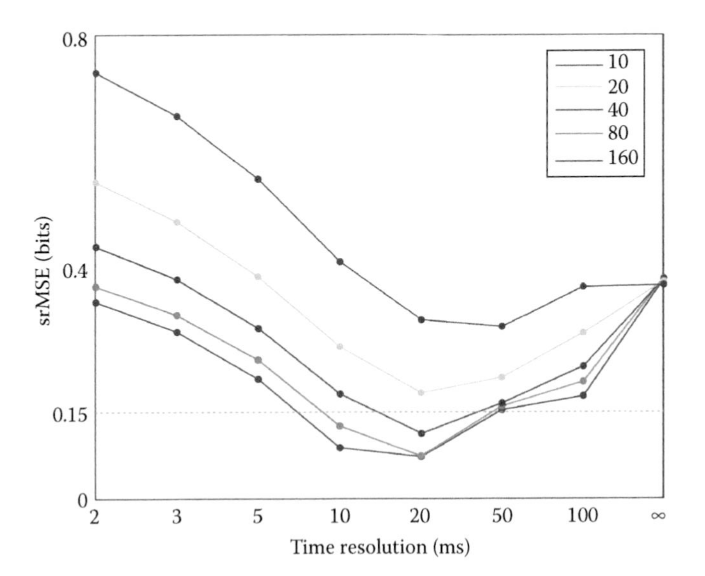

FIGURE 5.6 Estimation quality dependence on temporal resolution. The RMSE of the estimation is plotted against the temporal resolution ($1/q$) used for the number of trials/stimulus (blue: 160; green: 80; red: 40; cyan: 20; purple: 10). The gray dashed line marks the level below which we considered the estimates reliable.

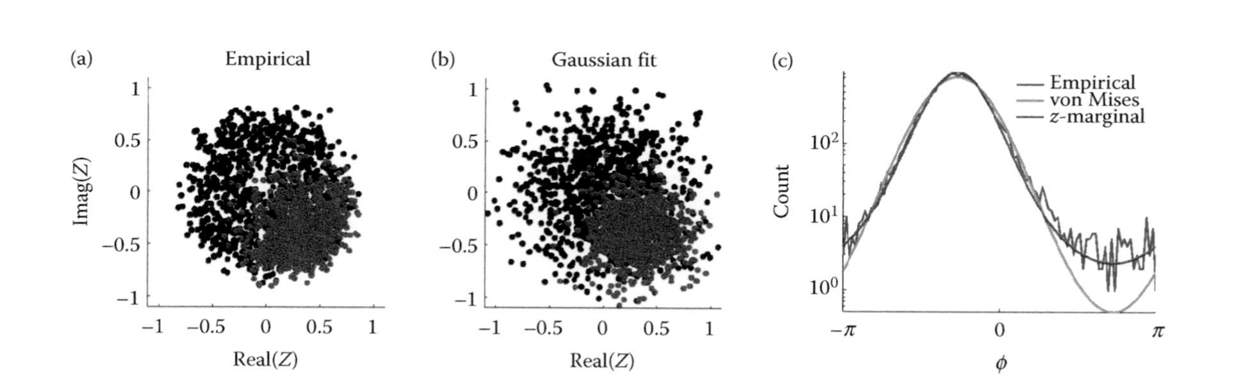

FIGURE 6.2 Fitting the correlation between the analytic reference signal and spikes for the one-dimensional case. (a) Distributions of raw ensemble ($p(Z)$, black) and STE ($p(Z|X = 1)$, red) of the analytic signal in the complex plane. (b) Complex circular-symmetric Gaussian fits of raw ensemble and STE. Restricting the fit to be circular-symmetric results in a distribution that is isotropic in the complex plane. (c) The spike-triggered phase distribution (blue), fit with von Mises distribution (green) and with marginal of complex Gaussian (red) from Equation 6.9.

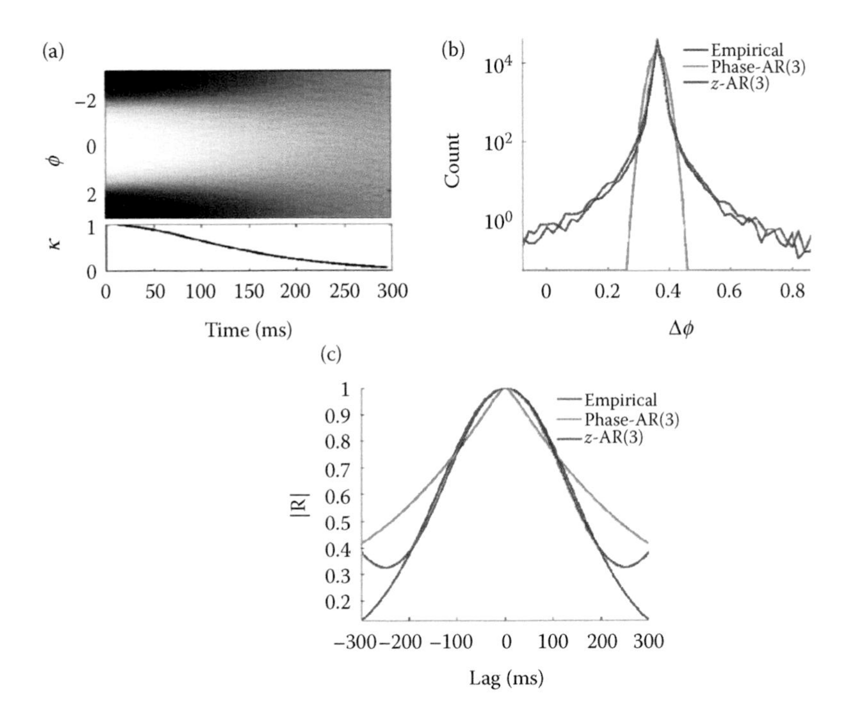

FIGURE 6.3 Autoregressive modeling of reference signal dynamics. (a) Following a spike, the phase distribution gradually broadens (upper panel), seen in the decrease of the concentration of the von Mises fit (lower panel). (b) Distribution of the phase difference in consecutive time steps (instantaneous frequency). The empirical distribution (blue) is well fitted by the AR model of the analytic signal (red) but only poorly fitted by the AR model of phase (green). (c) The autocorrelation of the analytic signal is well approximated by the z-AR model for 200 ms; here, the magnitude of the autocorrelation is shown for clarity.

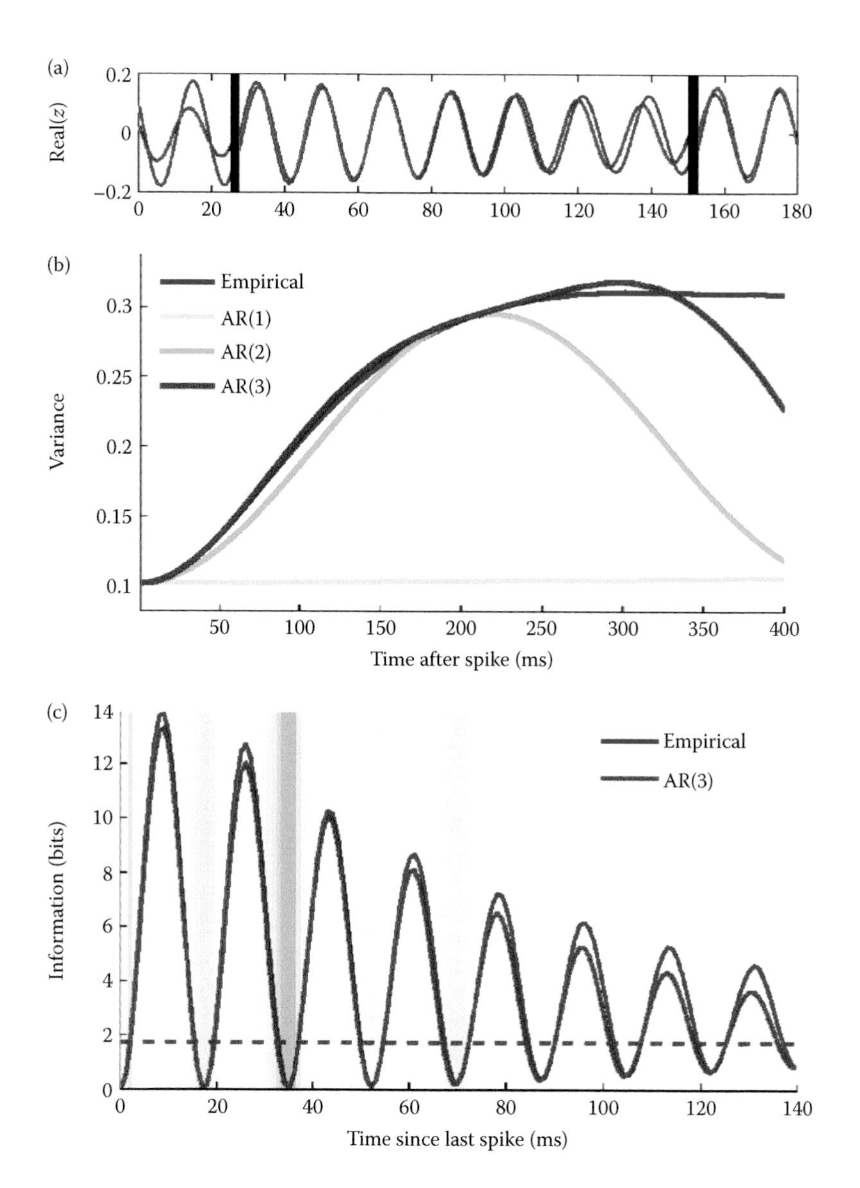

FIGURE 6.4 Estimating the variance of the analytic signal after a spike as a function of the interval *ISI* since the spike. (a) When a spike is observed (black bars), the AR estimate of the reference signal is updated by the spike-triggered average. This estimate is propagated through the AR model but gradually diverges from the actual signal. (b) The variance of the analytic signal increases following a spike, and is best approximated by of the AR(3) model. (c) The information (Equation 6.12) provided by a spike following different interspike intervals (ISIs), as measured by the KL divergence of the empirical distribution (blue), or the AR(3) model (red), from the STE. Dashed line indicates average information per spike. Darker green strips in the background mark ISIs where more spikes fall.

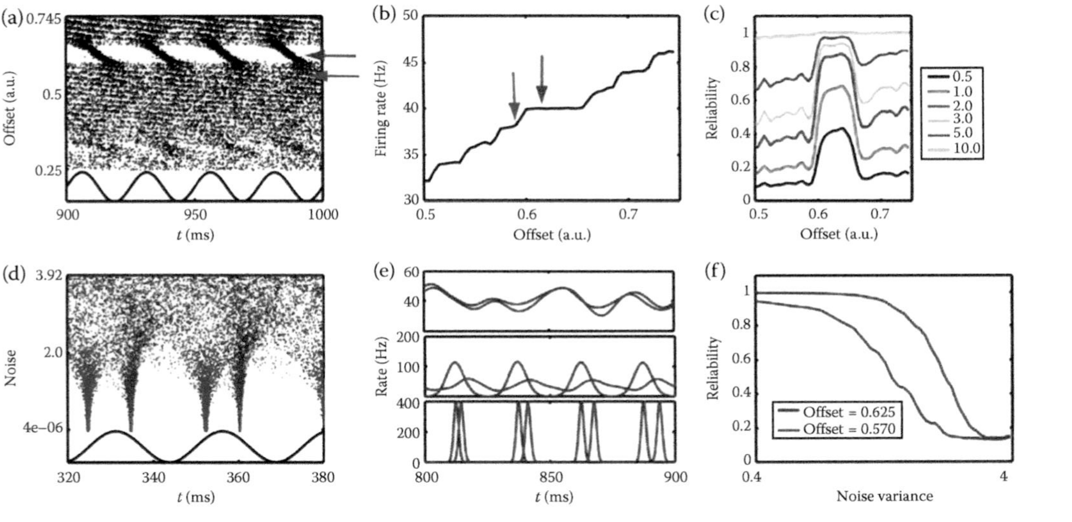

FIGURE 7.3 Phase locking to a sinusoidal drive leads to a resonance in the precision; resonant states are the most robust against noise. (a) The compound rastergram is used to show the effect of varying the level of the depolarizing current (offset). For each value, we plot 50 trials, with the offset increasing from bottom to top. At the bottom, we show the sinusoidal driving current. The blue and red arrows in panels (a) and (b) indicate an offset of $I = 0.57$ μA/cm² and $I = 0.625$ μA/cm², respectively. (b) The firing rate generally increases with the value of the offset, but the curve also shows plateaus where the firing rate does not change with the offset. (c) The plateau at which the firing rate is equal to the frequency of the driving current (40 Hz) also corresponds to a peak in the R-value. The R-value is a compound measure that increases both with precision and reliability and takes a temporal resolution σ as parameter. When σ increases, the measure becomes less sensitive to time differences between the spike trains. For $\sigma = 10$ ms, the R curve does not display a resonance anymore. (d) A compound rastergram with spike trains on multiple trials for different values of the noise variance. For clarity, only five trials are shown per noise value, but the reliability calculations were performed using all 50 trials. The spike trains with offset $I = 0.57$ μA/cm² (blue dots) are more noise sensitive than those for 0.625 μA/cm² (red dots), which is on the resonance step. (e) We show smoothed histograms representing the instantaneous firing rate for noise variances $D = 4 \times 10^{-6}$, 2.24, and 3.84 mV²/ms. The colors represent the two offset values as in the preceding panels. (f) This difference in sensitivity is visible as a shift in the R-reliability versus noise curve shown with the same color code.

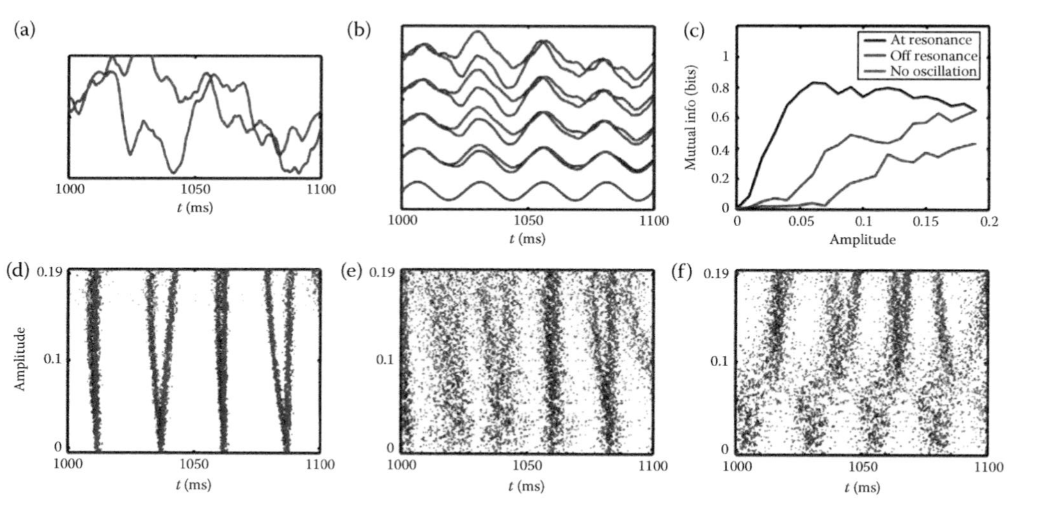

FIGURE 7.4 Neurons modulated by an oscillatory current encode stimulus information more efficiently than those without an oscillatory current, with the maximum benefit being obtained during resonance conditions. (a) Two different aperiodic stimulus waveforms, red (stimulus 1) and blue (stimulus 2), respectively, were injected on 50 trials into the neuron on top of either a constant depolarizing current or a sinusoidal current. (b) The amplitude of the stimulus was varied. We show the case with a periodic current present; for clarity, only a small segment of the 5 s long stimulus is shown and, for each amplitude, the curve is offset by a fixed amount. From bottom to top, the amplitude is increased and the difference between red and blue curves becomes more pronounced. (c) We used a clustering algorithm (Toups et al., 2011) to find two clusters of spike trains and determined the match between the clustering-determined labeling and the stimulus number (1 or 2) from which the spike train was obtained, in terms of the mutual information. We compare three cases: with periodic current injection and with a level of depolarizing current corresponding to either (black) a resonance or (blue) outside the resonance range, or (red) without a periodic current. The mutual information is plotted as a function of the amplitude of the stimulus. (d–f) The behavior of the mutual information is visualized by the rastergram. The two sets of trials are plotted on top of each other, with the stimulus number represented by the color of the dots as before. (d) For an on-resonance neuron, the precision is high and the difference in spike times between the two drives becomes immediately visible for nonzero amplitude. (e) Off-resonance, the precision is reduced and only for the highest amplitudes is a systematic difference between spike times visible. (f) When there is no periodic current, the response is imprecise for low amplitudes and the difference between spike times due to the two different stimuli is not visible.

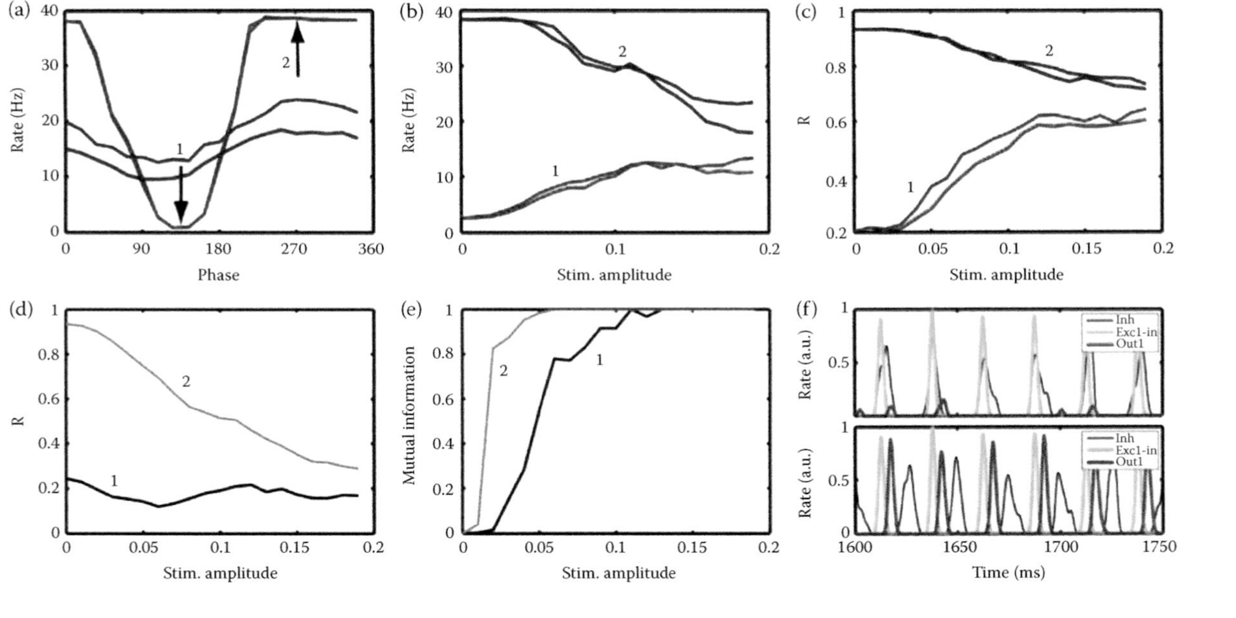

FIGURE 7.6 The transmission of E volleys is modulated by the phase of local inhibition. (a) A neuron received phase-locked E volleys as shown in Figure 7.4 together with periodic inhibitory volleys of which the phase was varied. The resulting firing rate as a function of inhibitory phase is shown for each of the two stimulus waveforms, in red and blue, respectively. Label 2 indicates the phase for which the E inputs arrive in the transmission window, whereas label 1 is for a phase outside the transmission window. We plot (b) the firing rate and (c) the R-reliability as a function of the stimulus amplitude for two phases, one inside the transmission window ("2") and one outside ("1"). (d) The similarity between the responses generated by stimulus 1 and stimulus 2 and (e) the distinguishability quantified as the mutual information as a function of stimulus amplitude for the two phases as labeled. (f) We plot (black) inhibitory volleys and (green) excitatory inputs and (red) the output spikes they generate. The histogram of each set of spike trains was convolved with a Gaussian with standard deviation of 1 ms and normalized by the maximum value across all stimulus conditions. For (top) a phase outside the transmission window, only a few output spikes result, whereas when the phase is inside the transmission window, there is a robust output just before the inhibitory volleys arrive.

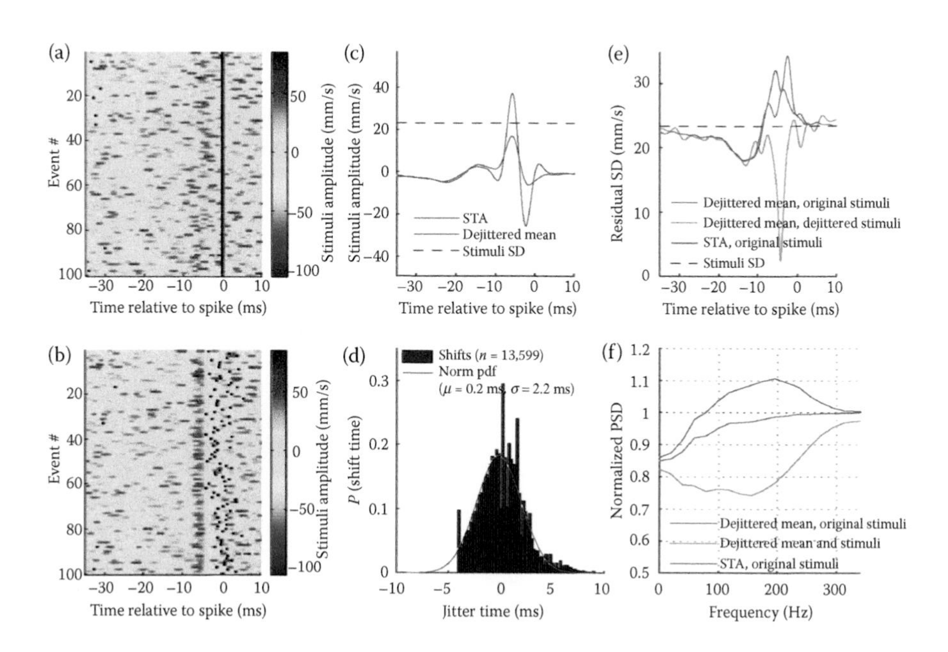

FIGURE 8.2 Characteristics of a dejittered mean stimulus. (a) Random representative subset of 100 sample segments of stimulus–response data aligned to the time of spike occurrence. The vertical black line at $t \approx 0$ is the raster plot of spikes superimposed on the color-coded traces of air-current velocity versus time. (b) The same random subset of 100 segments of stimulus–response data from (a) but now dejittered so that their alignment is based on minimal variance of the stimuli. (c) STA (red trace) and dejittered mean (blue trace). The standard deviation (STD) of the stimulus is indicated with the dashed black line. (d) Histogram of shift times (blue) compared with a Gaussian having an STD of 2.2 ms (red line). (e) STD of the residual under different dejittering conditions. The fully dejittered mean (green), which we assume is how the system interpreting the activity has the smallest residual. (f) Power spectra of the residuals, normalized by the power spectra of stimulus segments. The colors and abbreviations are the same as in (e).

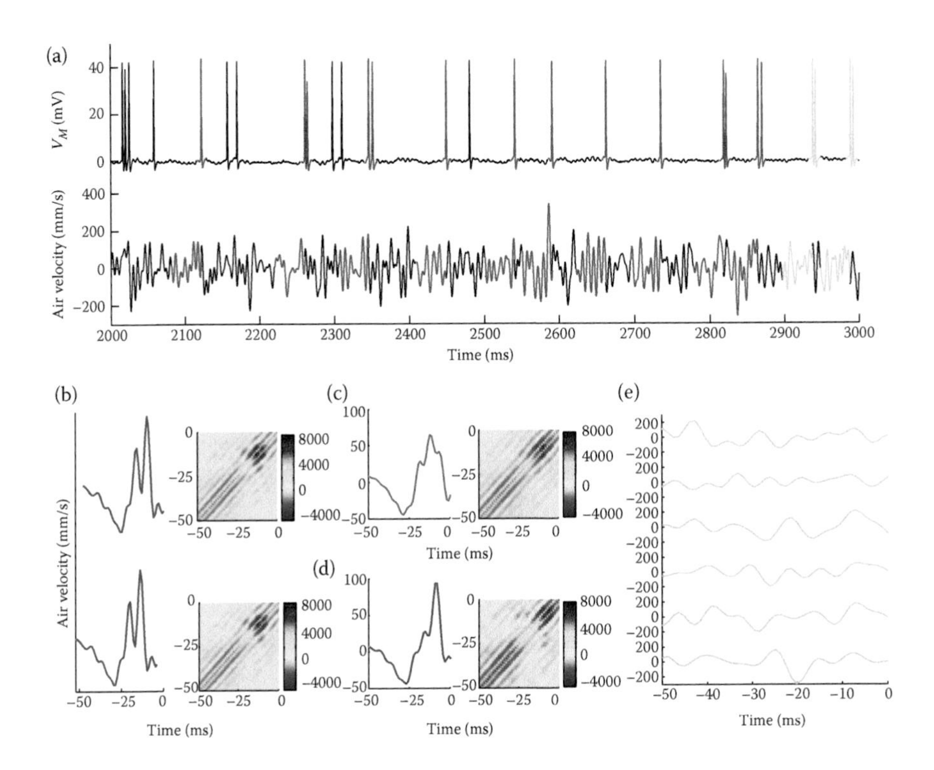

FIGURE 8.8 Schematic of modeling event-conditioned stimuli. (a) Simultaneous recording of 1 s of white noise wind stimulus (bottom trace) and intracellular membrane potential (upper trace) from the same interneuron as in Figure 8.6. Well-isolated response patterns are divided into isolated single-spike responses (blue) and ~2 ms doublets (red and cyan). Response patterns that are not sufficiently isolated are not considered in subsequent analysis (black). The 50 ms of the stimulus preceding the second spike of the response pattern is highlighted in matching colors (bottom trace). (b) Upper panel: Gaussian model of 50 ms of stimulus preceding an isolated single-spike response, consisting of a mean (blue, left panel) and covariance (right panel, color scale in mm²/s²) of the entire single-spike-conditioned stimulus ensemble. (b) Lower panel: Same Gaussian model as in the upper panel, offset by 2 ms. (c) Synthetic Gaussian model of stimulus preceding 2 ms doublets, obtained by summing the means from panel (b) (cyan, left panel), and summing and then constraining the covariances (Equation A.3). (d) Gaussian model (mean, red, and covariance) of 50 ms of stimulus preceding isolated doublet response patterns with 2 ms ISIs, based on 90% of the doublet-conditioned stimulus ensemble. (e) Selection of six of the stimulus samples that elicited a 2 ms doublet response and that were not used to build the Gaussian model in panel (d), later to be used for likelihood testing.

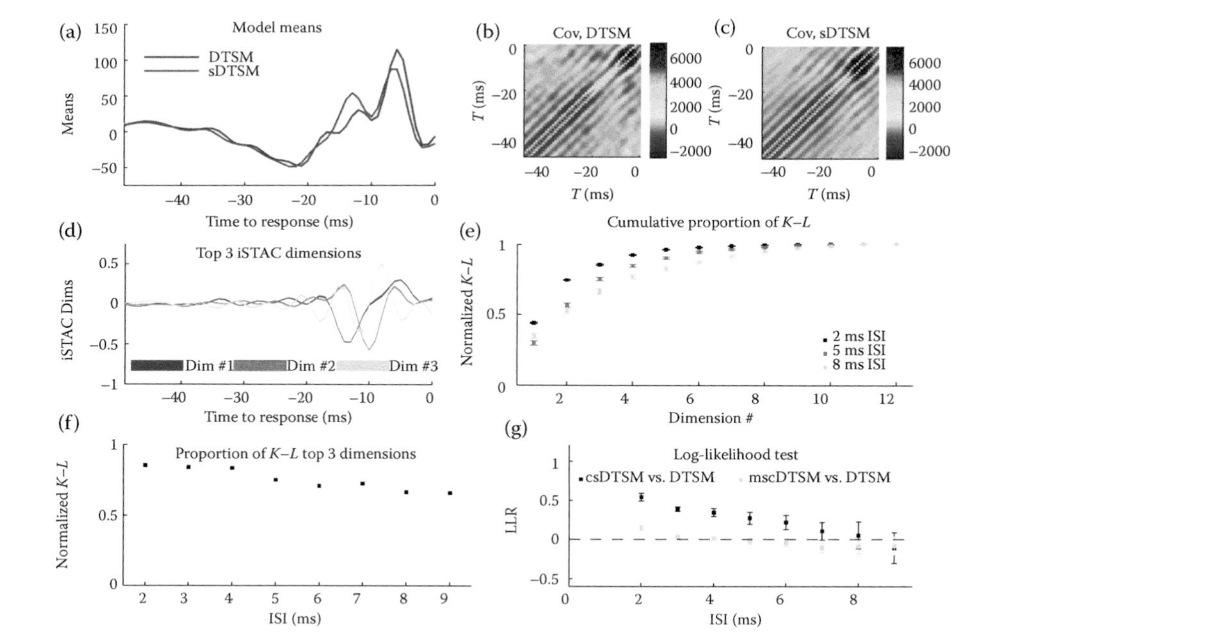

FIGURE 8.10 iSTAC analysis of data-based and synthetic models. (a) Mean of data-based (red) and synthetic (purple) multivariate Gaussian models for stimulus preceding a 2 ms doublet, from the same cell as in Figure 8.8. The covariance of data-based and synthetic models is shown in panels (b) and (c), respectively (color scale in mm^2/s^2). (d) Estimate of the three most informative iSTAC dimensions (shaded area indicates the mean ± SD across 10× validation). (e) Measure of the total normalized *K–L* divergence between data-based and synthetic models for 2, 5, and 8 ms, as a function of subspace dimensionality. The mean ± SD across 10× validation is shown with error bars, which are on the order of the size of the markers for the points. (f) Measure of the portion of the total *K–L* divergence explained by the subspace containing the three largest iSTAC vectors, as a function of ISI in the model. (g) Improvement of the synthetic model performance in LLR tests from Figure 8.9c (black makers) by modification along the three-dimensional subspace shown in panel (d) (cyan markers). The error bars represent ±95% CIs on the mean.

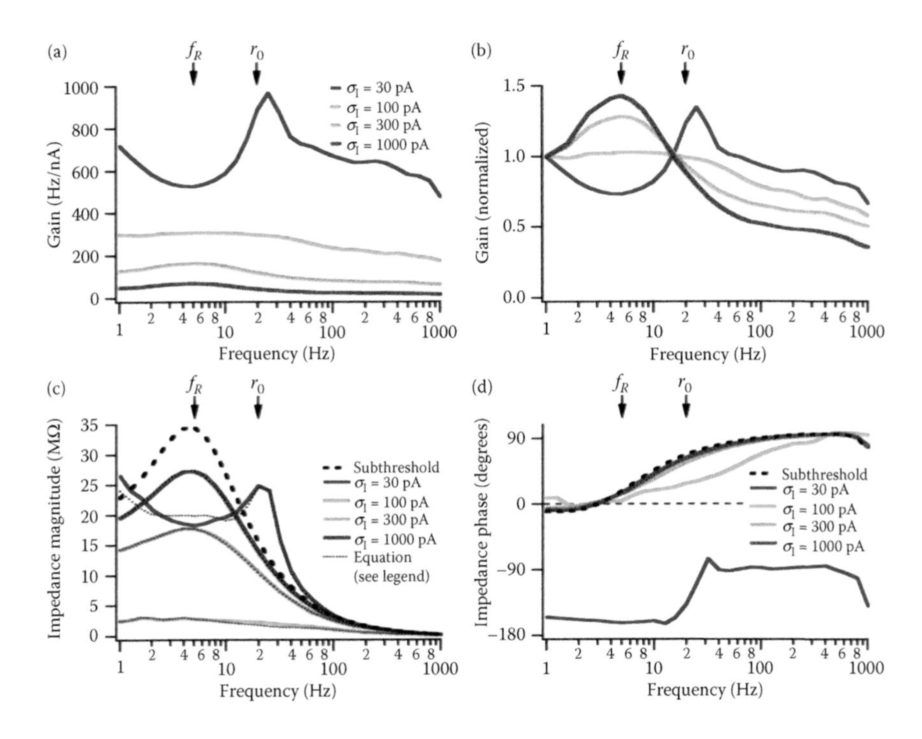

FIGURE 9.5 Generalized integrate-and-fire (GIF) model. Simulations were performed using the model of Richardson et al. (2003). The subthreshold response is described by $C\,dV/dt = -gV - g_1 w + I_{app}(t)$. The feedback variable w follows V with a first-order relaxation: $\tau_1\,dw/dt = V - w$. A spike is emitted when V reaches θ; V is then reset to V_r, but w is not reset. For the illustrations, $C = 500$ pF, $g = 25$ nS, $g_1 = 25$ nS, $\tau_1 = 100$ ms, $\theta = 20$ mV, and $V_r = 14$ mV. The noise had a relaxation time constant $\tau_{noise} = 5$ ms. (a) $G(f)$ for four levels of noise ($\sigma_I = 30$, 100, 300, and 1000 pA). To produce a mean firing rate $r_0 = 20$ Hz, I_0 was 971, 943, 782, and −89 pA, respectively. $G(f)$ was estimated by correlation analysis. (b) The same $G(f)$ curves, each normalized by the gain at 1 Hz. (c) Impedance magnitude for subthreshold response (dashed black line) and for $V(t)$ during each spiking response (colored lines), illustrating how the spike-and-reset mechanism disrupts the impedance resonance when σ_I is low. The equation predicting the impedance magnitude during spiking is $|Z_p(f)| = |Z(f)|\,[1 - G(f)\,C\,(\theta - V_r)\cos \Phi(f)]$. (d) Impedance phase.

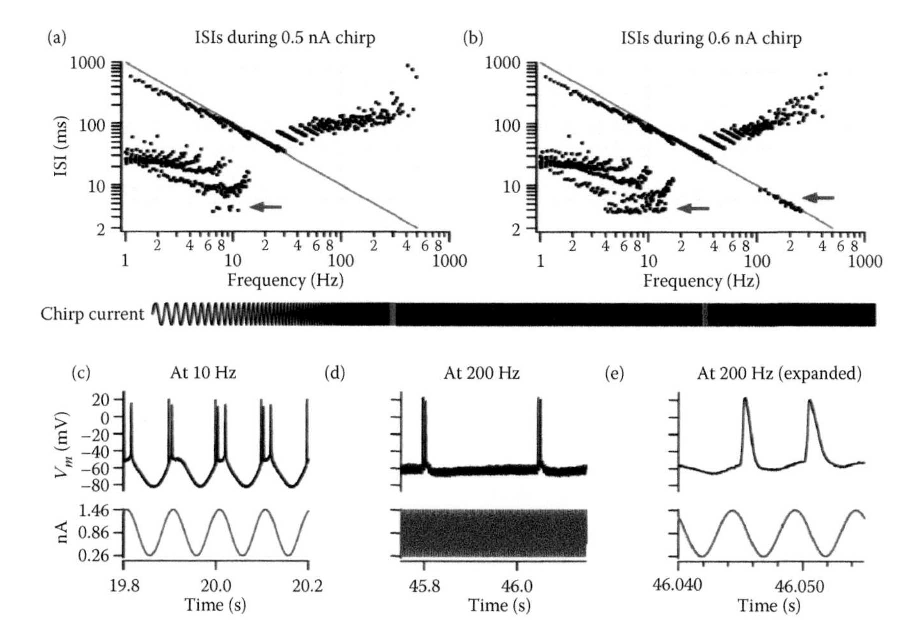

FIGURE 9.7 Bursts triggered by low- and high-frequency sine wave input. (a) ISIs in a cortical layer 2/3 PC during a chirp current stimulus with an amplitude of 0.5 nA and an exponential scan from 1 to 1000 Hz. The blue arrow indicates intraburst intervals at a chirp frequency near 10 Hz. (b) ISIs in the same cell when the chirp amplitude was increased to 0.6 nA. The red arrow indicates intraburst intervals at a chirp frequency of approximately 200 Hz. (c) Firing pattern during the 0.6 nA chirp at 10 Hz (blue region in chirp current trace above). Top plot shows V_m, and bottom plot shows the applied current. (d) Firing pattern at 200 Hz (red region in chirp current trace) on the same timescale. (e) Firing pattern at 200 Hz on an expanded timescale. (From Higgs MH, Spain WJ 2009. *J Neurosci* 29:1285–99.)

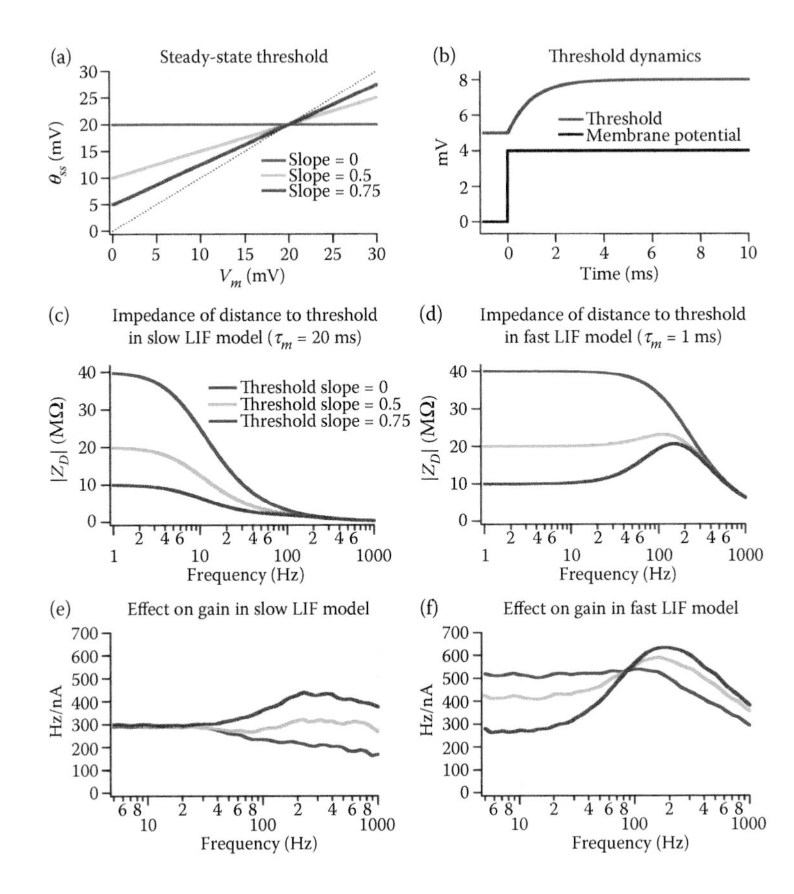

FIGURE 9.13 Functional effects of a dynamic threshold. (a) Steady-state threshold $\theta_{SS}(V)$. For comparison, each model was set up such that $\theta_{SS}(V)$ intersects V (dashed line) at 20 mV above the resting potential. (b) Threshold dynamics (first-order relaxation) illustrated by the response to a voltage step. (c) Calculated impedance of distance to threshold (Z_D) in slow LIF model ($\tau_m = 20$ ms). (d) Z_D in fast LIF model ($\tau_m = 1$ ms). In both cases, the dynamic threshold reduces Z_D at low frequencies. (e) Effect of the dynamic threshold on $G(f)$ in the slow LIF model. Simulations were performed with $\sigma_I = 100$ pA, $\tau_{noise} = 5$ ms, and I_0 set to give $r_0 = 20$ Hz. Blue, green, and red curves indicate steady-state threshold slopes of 0, 0.50, and 0.75. (f) $G(f)$ in fast model ($\tau_m = 1$ ms) with faster-fluctuating noise input ($\sigma_I = 100$ pA, $\tau_{noise} = 1$ ms).

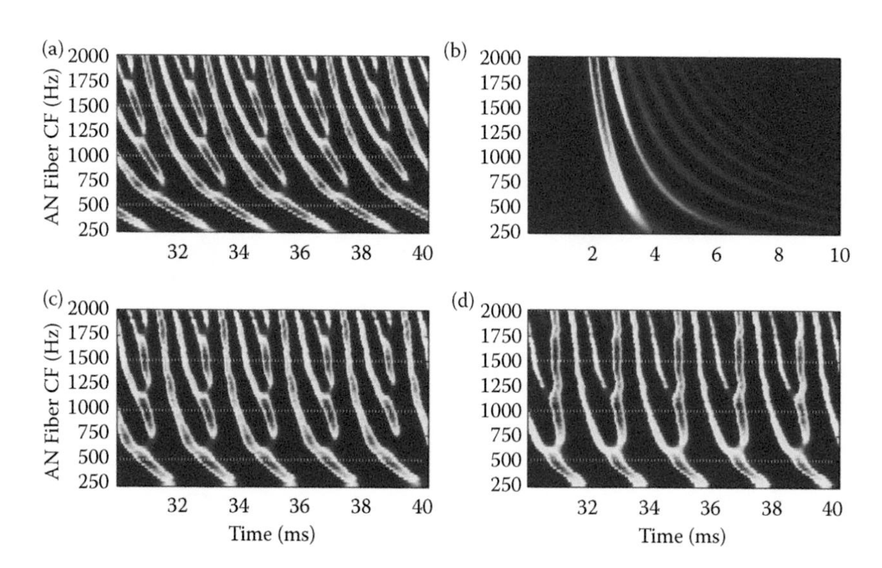

FIGURE 10.6 Spatiotemporal patterns of model auditory-nerve fiber responses to harmonic tone complexes (F0 = 500 Hz, 55 dB SPL). Responses were computed using the Zilany et al. (2009) auditory-nerve model. (a) Upper left: Responses of model auditory-nerve fibers to a steady-state portion of the harmonic tone complex before latency compensation. Each row shows the response rate as a function of time for a fiber with a given characteristic frequency (CF). Responses are normalized to the overall maximum in the image. Steep transitions in the spatiotemporal pattern occur near the fundamental frequency (500 Hz) and the harmonic frequencies (1000, 1500, 2000 Hz). (b) Population response to a 10 μs click. The peak response time of each frequency channel determined the latency used in the compensation. (c) The 55 dB SPL population response from (a) after compensation with the click latencies shown in (b). After compensation, patterns are more "vertical" (e.g., more coincident across frequency channels) near the harmonic frequencies. (d) A population response to a 65 dB SPL harmonic tone complex, including latency compensation. Spatiotemporal patterns near the harmonics become increasingly coincident as level increases.

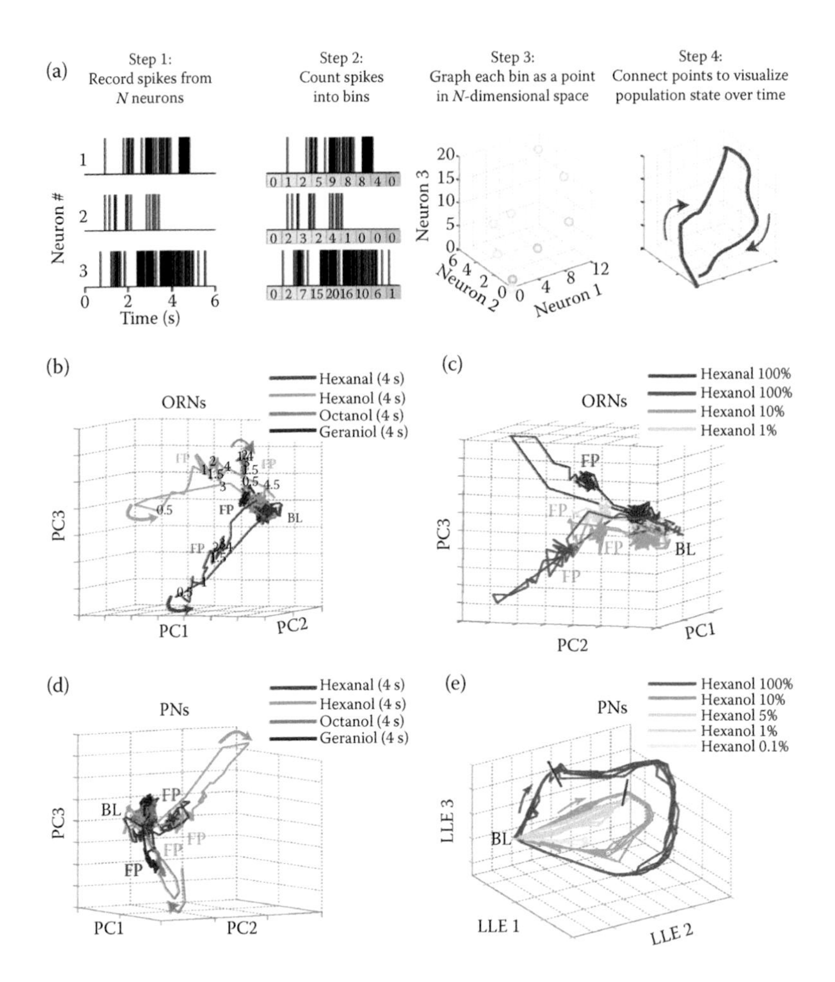

FIGURE 11.3 Trajectory plots illustrate spiking patterns of populations of ORNs and projection neurons (PNs) over time. (a) Method used to construct these plots (see text). (b) Spiking responses of 53 ORNs to four odorants each presented for 4 s. The first three principal components (PC1–3, see text) reveal that the population response departs from baseline activity (BL), undergoes an odor specific trajectory, settles into a fixed point (FP), and then returns to baseline via another pathway when the odor is removed. Numbers show time (seconds) after the odor's onset. (c) Trajectory responses of 43 ORNs responding to the odorant Hexanal, and to three different concentrations of a different odorant with a similar structure, Hexanol. Increasing concentrations of an odor elicit responses within the same area of response space (within the same manifold) but with trajectories of monotonically increasing length. (d) Trajectory responses of 94 PNs tested with the same odorants as in panel (b) reveal broadly similar time-varying response features. (e) Trajectory responses of 110 PNs responding to 1 s presentations of five different concentrations of Hexanol. Here, the trajectories were visualized using local linear embedding (LLE, see Roweis and Saul 2000), a dimensionality-reduction technique. Like the responses of ORNs shown in panel (c), PN trajectory magnitude increases monotonically within a manifold with odor concentration. (Panels (a) and (e): adapted from Stopfer, M., Jayaraman, V., and Laurent, G. 2003. *Neuron* 39, 991–1004.; Panels (b) through (d): adapted from Raman, B. et al. 2010. *The Journal of Neuroscience* 30, 1994–2006.)

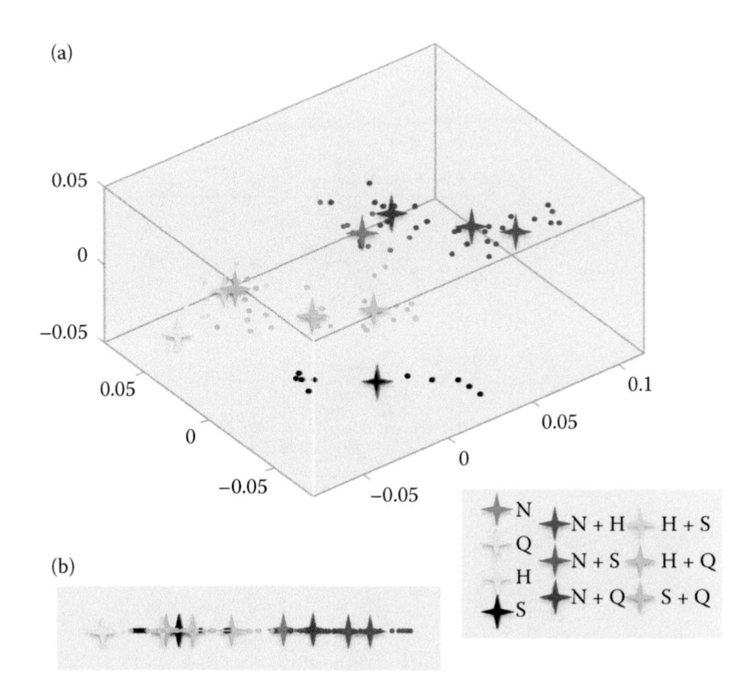

FIGURE 12.2 Multidimensional scaling (MDS) using differences in spike timing as a measure of similarity among responses. Dots indicate the location of individual responses and asterisks indicate the centroid of the clusters of responses to a given taste stimulus. Axes are labeled in arbitrary units. (a) The three-dimensional response space created by MDS of the spike time distances $D^{spike}[q_{max}]$, that is, the difference in spike timing calculated at the level of temporal precision (q) that conveys the maximum amount of information. For this cell, $q_{max} = 2.83$. (b) The one-dimensional response space created by MDS of the spike count distances, D_{count}. N, 0.1 M NaCl; Q, 0.01 M quinine HCl; H, 0.01 M HCl; S, 0.5 M sucrose; N + H, 0.1 M NaCl plus 0.01 M HCl; N + S, 0.1 M NaCl plus 0.5 M sucrose; N + Q, 0.1 M NaCl plus 0.01 M quinine HCl; H + S, 0.01 M HCl plus 0.5 M sucrose; H + Q, 0.01 M HCl plus 0.01 M quinine HCl; S + Q, 0.5 M sucrose plus 0.01 quinine HCl. (Reprinted from Di Lorenzo PM, Chen JY, and Victor JD. *J Neurosci* 29(29): 9227–38, 2009a.)

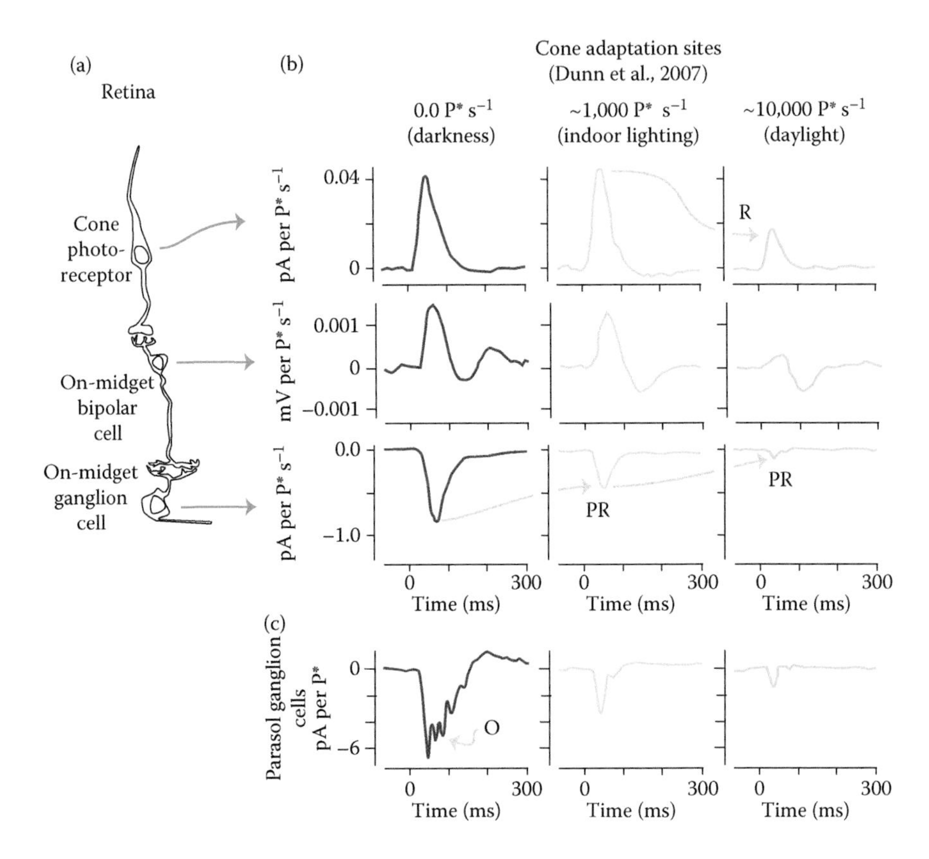

FIGURE 14.4 Influence of receptor and postreceptor circuits on response timing in the retina. (a) Schematic of midget–parasol pathway connections in monkey retina. The cone photoreceptors connect via midget–bipolar cells to midget–ganglion cells. (b) Light-evoked responses (membrane current flow or membrane potential changes) in these three cell classes. Each row shows the responses of one cell type at three different background light intensities. The sensitivity of individual cone receptors (upper row) shows little change between darkness and indoor light levels but receptor adaptation (R) is apparent with increases to outdoor light levels. Sensitivity of midget–bipolar cells (center row), parallels cone sensitivity, indicating that bipolar adaptation is inherited from the receptors. By contrast, ganglion cells show signs of postreceptor adaptation (PR) on transition from darkness to indoor levels (center column) and from indoor to outdoor lighting levels (right column). (c) The recordings from parasol cells show oscillatory potentials (O) attributable to reciprocal and feed-forward synaptic circuits in the IPL. (Modified from Dunn, FA et al. 2007. *Nature* 449, 603–07.)

8.2.6 Is It Just a Compressive Nonlinearity?

The analysis in the previous section demonstrates that the stimuli preceding patterns of spikes differ significantly from the predictions of linear stimulus reconstruction. However, the specific deviations remains unclear. For instance, one potential explanation for the results seen in Figure 8.8 is a form of compressive nonlinearity, where the stimuli preceding doublets of a specific ISI are modeled better by a sDTSM with a shorter ISI. This would be a natural way for a neuron to adjust its operational range within the limits imposed by biophysical constraints, allowing it to encode stimuli that "should" be represented by an ISI smaller than the cell's refractory period. Such an encoding mechanism would be the representational correlate of the "free firing rate" characterized by Berry and Meister [5]. To determine whether such a mechanism could explain the difference between the sDTSM and the DTSM, we carried out a likelihood test. We asked which of several sDTSMs (each having a different ISI) best explained the data. We built these models as in the previous section (using Equations A.3 and A.4), but with variable offset values of –3 to 29 ms (in this case, a –3 ms offset would be equivalent to a 3 ms offset, but with an additional 3 ms latency prior to the response), and tested them with doublet data containing ISIs from 2 to 26 ms.

The results of this analysis are shown in Figure 8.9, pooled across nine cells. Figure 8.9a shows the probability of each sDTSM (y axis) explaining the data for each ISI (x axis), averaged across all cells. For ISIs >2 ms, the clear peak lies along the diagonal of the image, indicating that, for these ISIs, the best offset between single-spike stimuli in the sDTSM is the actual ISI of the data being modeled. However, for doublets with an ISI of 2 ms, the best-match sDTSM was actually the one with two single spikes at 0 offset (i.e., completely superimposed). To see how consistent this relationship was across the cells in our data set, we found the peak probability for each ISI. The mean ±1 SD of this value across the cells in our population is plotted in Figure 8.9b, showing that this relationship is indeed consistent across this population of neurons. Having established that a compressive nonlinearity exists in the encoding scheme of these cells, we repeated the likelihood analysis with the best-fit synthetic model. We refer to this best-fit synthetic model as the compressed synthetic doublet-triggered stimulus model (csDTSM). The results of comparing the DTSM and the csDTSM are shown with gray markers in Figure 8.9b, superimposed on the original comparison between the DTSM and sDTSM (black markers). Although accounting for the nonlinearity significantly increases the predictive power of the 2 ms synthetic model, this improvement still explains only a fraction of the difference in predictive power between the sDTSM and the DTSM for short ISIs. This indicates that the mismatches between observed and linearly synthesized waveforms, as seen in Figure 8.8, cannot be explained solely by the refractory behavior of neurons.

8.2.7 What Are the Deviations from Linearity?

Having just shown that modifications accounting for refractory periods do not explain the differences in our models, we sought to fully quantify these differences in a rigorous manner. To do this, we used iSTAC analysis [79] as adapted to multivariate inference [24]. See Appendix Section "iSTAC Analysis" for further details. iSTAC is a

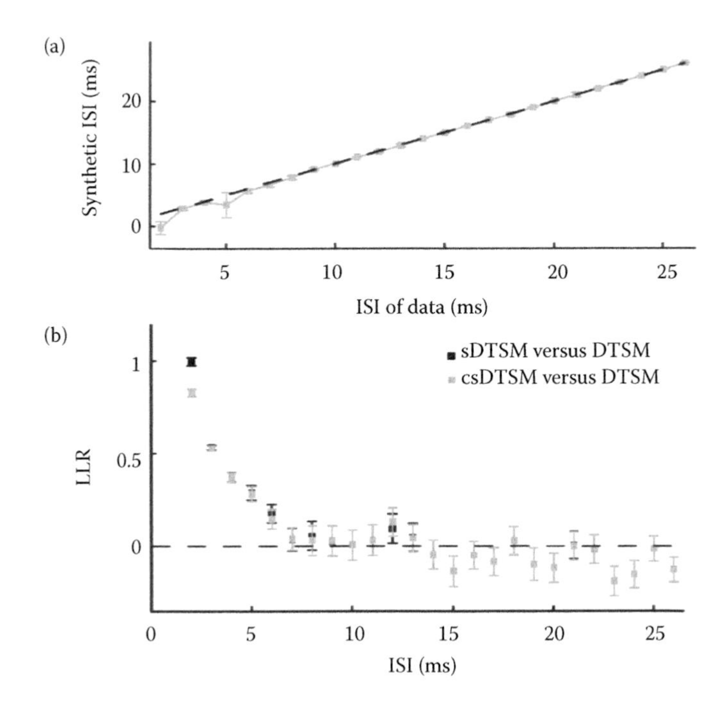

FIGURE 8.9 Nonlinear compression is not sufficient to explain the observations. (a) Nonlinear mapping between input ISI (x axis) and best-matched synthetic ISI (y axis), determined from peaks in likelihood. (b) Effects of nonlinear compression on estimates of log-likelihood ratios (LLRs). The black points show LLR between synthetic ("synth mod 1") and data-based doublet models, gray points show LLR between synthetic model modified by nonlinear compression ("synth mod 2") and data-based doublet models.

form of dimensional reduction, conceptually similar to principal components analysis (PCA). PCA has been used previously to examine the difference between burst- and single-spike-triggered stimulus ensembles in model neurons [56]. The difference (and, for this application, the distinct benefit) of iSTAC is that it is guaranteed to preserve the most information about the distinction between the two spaces, as assessed using Kullbach–Leibler (KL) divergence, for any given dimensionality of the subspace. The maximally informative subspace of the specified dimensionality provides the most compact description of the difference between the two models when the iSTAC assumptions are met [24]. When stimulus distributions are not well characterized by normal models, the iSTAC will provide an upper bound of the relevant space dimensionality, computing a low-dimensional linear subspace that contains an embedding of the relevant structure, which may be even lower-dimensional nonlinear manifolds. We do not search for such additional structures here.

In our case, the two multidimensional Gaussian models we wished to compare were the sDTSM and the DTSM. We were interested only in quantifying model differences that were potentially important in decoding responses. Since data-based and synthetic models for ISIs in which the LLR was not significantly different from zero were (by definition) equally good at decoding responses, we focused the comparison on the range for which the data-based model outperformed the synthetic model. For the cell analyzed here, this corresponded to ISI <8 ms. iSTAC analysis allowed us to characterize the difference in models using a small number of dimensions, ranging from a single dimension (i.e., a single vector representing the axis of greatest divergence between the two-model distributions) up to the full dimensionality of the original models. With iSTAC, we could also quantify in bits how much of the difference between the models was captured at each level of reduction.

Figure 8.10 shows the results for iSTAC analysis. The ISIs shown here were chosen from the region where the doublet outperformed the synthetic model in the LLR test, as represented by the 2 ms ISI models in Figure 8.10a–d. For the sake of visual clarity, the means, covariances, and iSTAC dimensions in Figure 8.10a–d are shown in the original 50-dimensional space, even though calculations were varied in the reduced-dimensionality space.

Figure 8.10a shows the mean of both the DTSM (red) and the csDTSM (purple). These represent the average predictions of the respective models for stimuli preceding a 2 ms response pattern. These two waveforms clearly differ, with the greatest visual difference occuring in the regions where the csDTSM overpredicts the stimulus (14–12 ms before the response pattern), and underpredicts the stimulus (6–3 ms before the response pattern). As shown previously, these differences are significant with respect to coding. The covariance for the DTSM and csDTSM is shown in Figure 8.10b and c, respectively. Once again, differences in the model are noticeable by the eye. Here, the diagonal elements from 21 to 9 ms prior to the response are overpredicted by the synthetic model.

The first three vectors describing the maximally informative subspace between the synthetic and doublet models are shown in Figure 8.10d. Note that instead of lines expressing the mean of the iSTAC vectors, we use shaded regions to depict the mean ± SD obtained from carring out iSTAC on every model from the 10× cross-validation (shaded area = 2SD), demonstrating the repeatability of the analysis. In the case of the 2 ms model, the power in the three most informative dimensions is concentrated in a continuous stimulus region from approximately 14 to 3 ms prior to the second spike in the response (Figure 8.10d). This space corresponds to the regions of dissimilarity from visual inspection, and can be subdivided into a region from 14 to 12 ms before the spike where the synthetic model overpredicts both the mean and variance relative to the data-based model, and a second region from 6 to 3 ms prior to the spike where the reverse is true (compare red and purple traces in Figure 8.10a). This means that to improve the predictive power of the synthetic model the most, we should increase the synthetic mean and covariance in the –3 to –6 ms region, and decrease them in the –12 to –14 ms region. This can be accomplished by scaling along the iSTAC dimensions.

In addition to showing which stimulus dimensions are most informative, we use the iSTAC analysis to quantify the extent to which the DTSM and linear sDTSM differ. This is accomplished by calculating the *K–L* divergence between the two models for each iSTAC dimension, which gives a measure of how well that dimension explains the difference between the two models in the information-theoretic units of bits. The normalized cumulative information recovered for using subspaces of various sizes up to 12 dimensions is shown in Figure 8.10e for models of 2, 5, and 8 ms ISIs. We see that dimensional reduction with the least loss of information

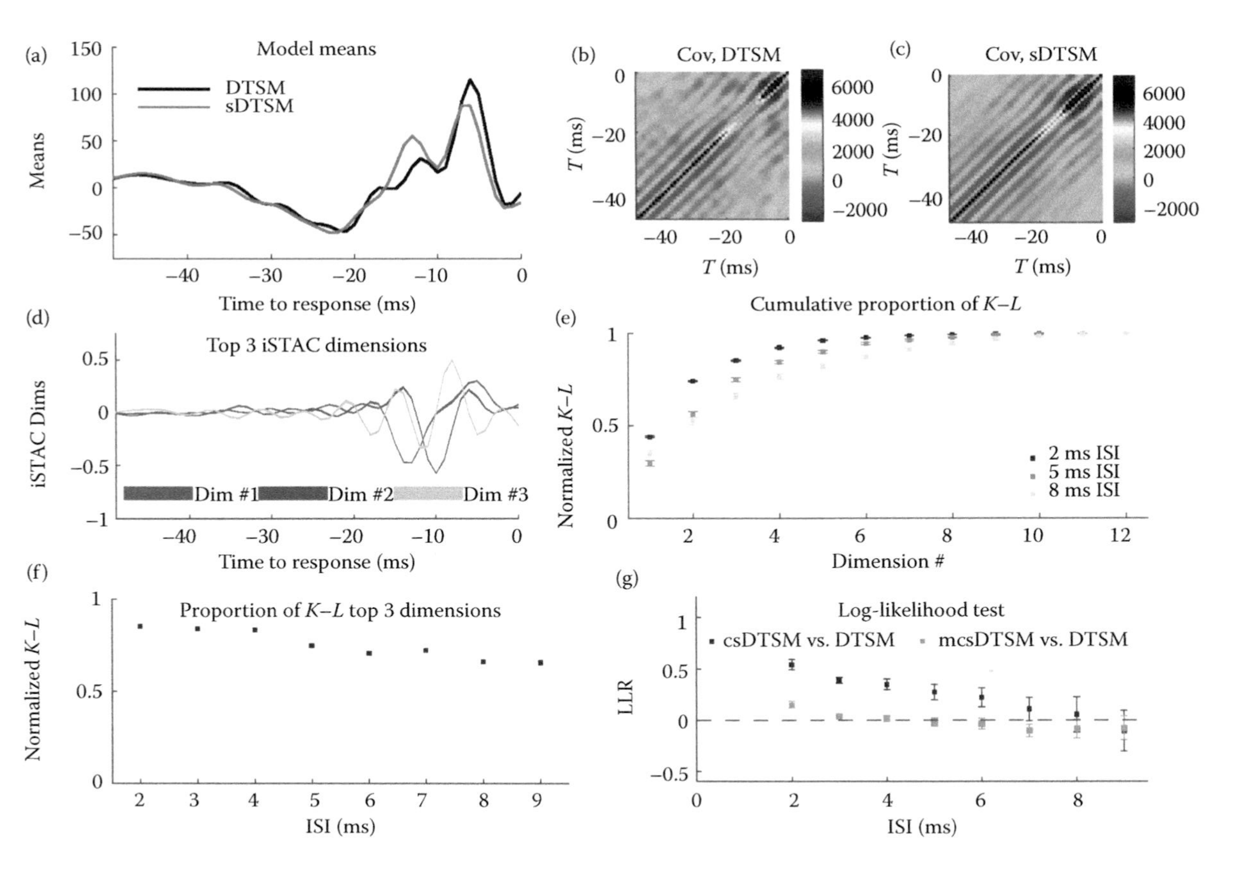
(a)
Model means
DTSM
sDTSM
Means
Time to response (ms)
(b)
Cov, DTSM
T (ms)
(c)
Cov, sDTSM
T (ms)
6000
4000
2000
0
−2000
(d)
Top 3 iSTAC dimensions
iSTAC Dims
Dim #1
Dim #2
Dim #3
Time to response (ms)
(e)
Cumulative proportion of K–L
Normalized K–L
2 ms ISI
5 ms ISI
8 ms ISI
Dimension #
(f)
Proportion of K–L top 3 dimensions
Normalized K–L
ISI (ms)
(g)
Log-likelihood test
csDTSM vs. DTSM
mcsDTSM vs. DTSM
LLR
ISI (ms)

FIGURE 8.10

is accomplished with the 2 ms models. In comparison, the longer ISIs require more dimensions to describe an equivalent amount of information about the differences between models. This point is underscored in Figure 8.10f, where we show the proportion of total information contained in the subspace containing iSTAC dimensions 1–3 for each ISI from 2 to 9 ms. Over 85% of the difference is captured by a three-dimensional subspace for the 2 ms ISI, whereas only ~65% of the difference is captured in the case of the 9 ms ISI.

Taken together, these results indicate that, unlike the compressive nonlinearity hypothesis, for short ISIs changes in a relatively small subspace of the synthetic model would cause substantial improvements in that model's performance. We tested this notion by modifying the synthetic model for each ISI so that they were identical to the corresponding DTSMs in a three-dimensional iSTAC subspace, but were unchanged along the remaining dimensions. We refer to such synthetic models as the modified compressed synthetic DTSM (mcsDTSM). Figure 8.10g shows the results of LLR analysis carried out for the csDTSM versus the DTSM, as well as for the mcsDTSM versus the DTSM, for ISIs between 2 and 9 ms. For each ISI from 2–7 ms, the LLR decreased significantly for the mcsDTSM in comparison with the csDTSM. In this case, the LLR stopped being significant at the 95% confidence level for all ISIs greater than 4 ms. This indicates that the iSTAC dimensions do indeed capture the differences between the models that are important for decoding neural activity. Note that although the three-dimensional subspace explains the greatest percentage of difference for the shortest ISIs (Figure 8.10f), these same ISIs have the greatest LLR difference between sDTSM and DTSM, and hence for the shortest ISIs, the mcsDTSM does not explain the data as well as the DTSM. These results further indicate that the deviations of these cells from linearity, previously shown in Figure 8.8, can be quantified using a dimensionally compact descriptor.

8.3 DISCUSSION

8.3.1 Temporal-Encoding Hypotheses

The nature of the neural code has long been studied. While early work such as that of Adrian showed that much of the information about a stimulus is contained

FIGURE 8.10 **(See color insert.)** iSTAC analysis of data-based and synthetic models. (a) Mean of data-based (red) and synthetic (purple) multivariate Gaussian models for stimulus preceding a 2 ms doublet, from the same cell as in Figure 8.8. The covariance of data-based and synthetic models shown in panels (b) and (c), respectively (color scale in mm^2/s^2). (d) Estimate of the three most informative iSTAC dimensions (shaded area indicates mean ± SD across 10× validation). (e) Measure of the total normalized *K–L* divergence between data-based and synthetic models for 2, 5, and 8 ms, as a function of subspace dimensionality. The mean ± SD across 10× validation is shown with error bars, which are on the order of the size of the markers for the points. (f) Measure of the portion of the total *K–L* divergence explained by the subspace containing the three largest iSTAC vectors, as a function of ISI in the model. (g) Improvement of the synthetic model performance in LLR tests from Figure 8.9c (black markers) by modification along the three-dimensional subspace shown in panel (d) (cyan markers). The error bars represent ±95% CIs on the mean.

in the firing rate of a neural response [1], more sophisticated analyses have demonstrated that information about the stimulus can be extracted from the timing of individual spikes in the neural response [7,20,87]. Additionally, it has been shown that neurons are capable of responding with as much temporal precision as 1 ms [4,6,10,65,85]. This has led to the hypothesis that neurons might use a temporal code, through which multispike patterns are used to represent stimuli that are distinct from those stimuli that could be predicted based on consideration of individual spikes [99]. Recent work in several systems has purported to show various types of temporal encoding with respect to this definition [9,61,73,75,77,84]. Our results are consistent with this temporal-encoding hypothesis, where high-frequency doublets (2–8 ms) are used to represent stimuli composed of frequencies less than 200 Hz. The results also indicate how the stimuli corresponding to these doublets differ from those stimuli that can be represented by sums of appropriately offset linear kernels.

8.3.2 Temporal Precision of Multispike Code Words

Several factors have been identified that would act to constrain the effectiveness of temporal codes. In particular, the upper bound on the duration of multiple-spike code words is imposed by the biophysical constraints on decoding and by selective pressure on the reaction time of the animal in making a decision based on sensory input. Similarly, the lower bound on the duration of multiple-spike code words is imposed by the refractory period of the cell and by the limiting temporal uncertainty in the stimulus–response relationship [5,20,75]. One specific factor contributing to the temporal uncertainty in stimulus–response relations is the inherent limiting noisiness or "jitter" in spike timing. While cells driven by dynamic, large-amplitude stimuli tend to minimize this jitter [4,6,11,13,21,38,45,65,82], the temporal variability of single spikes must limit the ability of a neuron to transfer information with precise patterns. This limit would become severe especially if noise from single spikes summed independently. We have recently shown that isolated single spikes recorded from cercal interneurons exhibit a stimulus-to-response jitter of ~2.2 ms around the mean latency [3]. Here, we have repeated that analysis, and further extended it in several ways. First, we show that doublet events with ISIs of less than 30 ms have event onset jitter considerably tighter than the onset variability of single spikes across trials (Figure 8.4e), which is in agreement with modeling studies [56]. Second, we show that this stimulus–response jitter does not independently affect spikes in short-interval doublets, and that the times of occurrence of spikes in doublets with average ISIs of 5 ms or less are tightly correlated, exceeding even the precision expected from consideration of a refractory period as shown by Berry and Meister [5]. Such correlations, where repeated stimuli elicit nearly identical patterns of spikes, are necessary for a temporal code to be able to efficiently transmit information about the stimulus. Such mechanisms have been theoretically implicated in models of visual cortex [56], and indeed, highly reproducible ISIs have been shown to exist in the presence of noise in several vertebrate sensory systems [5,75,102], suggesting that this form of temporal encoding is not restricted to the insect realm. The analysis reported here also derives an estimate of the minimum "word length" of temporal patterns distinct from single spikes in this set of neurons (≤8 ms), and should be used as the first step in determining parameters for analyses of dynamical neural coding [24,26,27].

8.3.3 Implications for Information Transmission

Information-theoretic analysis has proven to be a useful tool in determining the coding schemes of many different sensory systems. Two of the most popular methods of information-theoretic analysis in neuroscience, the direct method and linear stimulus reconstruction, each have distinct advantages. These methods of estimating information make different assumptions about how information is represented. The reconstruction method assumes that each spike in a group can be decoded independently, and provides the decoding. The direct method can capture most of the information but does not make specific coding assumptions, hence does not provide important functional details. We showed that in our case, these methods identify different amounts of information; hence, we can rigorously conclude that spike patterns have a meaning beyond that of their components.

One of the possible reasons for this discrepancy in information rates is that any aspects of the stimuli encoded by temporal patterns in the nervous system would not be accounted for in calculations using stimulus reconstruction [8,11,78]. Our analysis shows that a significant proportion of this information gap can be attributed to assumptions about the temporal variability implicit in reconstruction methodology (e.g., Figures 8.7, 8.6, and Table 8.1). This result is in agreement with the work of Bialek and colleagues, as well as several other studies that have measured the information contained in specific patterns of spikes [9,20,22,33,39].

In particular, de Ruyter van Steveninck and Bialek showed that in the H1 neurons of flies, ISIs of 10 ms and less transmitted the most information about the stimulus (their Figure 8.7). Similarly, in the same system, Brenner et al. showed that ISIs of 6 ms and less provided greater "event information" than longer ISIs (their Figure 8.3). In both studies, the authors attempted to estimate the mutual information tied to specific response events. Here, we use a complementary approach, characterizing only the conditional entropy for specific response events (Figures 8.6h and 8.7b), and then relating that to estimates of the total mutual information (Figure 8.7d and e). Our results are consistent with that work, showing that ISIs less than 10 ms are capable of carrying more information about the stimulus than longer-interval doublets.

The previous studies have shown that coherence-based estimates of information rates are higher in neurons at the sensory periphery than in neurons at higher processing layers [15,18,63,90,100]. Here, we have extended that analysis by examining various estimates of the temporal precision of neurons in afferent layers of different insect mechanosensory systems as well as first-order projecting interneurons. We showed that the afferent neurons respond to stimuli with greater temporal precision, whether that precision is based on measurements to repeated stimuli or by attempting to remove jitter from spike-triggered ensembles. This finding of decreasing stimulus-locked precision at higher layers is similar to findings in the vertebrate visual pathway [54,103]. In these studies simultaneous recordings from retina, thalamus, and cortex showed an increasing variability in the spike rate in small counting windows at successive processing layers within the system, while our results speak to the timing of individual spikes at successive layers. Our results further suggest that the reliable temporal patterning of short ISI "bursts" in the interneurons arise from processing of the postsynaptic afferents, rather than simply being a feed-forward phenomenon. Finally, our results show that in spite of their relatively large temporal uncertainty, the interneurons are at least as selective as the afferents in which features of the stimulus they respond to, once temporal invariances (such as those removed by dejittering) have been taken into account.

8.3.4 Biological Relevance

It is important to consider possible neural-coding schemes within a broader neuroethological context. The cercal system of crickets has been shown to be responsive to acceleration due to gravity [91], to the touch of approaching predators [29,72], to air movement caused by the approach of predators [14,41,42,71], and to air movement generated by the stridulation of nearby conspecifics [51–53]. All these types of stimuli activate the cercal filiform mechanosensors, which synapse onto the interneurons studied here. Although the precise synaptic connectivity with higher-order neurons is unknown in the cricket cercal system, it is known that these giant interneurons have axonal arborizations in the thoracic ganglia that connect to motor nerves, as well as arborizations in the mechanosensory centers in the protocerebrum [46,69]. The neurons with multimodal sensitivities (including sensitivity to airflow) project out of these areas and can effect behavior related to locomotion [44,47,94]. It is unclear what role the cercal system plays in specific behaviors such as phonotaxis and courtship [80], although at the very least, the relatively few cercal filiform interneurons must carry enough information to allow the animal to distinguish between the signature of an approaching predator and the infrasound components of conspecific calling songs. A myriad of different encoding schemes for representing this information can be imagined, including one where different postsynaptic neurons use short-term depression and facilitation to selectively filter for specific ISI durations in bursts [48]. There is, in fact, a strong evidence that crickets specifically use short bursts at the interneuron level to trigger evasive responses [66]. In addition, there is evidence that thoracic motorneurons that receive input from the cercal system undergo facilitation, as short presynaptic bursts trigger spiking response in the motorneurons, whereas presynaptic single spikes do not [88]. Further, it has been shown that direct intracellular current injection into neurons 10-2a and 10-3a can elicit escape-like running responses within tens of milliseconds [46]. Here, we show that stimuli that include the frequency content of both predatory and conspecific stimuli elicit both single-spike and doublet-spiking responses, and that these two response types represent distinct information about the stimuli.

A plausible working hypothesis is that the short-interval spike doublets we characterize here are the symbolic correlate of a component of the animal's evasive response, whereas single isolated spikes mediate detection of other sensory signals.

APPENDIX

Information-Theoretic Calculations

The models of ISI variability and onset precision were used to calculate information rates relating to various assumptions on the correlation between nearby spikes. This was done by assuming independence between onset jitter and ISI variability. For each ISI in each model, the probability of a response pattern conditioned on a stimulus was approximated by determining the temporal correlation between spikes for the ISI/model pair, determining the variance of the corresponding ISI using Equation 8.6 (representing the variability of ISIs conditioned on a stimulus), and then adding the onset jitter squared appropriate to each model (representing the variance in latency of patterns conditioned on a stimulus, see the previous section). The square root of the resulting sum was used as the SD to generate a normal probability density function representing the total pattern variability conditioned on the occurrence of a stimulus. The conditional entropy for the ISI, H_C, was calculated according to

$$H_C(\text{ISI}) = -\sum_{\text{ISI}} p(\text{ISI})\log_2[p(\text{ISI})] \tag{A.1}$$

This yielded the conditional entropy per stimulus event. To transform this into a rate we weighted by the probability of each ISI occurring in our *data set* (we used the ISI histogram as a surrogate, representing the probability of our white-noise stimulus eliciting a given pattern), and then multiplied this value by the rate of occurrence of ISIs in the recording (number of spikes in the recording divided by the length of the recording). The unconditional or total response entropy rate was calculated using only the ISI histogram plugged into Equation A.1, multiplied by the rate of occurrence of ISIs. The mutual information of the models was estimated as the difference in the two entropy rates. These model values were compared with entropy and information rates calculated from our data using the CTW method [97], as well as information rates using stimulus reconstruction methods [7,87] obtained through a multitaper calculation of the coherence function [50,100].

Response-Conditioned Stimulus Models

Three distinct response-conditioned stimulus models were developed: two doublet-conditioned models and a singlet-conditioned model. For the first doublet-conditioned model, all the well-isolated doublets with a given interspike interval, which were neither preceded nor followed by other spikes within a 20 ms window, were located. Note that this definition of doublets necessarily differs from the definition used in finding doublet responses from repeated presentations of a single stimulus described in the previous section. The stimulus segments starting 50 ms prior to the second spike of all doublet events were collected to form the doublet-triggered stimulus ensemble (DTSE). 10% of these stimuli were held out for later cross-validation as the test doublet-triggered stimulus ensemble (tDTSE), whereas the remaining

90% of the ensemble was used to build the DTSM. This consisted of the mean, μ_D, and the covariance matrix of the ensemble, C_D, both sampled at 1 kHz.

To build the singlet model and the second doublet model, we identified all the single spikes isolated by the same criteria used for the doublets (i.e., no other spikes in a 20 ms window around the spike). All the stimulus segments preceding the isolated spikes were collected, extending from 50 to 1 ms prior to the spike. The entire singlet-triggered stimulus ensemble (STSE) was then used to build a STSM with mean μ_s and covariance C_s. To form a synthetic mean of the stimulus for the doublet (μ_{sD}), μ_s was replicated, shifted in time, and summed according to

$$\mu_{sD}(t) = \mu_s(t) + \mu_s(t + \mathrm{ISI}) \tag{A.2}$$

where ISI represents the interspike interval of the desired model. The synthetic covariance of the stimulus for the doublet (C_{sD}) was calculated according to

$$C_{sD}(t_1,t_2) = \alpha\left(C_s(t_1,t_2) + C_s(t_1 + \mathrm{ISI}, t_2 + \mathrm{ISI})\right) \tag{A.3}$$

where

$$\alpha = \left(\frac{det(C_S)}{det\left(C_S(t_1,t_2) + C_s(t_1 + \mathrm{ISI}, t_2 + \mathrm{ISI})\right)}\right)^{1/n} \tag{A.4}$$

ensured that the sDTSM operated over approximately the same volume in stimulus space as the DTSM. The determinant is taken over all eigenvalues of C, and is implemented as the equivalent MATLAB expression *sum(log(eig(C)))* for the respective C_S, to avoid multiplication of a large set of numbers. See Ref. [2] for derivations of Equations A.3 and A.4.

Likelihood Tests

The relative abilities of the STSM, the sDTSM, and the DTSM to predict the stimuli preceding a doublet response were tested with a simple log-likelihood test. The log likelihood L for each sample x of the tDTSE coming from each model was calculated as

$$L = -\frac{1}{2}\left((x-\mu)C^{-1}(x-\mu)^T - n\log(2\pi) - \log\left(|C|\right)\right) \tag{A.5}$$

where n is the dimensionality of the model, μ and C are the mean and covariance of the model being tested (the STSM, DTSM, or sDTSM), log is the natural logarithm, $(\bullet)^T$ represents the transpose of the matrix, and $|\bullet|$ represents the determinant of the matrix. The difference of log-likelihood values, LDTSM-LSTSM, and LDTSM-LsDTSM were then calculated to determine the log-likelihood ratios (LLRs). Samples with LLRs greater than zero were more likely to have been elicited by the data-based model, whereas samples with LLRs of zero were equally likely to have been elicited by either model in the test. Prior to carringout likelihood tests, all models and test samples were projected into a reduced space to overcome spurious effects due to band-limited stimuli [24]. Owing to the large data demands of the multivariate models, we removed ISIs from experiments that had less than 80 samples. To avoid biasing due to large outliers, we also removed LLR values with absolute values greater than three SDs from the mean. This typically amounted to less than 2% of the available samples.

When visualizing the LLR distribution versus the ISI of the respective models, we modeled the observed decay with a sum of exponentials identical to Equation 8.4:

$$\mathbf{LLR(ISI)} = ae^{-(\mathbf{ISI}/\tau)} + b \tag{A.6}$$

We obtain the parameters of the model as a least-squared fit in a manner identical to the analysis of Equation 8.5. This functional relation was selected among several by again using the AIC.

iSTAC Analysis

We employed information-theoretic Spike-Triggered Average and Covariance (iSTAC) [79] to compactly describe the difference between the DTSM and the sDTSM. iSTAC finds a subspace that maintains as much as possible of the KL divergence between two distributions. We briefly summarize the method of Pillow and Simoncelli [79]:

iSTAC assumes that the probability of a stimulus x given a certain condition is a multivariate normal distribution:

$$P(x) = \mathcal{N}(x;\mu,C) \tag{A.7}$$

In this probabilistic formulation, the difference between two data sets is characterized by the KL divergence

$$D(P,P') = E_P \log_2 \frac{P(x)}{P'(x)} \tag{A.8}$$

where $P(x)$ is the base probability against which differences are sought (in our case, either the sDTSM or the sDTSM + DTSM), $P'(x)$ is the probability that needs to be discriminated (either the DTSM or the raw stimulus), and E_P is the expectation (integral) over the base probability. $D(P, P')$ is an information-theoretic quantity characterizing the difference between the two distributions in bits.

Since we were only interested in relative comparison between two distributions, the base probability $P(x)$ was rescaled to have zero mean and an identity–covariance matrix. Therefore, we let

$$\mu_W = C^{-1/2}(\mu - \mu') \tag{A.9}$$

and

$$C_W = C^{-1/2}\ C'C^{-1/2} \tag{A.10}$$

be the DTSM mean and covariance, respectively whitened against the base–stimulus probability. This allows us to simplify Equation A.8 to

$$D(P,P') = \frac{1}{2}\left(Tr(C_W) - \log|C_W| + \mu_W^T\mu_W - n\right) \tag{A.11}$$

where $Tr(\bullet)$ represents the trace of the matrix. We then specify an m-dimensional linear subspace defined by an orthonormal basis B in which $D(P, P')$ satisfies:

$$D_{[B]}(P,P') = \frac{1}{2}\left(Tr\left[B^T(C_W + \mu_W\mu_W^T)B\right] - \log|B^TC_WB| - m\right) \tag{A.12}$$

The most informative subspace is described by the matrix B that maximizes Equation A.12. We analyzed a variety of dimensionalities m, ranging from single-dimensional to the full dimensionality of the models.

REFERENCES

1. E. D. Adrian and Y. Zotterman. The impulses produced by sensory nerve-endings: Part II. The response of a single end-organ. *J Physiol*, 61(2):151–171, 1926.
2. Z. N. Aldworth, A. G. Dimitrov, G. I. Cummins, T. Gedeon, and J. P. Miller. Temporal encoding in a nervous system. *PLoS Comp Biol*, (7(5):e1002041.doi:10.1371/journal.pcbi.1002041), 2011.
3. Z. N. Aldworth, J. P. Miller, T. Gedeon, G. I. Cummins, and A. G. Dimitrov. Dejittered spike-conditioned stimulus waveforms yield improved estimates of neuronal feature selectivity and spike-timing precision of sensory interneurons. *J Neurosci*, 25(22):5323–5332, 2005.
4. W. Bair and C. Koch. Temporal precision of spike trains in extrastriate cortex of the behaving macaque monkey. *Neur Comp*, 8(6):1185–1202, 1996.
5. M. J. Berry 2nd and M. Meister. Refractoriness and neural precision. *J Neurosci*, 18(6):2200–2211, 1998.
6. M. J. Berry 2nd, D. K. Warland, and M. Meister. The structure and precision of retinal spike trains. *Proc Natl Acad Sci USA*, 94(10):5411–5416, 1997.
7. W. Bialek, F. Rieke, R. R. de Ruyter van Steveninck, and D. Warland. Reading a neural code. *Science*, 252:1854–1857, 1991.
8. A. Borst and F. E. Theunissen. Information theory and neural coding. *Nat Neurosci*, 2(11):947–957, 1999.
9. N. Brenner, S. P. Strong, R. Koberle, W. Bialek, and R. R. de Ruyter van Steveninck. Synergy in a neural code. *Neur Comp*, 12(7):1531–1552, 2000.
10. H. L. Bryant and J. P. Segundo. Spike initiation by transmembrane current: A white-noise analysis. *J Physiol*, 260:279–314, 1976.
11. G. T. Buracas, A. M. Zador, M. R. DeWeese, and T. D. Albright. Efficient discrimination of temporal patterns by motion-sensitive neurons in primate visual cortex. *Neuron*, 20(5):959–969, 1998.

12. K. P. Burnham and D. R. Anderson. *Model Selection and Multi-Model Inference: A Practical Information-Theoretic Approach*. Springer, New York, second edition, 2002.
13. D. A. Butts, C. Weng, J. Jin, C.-I. Yeh, N. A. Lesica, J.-M. Alonso, and G. B. Stanley. Temporal precision in the neural code and the timescales of natural vision. *Nature*, 449(7158):92–95, 2007.
14. J. M. Camhi, W. Tom, and S. Volman. The escape behavior of the cockroach *Periplaneta americana* II: Detection of natural predators by air displacement. *J Comp Physiol [A]*, 128(3):203–212, 1978.
15. M. J. Chacron. Nonlinear information processing in a model sensory system. *J Neurophys*, 95(5):2933–2946, 2006.
16. T. R. Chang, T. W. Chiu, P. C. Chung, and P. W. F. Poon. Should spikes be treated with equal weightings in the generation of spectro-temporal receptive fields? *J Physiol-Paris*, 104(3–4):215–222, 2010.
17. T.-R. Chang, P.-C. Chung, T.-W. Chiu, and P. W.-F. Poon. A new method for adjusting neural response jitter in the {STRF} obtained by spike-trigger averaging. *BioSystems*, 79:213–222, 2005.
18. H. Clague, F. E. Theunissen, and J. P. Miller. Effects of adaptation on neural coding by primary sensory interneurons in the cricket cercal system. *J Neurophys*, 77(1):207–220, 1997.
19. T. Daniel, Z. Aldworth, A. Hinterwirth, and J. Fox. *Insect Inertial Measurement Units: Gyroscopic Sensing of Body Rotation*, page in press. Springer, Wien-New York, 2011.
20. R. R. de Ruyter van Steveninck and W. Bialek. Real-time performance of a movement-sensitive neuron in the blowfly visual system: Coding and information transfer in short spike sequences. *Proc Roy Soc [B]*, 234:379–414, 1988.
21. R. R. de Ruyter van Steveninck, G. D. Lewen, S. P. Strong, R. Koberle, and W. Bialek. Reproducibility and variability in neural spike trains. *Science*, 275:1805–1808, 1997.
22. K. S. Denning and P. Reinagel. Visual control of burst priming in the anesthetized lateral geniculate nucleus. *J Neurosci*, 25(14):3531–3538, 2005.
23. M. H. Dickinson. Directional sensitivity and mechanical coupling dynamics of campaniform sensilla during chordwise deformations of the Fly wing. *J Exp Biol*, 169(1):221–233, 1992.
24. A. G. Dimitrov, G. I. Cummins, A. Baker, and Z. N. Aldworth. Characterizing the fine structure of a neural sensory code through information distortion. *J Comp Neurosci*, 30(1):163–179, 2011.
25. A. G. Dimitrov and T. Gedeon. Effects of stimulus transformations on estimates of sensory neuron selectivity. *J Comp Neurosci*, 20(3):265–283, 2006.
26. A. G. Dimitrov and J. P. Miller. Neural coding and decoding: Communication channels and quantization. *Netw: Comput Neural Systems*, 12:441–472, 2001.
27. A. G. Dimitrov, J. P. Miller, T. Gedeon, Z. Aldworth, and A. E. Parker. Analysis of neural coding through quantization with an information-based distortion measure. *Netw: Comput Neural Systems*, 14:151–176, 2003.
28. U. Dombrowski. *Untersuchungen zur funktionellen Organisation des Flugsystems von Manduca sexta (L.)*. PhD thesis, University of Cologne, 1991.
29. K. Dumpert and W. Gnatzy. Cricket combined mechanoreceptors and kicking response. *J Comp Physiol [A]*, 122(1):9–25, 1977.
30. O. J. Dunn. Multiple comparisons among means. *J Am Stat Assoc*, 56(293):52–64, 1961.
31. R. C. Elson. Flight motor neuron reflexes driven by strain-sensitive wing mechanoreceptors in the *locust*. *J Comp Physiol [A]*, 161(5):747–760, 1987.
32. M. A. Escabi, R. L. Nassiri, M. Miller, C. E. Schreiner, and H. L. Read, The contribution of spike threshold to acoustic feature selectivity, spike information content, and information throughput. *J Neurosci*, 25(41):9524–9534, 2005.

33. H. G. Eyherabide, A. Rokem, A. V. Herz, and I. Samengo. Burst firing is a neural code in an insect auditory system. *Front Comp Neurosci*, 2:3, 2008.
34. J. L. Fox and T. L. Daniel. A neural basis for gyroscopic force measurement in the halteres of *Holorusia*. *J Comp Physiol [A]*, 194(10):887–897, 2008.
35. J. L. Fox, A. L. Fairhall, and T. L. Daniel. Encoding properties of haltere neurons enable motion feature detection in a biological gyroscope. *Proc Natl Acad Sci USA*, 107(8):3840–3845, 2010.
36. F. Gabbiani, W. Metzhner, R. Wessel, and C. Koch. From stimulus encoding to feature extraction in weakly electric fish. *Nature*, 384:564–567, 1996.
37. R. L. Gaines. Development of campaniform sensilla on the wing of the *tobacco hornworm*, manduca sexta. Master's thesis, Kansas State University, Manhattan, 1979.
38. R. F. Galan, G. B. Ermentrout, and N. N. Urban. Optimal time scale for spike-time reliability: Theory, simulations, and experiments. *J Neurophysiol*, 99(1):277–283, 2008.
39. K. S. Gaudry and P. Reinagel. Information measure for analyzing specific spiking patterns and applications to LGN bursts. *Netw: Comput Neural Syst*, 19(1):69–94, 2008.
40. E. Gettrup. Sensory regulation of wing twisting in *Locusts*. *J Exp Biol*, 44(1):1–16, 1966.
41. W. Gnatzy and R. Heusslein. Digger *Wasp* against *Crickets* I: Receptors involved in the antipredator strategies of the prey. *Naturwissenschaften*, 73(4):212–215, 1986.
42. W. Gnatzy and G. Kämper. Digger *Wasp* against *Crickets* II: An airborne signal produced by a running predator. *J Comp Physiol [A]*, 167(4):551–556, 1990.
43. T. Gollisch. Estimating receptive fields in the presence of spike-time jitter. *Netw: Comput Neural Syst*, 17:103–129, 2006.
44. H. Gras, M. Hörner, L. Runge, and F. W. Schürmann. Prothoracic DUM neurons of the cricket *Gryllus bimaculatus*—Responses to natural stimuli and activity in walking behavior. *J Comp Physiol [A]*, 166(6):901–914, 1990.
45. J. Haag and A. Borst. Encoding of visual motion information and reliability in spiking and graded potential neurons. *J Neurosci*, 17(12):4809–4819, 1997.
46. K. Hirota, Y. Sonoda, Y. Baba, and T. Yamaguchi. Distinction in morphology and behavioral role between dorsal and ventral groups of cricket giant interneurons. *Zoolog Sci*, 10(4):705–709, 1993.
47. M. Hörner. Wind-evoked escape running of the cricket *Gryllus bimaculatus*: II. Neurophysiological analysis. *J Exp Biol*, 171(1):215–245, 1992.
48. E. M. Izhikevich, N. S. Desai, E. C. Walcott, and F. C. Hoppensteadt. Bursts as a unit of neural information: Selective communication via resonance. *Trends Neurosci*, 26(3):161–167, 2003.
49. G. A. Jacobs, J. P. Miller, and Z. Aldworth. Computational mechanisms of mechanosensory processing in the cricket. *J Exp Biol*, 211(11):1819–1828, 2008.
50. M. R. Jarvis and P. P. Mitra. Sampling properties of the spectrum and coherency of sequences of action potentials. *Neur Comp*, 13(4):717–749, 2001.
51. G. Kämper. Abdominal ascending interneurons in crickets: Responses to sound at the 30-Hz calling-song frequency. *J Comp Physiol [A]*, 155(4):507–520, 1984.
52. G. Kämper and M. Dambach. Response of the cercus-to-giant interneuron system in crickets to species-specific song. *J Comp Physiol [A]*, 141(3):311–317, 1981.
53. G. Kämper and M. Dambach. Low frequency airborne vibrations generated by crickets during singing and aggression. *J Insect Physiol*, 31:925–929, 1985.
54. P. Kara, P. Reinagel, and R. C. Reid. Low response variability in simultaneously recorded retinal, thalamic, and cortical neurons. *Neuron*, 27(3):635–646, 2000.
55. T. A. Keil. Comparative morphogenesis of sensilla: A review. *Int J Insect Morphol Embryol*, 26(3–4):151–160, 1997.
56. A. Kepecs and J. Lisman. Information encoding and computation with spikes and bursts. *Netw: Comput Neural Systems*, 14(1):103–118, 2003.

57. P. Kloppenburg and M. Hörner. Voltage-activated currents in identified giant interneurons isolated from adult crickets *Gryllus bimaculatus*. *J Exp Biol*, 201(17):2529–2541, 1998.
58. R. Krahe and F. Gabbiani. Burst firing in sensory systems. *Nat Rev Neurosci*, 5:13–23, 2004.
59. M. Landolfa and G. A. Jacobs. Direction sensitivity of the filiform hair population of the cricket cercal system. *J Comp Physiol [A]*, 177:759–766, 1995.
60. M. A. Landolfa and J. P. Miller. Stimulus–response properties of cricket cercal filiform hair receptors. *J Comp Physiol [A]*, 177:749–757, 1995.
61. G. Laurent. A systems perspective on early olfactory coding. *Science*, 286(5440):723, 1999.
62. G. Laurent and W. Davidowitz. Encoding of olfactory information with oscillating neural assemblies. *Science*, 265:1872–1875, 1994.
63. J. E. Levin and J. P. Miller. Broadband neural encoding in the cricket cercal sensory system enhanced by stochastic resonance. *Nature*, 380(6570):165–168, 1996.
64. R. C. Liu, S. Tzonev, S. Rebrik, and K. D. Miller. Variability and information in a neural code of the cat lateral geniculate nucleus. *J Neurophys*, 86:2789–2896, 2001.
65. Z. G. Mainen and T. J. Sejnowski. Reliability of spike timing in neocortical neurons. *Science*, 268(5216):1503–1506, 1995.
66. G. Marsat and G. S. Pollack. A behavioral role for feature detection by sensory bursts. *J Neurosci*, 26(41):10542–10547, 2006.
67. P. Maršálek, C. Koch, and J. Maunsell. On the relationship between synaptic input and spike output jitter in individual neurons. *Proc Natl Acad Sci USA*, 94:735–740, 1997.
68. M. Meister and M. J. Berry. The neural code of the retina. *Neuron*, 22(3):435–450, 1999.
69. B. Mendenhall and R. K. Murphey. The morphology of cricket giant interneurons. *J Neurobiol*, 5(6):565–580, 1974.
70. J. P. Miller, G. A. Jacobs, and F. E. Theunissen. Representation of sensory information in the cricket cercal sensory system. {I. Response} properties of the primary interneurons. *J Neurophys*, 66:1680–1689, 1991.
71. J. Mulder-Rosi, G. I. Cummins, and J. P. Miller. The cricket cercal system implements delay-line processing. *J Neurophys*, 103:1823–1832, 2010.
72. R. K. Murphey. A second cricket cercal sensory system: Bristle hairs and the interneurons they activate. *J Comp Physiol [A]*, 156(3):357–367, 1985.
73. I. Nemenman, G. D. Lewen, W. Bialek, and R. R. de Ruyter van Steveninck. Neural coding of natural stimuli: Information at sub-millisecond resolution. *PLoS Comp Biol*, 4(3):e1000025, 2008.
74. H. Ogawa, G. I. Cummins, G. A. Jacobs, and K. Oka. Dendritic design implements algorithm for synaptic extraction of sensory information. *J Neurosci*, 28(18):4592–4603, 2008.
75. A. M. Oswald, B. Doiron, and L. Maler. Interval coding. I. Burst interspike intervals as indicators of stimulus intensity. *J Neurophys*, 97(4):2731–2743, 2007.
76. L. Paninski. Estimation of entropy and mutual information. *Neur Comp*, 15:1191–1253, 2003.
77. S. Panzeri and S. R. Schultz. A unified approach to the study of temporal, correlational and rate coding. *Neur Comp*, 13(6):1311–1349, 2001.
78. C. L. Passaglia and J. B. Troy. Information transmission rates of cat retinal ganglion cells. *J Neurophysiol*, 91(3):1217–1229, 2004.
79. J. W. Pillow and E. P. Simoncelli, Dimensionality reduction in neural models: An information-theoretic generalization of spike-triggered average and covariance analysis. *J Vis*, 6(4):414–428, 2006.
80. G. S. Pollack, V. Givois, and R. Balakrishnan. Air-movement "signals" are not required for female mounting during courtship in the cricket *Teleogryllus oceanicus*. *J Comp Physiol [A]*, 183(4):513–518, 1998.

81. D. S. Reich, F. Mechler, K. P. Purpura, and J. D. Victor. Interspike intervals, receptive fields, and information encoding in primary visual cortex. *J Neurosci*, 20:1964–1974, 2000.
82. D. S. Reich, J. D. Victor, B. W. Knight, T. Ozaki, and E. Kaplan. Response variability and timing precision of neuronal spike trains *in vivo*. *J Neurophysiol*, 77(5):2836–2841, 1997.
83. P. Reinagel, D. Goodwin, M. Sherman, and C. Koch. Encoding of visual information by {LGN} bursts. *J Neurophysiol*, 81:2558–2569, 1999.
84. P. Reinagel and R. C. Reid. Temporal coding of visual information in the thalamus. *J Neurosci*, 20(14):5392–5400, 2000.
85. P. Reinagel and R. C. Reid. Precise firing events are conserved across neurons. *J Neurosci*, 22(16):6837–6841, 2002.
86. B. J. Richmond, L. M. Optican, M. Podel, and H. Spitzer. Temporal encoding of two-dimensional patterns by single units in the primate inferior temporal cortex. {I. Response} characteristics. *J Neurophysiol*, 57:132–146, 1987.
87. F. Rieke, D. Warland, R. R. de Ruyter van Steveninck, and W. Bialek. *Spikes: Exploring the Neural Code*. The MIT Press, Cambridge, 1997.
88. R. E. Ritzmann and J. M. Camhi. Excitation of leg motor neurons by giant interneurons in the cockroach *Periplaneta americana*. *J Comp Physiol [A]*, 125(4):305–316, 1978.
89. J. C. Roddey, B. Girish, and J. P. Miller. Assessing the performance of neural encoding models in the presence of noise. *J Comp Neurosci*, 8:95–112, 2000.
90. J. C. Roddey and G. A. Jacobs. Information theoretic analysis of dynamical encoding by filiform mechanoreceptors in the cricket cercal system. *J Neurophysiol*, 75:1365–1376, 1996.
91. D. S. Sakaguchi and R. K. Murphey. The equilibrium detecting system of the cricket: Physiology and morphology of an identified interneuron. *J Comp Physiol [A]*, 150(2):141–152, 1983.
92. K.-H. Schäffner and U. T. Koch. A new field of wing campaniform sensilla essential for the production of the attractive calling song in crickets. *J Exp Biol*, 129(1):1–23, 1987.
93. K.-H. Schäffner and U. T. Koch. Effects of wing campaniform sensilla lesions on stridulation in crickets. *J Exp Biol*, 129(1):25–40, 1987.
94. K. Schildberger. Multimodal interneurons in the cricket brain: Properties of identified extrinsic mushroom body cells. *J Comp Phsyiol [A]*, 154(1):71–79, 1984.
95. M. N. Shadlen and W. T. Newsome. Noise, neural codes and cortical organization. *Curr Opin Neurobiol*, 4:569–579, 1994.
96. T. Shimozawa and M. Kanou. Varieties of filiform hairs: Range fractionation by sensory afferents and cercal interneurons of a cricket. *J Comp Physiol [A]*, 155:485–493, 1984.
97. J. M. Shlens, B. Kennel, H. D. I. Abarbanel, and E. J. Chichilnisky, Estimating information rates with confidence intervals in neural spike trains. *Neur Comp*, 19(7):1683–1719, 2007.
98. G. K. Taylor and H. G. Krapp. Sensory systems and flight stability: What do insects measure and why? *Insect Mech Control: Adv Insect Physio*, 34:231–316, 2008.
99. F. Theunissen and J. P. Miller. Temporal encoding in nervous systems: A rigorous definition. *J Comp Neurosci*, 2:149–162, 1995.
100. F. Theunissen, J. C. Roddey, S. Stufflebeam, H. Clague, and J. P. Miller. Information theoretic analysis of dynamical encoding by four identified primary sensory interneurons in the cricket cercal system. *J Neurophysiol*, 75(4):1345–1364, 1996.
101. V. Uzzell and E. J. Chichilniski. Precision of spike trains in primate retinal ganglion cells. *J Neurophys*, 92:780–789, 2004.
102. J. D. Victor. Temporal aspects of neural coding in the retina and lateral geniculate: A review. *Network* 10(4):R1–66, 1999.

103. X. Wang, V. Vaingankar, C. Soto Sanches, F. T. Sommer, and J. A. Hirsch. Thalamic interneurons and relay cells use complementary synaptic mechanisms for visual processing. *Nat Neurosci*, 14:224–231, 2011.
104. D. Warland, M. A. Landolfa, J. P. Miller, and W. Bialek. Reading between the spikes in the cercal filiform hair receptors of the cricket. In F. H. Eeckman, ed., *Analysis and Modeling of Neural Systems*, pp. 327–333. Kluwer Academic Publishers, Boston, MA, 1991.
105. D. K. Warland, P. Reinagel, and M. Meister. Decoding visual information from a population of retinal ganglion cells. *J Neurophysiol*, 78(5):2336–2350, 1997.
106. D. Zoccolan, G. Pinato, and V. Torre. Highly variable spike trains underlie reproducible sensorimotor responses in the medicinal leech. *J Neurosci*, 22:10790–10800, 2002.

9 Neural Encoding of Dynamic Inputs by Spike Timing

Matthew H. Higgs and William J. Spain

CONTENTS

Oscillations are virtually ubiquitous in neural signals and play a wide variety of roles in information coding, signal transmission, coordination of neuronal ensembles, and pathophysiology, including epilepsy. This chapter begins with some general principles about how neurons respond to oscillating input, proceed to selected results from model neurons, and end with some experimental data from neurons in the cerebral cortex and the auditory brainstem. While the chapter focuses on the analysis of responses to sinusoidal inputs, many of the results have more general relevance, as linear systems analysis is an important first step in understanding the dynamics of neuronal responses to any type of input.

9.1 BASIC PRINCIPLES

9.1.1 Subthreshold Voltage Responses

Although our major focus is on action potential (AP) responses, we begin by summarizing some principles about subthreshold voltage responses to alternating current (AC) input, to lay a foundation for understanding the AP response. If we consider the input to a neuron as the summed synaptic current, the subthreshold input–output relationship may be defined as $\Sigma I_{synaptic}(t) \rightarrow V(t)$. In experiments using applied current, we have $I_{app}(t) \rightarrow V(t)$. Here, we focus on cases where $I_{app}(t)$ has an AC component or is decomposed into sine waves by Fourier analysis.

The linear component of the transformation from $I_{app}(t)$ to $V(t)$ can be quantified by the impedance (Z), which is a function of frequency (f). The stimulus most often used to measure $Z(f)$ in neurons is a chirp, or sine wave of varying frequency (Hutcheon and Yarom, 2000). If the stimulus amplitude is not too large, the response is primarily linear. In other words, each frequency of current causes a voltage oscillation of the same frequency. Analysis of the response provides the impedance magnitude, $|Z(f)|$, which is the amplitude of the voltage response divided by the amplitude of the sine wave current. When f approaches zero, $|Z(f)|$ is equal to the input resistance of the neuron. Impedance data show that some neurons function as low-pass filters, with maximal impedance at $f = 0$, whereas others are band-pass, or resonant, with greatest impedance at a nonzero frequency, f_R (Figure 9.1). The impedance phase also provides information about a resonance, generally showing a rapid change from phase lead for $f < f_R$ to phase lag for $f > f_R$. In the time domain, a resonant neuron recovers from a current pulse with an overshoot or damped oscillation. Stronger resonance results in more cycles of oscillation and a sharper peak in the impedance plot.

The essential mechanism of subthreshold resonance is slow negative feedback; stronger resonance can occur when the neuron also has fast positive feedback (Hutcheon et al., 1996a,b; Hutcheon and Yarom, 2000). Common negative feedback mechanisms include voltage-gated K^+ channels, which produce an outward current that is activated by depolarization, and hyperpolarization-activated, cyclic nucleotide-gated (HCN) channels, which produce an inward current with the opposite voltage dependence. Kv7 channels (M current) commonly give rise to theta frequency (~4–7 Hz) resonance at potentials near spike threshold, whereas HCN channels often cause theta resonance at more hyperpolarized levels (Gutfreund et al., 1995; Hutcheon et al., 1996a,b; Hu et al., 2002, 2009; Peters et al., 2005). In contrast, fast delayed rectifiers such as Kv1 channels contribute to resonance at much higher frequency in some neurons (Hsiao et al., 2009) and in axons (Hodgkin and Huxley, 1952; Mauro et al., 1970).

Resonance at near-threshold potentials can be enhanced by fast positive feedback via persistent sodium current (Carnevale and Wachtel, 1980; Amitai, 1994; Gutfreund et al., 1995; Hutcheon et al., 1996a,b; Tennigkeit et al., 1999; Hutcheon and Yarom, 2000). This can be understood by considering a single voltage oscillation in a

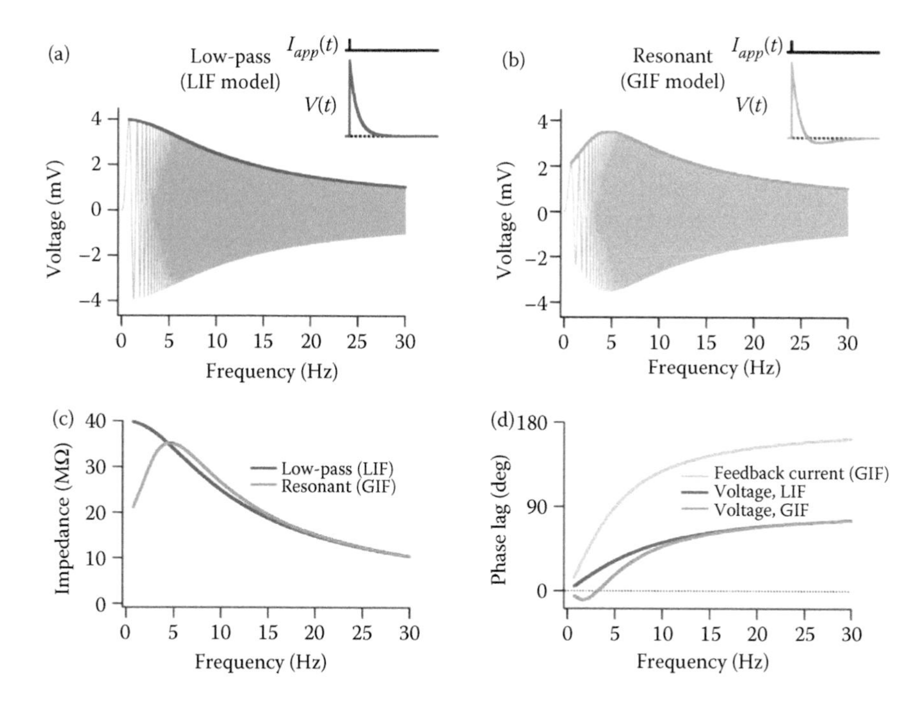

FIGURE 9.1 Subthreshold voltage responses in low-pass and resonant models. Simulations were performed using the generalized integrate-and-fire (GIF) model of Richardson et al. (2003), which reduces to the leaky integrate-and-fire (LIF) model when the scale of the feedback current (g_1) is zero. (a) LIF model response to a chirp stimulus (linear scan from 0 to 30 Hz in 30 s). Black envelope indicates the peak response on each cycle. Inset is the response to a current impulse. (b) Response of GIF model with subthreshold dynamics described by $C\,dV/dt = -gV - g_1 w + I_{app}(t)$ and $\tau_1\,dw/dt = V - w$, where w is the feedback variable, τ_1 is the time constant of the feedback, $C = 500$ pF, $g = 25$ nS, $g_1 = 25$ nS, and $\tau_1 = 100$ ms. (c) Impedance magnitude of each model. (d) Phase lag of the voltage responses and phase lag of the feedback current in the GIF model, $g_1 w(t)$. The increasing phase lag of the feedback current accounts for the greater impedance of the GIF model compared to the LIF model at intermediate frequencies (panel C).

neuron with Kv7 channels and persistent sodium current. As the neuron depolarizes, the sodium current activates, increasing the amplitude of the depolarization. After a delay, this results in greater Kv7 activation, causing a larger hyperpolarization. This cycle of oscillation increases resonance for well-matched frequencies of applied current. Resonance also depends on the passive membrane dynamics of the neuron, which are governed by the membrane time constant as well as charge redistribution between neuronal compartments (e.g., dendrites). For resonance to occur, part of the membrane potential relaxation must be fast relative to the activation of the negative feedback current. Subthreshold resonance can also depend on what parts of the neuron receive synaptic input, largely because of differences in ion channel expression between the soma and different subdivisions of the dendritic tree (Narayanan and Johnston, 2007; Hu et al., 2009).

9.1.2 Action Potential Responses

While the subthreshold voltage response of a neuron is of interest, its primary importance is the effect on AP output. For most purposes, we consider each AP as an impulse whose meaning is determined by its timing. An AP response, $r(t)$, consisting of n spikes can then be represented as a sum of delta function impulses, $\delta(t - t_i)$, where t_i is the time of each spike:

$$r(t) = \sum_{i=1}^{n} \delta(t - t_i)$$

The input–output relationship that we wish to characterize is $I_{app}(t) \rightarrow r(t)$. A fundamental difference between the subthreshold voltage response and the AP response is that the latter is much more strongly nonlinear. Clearly, even when $I_{app}(t)$ is a sine wave, $r(t)$ is not. However, the AP response has a linear component that can be quantified by measures analogous to the impedance. The sections below discuss AP responses to sine wave input in two general conditions: (1) low noise, where the spikes phase-lock to the stimulus, and (2) high noise, where the AP probability varies sinusoidally as a function of the input phase.

9.1.3 Phase-Locking

In some cases, neurons receiving periodic input generate AP output restricted to a narrow range of input phases. Such "phase-locking" does not require that the mean firing rate of the neuron (r_0) be equal to the input frequency (f); rather, the spikes can be highly phase-locked even when cycles are skipped. In general, neurons generate phase-locked AP responses when their input contains a large periodic component and relatively little "noise," or power at frequencies other than multiples of f. Depending on r_0, f, and the amplitude of the periodic input, there may be one or more spikes per input cycle, cycles may be skipped, or different numbers of spikes may occur on alternating cycles (Brumberg, 2002).

The distribution of AP phases is commonly analyzed by constructing a histogram of spike count versus phase ($\phi = 0$ to 360° or 0 to 2π radians). With appropriate normalization, the histogram represents the mean firing rate at each phase and may be designated $r(\phi)$. In a low-noise regime, $r(\phi)$ generally has one or more narrow peaks rising from a near-zero baseline. If $r(\phi)$ is unimodal, we can quantify the precision of output spike timing by the jitter (standard deviation) in units of phase or time. More generally, the strength of phase-locking can be measured by the normalized sum of unit-length phase vectors, or "vector strength" (V_s) (Goldberg and Brown, 1969), which varies from 0 for a uniform distribution to 1 when every AP occurs at the same phase.

9.1.4 Linearization of the AP Response by Noise

As discussed above, AP responses are not generally linear. A phase-locked spike response to a sine wave input is highly nonlinear because it has power at a series

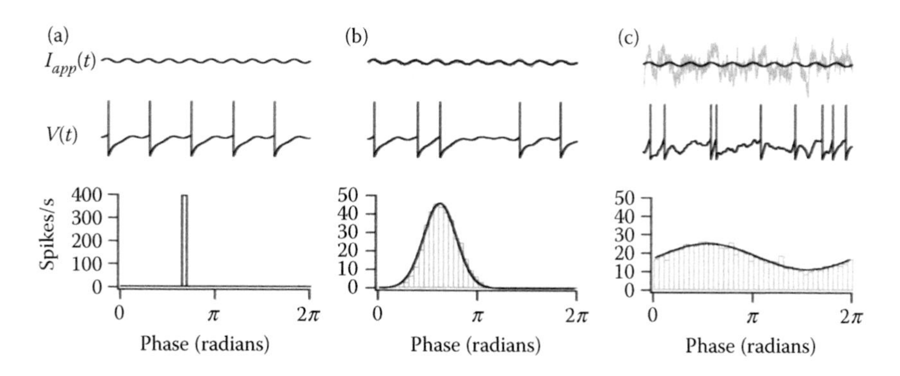

FIGURE 9.2 The effect of increasing noise on $r(\phi)$ in the leaky integrate-and-fire (LIF) model. In each simulation, $C = 500$ pF, $R = 40$ MΩ, $V_{rest} = 0$ mV, threshold = 20 mV, and reset = 14 mV. The applied current is $I_{app}(t) = I_0 + I_1 \sin(2\pi ft) + I_{noise}(t)$, where $I_0 = 494$ pA, $I_1 = 25$ pA, $f = 20$ Hz, and $I_{noise}(t)$ is Gaussian noise with a relaxation time constant $\tau_{noise} = 5$ ms. Top traces show the total $I_{app}(t)$ (gray) and the sine wave component (black), middle traces are $V(t)$, and bottom graphs are $r(\phi)$. (a) No noise. (b) Noise standard deviation (σ_I) = 10 pA. (c) $\sigma_I = 100$ pA.

of higher harmonics, or multiples of the input frequency. However, under certain conditions, $r(\phi)$ is approximately sinusoidal, and in this sense the response is linear. Although the response of one neuron to one cycle of input is not linear, the average response across many cycles may be linear. Similarly, the total AP response of a large population of neurons on one cycle can be linear. The essential condition required for a linear AP response is high noise or variability, relative to the amplitude of a periodic input. In each neuron, noise arises from stochastic ion channel gating and random presynaptic vesicle release. The noise is greater when presynaptic neurons fire synchronously, producing large compound synaptic currents. In a population of neurons, additional variability results from heterogeneous cell properties and synaptic connections. Noise and heterogeneity disrupt phase-locking, and with sufficient noise $r(\phi)$ becomes sinusoidal (Figure 9.2).

9.1.5 Gain and Phase Shift of the AP Response

In a high-noise, linear response regime, the sensitivity of a neuron's AP response to sine wave input can be quantified by measures analogous to the subthreshold impedance: the frequency-dependent gain, $G(f)$, and the average phase shift, $\Phi(f)$. These measures do not provide a general description of the neuron's input–output relationship but simply quantify the response under given stimulus parameters. In the discussion below, we consider the response to an applied current with direct current (DC), sine wave, and noise components:

$$I_{app}(t) = I_0 + I_1 \sin(2\pi ft) + I_{noise}(t)$$

To restrict our analysis to the high-noise regime, we consider the case where the standard deviation of the noise (σ_I) is larger than I_1, producing a sinusoidal $r(\phi)$. We can then analyze $r(\phi)$ by least-squares fitting or Fourier transformation to determine the amplitude (r_1) and the phase shift (Φ) of the firing rate modulation. The gain is the ratio of response amplitude to current amplitude:

$$G(f) = r_1/I_1$$

In the limit of zero frequency, $G(f)$ is the slope of the steady-state firing frequency–current relationship (r_0 vs. I_0) around a given I_0 with a particular I_{noise}. At nonzero frequencies, $G(f)$ shows a number of interesting phenomena that are not intuitively obvious, some of which are discussed below.

9.1.6 Gain Estimation Using Noise Stimulation and Correlation Analysis

If recordings are extremely stable, $G(f)$ can be determined at many frequencies using individual sine wave inputs, each added to the same I_0 and I_{noise} (Köndgen et al., 2008). However, this method is time consuming because hundreds of spikes must be collected for each input frequency. Fortunately, $G(f)$ can be estimated more efficiently using correlation methods (Lee and Schetzen, 1965; Guttman et al., 1974; Bryant and Segundo, 1976). In the high-noise, linear response regime ($\sigma_I \gg I_1$), we expect that $G(f)$ will be independent of I_1. Thus, we can apply a stimulus with only the DC and noise components:

$$I_{app}(t) = I_0 + I_{noise}(t)$$

We then use stimulus–response correlation and Fourier analysis to measure $G(f)$ and $\Phi(f)$ for the AP response, $r(t)$, with respect to each frequency component of the time-varying stimulus, $I_{noise}(t)$. To simplify the notation, we define $s(t) = I_{noise}(t)$. Based on $s(t)$ and $r(t)$, we obtain the stimulus–response correlation, $c_{SR}(\tau)$, and the stimulus autocorrelation, $c_{SS}(\tau)$, where τ is the time difference:

$$c_{SR}(\tau) = \langle s(t)\, r(t+\tau) \rangle \qquad c_{SS}(\tau) = \langle s(t)\, s(t+\tau) \rangle$$

To move to the frequency domain, we calculate the Fourier transform (FT) of each correlation over an appropriate time window (typically via the fast Fourier transform algorithm). The window width (W) must be greater than the longest cycle of interest, and determines the frequency resolution of the results ($\Delta f = 1/W$).

$$C_{SR}(f) = \mathrm{FT}[c_{SR}(\tau)] \qquad C_{SS}(f) = \mathrm{FT}[c_{SS}(\tau)]$$

$G(f)$ and $\Phi(f)$ are then determined based on the magnitude and phase of the two correlations:

$$G(f) = |C_{SR}(f)|/|C_{SS}(f)| \qquad \Phi(f) = \mathrm{atan}[C_{SR}(f)]$$

A refinement of this method uses a variable integration window for Fourier transformation, where the window is a Gaussian envelope with a standard deviation that scales with the period, $1/f$. This technique was found to reduce the error at high frequencies, where meaningful stimulus–response correlations are restricted to a narrow time window, and was used to estimate the gain of cortical neurons over a wide range of frequencies (Higgs and Spain, 2009).

9.2 RESULTS IN MODELS

While the principles that govern subthreshold frequency response properties (e.g., impedance) are relatively well understood, AP responses show differences in frequency dependence that are not intuitively obvious. Differences between the subthreshold impedance, $Z(f)$, and $G(f)$ arise from the spike threshold nonlinearity as well as the perturbation of subthreshold voltage and/or threshold caused by each spike (sometimes called "spike history"). Unlike a linear system, a neuron in which each spike results from an upward threshold crossing has a sensitivity to dV/dt that greatly increases the impact of high-frequency input. In addition, regular spiking can introduce a natural frequency r_0 (the mean firing rate) to the system. Thus, a neuron with subthreshold resonance has two natural frequencies: the subthreshold resonant frequency f_R (which is relatively fixed) and r_0 (which is highly sensitive to the input current). Both frequencies can affect the profile of $G(f)$, as discussed in Section 9.2.3. To investigate these factors, a number of analytical and computational studies have been performed using simple, single-compartment models.

9.2.1 Effects of a Spike Threshold and Voltage Reset: Leaky Integrate-and-Fire Model

The leaky integrate-and-fire model (LIF) model is a parallel resistance–capacitance (RC) circuit with a spike threshold and a voltage reset. The model has no subthreshold resonance, acting as a low-pass filter below the threshold voltage (θ). When V reaches θ, the model is said to "emit" a spike and V is reset to a specified V_r. The AP is treated as an instantaneous pulse, and its dynamics are not modeled. Thus, $V(t)$ is entirely subthreshold, and the only dynamical effect of the AP is the reset, which enforces the condition that each spike must result from an upward threshold crossing.

Although the LIF model has a linear subthreshold response, characterized by a fixed subthreshold $Z(f)$, the profile of $G(f)$ is highly dependent on the stimulus statistics, including the mean current (I_0), the standard deviation of the noise (σ_I), and the relaxation time of the noise (τ_{noise}). In general, we can distinguish two firing regimes based on the rheobase, defined as the current threshold when $\sigma_I = 0$. In the "low-noise regime," I_0 is above the rheobase and is primarily responsible for spike generation; in contrast, in the "high-noise regime," I_0 is below the rheobase and spikes are triggered by noise. In the low-noise regime, the spike response is regular (the interspike intervals (ISIs) have a low coefficient of variation) and $G(f)$ has a peak at r_0. In contrast, in the high-noise regime, the spike response is irregular and this "firing rate resonance" is eliminated (Knight, 1972; Gerstner, 2000) (Figure 9.3). We can understand this effect by considering a single sine wave component of the stimulus: $I_{sine} = I_1 \sin(2\pi ft)$. When the level of noise is low and firing is regular, the spike response can align with I_{sine} when $f = r_0$, producing a deep modulation of the spike phase histogram and a peak in $G(f)$ (Figure 9.3c,

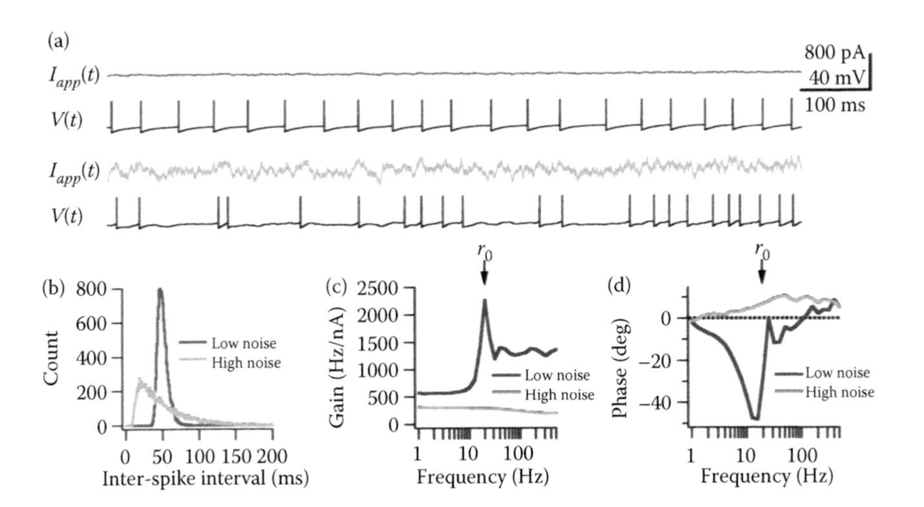

FIGURE 9.3 Interspike intervals, $G(f)$, and $\Phi(f)$ in the LIF model with low and high noise. (a) Examples of stimulus current and response with low noise ($\sigma_I = 10$ pA) and high noise ($\sigma_I = 100$ pA). To obtain matched firing rates ($r_0 = 20$ Hz), I_0 was 514 pA with low noise and 500 pA with high noise. (b) ISI histograms. (c) $G(f)$ estimated by correlation analysis with frequency-dependent windowing, as described above, based on 300 s of simulated firing. With low noise, there is a large peak at $f = r_0$, whereas with high noise, the gain is lower at all frequencies and the peak is absent. (d) $\Phi(f)$ from the same analysis. The rapid phase change seen at $f = r_0$ with low noise is a consequence of the firing rate resonance.

gray line). In contrast, when the level of noise is high and thus firing is irregular, the broad distribution of ISIs prevents consistent alignment of output spikes with any I_{sine}, where f is on the order of r_0, and this reduces the firing rate resonance (Figure 9.3c).

9.2.2 AP Generation and Sensitivity to High-Frequency Input

The profile of $G(f)$ in the LIF model is remarkably flat at high frequencies in contrast to the subthreshold impedance magnitude, which varies with $1/f$ where $f \gg 1/(2\pi\tau_m)$, or ~8 Hz for the illustrated model (see Figure 9.1c). The flat gain spectrum shows that high-frequency input can modulate the probability of AP generation even when the subthreshold voltage response is small. This result depends on the power spectrum of the noise. With white noise (flat power spectrum), $G(f)$ in the LIF model was shown to decrease at high frequency; however, with low-pass filtered noise, which simulates a synaptic time constant, $G(f)$ does not fall off (Brunel et al., 2001). Thus, in the time domain, the AP probability of the LIF model can respond instantaneously to a fast change in the input signal (provided that this signal is not subjected to the same filtering). In simple terms, the reason that the gain does not fall at high frequency is that the threshold has a differentiator-like effect. Because V_m is reset after each spike, each threshold crossing must be upward ($dV/dt > 0$). While $Z(f)$ decreases at high frequencies, the effect of a sine wave current on dV/dt remains constant.

Thus, high-frequency input can influence spike timing and thereby modulate the spike phase distribution.

Real neurons also have high gain above the subthreshold corner frequency; however, their gain falls at very high frequencies (Köndgen et al., 2008) because of the dynamics of spike generation. This effect was investigated using a variant of the LIF model, the exponential integrate-and-fire (EIF) model (Fourcaud-Trocmé et al., 2003). The theoretical basis of this model is that the voltage dependence of the AP-generating sodium current is exponential near threshold. Ignoring the time dependence of sodium current activation, the instantaneous current–voltage (I–V) relationship of a leaky integrator with a sodium current is approximated by a leak (a linear function of V) plus an exponential function of V. Taking V as the membrane potential relative to the leak reversal potential, g as the leak conductance, A as the amplitude of the exponential current at $V = 0$, and K as the sharpness, or "slope factor" of the exponential current activation curve, the dynamics of the EIF model are described by $C\,\mathrm{d}V/\mathrm{d}t = I(V) = gV - A\exp(V/K) + I_{app}(t)$. Because of these dynamics, the AP will increase to an arbitrarily high V in finite time, and in practice, V is reset from an arbitrary level that is not critical, provided that it is sufficiently high. The EIF model does not have a fixed AP threshold. At steady state, the threshold may be defined as the voltage where $I(V)$ reaches a maximum, but $V(t)$ can transiently exceed the steady-state threshold without producing a spike.

The critical parameter affecting the high-frequency responses of the EIF model is K, the sharpness of the exponential current activation curve, which determines the dynamics of AP generation. The spike onset is slow when K is large and fast when K is small; when K approaches 0, the EIF model becomes equivalent to the LIF. For $K > 0$, the high-frequency gain of the EIF model is proportional to $1/f$ (irrespective of the power spectrum of the noise), and increasing K causes $G(f)$ to fall off at a lower frequency (Fourcaud-Trocmé et al., 2003). Figure 9.4 illustrates this effect, comparing EIF models with $K = 1$, 3, and 6 mV to a matched LIF model. These simulations show that the dynamics of AP generation are crucial for the sensitivity to high-frequency input. In real neurons, a parameter analogous to K is the sharpness of the

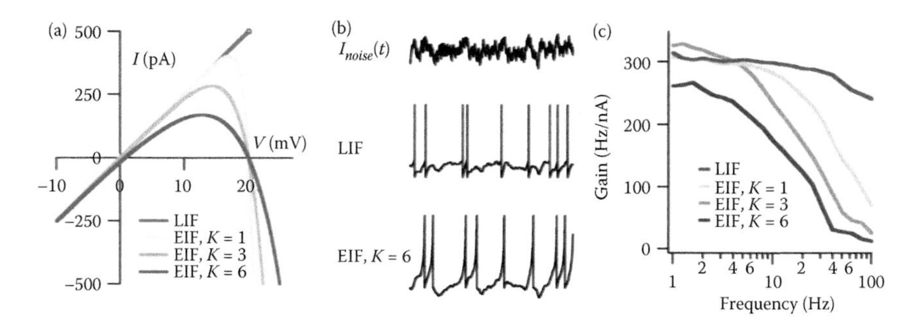

FIGURE 9.4 Exponential integrate-and-fire (EIF) model. (a) Current–voltage (I–V) relationships of the LIF and EIF models. (b) Responses to a high-noise stimulus ($\sigma_I = 100$ pA, $\tau_{noise} = 5$ ms) in the LIF model and in the EIF model with $K = 6$ mV. For each model, I_0 was set to give $r_0 = 20$ Hz. Trace length = 500 ms. Note the gradual AP onsets in the EIF model. (c) $G(f)$ for each model, estimated by correlation analysis based on 300 s of simulated firing.

sodium current activation curve. In addition, because real sodium channels activate extremely rapidly but not instantaneously, the kinetics of sodium channel activation might affect the AP response to very high-frequency input.

In real neurons, additional factors affecting spike generation are also expected to influence $G(f)$, particularly including sodium channel inactivation and fast voltage-gated K^+ channels that activate at subthreshold potentials (e.g., Kv1). Both these mechanisms are slower than sodium channel activation but can reduce the net inward current when the depolarization is relatively slow. In essence, the *I–V* curve varies dynamically, shifting upward with depolarization. This generally has a high-pass effect on the spike response. In a single-compartment neuron model, these dynamic mechanisms are apparent at the level of membrane potential. In a neuron with an electrically isolated axonal spike generation zone, similar mechanisms can regulate spike threshold with only minimal effects on the somatic voltage, as discussed later in this chapter.

9.2.3 Models with Subthreshold Resonance

While suprathreshold resonance at the mean firing rate (r_0) can arise without any subthreshold resonance, r_0 depends on the current input to a neuron. Thus, the r_0 resonance may play a relatively limited role in synchronous oscillations among neurons with heterogeneous synaptic input and/or intrinsic properties. Because subthreshold impedance resonances generally show a less variable resonant frequency (f_R), they might play a greater role in synchronization. Thus, it is of great interest to understand how a subthreshold resonance can affect the AP response.

Richardson et al. (2003) investigated this question using model neurons with subthreshold resonance. The simplest was a "generalized integrate-and-fire" (GIF) model constructed by adding a slowly activating negative feedback current to the LIF model. Simulations were performed with different combinations of mean current (I_0) and noise standard deviation (σ_I) adjusted to elicit the same r_0. The data showed that with low noise the r_0 resonance was dominant, whereas with high noise the f_R resonance was unmasked. For illustration, we reproduced these results in Figure 9.5, using the GIF model of Richardson et al. (2003) and the methods of gain analysis described above. In absolute terms, high noise produces lower gain at all frequencies in the GIF model (Figure 9.5a). However, the relative suppression of gain is least near f_R, resulting in a peak at this frequency (Figure 9.5b, orange and red traces).

To illustrate how spiking disrupts the f_R resonance in the low-noise condition, Figure 9.5c and d compares the subthreshold impedance of the GIF model to the "perturbed impedance" during spiking. The impedance magnitude and phase were estimated by correlation analysis identical to that described above, except that the analyzed response was $V(t)$ rather than the spike response $r(t)$. We note that $V(t)$ does not actually contain spikes, but is reset to V_r upon reaching the threshold.

For pure subthreshold input, the impedance of the model peaks at $f_R \approx 5$ Hz, and the phase lag increases most rapidly near f_R and reaches a maximum of 90° (Figure 9.5d). In contrast, the "perturbed impedance" during spiking with low noise has a peak magnitude near the mean firing rate ($r_0 = 20$ Hz) and a very different phase behavior, where the voltage response is nearly 180° out of phase with the current

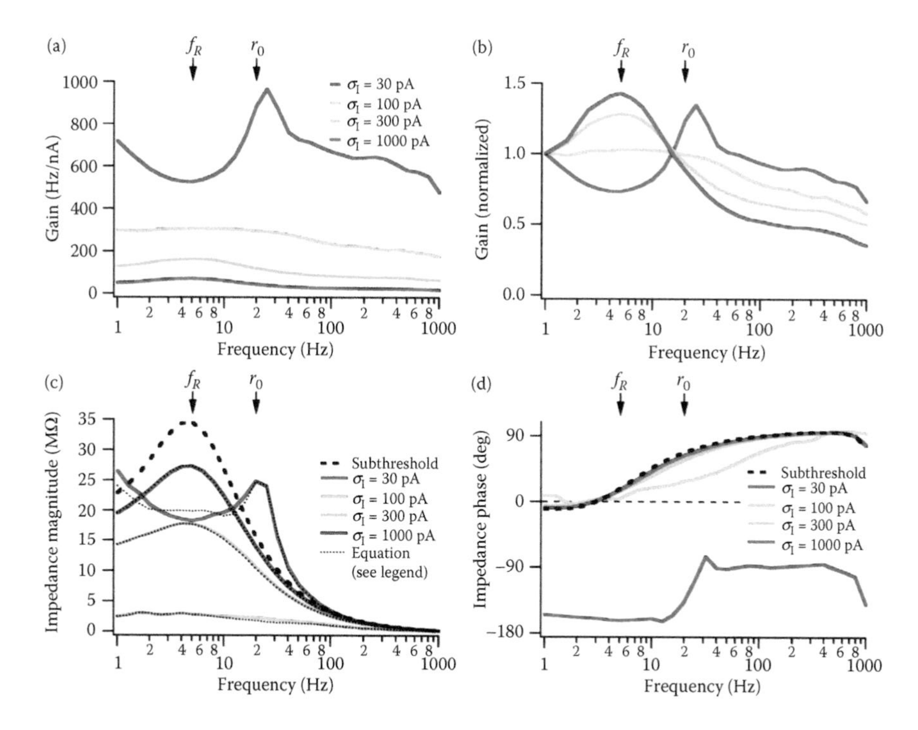

FIGURE 9.5 **(See color insert.)** Generalized integrate-and-fire (GIF) model. Simulations were performed using the model of Richardson et al. (2003). The subthreshold response is described by $C\ dV/dt = -gV - g_1 w + I_{app}(t)$. The feedback variable w follows V with a first-order relaxation: $\tau_1\ dw/dt = V - w$. A spike is emitted when V reaches θ; V is then reset to V_r, but w is not reset. For the illustrations, $C = 500$ pF, $g = 25$ nS, $g_1 = 25$ nS, $\tau_1 = 100$ ms, $\theta = 20$ mV, and $V_r = 14$ mV. The noise had a relaxation time constant $\tau_{noise} = 5$ ms. (a) $G(f)$ for four levels of noise ($\sigma_I = 30$, 100, 300, and 1000 pA). To produce a mean firing rate $r_0 = 20$ Hz, I_0 was 971, 943, 782, and −89 pA, respectively. $G(f)$ was estimated by correlation analysis. (b) The same $G(f)$ curves, each normalized by the gain at 1 Hz. (c) Impedance magnitude for subthreshold response (dashed black line) and for $V(t)$ during each spiking response (colored lines), illustrating how the spike-and-reset mechanism disrupts the impedance resonance when σ_I is low. The equation predicting the impedance magnitude during spiking is $|Z_p(f)| = |Z(f)|\ [1 - G(f)\ C\ (\theta - V_r) \cos \Phi(f)]$. (d) Impedance phase.

input at frequencies below r_0. With increasing noise, the impedance peak at r_0 is lost and the peak at f_R recovers. Spiking affects the impedance via the voltage reset (or "afterhyperpolarization (AHP)"), which perturbs $V(t)$ each time it reaches threshold. The size of the effect depends on the gain of the neuron (G), the membrane capacitance (C), and the depth of the reset ($\theta - V_r$). At steady state, it can be shown that the "perturbed resistance" is given by $R_p = R[1 - G\ C\ (\theta - V_r)]$. For sine wave input, we must account for the phase of the spike response relative to the current, and the perturbed impedance magnitude is $|Z_p(f)| = |Z(f)|\ [1 - G(f)\ C\ (\theta - V_r) \cos \Phi(f)]$. This equation provides good fits to the perturbed impedance data shown in Figure 9.5c (dotted lines).

A qualitative interpretation of this analysis is that firing perturbs the subthreshold impedance when G is high, when C is large (producing a slow membrane time constant that shapes the AHP), and when $\theta - V_r$ is large (giving a deep AHP). The statistics of the noisy current (I_0 and σ_I) enter the equation via their effects on $G(f)$ and $\Phi(f)$. In the GIF model, for a given r_0, increasing σ_I lowers $G(f)$ across the entire frequency spectrum, and this lessens the perturbation of subthreshold resonance. This analysis shows how subthreshold feedback currents and spike-dependent AHPs interact with stimulus statistics to determine the frequency response properties of a simple model neuron. The following sections discuss these mechanisms in more detail.

9.3 AP-DEPENDENT IONIC MECHANISMS

9.3.1 Afterhyperpolarizations

As discussed above, the AHP after each spike can influence the frequency response properties of a neuron. In real neurons, the AHPs are not simple resets but have distinctive waveforms produced by specific collections of channels. For example, cortical pyramidal cells (PCs) have AHPs with fast, medium, and slow components (the fast afterhyperpolarization (fAHP), medium-duration afterhyperpolarization (mAHP), and slow afterhyperpolarization (sAHP), respectively). The fAHP is the final phase of AP repolarization and is mediated primarily by voltage-activated K^+ channels and large-conductance calcium-activated K^+ (BK) channels (Barrett and Barrett, 1976; Storm, 1987; Schwindt et al., 1988b). The mAHP is the scooped voltage trajectory lasting tens of milliseconds after each AP or burst, and can be produced by small-conductance calcium-activated K^+ (SK) channels in cortical PCs (Schwindt et al., 1988b), or Kv7 channels in hippocampal CA1 PCs (Storm, 1989; Peters et al., 2005; Gu et al., 2005). The sAHP lasts for seconds and is caused by calcium- and sodium-dependent K^+ channels (Lancaster and Adams, 1986; Schwindt et al., 1988a, 1989).

The function of each AHP component depends on its kinetics and the ISIs. The fAHP is closest to the reset of the LIF model, and is relatively ineffective at producing regular firing and r_0 resonance unless the ISIs are short or the noise level is low (Figure 9.6a). In contrast, a medium AHP conductance with a deactivation time course similar to the ISI produces a scooped voltage trajectory that lowers the AP probability at intermediate times, regularizing firing. Thus, the mAHP can produce a strong r_0 resonance at a relatively low firing rate (Figure 9.6b). The sAHP current generally has little effect on firing regularity because of its slow time course and the very small change in sAHP conductance caused by each spike. Thus, the sAHP does not cause a strong r_0 resonance. However, the sAHP conductance accumulates over multiple ISIs and thereby limits the gain in response to low-frequency input (Lorenzon and Foehring, 1992) (Figure 9.6c).

9.3.2 Fast Afterdepolarization and Burst Firing

In many neurons, spikes activate depolarizing currents that give rise to the fADP and burst firing. The ionic basis of bursting is complex and cannot be adequately

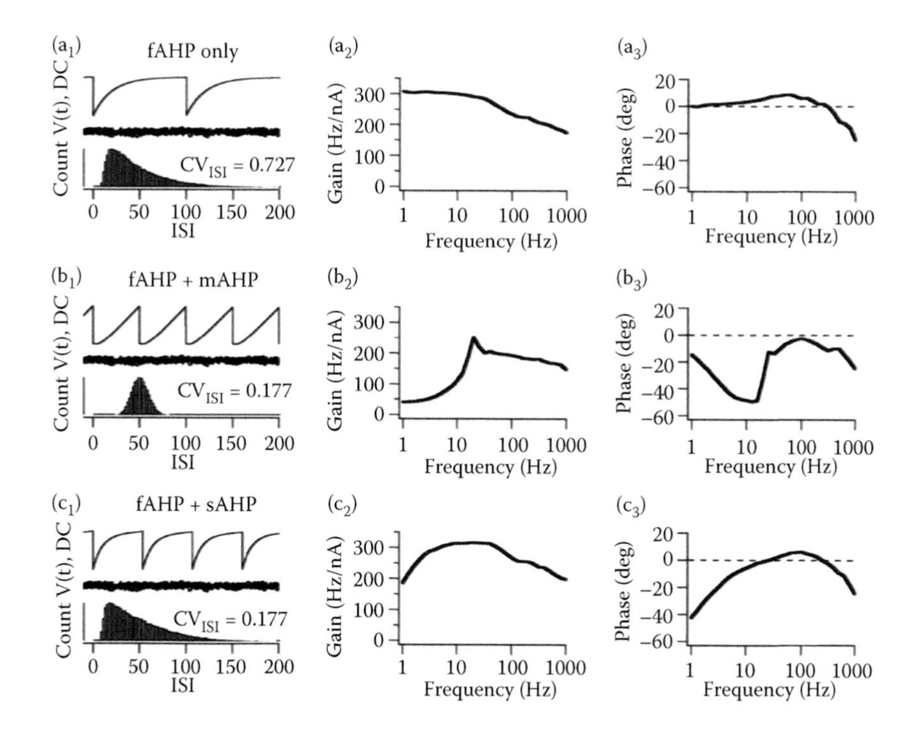

FIGURE 9.6 Distinct effects of fast, medium, and slow AHPs in the LIF model. Panels (a_1–c_1) show the interspike $V(t)$ with DC input only (top), several samples of the noise current (middle), and the ISI histogram with noise input (bottom), (a_2–c_2) show $G(f)$, and (a_3–c_3) show $\Phi(f)$. (a) Standard LIF model with "fAHP" only (reset to 6 mV below threshold); $r_0 = 20$ Hz with $I_0 = 501$ pA. Note the broad ISI distribution, the ISI shortening caused by noise (from ~100 ms with DC only to a mean of 50 ms with noise), and the lack of r_0 resonance. (b) Model with fAHP and mAHP ($\Delta G = 25$ nS per spike, $E_{rev} = V_{rest}$, time constant = 50 ms); $r_0 = 20$ Hz with $I_0 = 867$ pA. Note the narrow ISI distribution and the r_0 resonance. (c) Model with fAHP and sAHP ($\Delta G = 1.25$ nS per spike, time constant = 1000 ms); $r_0 = 20$ Hz with $I_0 = 961$ pA. Note the broad ISI distribution, lack of r_0 resonance, and reduced gain at low f (compared to the fAHP-only model).

summarized in this chapter. However, as an example, two mechanisms that can produce the fADP in PCs are calcium-dependent dendritic spikes (Traub, 1979; Schwindt and Crill, 1999; Larkum et al., 1999; Magee and Carruth, 1999; Williams and Stuart, 1999; Bekkers and Häusser, 2007) and persistent sodium current located more proximally (Yue et al., 2005), and particularly in the first node of Ranvier (Kole, 2011). In most bursting neurons, the fADP rises quickly from the fAHP and either produces the next AP or is terminated by the mAHP. The fADP trajectory can also be influenced by the input current; thus, many "conditional bursting" neurons can fire single spikes and/or bursts, depending on the input.

Because of the distinct timescales of interburst and intraburst intervals, either low- or high-frequency input can trigger bursts. This may be seen by examining the suprathreshold responses of a cortical layer 2/3 PC to a chirp current. Figure 9.7a and b shows the ISIs in response to chirps of different amplitudes. This cell generated two-spike bursts with intraburst ISIs of approximately 3–6 ms, which were most readily evoked at a current frequency of ~10 Hz. When the chirp amplitude was increased, bursts also appeared at a current frequency of ~200 Hz. At 10 Hz, a burst occurred near the crest of each current oscillation (Figure 9.7c), whereas at 200 Hz, the two spikes comprising each burst were triggered on consecutive cycles (Figure 9.7e).

The fADP and bursting can have major effects on $G(f)$, as illustrated in Figure 9.8. At low frequencies, the fADP can increase gain by offsetting some of the AHP current activated by each spike. In real neurons, an additional increase in gain may result from sublinear summation of the mAHP current at short burst ISIs. In addition, the interaction between the fADP, bursting, and the mAHP can strengthen the firing rate-dependent resonance, and bursts can produce a second peak of $G(f)$ at a high frequency corresponding to the intraburst ISIs (Schindler et al., 2006; Higgs and Spain, 2009; Figure 9.8b).

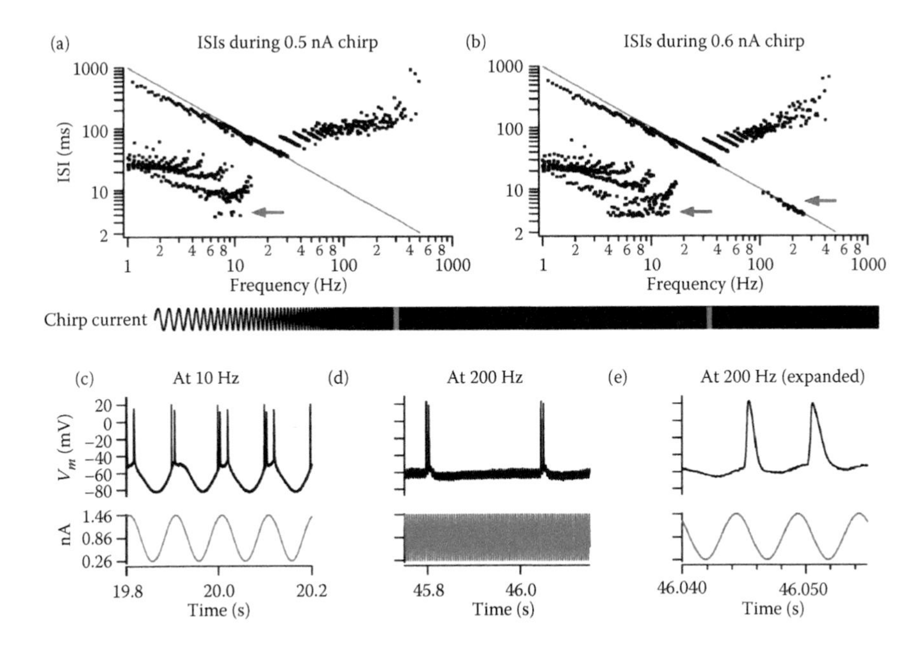

FIGURE 9.7 (**See color insert.**) Bursts triggered by low- and high-frequency sine wave input. (a) ISIs in a cortical layer 2/3 PC during a chirp current stimulus with an amplitude of 0.5 nA and an exponential scan from 1 to 1000 Hz. The blue arrow indicates intraburst intervals at a chirp frequency near 10 Hz. (b) ISIs in the same cell when the chirp amplitude was increased to 0.6 nA. The red arrow indicates intraburst intervals at a chirp frequency of approximately 200 Hz. (c) Firing pattern during the 0.6 nA chirp at 10 Hz (blue region in chirp current trace above). Top plot shows V_m, and bottom plot shows the applied current. (d) Firing pattern at 200 Hz (red region in chirp current trace) on the same timescale. (e) Firing pattern at 200 Hz on an expanded timescale. (From Higgs MH, Spain WJ 2009. *J Neurosci* 29:1285–99.)

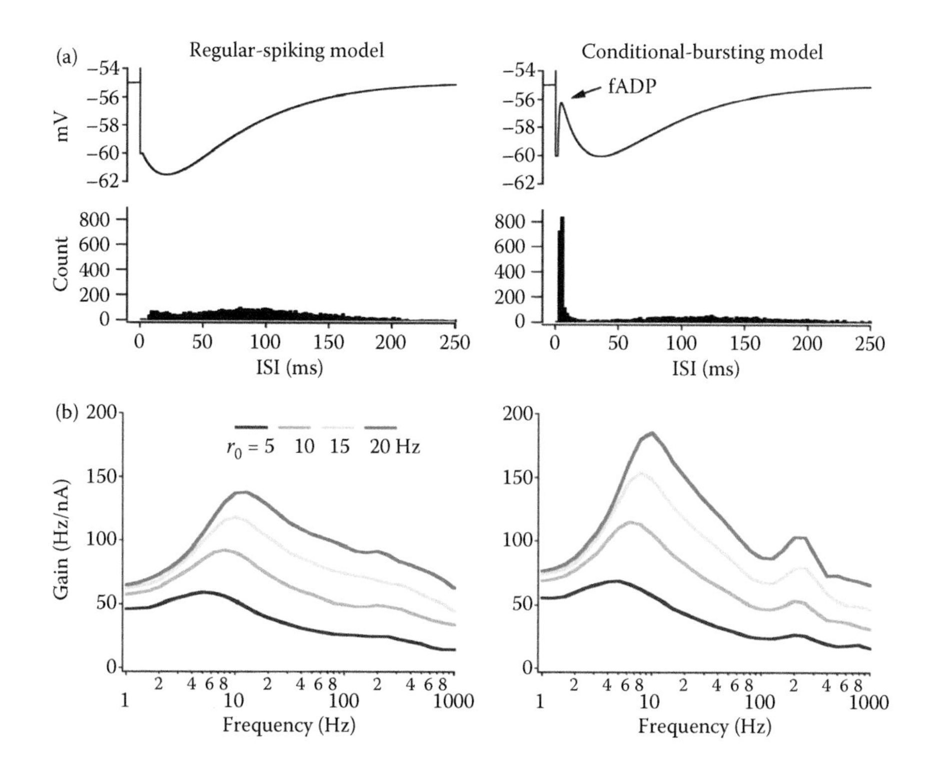

FIGURE 9.8 Effects of the fADP and conditional bursting in the LIF model. Left panels are for regular-spiking model with fAHP and mAHP only; right panels are for model with additional fADP current. (a) Top: AHP trajectory with just-suprathreshold DC input. Bottom: ISI histogram with noise input. (b) $G(f)$ for each model. I_0 was adjusted to give mean firing rates of $r_0 = 5$, 10, 15, and 20 Hz (from light gray to black) with $\sigma_I = 100$ pA and $\tau_{noise} = 5$ ms. (Modified from Higgs MH, Spain WJ 2009. *J Neurosci* 29:1285–99.)

9.4 FAST SIGNALS AND HIGH-FREQUENCY RESONANCE

Thus far, our discussion has primarily considered the responses of neurons to relatively low input frequencies, which are relevant to the coarse synchronization of neuronal ensembles. However, there is increasing interest in the physiological and pathological roles of higher-frequency signals that affect precise spike timing. Some neurons appear to operate primarily on one fast timescale. These cells have a large membrane conductance and a short time constant, and often express high densities of rapidly activating voltage-gated K^+ channels, giving rise to subthreshold resonance at a high frequency. In contrast, other neurons appear to operate simultaneously on fast and slow timescales, in part because of somatodendritic compartments with slow dynamics and an axonal AP generator with fast dynamics of membrane potential and spike threshold. Examples of these types of behaviors are described below.

One mechanism discussed above that can create or modify a high-frequency resonance is burst firing (Schindler et al., 2006; Higgs and Spain, 2009). In general, bursting can create a strong resonance when the occurrence of each burst spike is stimulus dependent ("conditional bursting") but the intraburst intervals are relatively constant. However, high-frequency resonance can also be generated by subthreshold mechanisms, with or without high-frequency firing, and in some systems, the burst firing rate may be directly related to the subthreshold resonance of the spike generator (e.g., Hsiao et al., 2009).

9.4.1 Fast Neurons in the Auditory Brainstem

Among neurons that encode high-frequency signals, those in the avian interaural time difference (ITD) pathway are perhaps the most studied (and mammalian auditory neurons have many similar mechanisms, reviewed in Ashida and Carr, 2011). The ITD pathway begins in the cochlea, which separates sound input into frequency bands. The cochlear hair cells synapse onto auditory nerve fibers, which fire spikes that are phase-locked to the sound wave (Rose et al., 1967; Köppl, 1997). The auditory nerve fibers synapse onto cochlear nucleus magnocellularis (NM) neurons, which maintain or enhance the phase-locking (Fukui et al., 2006). The NM axons project to the ipsilateral and contralateral nucleus laminaris (NL) (Figure 9.9a), which forms a map of characteristic sound frequency (CF) and sound location. The NM axons function as delay lines, whose conduction time varies along the sound location axis of NL (Jeffress, 1948; Carr and Konishi, 1990; Overholt et al., 1992). At each position, the axonal delay cancels a particular ITD, resulting in coincident synaptic input. The NL neurons act as coincidence detectors, converting the time delay to firing rate (Sullivan and Konishi, 1986; Overholt et al., 1992; Joseph and Hyson, 1993; Reyes et al., 1996; Funabiki et al., 2011), similar to the function of mammalian medial superior olivary (MSO) neurons (Goldberg and Brown, 1969; Yin and Chan, 1990).

To encode small ITDs, NM neurons must have great timing precision, and NL cells must convert small changes in input timing to large changes in firing rate. These functions depend on highly specialized synapses and intrinsic properties. Cells in NM and the high-CF region of NL receive extremely fast excitatory postsynaptic currents (EPSCs) mediated by specific types of AMPA receptor subunits (Zhang and Trussell, 1994; Otis et al., 1995; Kuba et al., 2005). These EPSCs are unlikely to be slowed by electrotonic filtering, as NM neurons receive their synaptic input on the cell body (Parks, 1981; Köppl, 1994) and high-CF NL cells have only short dendrites (Smith and Rubel, 1979). In contrast, low-CF NL cells have slower EPSCs, which might result from electrotonic filtering in their longer dendrites. The intrinsic properties of NM and NL neurons include a large low-threshold K^+ current mediated by Kv1 channels (Reyes et al., 1994; Rathouz and Trussell, 1998; Kuba et al., 2002, 2005; Fukui and Ohmori, 2004) and a restricted distribution of voltage-gated sodium channels, which are found in the axon initial segment but show little expression on the soma (Kuba et al., 2006; Kuba and Ohmori, 2009; Ashida et al., 2007). Together, these properties provide great temporal precision, while allowing spike generation only in response to rapidly fluctuating input (Reyes et al., 1994, 1996). This "DC filtering" property arises from strong outward rectification of the somatic membrane current, combined with a

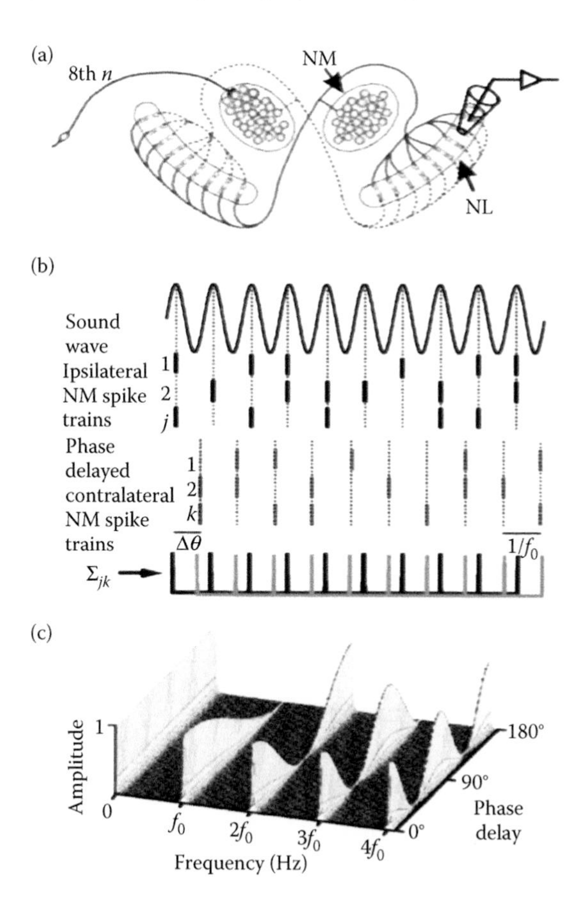

FIGURE 9.9 Phase-locked signals in the chicken auditory brainstem. (a) Diagram of nucleus magnocellularis (NM), nucleus laminaris (NL), and the ipsilateral and contralateral projections (delay lines) from NM to NL. (b) Schematic of phase-locked spike input from ipsilateral and contralateral NM neurons. (c) Fourier analysis of a hypothetical, perfectly phase-locked spike input to NL. The signal amplitude is largest at harmonics of the sound frequency, and the amplitude of each harmonic varies as a function of the IPD (right axis). Only the sound frequency ("f_0") component has an amplitude that varies monotonically from 0 to 180° IPD. (Modified from Slee SJ, Higgs MH, Fairhall AL, Spain WJ 2010. *J Neurophysiol* 103:2857–75.)

dynamic spike threshold (Howard and Rubel, 2010) that cannot be reached by slow depolarization. Interestingly, the spike timing precision of NM neurons is reduced during adaptation to long-duration input, as a consequence of slow inactivation of Kv1 channels (Kuznetsova et al., 2008). This novel form of adaptation is predicted to reduce the response to ITD in postsynaptic NL neurons.

The function of each NL neuron is to convert the interaural phase difference (IPD) to firing rate. To do this, each neuron must distinguish a signal at the CF (or "f_0"), which decreases monotonically as a function of IPD, from a signal at $2f_0$ that varies nonmonotonically (Figure 9.9b and c). Input components below f_0 do not contribute directly to IPD encoding but might add noise that interferes with this function.

Because of these input properties, each NL neuron should (1) respond to signals at f_0, (2) filter out signals at $2f_0$ or higher, and (3) filter out signals below f_0. The synaptic and intrinsic properties of NL cells meet these requirements (Slee et al., 2010). Near-optimal synaptic filtering results from EPSCs having a width of approximately 1/2 sound cycle at half height; the width varies across the tonotopic axis (Figure 9.10a–c) preserving this relationship for each CF. Filtering of the spike input by this EPSC waveform preserves signals at f_0 and attenuates signals at $2f_0$.

Input encoding by NL neurons also depends on their intrinsic properties. $G(f)$ data obtained using noise stimulation showed a band-pass profile (Figure 9.10d), with low gain at $f = 0$, a peak well below f_0 (varying nonmonotonically across the tonotopic axis), and a fall-off from f_0 to $2f_0$ and beyond. $G(f)$ does not directly predict a change in mean firing rate (r_0) caused by an AC input; however, linear–nonlinear models constructed based on the gain data approximately reproduced experimental r_0 versus IPD curves. While a peak gain below f_0 may appear "non-optimal" for IPD encoding, an alternative interpretation is that complete removal

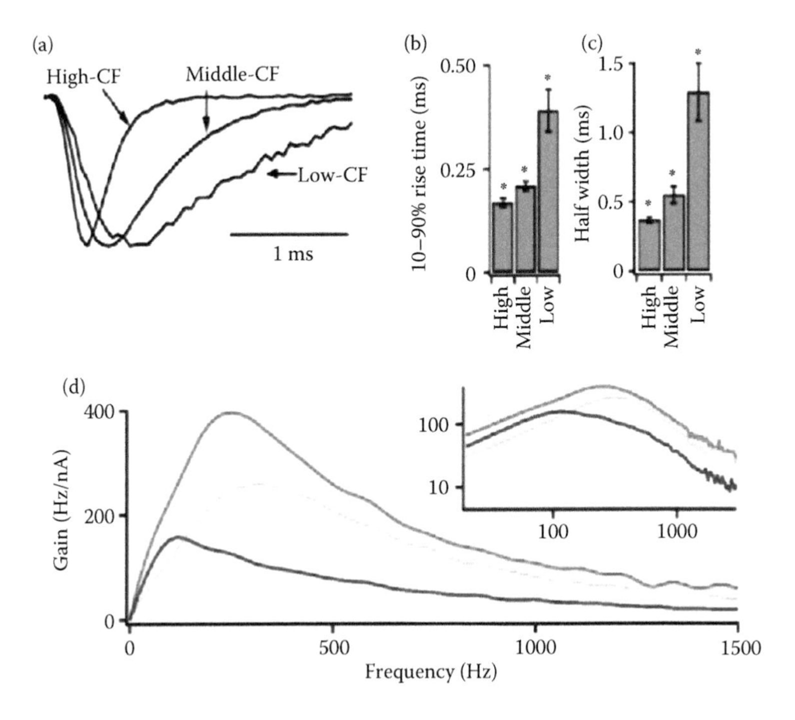

FIGURE 9.10 Variation of NL EPSC kinetics and frequency-dependent gain across the tonotopic axis. (a) EPSCs in low-, middle-, and high-CF regions of NL. For comparison, each trace is normalized by its peak amplitude. (b) 10–90% rise time of EPSCs in each CF region. (c) Width at half height of EPSCs in each region. (d) $G(f)$ in NL neurons, estimated by noise stimulation and correlation analysis. (From Slee SJ, Higgs MH, Fairhall AL, Spain WJ 2010. *J Neurophysiol* 103:2857–75.)

of the $2f_0$ signal is more essential than filtering of components below f_0, whose magnitude is independent of IPD.

9.4.2 Responses to Fast Signals in Cortical Pyramidal Cells

In contrast to auditory brainstem neurons, cortical PCs have relatively slow EPSCs, excitatory postsynaptic potentials (EPSPs), and membrane time constants, and most can fire repetitively in response to DC input. Thus, it is commonly thought that these cells function as rate coding integrators. However, cortical responses with a temporal precision of less than 10 ms have been reported (e.g., Elhilali et al., 2004; Mechler et al., 1998; Cardin et al., 2010; Kayser et al., 2010), and some investigators believe that precise spike timing plays a major role in cortical function. Despite their low-pass somatodendritic membrane properties (modified in some cases by low-frequency resonance), cortical PCs generate highly reproducible spike trains in response to repeated "frozen" noise stimuli (Mainen and Sejnowski, 1995), and studies using noisy current stimuli have shown that cortical PCs have high gain at frequencies up to hundreds of hertz (Köndgen et al., 2008; Higgs and Spain, 2009).

Gain data also suggest that some cortical PCs have a high-frequency resonance, producing a maximum $G(f)$ at ~400 Hz. This high-frequency peak showed only subtle effects of burst firing, suggesting that it is primarily a subthreshold resonance rather than a firing rate resonance, and the peak was greatly reduced by blocking Kv1 channels (Higgs and Spain, 2009, 2011; Figure 9.11).

9.4.3 High-Frequency Resonance Produced by a Dynamic Spike Threshold

As discussed above, the profile of $G(f)$ can be quite different from that of the subthreshold $Z(f)$, and a major reason is the spike threshold. In addition to its basic

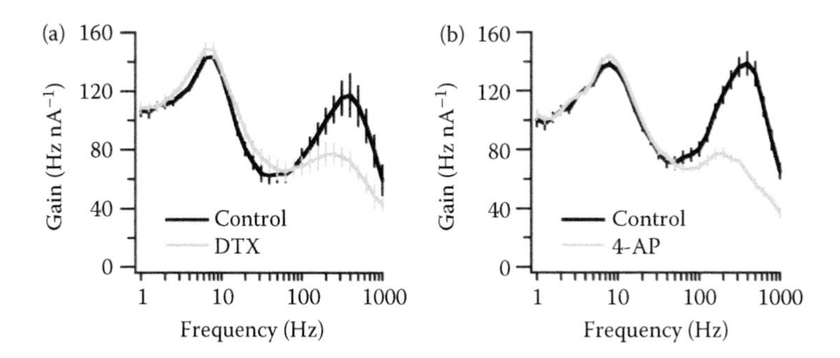

FIGURE 9.11 High-frequency resonance in cortical layer 2/3 PCs. Plots show $G(f)$ in response to noise current ($\sigma_I = 100$ pA, $\tau_{noise} = 5$ ms). I_0 was adjusted to produce a mean firing rate $r_0 = 5$ Hz in each cell. (a) Effect of Kv1 potassium channel blockade by 100–200 nM α-dendrotoxin (DTX). (b) Effect of less selective Kv1 blockade by 100 μM 4-aminopyridine (4-AP). (From Higgs MH, Spain WJ 2011. *J Physiol* 589(Pt 21):5125–42.)

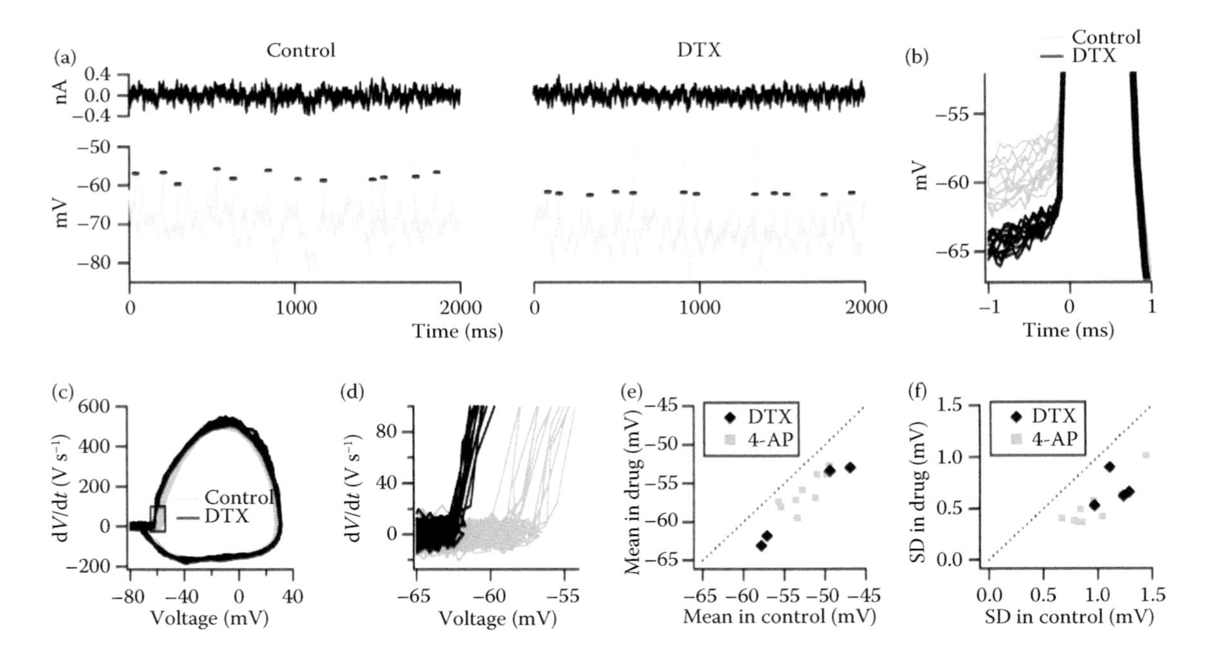

FIGURE 9.12 Contribution of Kv1 potassium channels to threshold variation in cortical layer 2/3 PCs. (a) Noise current (top) and voltage response (bottom) in control solution (left) and in the presence of the Kv1 blocker α-dendrotoxin (DTX) (right). Horizontal marks indicate spike thresholds, measured at $dV/dt = 20\ V\ s^{-1}$. Note the lesser variation of threshold in the presence of DTX. (b) Expanded view spike onsets from A. Gray traces are control data, and black traces were obtained in DTX. (c) Phase plot showing dV/dt versus V for each AP. (d) Expanded view of the rapidly rising portion of the phase plot (AP onset). (e) Scatter plot showing the mean threshold during Kv1 blockade by DTX or 4-aminopyridine (4-AP) versus the mean threshold in control solution. (f) Scatter plot showing the standard deviation (SD) of threshold during Kv1 blockade versus the SD in control solution. (From Higgs MH, Spain WJ 2011. *J Physiol* 589(Pt 21):5125–42.)

nonlinearity, the threshold can vary in response to the membrane potential. A dynamic threshold can result from sodium channel inactivation (Platkiewicz and Brette, 2011) and/or K^+ channel activation (Higgs and Spain, 2011; Figure 9.12), and appears to depend on channels localized near the spike initiation zone (SIZ) in the proximal axon. Threshold changes can be rapid, even in a neuron with a slow somatic membrane time constant, and can have major effects on neuronal frequency response properties.

To understand the effects of a dynamic threshold, it is helpful to consider the threshold as a dynamical variable, $\theta(t)$, where spikes initiate when $V(t) = \theta(t)$. We note that this concept presents some subtle issues. Most fundamentally, $\theta(t)$ is not uniquely defined by a discrete set of threshold points. In addition, $V(t)$ differs between the soma and the SIZ (Yu et al., 2008; Kole and Stuart, 2008). Thus, we must distinguish between the biophysical threshold in the SIZ (a difficult quantity to measure or even define in a cable with nonuniform voltage and currents) and the somatic voltage required to produce a spike. To sidestep these issues, we focus on the functional somatic threshold, which can be measured directly (e.g., by determining the largest EPSP that does not cause a spike) and is generally well correlated with the voltage at the AP onset (at a specified dV/dt), although deviations can occur when the subthreshold voltage changes rapidly.

In most cases, spikes preceded by rapid depolarization have a lower threshold (Azouz and Gray, 2000, 2003; Farries et al., 2010), suggesting that θ increases in response to maintained depolarization, or "accommodates" (Hill, 1936). θ can be modeled by a steady-state function of voltage, $\theta_{SS}(V)$ (Figure 9.13a), and a description of the dynamics (Figure 9.13b). If there is a point where $\theta_{SS}(V) = V$, the neuron can spike in response to slow depolarization; in contrast, in "DC filtering" cells such as the auditory neurons discussed above, $\theta_{SS}(V) > V$ for all V. To understand the function of a dynamic threshold, we consider the "distance to threshold," $D(t) = V(t) - \theta(t)$. If the dynamics of $\theta(t)$ are low-pass with respect to $V(t)$, then $D(t)$ is high-pass filtered, as we can show by calculating the impedance of D with respect to the applied current, or $Z_D(f)$ (Figure 9.13c and d). While the relationship between impedance and gain is complex, as discussed above, the change in effective impedance caused by the dynamic threshold can produce a resonance at the level of $G(f)$ (Figure 9.13e and f), and is likely to account for the Kv1-dependent, high-frequency resonance observed in cortical PCs (Higgs and Spain, 2011; Figure 9.12).

9.5 CONCLUSIONS

The topics reviewed in this chapter illustrate the great complexity of single neuron spike responses to oscillating input, showing that the frequency-dependent gain depends on the input statistics, on intrinsic currents that give rise to subthreshold resonance, on axonal mechanisms that control the dynamics of spike generation and lead to variation of spike threshold, and on spike-dependent currents that produce the diversity of AHPs observed in different types of neurons. All these mechanisms can potentially influence the synchronization of a neuronal ensemble receiving a shared oscillatory input, as well as the dynamics of responses to time-varying inputs in general.

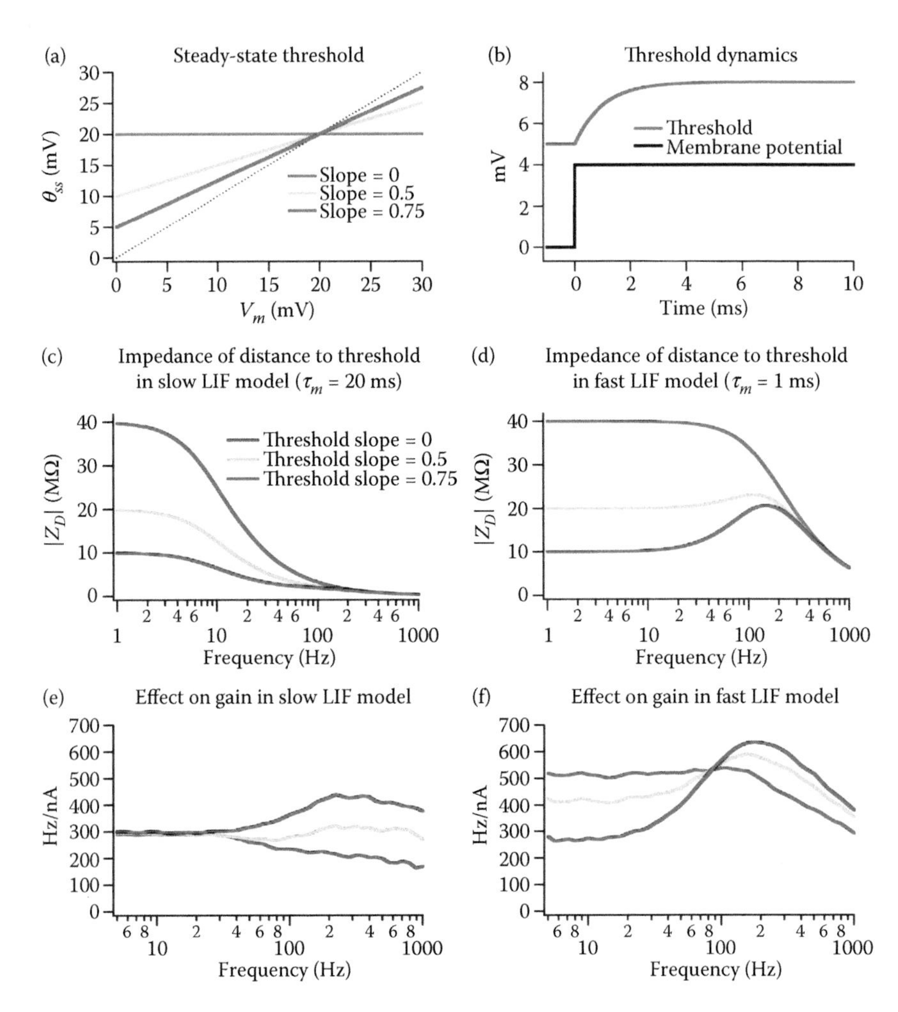

FIGURE 9.13 **(See color insert.)** Functional effects of a dynamic threshold. (a) Steady-state threshold $\theta_{SS}(V)$. For comparison, each model was set up such that $\theta_{SS}(V)$ intersects V (dashed line) at 20 mV above the resting potential. (b) Threshold dynamics (first-order relaxation) illustrated by the response to a voltage step. (c) Calculated impedance of distance to threshold (Z_D) in slow LIF model (τ_m = 20 ms). (d) Z_D in fast LIF model (τ_m = 1 ms). In both cases, the dynamic threshold reduces Z_D at low frequencies. (e) Effect of the dynamic threshold on $G(f)$ in the slow LIF model. Simulations were performed with σ_I = 100 pA, τ_{noise} = 5 ms, and I_0 set to give r_0 = 20 Hz. Blue, green, and red curves indicate steady-state threshold slopes of 0, 0.50, and 0.75. (f) $G(f)$ in fast model (τ_m = 1 ms) with faster-fluctuating noise input (σ_I = 100 pA, τ_{noise} = 1 ms).

SYMBOLS AND ABBREVIATIONS

AC	Alternating current
AHP	Afterhyperpolarization following a spike
AP	Action potential
BK	Large ("big")-conductance calcium-activated potassium channels

C	Membrane capacitance
CF	Characteristic frequency of sound to which an auditory neuron is most sensitive
$C_{SR}(f)$	Fourier transform of stimulus–response correlation (a function of frequency)
$c_{SR}(\tau)$	Stimulus–response correlation (a function of the time difference)
$C_{SS}(f)$	Fourier transform of the stimulus autocorrelation (a function of frequency)
$c_{SS}(\tau)$	Stimulus autocorrelation (a function of the time difference)
D	"Distance to threshold"; the spike threshold minus the membrane voltage
DC	Direct current
DTX	Dendrotoxin, a blocker of Kv1 potassium channels
EIF	Exponential integrate-and-fire model
EPSC	Excitatory postsynaptic current
EPSP	Excitatory postsynaptic potential
E_{rev}	Reversal potential of an ionic current
f	Frequency
f_0	The fundamental frequency of a periodic input
fAHP	Fast afterhyperpolarization following a spike
FT	Fourier transform
f_R	Resonant frequency (subthreshold)
g	Membrane conductance (passive)
$G(f)$	Gain versus frequency
g_1	Feedback conductance of GIF model
GIF	Generalized integrate-and-fire model
HCN	Hyperpolarization-activated, cyclic nucleotide-gated channels
I	Current
I_{app}	Applied current
I_{noise}	Random component of applied current
IPD	Interaural phase difference
ISI	Interspike interval
ITD	Interaural time difference
I_0	Mean applied current
I_1	Amplitude of sine wave component of applied current
K	Sharpness of exponential current voltage dependence in the EIF model
Kv	Voltage-gated potassium channels
LIF	Leaky integrate-and-fire model
mAHP	Medium-duration afterhyperpolarization
MSO	Medial superior olivary nucleus
NL	Nucleus laminaris in the avian auditory brainstem
NM	Nucleus magnocellularis in the avian auditory brainstem
PC	Pyramidal cell
R	Membrane resistance
r_0	Mean firing rate
r_1	The amplitude of a sinusoidal modulation of the instantaneous firing rate
$r(t)$	A time-varying response; the spike output represented as a sum of impulses
$r(\varphi)$	Firing rate versus phase of a periodic input
sAHP	Slow afterhyperpolarization following one or more spikes

SD	Standard deviation
SIZ	Spike initiation zone
SK	Small-conductance calcium-activated potassium channels
$s(t)$	A time-varying stimulus
t	Time
t_i	The time of spike i
V	Transmembrane voltage
V_r	"Reset" voltage of LIF, GIF, or EIF model after each spike
V_{rest}	Resting transmembrane voltage in the absence of current input
V_s	Vector strength, a measure of phase-locking of spikes to a periodic stimulus
w	Feedback variable of GIF model
W	The width of a data analysis window
$Z(f)$	Impedance (a function of frequency)
$Z_D(f)$	Impedance of "distance to threshold"
δ	Dirac's delta function; an impulse with infinitesimal width and unit area
φ	Phase with respect to a periodic input
Φ	The average phase shift of the response with respect to a periodic input
σ_I	Standard deviation of applied noise current
θ	Threshold voltage for spike generation
$\theta_{SS}(V)$	The steady-state value of a voltage-dependent, dynamic spike threshold
τ_{noise}	Time constant of a "colored" noise current with exponential relaxation
τ	A time difference
τ_m	Membrane time constant
τ_1	Time constant of feedback in GIF model
4-AP	4-Aminopyridine

REFERENCES

Amitai Y 1994. Membrane potential oscillations underlying firing patterns in neocortical neurons. *Neuroscience* 63:151–61.

Ashida G, Abe K, Funabiki K, Konishi M 2007. Passive soma facilitates submillisecond coincidence detection in the owl's auditory system. *J Neurophysiol* 97:2267–82.

Ashida G, Carr CE 2011. Sound localization: Jeffress and beyond. *Curr Opin Neurobiol* 21:745–51.

Azouz R, Gray CM 2000. Dynamic spike threshold reveals a mechanism for synaptic coincidence detection in cortical neurons *in vivo*. *Proc Natl Acad Sci USA* 97:8110–15.

Azouz R, Gray CM 2003. Adaptive coincidence detection and dynamic gain control in visual cortical neurons *in vivo*. *Neuron* 37:513–23.

Barrett EF, Barrett JN 1976. Separation of two voltage-sensitive potassium currents, and demonstration of a tetrodotoxin-resistant calcium current in frog motoneurones. *J Physiol* 255:737–74.

Bekkers JM, Häusser M 2007. Targeted dendrotomy reveals active and passive contributions of the dendritic tree to synaptic integration and neuronal output. *Proc Natl Acad Sci USA* 104:11447–52.

Brumberg JC 2002. Firing pattern modulation by oscillatory input in supragranular pyramidal neurons. *Neuroscience* 114:239–46.

Brunel N, Chance FS, Fourcaud N, Abbott LF 2001. Effects of synaptic noise and filtering on the frequency response of spiking neurons. *Phys Rev Lett* 86:2186–9.

Bryant HL, Segundo JP 1976. Spike initiation by transmembrane current: A white noise analysis. *J Physiol* 260:279–314.

Cardin JA, Kumbhani RD, Contreras D, Palmer LA 2010. Cellular mechanisms of temporal sensitivity in visual cortex neurons. *J Neurosci* 30:3652–62.

Carnevale NT, Wachtel H 1980. Two reciprocating current components underlying slow oscillations in Aplysia bursting neurons. *Brain Res* 203:45–65.

Carr CE, Konishi M 1990. A circuit for detection of interaural time differences in the brain stem of the barn owl. *J Neurosci* 10:3227–46.

Elhilali M, Fritz JB, Klein DJ, Simon JZ, Shamma SA 2004. Dynamics of precise spike timing in primary auditory cortex. *J Neurosci* 24:1159–72.

Farries MA, Kita H, Wilson CJ 2010. Dynamic threshold and zero membrane slope conductance shape the response of subthalamic neurons to cortical input. *J Neurosci* 30:13180–91.

Fourcaud-Trocmé N, Hansel D, van Vreeswijk C, Brunel N 2003. How spike generation mechanisms determine the neuronal response to fluctuating inputs. *J Neurosci* 23:11628–40.

Fukui I, Ohmori H 2004. Tonotopic gradients of membrane and synaptic properties for neurons of the chicken nucleus magnocellularis. *J Neurosci* 24:7514–23.

Fukui I, Sato T, Ohmori H 2006. Improvement of phase information at low sound frequency in nucleus magnocellularis of the chicken. *J Neurophysiol* 96:633–41.

Funabiki K, Ashida G, Konishi M 2011. Computation of interaural time difference in the owl's coincidence detector neurons. *J Neurosci* 31:15245–56.

Gerstner W 2000. Population dynamics of spiking neurons: Fast transients, asynchronous states, and locking. *Neural Comput* 12:43–89.

Goldberg JM, Brown PB 1969. Response of binaural neurons of dog superior olivary complex to dichotic tonal stimuli: Some physiological mechanisms of sound localization. *J Neurophysiol* 32:613–36.

Gu N, Vervaeke K, Hu H, Storm JF 2005. Kv7/KCNQ/M and HCN/h, but not KCa2/SK channels, contribute to the somatic medium after-hyperpolarization and excitability control in CA1 hippocampal pyramidal cells. *J Physiol* 566:689–715.

Gutfreund Y, Yarom Y, Segev I 1995. Subthreshold oscillations and resonant frequency in guinea-pig cortical neurons: Physiology and modelling. *J Physiol* 483:621–40.

Guttman R, Feldman L, Lecar H 1974. Squid axon membrane response to white noise stimulation. *Biophys J* 14:941–55.

Higgs MH, Spain WJ 2009. Conditional bursting enhances resonant firing in neocortical layer 2–3 pyramidal neurons. *J Neurosci* 29:1285–99.

Higgs MH, Spain WJ 2011. Kv1 channels control spike threshold dynamics and spike timing in cortical pyramidal neurones. *J Physiol* 589(Pt 21):5125–42.

Hill AV 1936. Excitation and accommodation in nerve. *Proc R Soc B* 119:305–55.

Hodgkin AL, Huxley AF 1952. A quantitative description of membrane current and its application to conduction and excitation in nerve. *J Physiol* 117:500–44.

Howard MA, Rubel EW 2010. Dynamic spike thresholds during synaptic integration preserve and enhance temporal response properties in the avian cochlear nucleus. *J Neurosci* 30:12063–74.

Hsiao CF, Kaur G, Vong A, Bawa H, Chandler SH 2009. Participation of Kv1 channels in control of membrane excitability and burst generation in mesencephalic V neurons. *J Neurophysiol* 101:1407–18.

Hu H, Vervaeke K, Graham LJ, Storm JF 2009. Complementary theta resonance filtering by two spatially segregated mechanisms in CA1 hippocampal pyramidal neurons. *J Neurosci* 29:14472–83.

Hu H, Vervaeke K, Storm JF 2002. Two forms of electrical resonance at theta frequencies, generated by M-current, h-current and persistent Na^+ current in rat hippocampal pyramidal cells. *J Physiol* 545:783–805.

Hutcheon B, Miura RM, Puil E 1996a. Subthreshold membrane resonance in neocortical neurons. *J Neurophysiol* 76:683–97.
Hutcheon B, Miura RM, Puil E 1996b. Models of subthreshold membrane resonance in neocortical neurons. *J Neurophysiol* 76:698–714.
Hutcheon B, Yarom Y 2000. Resonance, oscillation and the intrinsic frequency preferences of neurons. *Trends Neurosci.* 23:216–22.
Jeffress LA 1948. A place theory of sound localization. *J Comp Physiol Psychol* 41:35–9.
Joseph AW, Hyson RL 1993. Coincidence detection by binaural neurons in the chick brain stem. *J Neurophysiol* 69:1197–211.
Kayser C, Logothetis NK, Panzeri S 2010. Millisecond encoding precision of auditory cortex neurons. *Proc Natl Acad Sci USA* 107:16976–81.
Knight BW 1972. Dynamics of encoding in a population of neurons. *J Gen Physiol* 59:734–66.
Kole MH 2011. First node of Ranvier facilitates high-frequency burst encoding. *Neuron* 71:671–82.
Kole MH, Stuart GJ 2008. Is action potential threshold lowest in the axon? *Nat Neurosci* 11:1253–5.
Köndgen H, Geisler C, Fusi S, Wang XJ, Lüscher HR, Giugliano M 2008. The dynamical response properties of neocortical neurons to temporally modulated noisy inputs *in vitro*. *Cereb Cortex* 18:2086–97.
Köppl C 1994. Auditory nerve terminals in the cochlear nucleus magnocellularis: Differences between low and high frequencies. *J Comp Neurol* 339:438–46.
Köppl C 1997. Phase locking to high frequencies in the auditory nerve and cochlear nucleus magnocellularis of the barn owl, Tyto alba. *J Neurosci* 17:3312–21.
Kuba H, Ohmori H 2009. Roles of axonal sodium channels in precise auditory time coding at nucleus magnocellularis of the chick. *J Physiol* 587:87–100.
Kuba H, Koyano K, Ohmori H 2002. Development of membrane conductance improves coincidence detection in the nucleus laminaris of the chicken. *J Physiol* 540:529–42.
Kuba H, Yamada R, Fukui I, Ohmori H 2005. Tonotopic specialization of auditory coincidence detection in nucleus laminaris of the chick. *J Neurosci* 25:1924–34.
Kuba H, Ishii TM, Ohmori H 2006. Axonal site of spike initiation enhances auditory coincidence detection. *Nature* 444:1069–72.
Kuznetsova MS, Higgs MH, Spain WJ 2008. Adaptation of firing rate and spike-timing precision in the avian cochlear nucleus. *J Neurosci* 28:11906–15.
Lancaster B, Adams PR 1986. Calcium-dependent current generating the afterhyperpolarization of hippocampal neurons. *J Neurophysiol* 55:1268–82.
Larkum ME, Zhu JJ, Sakmann B 1999. A new cellular mechanism for coupling inputs arriving at different cortical layers. *Nature* 398:338–41.
Lee YW, Schetzen M 1965. Measurement of the Wiener kernels of a non-linear system by cross-correlation. *Int J Control* 2:237–54.
Lorenzon NM, Foehring RC 1992. Relationship between repetitive firing and afterhyperpolarizations in human neocortical neurons. *J Neurophysiol* 67:350–63.
Magee JC, Carruth M 1999. Dendritic voltage-gated ion channels regulate the action potential firing mode of hippocampal CA1 pyramidal neurons. *J Neurophysiol* 82:1895–901.
Mainen ZF, Sejnowski TJ 1995. Reliability of spike timing in neocortical neurons. *Science* 268:1503–6.
Mauro A, Conti F, Dodge F, Schor R 1970. Subthreshold behavior and phenomenological impedance of the squid giant axon. *J Gen Physiol* 55:497–523.
Mechler F, Victor JD, Purpura KP, Shapley R 1998. Robust temporal coding of contrast by V1 neurons for transient but not for steady-state stimuli. *J Neurosci* 18:6583–98.
Narayanan R, Johnston D 2007. Long-term potentiation in rat hippocampal neurons is accompanied by spatially widespread changes in intrinsic oscillatory dynamics and excitability. *Neuron* 56:1061–75.

Otis TS, Raman IM, Trussell LO 1995. AMPA receptors with high Ca^{2+} permeability mediate synaptic transmission in the avian auditory pathway. *J Physiol* 482:309–15.

Overholt EM, Rubel EW, Hyson RL 1992. A circuit for coding interaural time differences in the chick brainstem. *J Neurosci* 12:1698–708.

Parks TN 1981. Morphology of axosomatic endings in an avian cochlear nucleus: Nucleus magnocellularis of the chicken. *J Comp Neurol* 203:425–40.

Peters HC, Hu H, Pongs O, Storm JF, Isbrandt D 2005. Conditional transgenic suppression of M channels in mouse brain reveals functions in neuronal excitability, resonance and behavior. *Nat Neurosci* 8:51–60.

Platkiewicz J, Brette R 2011. Impact of fast sodium channel inactivation on spike threshold dynamics and synaptic integration. *PLoS Comput Biol* 7:e1001129.

Rathouz M, Trussell L 1998. Characterization of outward currents in neurons of the avian nucleus magnocellularis. *J Neurophysiol* 80:2824–35.

Reyes AD, Rubel EW, Spain WJ 1994. Membrane properties underlying the firing of neurons in the avian cochlear nucleus. *J Neurosci* 14:5352–64.

Reyes AD, Rubel EW, Spain WJ 1996. *In vitro* analysis of optimal stimuli for phase-locking and time-delayed modulation of firing in avian nucleus laminaris neurons. *J Neurosci* 16:993–1007.

Richardson MJE, Brunel N, Hakim V 2003. From subthreshold to firing-rate resonance. *J Neurophysiol* 89:2538–54.

Rose JE, Brugge JF, Anderson DJ, Hind JE 1967. Phase-locked response to low-frequency tones in single auditory nerve fibers of the squirrel monkey. *J Neurophysiol* 30:769–93.

Schindler KA, Goodman PH, Wieser HG, Douglas RJ 2006. Fast oscillations trigger bursts of action potentials in neocortical neurons *in vitro*: A quasi-white-noise analysis study. *Brain Res* 1110:201–10.

Schwindt P, Crill W 1999. Mechanisms underlying burst and regular spiking evoked by dendritic depolarization in layer 5 cortical pyramidal neurons. *J Neurophysiol* 81:1341–54.

Schwindt PC, Spain WJ, Foehring RC, Chubb MC, Crill WE 1988a. Slow conductances in neurons from cat sensorimotor cortex *in vitro* and their role in slow excitability changes. *J Neurophysiol* 59:450–67.

Schwindt PC, Spain WJ, Foehring RC, Stafstrom CE, Chubb MC, Crill WE 1988b. Multiple potassium conductances and their functions in neurons from cat sensorimotor cortex *in vitro*. *J Neurophysiol* 59:424–49.

Schwindt PC, Spain WJ, Crill WE 1989. Long-lasting reduction of excitability by a sodium-dependent potassium current in cat neocortical neurons. *J Neurophysiol* 61:233–44.

Slee SJ, Higgs MH, Fairhall AL, Spain WJ 2010. Tonotopic tuning in a sound localization circuit. *J Neurophysiol* 103:2857–75.

Smith DJ, Rubel EW 1979. Organization and development of brain stem auditory nuclei of the chicken: Dendritic gradients in nucleus laminaris. *J Comp Neurol* 186:213–39.

Storm JF 1987. Intracellular injection of a Ca^{2+} chelator inhibits spike repolarization in hippocampal neurons. *Brain Res* 435:387–92.

Storm JF 1989. An after-hyperpolarization of medium duration in rat hippocampal pyramidal cells. *J Physiol* 409:171–90.

Sullivan WE, Konishi M 1986. Neural map of interaural phase difference in the owl's brainstem. *Proc Natl Acad Sci USA* 83:8400–4.

Tennigkeit F, Schwarz DW, Puil E 1999. Modulation of frequency selectivity by Na+ - and K^+-conductances in neurons of auditory thalamus. *Hear Res* 127:77–85.

Traub RD 1979. Neocortical pyramidal cells: A model with dendritic calcium conductance reproduces repetitive firing and epileptic behavior. *Brain Res* 173:243–57.

Yin TC, Chan JC 1990. Interaural time sensitivity in medial superior olive of cat. *J Neurophysiol* 64:465–88.

Yu Y, Shu Y, McCormick DA 2008. Cortical action potential backpropagation explains spike threshold variability and rapid-onset kinetics. *J Neurosci* 28:7260–72.

Yue C, Remy S, Su H, Beck H, Yaari Y 2005. Proximal persistent Na^+ channels drive spike afterdepolarizations and associated bursting in adult CA1 pyramidal cells. *J Neurosci* 25:9704–20.

Williams SR, Stuart GJ 1999. Mechanisms and consequences of action potential burst firing in rat neocortical pyramidal neurons. *J Physiol* 521:467–82.

Zhang S, Trussell LO 1994. Voltage clamp analysis of excitatory synaptic transmission in the avian nucleus magnocellularis. *J Physiol* 480:123–36.

10 Relating Spike Times to Perception

Auditory Detection and Discrimination

Laurel H. Carney

CONTENTS

10.1 INTRODUCTION

A key question for the study of spike times is whether or not they are actually important. That is, are they an interesting phenomenon or simply an epiphenomenon? Every spike must have a time associated with it, after all. A critical test for the significance of spike timing is to relate it to function, and in the case of the auditory system, an important function is perception, including detection and discrimination of sounds. In particular, the role of spike timing in the ability of listeners to discriminate the most basic aspects of simple stimuli is of interest, such as the amplitude and frequency of tonal stimuli.

In the auditory system, sensitivity to sound frequency is spatially mapped along the length of the inner ear (cochlea) due to changes in the mechanical properties, especially the stiffness, of the sensory apparatus along its length. The frequency map set up by the inner ear, referred to as tonotopy, is preserved throughout the central nervous system, providing a spatial map of sound frequency at each level of the ascending auditory pathway. Thus, spectral information could potentially be represented by the average discharge rates across the spatial frequency map, the so-called rate-place code. Spike timing is an alternative, or additional, strategy for spectral coding. This chapter will present a series of studies of coding in the auditory system that have tested both rate-place and timing coding strategies and compared the results to perceptual abilities in a variety of psychophysical discrimination tasks. The focus will be on studies of coding of simple and complex sounds presented to one ear (i.e., monaural stimulation). Responses of neurons in the auditory periphery and of coincidence detectors that receive convergent inputs that are stimulated by one ear will be considered. Spatiotemporal representations, which combine rate-place and timing information, will also be discussed.

10.2 PREDICTING PSYCHOPHYSICAL RESULTS USING AUDITORY-NERVE POPULATION MODELS

Siebert (1965, 1968, 1970) tested the hypotheses that the information in rate-place and timing in the responses of the population of auditory-nerve fibers could quantitatively predict psychophysical performance in the discrimination of tone frequencies and levels. An important aspect of his hypotheses was that the randomness inherent in the Poisson firing statistics of auditory-nerve fibers, the so-called internal noise, was the limiting factor in these psychophysical discriminations. Thus, he assumed Poisson variance for the auditory-nerve responses, and that the responses of the fibers were independent from each other. These reasonable assumptions allowed the derivation of a just-noticeable difference (JND) for tone frequency and level using a Cramer–Rao bound (van Trees, 1968). The Cramer–Rao inequality describes the upper bound of the variance of an estimate of a stimulus parameter value, and thus provides an estimate of the discrimination threshold for that parameter. The Cramer–Rao bound is calculated based on the change in the system response to stimuli that vary (slightly) in the parameter of interest.

The process for computing the discrimination threshold for a change in the amplitude of a tonal stimulus is illustrated in Figure 10.1. The acoustic stimulus (a pressure waveform in pascals) is the input signal to the population model for auditory-nerve responses (Figure 10.1, top). The model output is a function that describes the time-varying discharge rate of each fiber, $r_i(t,f,L)$, which depends upon stimulus frequency, f, level, L, and the fiber's characteristic frequency (CF, the stimulus frequency to which a fiber is most sensitive), where i indicates the index (or "place") of the fiber within the frequency-mapped population of fibers. The left-hand side of Figure 10.1 illustrates the procedure for estimating level discrimination based on the entire response, or spike count, of each fiber over the entire time course of the stimulus, the so-called rate-place model. The procedure illustrated on the right-hand side of Figure 10.1 uses both discharge rate and temporal information by working with the time-varying rate function; this model is referred to as the "all-information" model. In both cases, Poisson variance is assumed for spike counts (either for the average count or for the instantaneous responses described by the rate function), and optimal decision theory, that is, the Cramer–Rao bound, is used to compute the discrimination threshold, as described below.

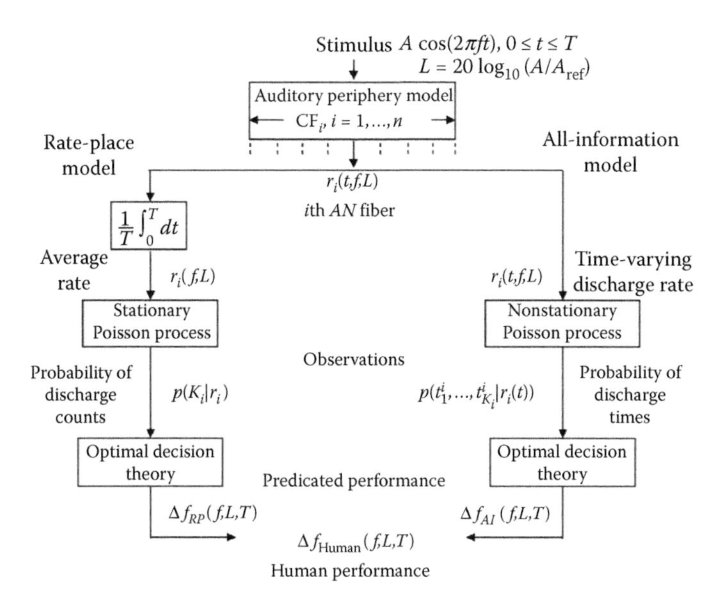

FIGURE 10.1 Schematic diagram illustrating Siebert's strategy for estimating just-noticeable differences (JNDs) in a discrimination task. The case illustrated is the discrimination of the sound level, L; f is the stimulus frequency; T is the stimulus duration, and the t's are discharge times. (Reprinted from Heinz, M.G., Colburn, H.S., and Carney, L.H. 2001a. *Neural Computat.* 13:2273–316, Figure 2. With permission.)

In Equation 10.1, the neural response is designated as X, and the change in this response can be computed based either on the average rate or on the time-varying rate function. This change is measured in response to stimuli that vary in one stimulus parameter, α (e.g., α may refer to stimulus frequency or to amplitude for a pure tone stimulus). Finally, the variance of the response, which according to the Poisson assumption is equal to the rate itself, ultimately limits how accurately the parameter value, α, can be estimated based on changes in the response, X. The expression for the upper bound of the variance in the estimate of the stimulus parameter α, based on the Cramer–Rao bound, is shown in Equation 10.1 (from Siebert 1968, Equation 4.5). (See Heinz et al., 2001a, for more detailed derivations and explanation of this procedure.)

$$\sigma^2\left[\hat{\alpha}(\mathrm{X})|\alpha\right] \ge \frac{1}{-\int\left[(\partial^2 \ln p(\mathrm{X}|\alpha))/\partial\alpha^2\right]p(\mathrm{X}|\alpha)\,d\mathrm{X}} \tag{10.1}$$

An advantage of Siebert's approach is that the JND derived from the Cramer–Rao bound can be directly compared to psychophysical thresholds, or JNDs, for the same stimulus parameter (Figure 10.1, bottom). Also, the JND can be estimated based either on a description of the response of a single fiber or on the response of a population of fibers. In the latter case, the contribution of each fiber, or frequency channel, can also be assessed.

The description of the auditory-nerve responses used by Siebert (1965, 1968) was relatively simple, but captured the key properties of auditory-nerve responses that had been described to date (e.g., Kiang et al. 1965). Frequency tuning of the auditory-nerve fibers was described by linear triangular filters; thus, the average rate depended on the distance between the stimulus frequency and the fiber's characteristic frequency. The average discharge rate (sp/s) was assumed to be linearly related to stimulus level (in dB SPL), and the average rate did not change over the time course of the stimulus; thus, rate adaptation was not included in the model. Predictions of JND based on rate-place properties only required filtering and rate-level descriptions. Predictions based on spike times required a description of the time-varying response properties of each fiber (Siebert 1970). For low stimulus frequencies, auditory-nerve fibers phase-lock, or synchronize, to the stimulus waveform; the rate varies within each stimulus cycle, peaking at a particular phase. Siebert (1970) used an exponentiated sinusoid to describe the rate function, r, which is the time-varying rate of auditory-nerve fibers to low-frequency pure tones:

$$r(t,f,A) = r_0 \exp\left\{ g\left[A\,H\left(\frac{f}{f_i}\right)\right] \cos(2\pi f t)\right\} \tag{10.2}$$

where the average rate, r_0, as a function of level is proportional to level (in dB SPL). The function, g, determines the strength of phase-locking and varies with stimulus amplitude, A, and stimulus frequency, f, relative to the fiber's CF, f_i, according to a triangular filter function, H (see Equation 1 in Siebert 1970, and Equation 9 in Colburn et al. 2003). Siebert's analog model for the auditory-nerve responses provides a reasonably good description for responses at moderate levels (above rate threshold and below rate saturation), and for stimulus frequencies below about 1250 Hz, above which the strength of phase-locking rolls off (Kiang et al. 1965, Johnson 1980). Thus, Siebert limited his predictions to the range of stimulus levels and frequencies where his simple analog model for auditory-nerve responses was appropriate.

The general conclusions of Siebert's predictions for frequency and level discrimination of low-frequency tones were that rate-place predictions were appropriate to explain perceptual thresholds, and that timing-based predictions were too sensitive. Furthermore, the simple extrapolation of the results to higher frequencies tends to argue that because the strength of phase-locking of auditory-nerve responses rolls off as stimulus frequency increases, the system *must* depend on rate-place information at high frequencies. A logical problem with this extrapolation is that perceptual thresholds indeed become less sensitive as stimulus frequency increases, but rate-place-based predictions remain basically unchanged as frequency increases. To extend this approach to a wider range of stimuli, including higher frequencies, requires a more accurate model for the auditory periphery. Recent studies that have addressed the problems of level and frequency discrimination are presented next.

10.3 PREDICTIONS OF SOUND LEVEL DISCRIMINATION

How does the auditory system encode stimuli over a wide dynamic range of at least 100 dB? This fundamental question is not unique to auditory level discrimination but applies to other sensory systems, as well. For many scenarios, Weber's law describes sensitivity over a wide range of stimulus levels, and sensitivity scales proportionally to stimulus level in log units. In the case of the auditory system, there is a so-called near-miss to Weber's law, or an improvement in sensitivity (for some types of stimuli), as stimulus level increases, making the dynamic range problem that much more challenging (see Florentine et al. 1987, Viemeister 1988a,b).

The predictions of Siebert's analytical model for level discrimination are illustrated by the dotted lines with open circles in Figure 10.2 (from Heinz et al. 2001a). These predictions meet or exceed human thresholds for all levels, durations, and frequencies shown. Because the predictions are for optimal detectors, and thus typically outperform actual listeners, it is instructive to look further at the *trends* in the predictions as a function of the stimulus parameters, rather than only comparing the absolute values of the predictions with the thresholds. The decrease in threshold predictions with increasing level agrees with the data (Figure 10.2a, filled stars), except above 40 dB SPL where predicted thresholds flatten due to limitations in the auditory-nerve model he used (the rate-place predictions made with a more accurate

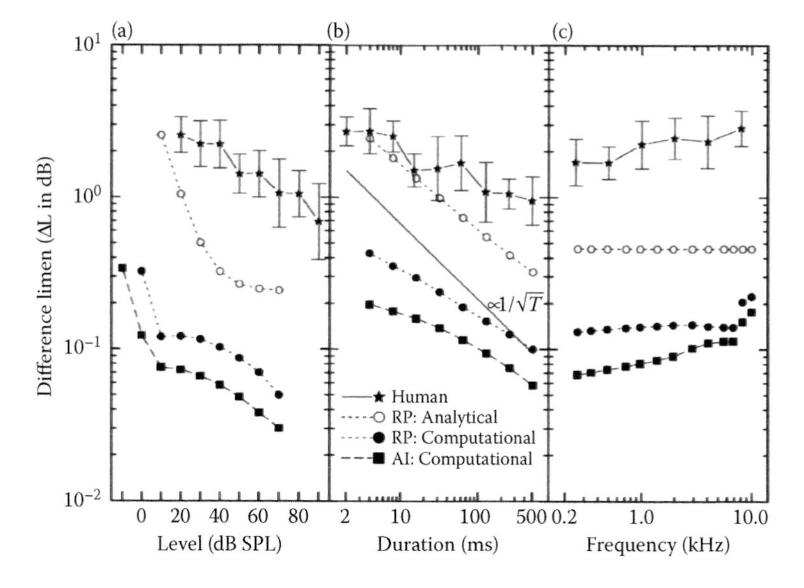

FIGURE 10.2 Estimates of level discrimination thresholds for an optimal detector based on population auditory-nerve model responses as a function of level, duration, and frequency. The analytical model is the rate-place model of Siebert (1968). (a) Frequency = 970 Hz, duration = 500 ms. Human data: Florentine et al. (1987). (b) Level = 40 dB SPL, frequency = 970 Hz. Human data: Florentine (1986). (c) Level = 40 dB SPL; duration = 500 ms. Human data: Florentine et al. (1987). (Reprinted from Heinz, M.G. et al. 2001a. *Neural Computat.* 13:2273–316, Figure 6. With permission.)

computation model [filled circles, see below] do not share this effect). The trends in predicted thresholds improve more quickly than actual thresholds as the stimulus duration increases (Figure 10.2b), as is true for all of the optimal-detector-based models illustrated here. Finally, the analytical rate-place model predicts uniform thresholds as a function of stimulus frequency (Figure 10.2c), which does not agree with the trend in human thresholds. Threshold predictions based on temporal information in the auditory-nerve responses will be described below.

Other strategies for relating psychophysical discriminations to neural responses have involved various neurometrics, including sensitivity measures similar to the discriminability measure, d', but based on the changes and variance of model auditory-nerve spike rates (e.g., Viemeister 1988a,b, Winter and Palmer 1991, Delgutte 1987, 1996). These studies have explored the potential contribution of a small population of fibers, the low-spontaneous-rate auditory-nerve fibers that have relatively large dynamic ranges. If this class of fibers is heavily weighted in a pooled representation of the population response, it is possible to explain a relatively wide dynamic range (Viemeister 1988a,b, Winter and Palmer 1991, Delgutte 1987, 1996). However, low-spontaneous-rate fibers only exhibit large dynamic ranges at mid to high characteristic frequencies (>1500 Hz, Winter and Palmer 1991), presumably because the wide dynamic range of the low-spontaneous-rate fibers requires a strong compressive nonlinearity in the inner-ear mechanics (see below); the amount of inner-ear compression is reduced at low frequencies (Cooper 1996).

Colburn et al. (2003) revisited the dynamic range problem using Siebert's approach, but included in the description of the auditory periphery a nonlinearity that had not been described at the time of Siebert's work: the level-dependent phase of the temporal responses. Tuning to audio frequencies in the cochlea is sharpest at low sound levels and broadens as the level increases. The increase in the bandwidth of tuning at higher levels is associated with a decrease in amplification of the cochlear response as the level increases, referred to as the compressive nonlinearity (Rhode 1971, Ruggero et al. 1997, reviewed in Robles and Ruggero 2001). As the gain and bandwidth of filters change as a function of the sound level, so do their phase properties (Figure 10.3). Thus, the phase-locked responses of auditory-nerve fibers have level-dependent phase properties, or timing, that provide potential information for level discrimination. The level-dependent timing is systematic: tone-evoked spike times of fibers tuned to frequencies below the stimulus frequency have increasing phase lags as level increases, whereas response times of fibers tuned above the stimulus frequency have increasing phase leads (Anderson et al. 1971, Geisler and Rhode 1982). The result is that responses of a population of fibers tuned to frequencies straddling the stimulus frequency have increasing coincidence as tone level increases (Carney 1994). Colburn et al. (2003) showed that inclusion of this cue in predictions of level discrimination can explain behavioral-level discrimination over a wide range of levels, without having to resort to heavy weighting of the small population of low-spontaneous-rate fibers used in earlier models (Winter and Palmer 1991, Delgutte 1987, 1996). These results support the hypothesis that spike timing information may play an important role in sound level coding.

Heinz (2000) extended Siebert's approach to higher stimulus frequencies and a wider range of stimulus levels by modernizing the peripheral model and adapting

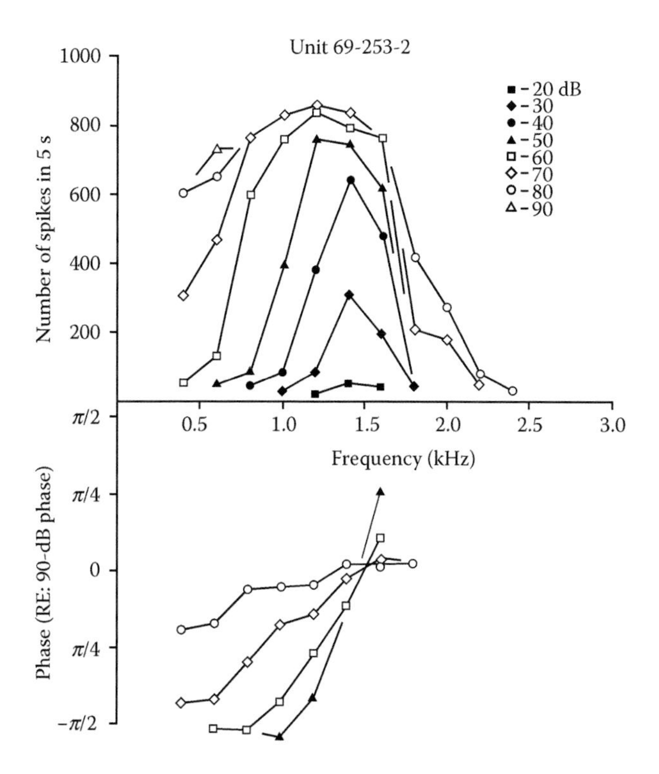

FIGURE 10.3 Discharge rates (top) and timing (bottom) for responses of an auditory-nerve fiber in squirrel monkey to tones across a range of frequencies and levels. Spike timing, or phase, is plotted with respect to the timing of responses to a 90 dB SPL tone at each frequency. Note the systematic changes in timing as the sound level is increased: responses of the fiber to tones below the fiber's CF (~1.4 kHz) are more delayed, or lagging, as the sound level is increased. The opposite is true for responses to frequencies above CF. These level-dependent changes in spike timing reflect changes that have been observed in the mechanical response of the basilar membrane in the inner ear (Geisler and Rhode 1982; Ruggero et al. 1997). (Reprinted with permission from Anderson, D.J. et al. *J. Acoust. Soc. Am.* 49:1131–9, Figure 5. Copyright 1971, Acoustical Society of America.)

the calculation of the Cramer–Rao bound to work with the response of a computational model. Computational models allow the inclusion of a number of nonlinearities in the peripheral responses that have been described since Siebert's work in the 1960s. These include a detailed description of the compressive nonlinearity that is associated with changes in sensitivity of the cochlea as a function of the sound level. These changes in gain are impressive, especially at high frequencies, where the sensitivity of the cochlea to sounds near threshold can be 50–60 dB greater than the sensitivity at high sound levels (Rhode 1971, Ruggero et al. 1997, Robles and Ruggero 2001). The changes in timing information with the sound level are more complex than the changes in strength of synchrony that were included in Siebert's (1970) model and in the simple model used in the Colburn et al. (2003) study.

Heinz et al. (2001a,b) used a modified version of the Carney (1993) model for auditory-nerve responses that included the compressive nonlinearity, the roll-off in strength of phase-locking with increasing stimulus frequency, smooth changes in average rate near threshold and saturation, and a detailed description of the level-dependent time-varying discharge rate within each cycle of the stimulus as well as over the course of the stimulus, including rate adaptation at stimulus onset and recovery after stimulus offset. The responses of the model were still assumed to be Poisson random variables, and for the purposes of Heinz et al.'s predictions, refractoriness was not directly included in the description of the time-varying rate function, though rates were adjusted to include the overall decrease in rate due to refractoriness.

Figure 10.2a–c illustrates Heinz et al.'s (2001a,b) predictions based on rate-place (filled circles) and all information (which included both rate and timing information, filled squares). The rate-place model predictions based on the more complete auditory-nerve model are generally similar to Siebert's results, with the improvement in the trend of predicted thresholds over a wide range of levels (Figure 10.2a) being the most notable difference. Thresholds predicted based on both rate and temporal information were lower (as expected, the optimal detector's performance improves when more information is provided). Overall, the trends in threshold versus level, duration, and frequency are better described by the all-information predictions. The most important difference between the two sets of predictions is the increasing trend of all-information thresholds as a function of frequency, which better matches the trend in the human data. Because one can posit the existence of suboptimal detectors that would simply shift the thresholds upward—by some form of information "loss," or inefficient processing of information—while maintaining the same trends across stimulus parameters, it is easier to imagine obtaining the actual human thresholds based on the all-information predictions than on the rate-place predictions. Thus, these results also support the hypothesis that temporal information plays a role in level discrimination.

10.4 PREDICTIONS OF FREQUENCY DISCRIMINATION

Predictions for frequency discrimination thresholds based on Siebert's (1965, 1968, 1970) and Heinz et al.'s (2001a) models are shown in Figure 10.4. As expected, the rate-place and rate-plus-timing (all information) predictions of the more modern computational model (filled circles and squares, respectively) generally agree with those of Siebert's analytical models (open circles and squares) at low frequencies, where the relatively simple model of Siebert (1970) does a good job of describing phase-locked responses of auditory-nerve fibers. At high frequencies, where phase-locking rolls off, the rate-place predictions are unaffected (as expected), whereas predictions based on the combination of rate and timing information increase. The increase in the model thresholds as the frequency increases is due to the reduction in the strength of phase-locking as the stimulus frequency increases. It is notable that the general trend in the timing-based predictions is similar to that seen in the human perceptual thresholds. One can imagine that a suboptimal detector, with thresholds that were uniformly elevated across frequency, could nearly match the perceptual results. However, there is no simple way to match the rate-place predictions to the

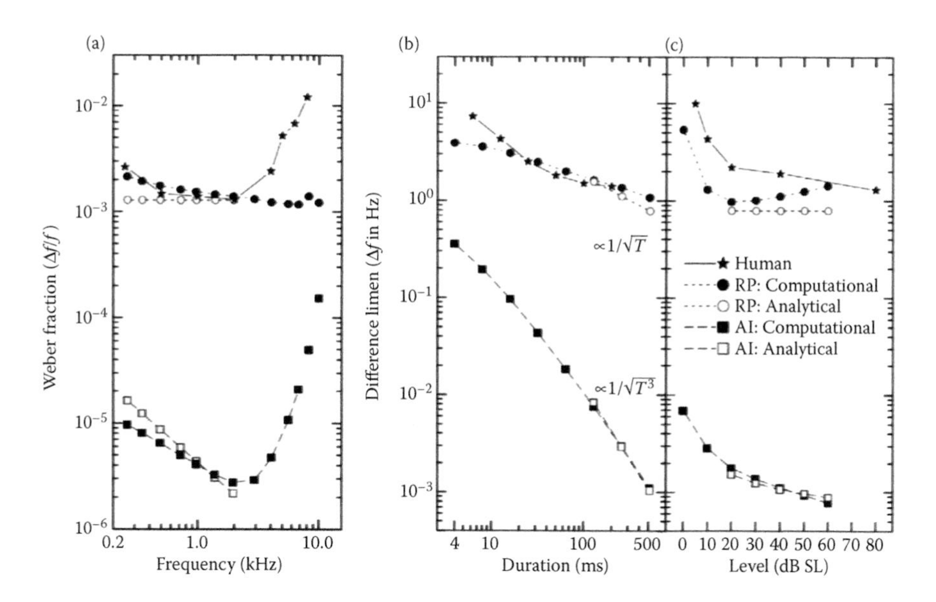

FIGURE 10.4 Estimates of frequency discrimination thresholds for an optimal detector based on population auditory-nerve model responses as a function of frequency, duration, and level. The analytical all-information predictions are based on Siebert (1970). (a) Level = 40 dB SPL; duration = 200 ms, Human data: Moore (1973). (b) Level = 40 dB SPL, frequency = 970 Hz. Human data: Moore (1973) for 60 dB SPL, frequency = 1000 Hz. (c) Frequency = 970 Hz; duration = 500 ms. Human data: Wier et al. (1977) for frequency = 1000 Hz, duration = 500 ms. (Reprinted from Heinz, M.G., Colburn, H.S., and Carney, L.H. 2001a. *Neural Computat.* 13:2273–316, Figure 4. With permission.)

perceptual results: these predictions require near-optimal use of information at low frequencies, and increasingly suboptimal use of information as frequency increases to explain the perceptual results. There is no reason to believe that a system that could optimally use average-rate information for one set of neurons (with low CFs) would not do so for another set of neurons (with higher CFs).

Heinz et al.'s (2001a) predicted thresholds for frequency discrimination thresholds have the correct trends as a function of frequency (Figure 10.4a), but they are still considerably lower than actual thresholds, suggesting that listeners do not make optimal use of the temporal information in the auditory periphery. This is not surprising, especially given that optimal use of the spike times in this task includes feats such as the precise use of the time differences between the very first and very last spike in the response (which explains the too-rapid improvement in predicted performance as stimulus duration increases, Figure 10.4b). Consideration of suboptimal processors opens a wide door for proposed processing schemes. For example, Goldstein and Srulovicz (1977) explored models that placed reasonable limits on the information available to timing-based predictions, by limiting the numbers of fibers and constraining the models to use first-order interspike intervals, and only intervals that are less than a certain duration; they demonstrated that models that make use of restricted temporal information cannot be ruled out. It should be noted that interspike intervals have been explored not only for the discrimination of pure tones, as discussed here, but also for more complex sounds such as harmonic tone complexes that evoke the perception of pitch (Cariani and Delgutte 1996a,b).

Heinz's and Siebert's approaches can also be applied to more complex stimuli. For example, Tan and Carney (2005) used this strategy to predict formant-frequency discrimination thresholds for synthetic single-formant vowel-like sounds that were used in a psychophysical study by Lyzenga and Horst (1995). The psychophysical study included sound levels that were randomly varied from trial to trial to obscure energy-based cues, and these "roving" levels were also included in the neural simulations. Tan and Carney (2005) found that if both rate-place and timing information were included, a relatively small number of auditory-nerve fiber responses was sufficient to predict the trends in the psychophysical thresholds across a wide range of stimulus parameters.

10.5 MONAURAL COINCIDENCE DETECTORS FOR DECODING THE TIMING AND RATE INFORMATION IN THE AUDITORY PERIPHERY

Consideration of neural mechanisms for extracting temporal information from the population response of the auditory periphery leads naturally to the exploration of suboptimal but physiologically realistic processing strategies. One such strategy that has been the subject of a number of studies is monaural coincidence detection. Binaural coincidence detectors have been studied more extensively, given their presumed role in low-frequency sound localization (e.g., Jeffress 1948, Yin and Chan 1990, reviewed by Joris et al. 1998). However, the same ion channels and nonlinear membrane properties that provide binaural neurons in the medial superior olive with exquisite sensitivity to interaural time differences (Smith 1995) are also present in the monaural bushy cells of the cochlear nucleus (Manis and Marx 1991, Rothman and Manis 2003a,b,c). The potential role of these neurons in performing monaural cross-frequency coincidence detection has been explored in a few physiological studies (Carney 1990, Jiang et al. 1996, Wang and Delgutte 2012) and models (e.g., Carney 1994, Heinz et al. 2001c, Zhang and Carney 2005).

Cross-frequency coincidence detectors receive convergent input from auditory-nerve fibers tuned across some range of characteristic frequencies. Responses of the coincidence detector are more probable when the input fibers have more similar, or coincident, spike times. These coincidence detectors could potentially decode the temporal cues associated with nonlinear phase properties of auditory-nerve fibers (Figure 10.3) because these level-dependent phases result in changes in the *relative* timing of neighboring fibers (Carney 1994). Increases in average discharge rate also lead to increased probabilities of coincidences, and combined changes in timing and rate can be decoded naturally by coincidence detectors. For example, as the sound level increases, the spike times of fibers tuned to neighboring frequencies become more similar *and* the number of spikes increase until the discharge rates of the fibers saturate. At even higher sound levels, for which the discharge rate is saturated for many auditory-nerve fibers, the level-dependent changes in spike timing continue to affect the responses, so cross-fiber coincidences continue to increase as the sound level increases, even though discharge rates no longer increase. Colburn et al. (2003) showed that level-dependent spike timing provides a strong cue for level coding at low frequencies. At higher frequencies, the strength of phase-locking to pure tones gradually decays, but level-dependent changes in cochlear gain (and thus average discharge rate) are more significant at high frequencies. A population of coincidence detectors can decode level-dependent phase cues, level-dependent discharge rate, and their combination, across a wide range of frequencies (Heinz et al. 2001c).

Heinz et al. (2001c) explored level discrimination using a two-input coincidence-detection mechanism based on a similar model used by Colburn (1969, 1973, 1977) for binaural coincidence detection (Figure 10.5a). The monaural coincidence-counting cell model discharges whenever the two auditory-nerve inputs discharge within a specified temporal window, and counts up these coincidences over the course of a stimulus. Their results showed that the same across-frequency coincidence-detection mechanism could explain level discrimination performance across a wide range of stimulus frequencies. At low frequencies (Figure 10.5b left), nonlinear phase cues dominated detection performance, and at high frequencies (Figure 10.5b right), level-dependent rates were dominant. Because the same mechanism works effectively for both cues, the model naturally transitions between these two cues at intermediate frequencies. Heinz et al.'s (2001c) model also explains two well-known characteristics of level discrimination: the "mid-level bump," or elevation in thresholds at mid-levels, which is explained by the change in slope of the basilar membrane input–output function related to the compressive nonlinearity, and the improvement in discrimination at high sound levels, which is referred to as the "near miss to Weber's law" (Florentine et al. 1987) (Figure 10.5b).

The studies described above have shown that the coincidence-detection mechanism is useful for making predictions based on population responses; however, relating the responses of simple coincidence detectors to actual neural responses is difficult. Model cells with realistic numbers of subthreshold auditory-nerve inputs and with temporal windows that are small enough to provide effective coincidence detection tend to have very low discharge rates (e.g., Joris et al. 1994). Zhang and Carney (2005) explored the parameter space of integrate-and-fire models for coincidence detectors and found that the response properties of some types of auditory brainstem cells could be explained by varying the numbers and amplitude of the inputs to integrate-and-fire-type model cells. Convergence of many small-amplitude inputs was required to explain the response properties of some auditory brainstem cells; others required model cells with inputs that had short-duration input potentials, or mixed-amplitude inputs.

A more general approach to the study of coincidence-detection and related mechanisms for converging both excitatory and inhibitory inputs was presented by Krips and Furst (2009a,b). They recognized the power of Siebert's approach to predicting discrimination ability based on responses of neural populations, and the potential benefit of extending this approach to psychophysical problems that require central processing, such as binaural detection. Their studies describe a family of model

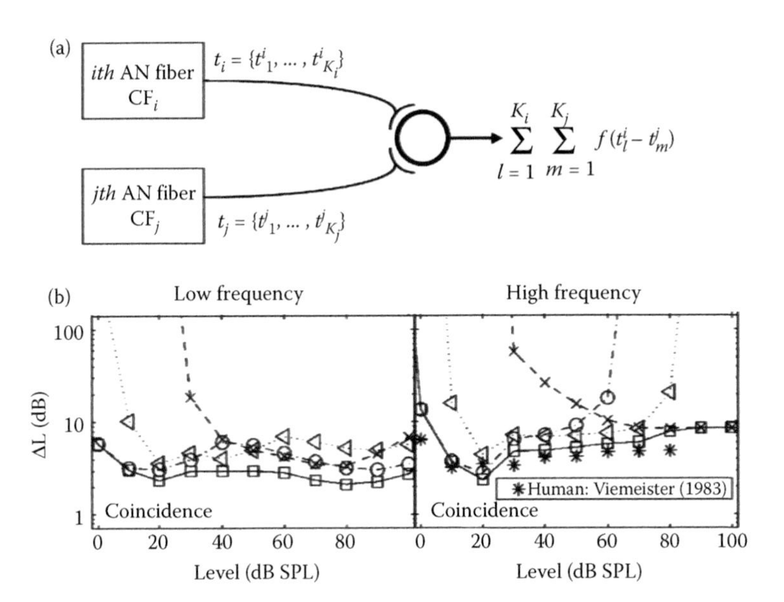

FIGURE 10.5 (a) Schematic illustration of two-input coincidence detector. (Reprinted with permission from Heinz, M.G., Colburn, H.S., and Carney, L.H. 2001c. *J. Acoust. Soc. Am.* 110:2065–84, Figure 2. Copyright 2001, Acoustical Society of America.) (b) Level-discrimination threshold estimates based on the coincidence-detection model in (a) with model auditory-nerve inputs. Thresholds for the optimal detector were based on individual and combined spontaneous-rate groups of model auditory-nerve fibers (HSR, MSR, and LSR are high, medium, and low spontaneous rates, respectively). Small populations of fibers tuned to frequencies near the stimulus frequency were included in the predictions. Predictions for a low-frequency (996 Hz) stimulus are in the left panel, and for a high-frequency (9874 Hz) stimulus on the right. TOT (total) indicates predictions based on combined populations of fibers. Human data: Viemeister (1983) for high-frequency noise band (6–14 kHz) in a notched noise; duration = 200 ms. (Reprinted with permission from Heinz, M.G., Colburn, H.S., and Carney, L.H. 2001c. *J. Acoust. Soc. Am.* 110:2065–84, Figure 7. Copyright 2001, Acoustical Society of America.)

excitatory–excitatory (coincidence-detection) cells, excitatory–inhibitory neurons, and cells that receive a combination of m excitatory inputs and n inhibitory inputs. Their model cells respond in a "rule-based" manner, for example, based on intervals between inputs of each type, as opposed to being governed by signal-processing mechanisms (cf. the integrate-and-fire model, Tuckwell 1988, Zhang and Carney 2005). Krips and Furst (2009a,b) proved that if their model cells receive inputs that are Poisson in nature, then the responses of these model neurons are also Poisson, and thus the Cramer–Rao bound can be used to estimate JND based on the responses of these model cells. Although their studies focused on binaural discriminations, their coincidence detectors and excitatory–inhibitory cells are also applicable to monaural coincidence-detection and spatiotemporal information-processing problems. A nice feature of these models is their utility in studying combinations of excitatory and inhibitory inputs, a universal feature of auditory brainstem neurons that is often excluded from coincidence-detection models.

10.6 DETECTION OF SIGNALS IN THE PRESENCE OF NOISE

The studies described above first focused on the problem of discriminations between two, typically periodic, stimuli that differed in frequency or amplitude. The problem of detection of signals in noise is probably more relevant to an understanding of hearing abilities in the real world, as well as problems for listeners with hearing loss. Siebert (1968) considered detection in noise for a rate-place coding scheme and recognized that the problem of rate saturation was a significant limiting factor. Predicting masked detection thresholds based on temporal information requires a strategy for handling the uncertainty in the noise waveform. When using timing information, the Cramer–Rao bound predicts an unreasonably low threshold for discrimination between a particular masker waveform and that masker waveform with an added tone because it takes advantage of any difference in the time waveforms between the two signals being compared; that is, the masker waveform is perfectly "known" to the processor.

To simulate the more realistic situation in which a signal is being detected in the presence of a random (unknown) masker, Heinz (2000) developed a strategy for basing predictions on responses of peripheral models that were averaged across a large set of masker waveforms. Tan and Carney (2006) applied this approach to a model predicting behavioral results for formant-frequency discrimination in the presence of noise. Their results suggested that combined rate and temporal information across a small population of model auditory-nerve fibers tuned near the format frequency could explain behavioral performance. Interestingly, a simple across-frequency coincidence-detection mechanism was not effective in extracting the information for this task, suggesting that the critical temporal information for this task lies not in the spatiotemporal pattern of the response but in some other aspect, such as temporal coding of the amplitude envelope fluctuations.

10.7 SPATIOTEMPORAL PATTERNS FOR COMPLEX SOUNDS

Most of the studies described above based predictions of discrimination or detection thresholds on changes in the rates and/or timing information of *single* fibers, or frequency channels, with population-based predictions that simply integrate the information across channels (e.g., Siebert 1965, 1968, 1970, Heinz et al. 2001a,b, Colburn et al. 2003). As mentioned above, cross-channel coincidence detectors introduce sensitivity to changes in the spatiotemporal patterns of the peripheral responses (e.g., Carney 1994, Heinz et al. 2001c). Consideration of spatiotemporal patterns has also been applied to more complex sounds, such as speech sounds (Shamma 1985a,b), resonant sounds (Carney and Yin 1988), Huffman sequences (Carney 1990, Wang and Delgutte 2012), and harmonic tone complexes use to study pitch perception (Cedolin and Delgutte 2010). Spatiotemporal patterns naturally combine rate-place and timing information by examining information in the temporal patterns distributed across fibers tuned to different audio frequencies.

Spatiotemporal patterns potentially convey a rich form of information to the central nervous system; however, there are interesting questions regarding the mechanisms for decoding information in this form. Coincidence detection is an obvious strategy for decoding changes in the relative timing of spikes across neighboring fibers (e.g., Carney and Yin 1988, Carney 1990, 1994, Heinz et al. 2001c). Carney (1994) pointed out that the degree of coincidence across neighboring fibers in response to a pure tone tends to increase as the sound level increases.

When the spatiotemporal patterns of auditory-nerve population responses to complex sounds are considered, responses near spectral features (such as harmonics in a complex tone) are associated with *transitions* in the spatiotemporal pattern (Figure 10.6a), which would result in *reduced* responses of coincidence detectors that received those patterns as inputs (Cedolin and Delgutte 2010). Shamma (1985a,b) described these transitions in population responses to speech sounds and suggested that lateral-inhibitory networks provide an appropriate mechanism for detecting the transitions in the spatiotemporal pattern associated with spectral peaks. A difficulty with this hypothesis is that there is no physiological or anatomical evidence for *lateral* inhibition along the ascending auditory pathway; instead, *on-frequency* inhibition dominates responses in auditory neurons of the brainstem and midbrain (Caspary et al. 1994, Palombi and Caspary 1996, Gai and Carney 2008a).

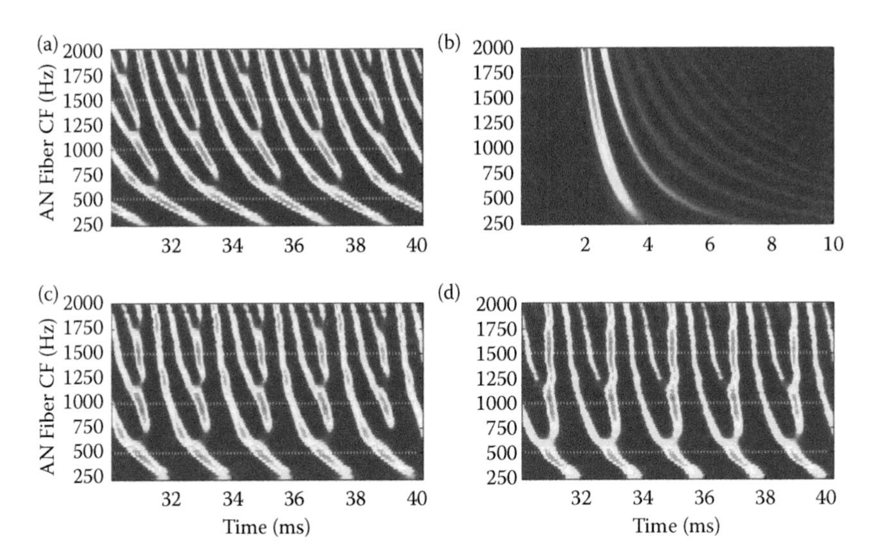

FIGURE 10.6 **(See color insert.)** Spatiotemporal patterns of model auditory-nerve fiber responses to harmonic tone complexes (F0 = 500 Hz, 55 dB SPL). Responses were computed using the Zilany et al. (2009) auditory-nerve model. (a) Upper left: Responses of model auditory-nerve fibers to a steady-state portion of the harmonic tone complex before latency compensation. Each row shows the response rate as a function of time for a fiber with a given characteristic frequency (CF). Responses are normalized to the overall maximum in the image. Steep transitions in the spatiotemporal pattern occur near the fundamental frequency (500 Hz) and the harmonic frequencies (1000, 1500, 2000 Hz). (b) Population response to a 10 μs click. The peak response time of each frequency channel determined the latency used in the compensation. (c) The 55 dB SPL population response from (a) after compensation with the click latencies shown in (b). After compensation, patterns are more "vertical" (e.g., more coincident across frequency channels) near the harmonic frequencies. (d) A population response to a 65 dB SPL harmonic tone complex, including latency compensation. Spatiotemporal patterns near the harmonics become increasingly coincident as level increases.

This dilemma may be solved by considering the possible role of axonal or dendritic delays that may affect the timing of auditory-nerve responses en route to higher-order neurons. The population response of the auditory-nerve to onsets or tones is not well aligned but is dispersed across time in a systematic way, reflecting the traveling-wave delay in the inner ear (Figure 10.6b). Yet, the strongest characteristic of nearly all auditory brainstem neurons is their well-timed, high-probability response to the onset of a tone or an acoustic click (transient). The percept of a click stimulus is also very compact in time. Both of these facts suggest the presence of a compensation mechanism for the disperse response in the cochlea. Compensation of the click response for the traveling-wave delay would result in well-timed patterns of activity that would provide strong onset and click responses in coincidence detectors. This temporal compensation would also result in increased coincidences (rather than transitions) associated with spectral peaks in complex stimuli (e.g., harmonics in complex tones or speech sounds) (Figure 10.6c); the vertical alignment of the peaks near harmonic frequencies increases as a function of the sound level (Figure 10.6d), consistent with the strengthening of the pitch percept as the level increases. These hypotheses have been developed mainly based on the knowledge of auditory-nerve responses for simple and complex sounds and on modeling studies of higher-level processing. Detailed models of the auditory periphery (e.g., Zilany et al. 2009) provide an excellent tool for exploring spatiotemporal patterns, which are inherently difficult to explore physiologically. However, the determination of whether lateral inhibition, coincidence detection, or neither, is involved in decoding spatiotemporal patterns will require physiological recordings in the central nervous system.

10.8 FUTURE DIRECTIONS

One strategy for exploring the role of spike times in the coding of complex sounds is to use psychophysical and behavioral tools to manipulate the temporal information in the sound itself. In this way, one can determine the importance for perception of temporal cues, which are often coded by spike times. Studies of the different temporal cues involved in detection have revealed that both fine structure and envelope cues, as well as nonlinear interactions between them, are involved in the detection of tones in noise by human listeners (Davidson et al. 2009). Interestingly, detection studies using the same acoustic stimuli in rabbits have shown that rabbits depend much more strongly on energy-based cues, rather than on temporal cues (Gai et al. 2007), even though substantial timing information in response to these stimuli is present in the responses of brainstem neurons (Gai and Carney 2006, 2008b). These results raise interesting issues regarding our studies of spike timing in animal models that may or may not use temporal information in the same way that human listeners do. Physiologists tend to assume that organisms make optimal, or nearly optimal, use of all available information in the spike responses, but this assumption may be inappropriate. Coordinated behavioral and physiological studies will be required to pursue general questions about the functional role of spike times in hearing, as well as to pin down the neural mechanisms for processing auditory information.

Another important future direction is to explore the role of timing information in explaining the difficulties of listeners with hearing loss. Sensorineural hearing loss is associated with a reduction in amplification in the inner ear, which has ramifications not only for sensitivity but also for the bandwidths and phase properties of the peripheral filters. Thus, the spike times in the responses of ears with hearing loss will be quantitatively and qualitatively different from those of the normal ear. Heinz et al. (2001c) explored changes in both rate and timing of peripheral responses as a result of reduction of cochlear amplification. These issues can be further explored using peripheral models that allow manipulations of both outer and inner hair-cell function to simulate different forms of hearing impairment (e.g., Zilany et al. 2009).

ACKNOWLEDGMENTS

Helpful comments were provided by Kris Abrams, David Cameron, Brian Flynn, Junwen Mao, and Muhammad Zilany. The preparation of this chapter was supported by NIH-NIDCD R01-001641.

REFERENCES

Anderson, D.J., Rose, J.E., Hind, J.E., and Brugge, J.F. 1971. Temporal position of discharges in single auditory nerve fibers within the cycle of a sinewave stimulus: Frequency and intensity effects. *J. Acoust. Soc. Am.* 49:1131–9.

Cariani, P.A. and Delgutte, B. 1996a. Neural correlates of the pitch of complex tones. I. Pitch and pitch salience. *J. Neurophysiol.* 76:1698–716.

Cariani, P.A. and Delgutte, B. 1996b. Neural correlates of the pitch of complex tones. II. Pitch shift, pitch ambiguity, phase invariance, pitch circularity, rate pitch, and the dominance region for pitch. *J. Neurophysiol.* 76:1717–34.

Carney, L.H. 1990. Sensitivities of cells in the anteroventral cochlear nucleus of cat to spatiotemporal discharge patterns across primary afferents. *J. Neurophysiol.* 64:437–56.

Carney, L.H. 1993. A model for the responses of low-frequency auditory nerve fibers in cat. *J. Acoust. Soc. Am.* 93:401–17.

Carney, L.H. 1994. Spatiotemporal encoding of sound level: Models for normal encoding and recruitment of loudness. *Hearing Res.* 76:31–44.

Carney, L.H. and Yin, T.C.T. 1988. Temporal coding of resonances by low-frequency auditory nerve fibers: Single fiber responses and a population model. *J. Neurophysiol.* 60:1653–77.

Caspary, D.M., Backoff, P.M., Finlayson, P.G., and Palombi, P.S. 1994. Inhibitory inputs modulate discharge rate within frequency receptive fields of anteroventral cochlear nucleus neurons. *J. Neurophysiol.* 72:2124–33.

Cedolin, L. and Delgutte, B. 2010. Spatiotemporal representation of the pitch of harmonic complex tones in the auditory nerve. *J. Neurosci.* 30:12712–24.

Colburn, H.S. 1969. *Some Physiological Limitations on Binaural Performance*. PhD dissertation, Massachusetts Institute of Technology, Cambridge, MA.

Colburn, H.S. 1973. Theory of binaural interaction based on auditory-nerve data. I. General strategy and preliminary results on interaural discrimination. *J. Acoust. Soc. Am.* 54, 1458–70.

Colburn, H.S. 1977. Theory of binaural interaction based on auditory-nerve data. II. Detection of tones in noise. *J. Acoust. Soc. Am.* 61, 525–33.

Colburn, H.S., Carney, L.H., and Heinz, M.G. 2003. Quantifying the information in auditory-nerve responses for level discrimination. *JARO*. 4:294–311.

Cooper, N.P. 1996. Two-tone suppression in cochlear mechanics. *J. Acoust. Soc. Am.* 99:3087–98.

Davidson, S.A., Gilkey, R.H., Colburn, H.S., and Carney, L.H. 2009. Diotic and dichotic detection with reproducible chimeric stimuli. *J. Acoust. Soc. Am.* 126:1889–905.

Delgutte, B. 1987. Peripheral auditory processing of speech information: Implications from a physiological study of intensity discrimination. In *The Psychophysics of Speech Perception*, ed. M. E. H. Schouten, pp. 333–353. Dordrecht: Nijhoff.

Delgutte, B. 1996. Physiological models for basic auditory percepts. In *Auditory Computation*, eds. H. L. Hawkins, T. A. McMullen, A. N. Popper, and R. R. Fay, pp. 157–220. New York: Springer-Verlag.

Florentine, M. 1986. Level discrimination of tones as a function of duration. *J. Acoust. Soc. Am.* 79:792–98.

Florentine, M., Buus, S., and Mason, C.R. 1987. Level discrimination as a function of level for tones from 0.25 to 16 kHz. *J. Acoust. Soc. Am.* 81:1528–41.

Gai, Y. and Carney, L.H. 2006. Temporal measures and neural strategies for detection of tones in noise based on responses in anteroventral cochlear nucleus. *J. Neurophysiol.* 96:2451–64.

Gai, Y. and Carney, L.H. 2008a. Influence of inhibitory inputs on rate and timing of responses in the anteroventral cochlear nucleus. *J. Neurophysiol.* 99:1077–95.

Gai, Y. and Carney, L.H. 2008b. Statistical analyses of temporal information in auditory brainstem responses to tones in noise: Correlation index and spike-distance metric. *JARO.* 9:373–87.

Gai, Y., Carney, L.H., Abrams, K.S., Idrobo, F., Harrison, J.M., and Gilkey, R.H. 2007. Detection of tones in reproducible noise maskers by rabbits and comparison to detection by humans. *JARO.* 8:522–38.

Geisler, C.D. and Rhode, W.S. 1982. The phases of basilar-membrane vibrations. *J. Acoust. Soc. Am.* 71:1201–3.

Goldstein, J.L. and Srulovicz, P. 1977. Auditory-nerve spike intervals as an adequate basis for aural frequency measurement. In *Psychophysics and Physiology of Hearing*, ed. E.F. Evans and J.P. Wilson, 337–47. New York: Academic Press.

Heinz, M.G. 2000. Quantifying the effects of the cochlear amplifier on temporal and average-rate information in the auditory nerve. PhD dissertation, Massachusetts Institute of Technology, Cambridge, MA.

Heinz, M.G., Colburn, H.S., and Carney, L.H. 2001a. Evaluating auditory performance limits: I. One-parameter discrimination using a computational model for the auditory nerve, *Neural Computat.* 13:2273–316.

Heinz, M.G., Colburn, H.S., and Carney, L.H. 2001b. Evaluating auditory performance limits: II. One-parameter discrimination with random level variation. *Neural Computat.* 13:2317–39.

Heinz, M.G., Colburn, H.S., and Carney, L.H. 2001c. Rate and timing cues associated with the cochlear amplifier: Level discrimination based on monaural cross-frequency coincidence detection. *J. Acoust. Soc. Am.* 110:2065–84.

Jeffress, L.A. 1948. A place theory of sound localization. *J. Comparative Physiol. Psychol.* 41:35–9.

Jiang, D., Palmer, A.R., and Winter, I.M. 1996. Frequency extent of two-tone facilitation in onset units in the ventral cochlear nucleus. *J. Neurophysiol.* 75:380–95.

Johnson, D.J. 1980. The relationship between spike rate and synchrony in responses of auditory-nerve fibers to single tones. *J. Acoust. Soc. Am.* 68:1115–22.

Joris, P.X., Carney, L.H., Smith, P.H., and Yin, T.C.T. 1994. Enhancement of neural synchronization in the anteroventral cochlear nucleus I: Responses to tones at the characteristic frequency. *J. Neurophysiol.* 71:1022–36.

Joris, P.X., Smith, P.H., and Yin, T.C.T. 1998. Coincidence detection in the auditory system: 50 years after Jeffress. *Neuron.* 21:1235–8.

Kiang, N.Y.S., Watanabe, T., Thomas, E.C., and Clark, L.F. 1965. *Discharge Patterns of Single Fibers in the Cat's Auditory Nerve.* Cambridge, MA: MIT Press.

Krips, R. and Furst, M. 2009a. Stochastic properties of auditory brainstem coincidence detectors in binaural perception. *J. Acoust. Soc. Am.* 125:1567–83.

Krips, R. and Furst, M. 2009b. Stochastic properties of coincidence-detector neural cells. *Neural Comput.* 21:2524–53.

Lyzenga, J. and Horst, J.W. 1995. Frequency discrimination of bandlimited harmonic complexes related to vowel formants. *J. Acoust. Soc. Am.* 98:1943–55.

Manis, P.B. and Marx, S.O. 1991. Outward currents in isolated ventral cochlear nucleus neurons. *J. Neurosci.* 11:2865–80.

Moore, B.C.J. 1973. Frequency difference limens for short-duration tones. *J. Acoust. Soc. Am.* 54:610–9.

Palombi, P.S. and Caspary, D.M. 1996. GABA inputs control discharge rate primarily within frequency receptive fields of inferior colliculus neurons. *J. Neurophysiol.* 75:2211–9.

Rhode, W.S. 1971. Observations of the vibration of the basilar membrane in squirrel monkeys using the Mössbauer technique. *J. Acoust. Soc. Am.* 64:158–76.

Robles, L. and Ruggero, M.A. 2001. Mechanics of the mammalian cochlea. *Physiol. Rev.* 81:1305–52.

Rothman, J.S. and Manis, P.B. 2003a. Differential expression of three distinct potassium currents in the ventral cochlear nucleus. *J. Neurophysiol.* 89:3070–82.

Rothman, J.S. and Manis, P.B. 2003b. Kinetic analyses of three distinct potassium conductances in ventral cochlear nucleus neurons. *J. Neurophysiol.* 89:3083–96.

Rothman, J.S. and Manis, P.B. 2003c. The roles potassium currents play in regulating the electrical activity of ventral cochlear nucleus neurons. *J. Neurophysiol.* 89:3097–113.

Ruggero, M.A., Rich, N.C., Recio, A., Narayan, S., and Robles, L. 1997. Basilar-membrane responses to tones at the base of the chinchilla cochlea. *J. Acoust. Soc. Am.* 101:2151–63.

Shamma, S.A. 1985a. Speech processing in the auditory system. I: The representation of speech sounds in the responses of the auditory nerve. *J. Acoust. Soc. Am.* 78:1612–21.

Shamma, S.A. 1985b. Speech processing in the auditory system. II: Lateral inhibition and the central processing of speech evoked activity in the auditory nerve. *J. Acoust. Soc. Am.* 78:1622–32.

Siebert, W.M. 1965. Some implication of the stochastic behavior of primary auditory neurons. *Kybernetik.* 2:206–15.

Siebert, W.M. 1968. Stimulus transformation in the peripheral auditory system. In *Recognizing Patterns*, eds. P.A. Kolers and M. Eden, pp. 104–133. Cambridge, MA: MIT Press.

Siebert, W.M. 1970. Frequency discrimination in the auditory system: Place or periodicity mechanisms? *Proc. IEEE.* 58:723–30.

Smith, P.H. 1995. Structural and functional differences distinguish principal from nonprincipal cells in the guinea pig MSO slice. *J. Neurophysiol.* 73:1653–67.

Tan, Q. and Carney, L. H. 2005. Encoding of vowel-like sounds in the auditory-nerve: Model predictions of discrimination performance. *J. Acoust. Soc. Am.* 117:1210–22.

Tan, Q. and Carney, L.H. 2006. Predictions of formant-frequency discrimination in noise based on model auditory-nerve responses. *J. Acoust. Soc. Am.* 120:1435–45.

Tuckwell, H.C. 1988. *Introduction to Theoretical Neurobiology*. Cambridge: Cambridge University Press.

van Trees, H.L. 1968. *Detection, Estimation, and Modulation Theory: Part I.* New York: Wiley, Chapter 2.

Viemeister, N.F. 1983. Auditory intensity discrimination at high frequencies in the presence of noise. *Science.* 221:1206–8.

Viemeister, N.F. 1988a. Intensity coding and the dynamic range problem. *Hear Res.* 34:267–74.

Viemeister, N.F. 1988b. Psychophysical aspects of intensity discrimination. In *Auditory Function: Neurobiological Bases of Hearing*, eds. G. M. Edelman, W. E. Gall, and W. M. Cowan, pp. 213–41. New York: Wiley.

Wang, G.I. and Delgutte, B. 2012. Sensitivity of cochlear nucleus neurons to spatio-temporal changes in auditory nerve activity. *J. Neurophysiol.* 108:3172–95.

Wier, C.C., Jesteadt, W., and Green, D.M. 1977. Frequency discrimination as a function of frequency and sensation level. *J. Acoust. Soc. Am.* 61:178–84.

Winter, I. M. and Palmer, A. R. 1991. Intensity-coding in low-frequency auditory-nerve fibers of the guinea pig. *J. Acoust. Soc. Am.* 90:1958–67.

Yin, T.C.T. and Chan, J.C.K. 1990. Interaural time sensitivity in medial superior olive of cat. *J. Neurophysiol.* 64:465–88.

Zhang, X. and Carney, L. H. 2005. Response properties of an integrate-and-fire model that receives subthreshold inputs. *Neural Computat.* 17:2571–601.

Zilany, M.S.A., Bruce, I.C., Nelson, P.C., and Carney, L.H. 2009. A phenomenological model of the synapse between the inner hair cell and auditory nerve: Long-term adaptation with power-law dynamics. *J. Acoust. Soc. Am.* 126:2390–412.

11 Spike Timing and Neural Codes for Odors

Sam Reiter and Mark Stopfer

CONTENTS

11.1 ONE BEST WAY TO ACHIEVE OLFACTION

The sense of smell informs animals about the world around them. The job of the olfactory system is challenging: it detects vast numbers of volatile chemicals in wide ranges of concentrations, and it translates these encounters into the spiking language of neural activity. The olfactory system provides researchers an important example of information processing achieved by well-defined populations of neurons. In this chapter, we review recent work investigating how olfactory systems accomplish this goal. We focus particularly on ways neural representations of an odorant's identity and concentration are transformed as this information moves through successive populations of neurons. These transformations can be understood as changes in the number and timing of spikes.

Relatively, simple animals such as insects have proved to be useful subjects for studies of neural coding in the olfactory system. Possessing relatively few but well-characterized neurons, these animals provide robust experimental preparations and are amenable to many technical approaches.

To date, most of these studies have focused on the first few stages of olfactory processing, as outlined in Figure 11.1. Insects can detect odors through several parts of their bodies, including the antennae, where olfactory receptor neurons (ORNs) come into contact with volatile chemicals in the environment. From the antenna, ORNs

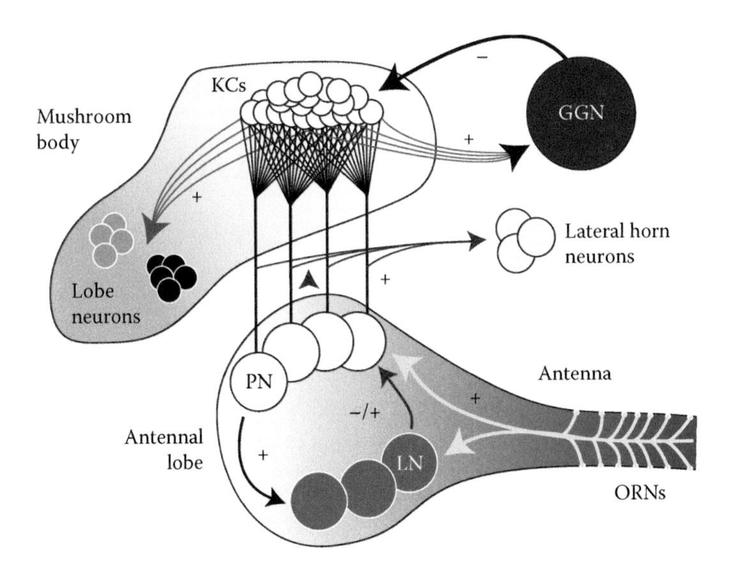

FIGURE 11.1 The locust olfactory system (see text for details). ORN, olfactory receptor neuron; LN, local neuron; PN, projection neuron; KCs, Kenyon cells; GGN, giant GABAergic neuron. " + " and " − " refer to excitatory (cholinergic) and inhibitory (GABAergic) synapses, respectively.

project to a structure called the antenna lobe (AL). In the AL, ORN outputs sort by receptor type and converge upon spherical neuropils known as glomeruli, where they make excitatory connections with projection neurons (PNs) and local neurons (LNs). Inhibitory and excitatory LNs also synapse onto PNs, and interact with each other through gap junctions. PNs project from the AL to the mushroom body (MB) and to the lateral horn (LH). The LH contains many types of odor-responsive neurons that project to several areas of the insect brain. Neurons of the MB, called Kenyon cells (KCs), have been studied in detail. In several species, including the locust, fly, and cockroach, KCs synapse en masse onto a single giant GABAergic neuron called, in the locust, GGN. This neuron then provides feedback inhibition to the entire population of KCs. KCs also make connections with a variety of cells within the MB, collectively known as lobe neurons. Some lobe neurons inhibit each other and have processes confined to the lobes of the MB. Others project to many areas of the brain, but their synaptic targets are unknown (Cassenaer and Laurent 2012; Couto et al. 2005; Fishilevich and Vosshall 2005; Gupta and Stopfer 2012; Homberg et al. 1989; Huang et al. 2010; Papadopoulou et al. 2011; Stocker 1994; Waldrop et al. 1987; Yaksi and Wilson 2010).

The vertebrate olfactory system is strikingly analogous to the invertebrate version (Hildebrand and Shepherd 1997). ORNs within the nose converge type-wise upon glomeruli within the olfactory bulb (OB), where excitatory mitral cells (MCs) interact with ORNs and several types of inhibitory interneurons that provide both inter- and intraglomerular connections (Kay and Stopfer 2006; Mombaerts et al. 1996; Ressler 1993; Schoppa 2003; Su et al. 2009; Vassar et al. 1993; Wilson and Mainen

2006). Mitral and tufted cells project broadly and divergently to other brain areas, including the piriform cortex.

The olfactory systems of both vertebrates and invertebrates utilize the precise timing of spikes to encode olfactory information (Kay and Stopfer 2006). That such diverse species have separately evolved similar anatomical and physiological solutions suggests there may be one best way to achieve olfaction. Here, we review how volatile chemicals in the environment are first encoded as patterns of spikes in ORNs, how these patterns are transformed by successive populations of neurons, and how these transformations benefit olfactory coding.

11.2 OLFACTORY RECEPTOR NEURONS

Olfactory information processing begins when volatile chemicals in the environment make their way through a protein-filled mucosa and then bind to olfactory receptor proteins (ORs) expressed on the dendrites of ORNs. In insects, the dendrites of ORNs are found within fluid-filled structures known as sensilla, which protrude from the antennae and maxillary palps. The number of ORNs varies widely from species to species. *Drosophila*, for example, possess ~1200 ORNs on the antenna and ~200 on the maxillary palp (Stocker 1994), and the honeybee *Apis mellifera* has ~120,000 in its antenna (Esslen and Kaissling 1976). By comparison, the mouse has ~2,000,000 ORNs within the olfactory epithelium of its nose (Pomeroy et al. 1990).

ORNs from a wide range of species display similar behaviors. Even when not stimulated by environmental odorants, these neurons spike at rates of ~1–30 Hz (Joseph et al. 2012). When an odor is presented to the nose or antenna, a subset of the population of ORNs responds with patterns of excitation and inhibition that vary with the odorant and its concentration (Duchamp-Viret 1999; Hallem and Carlson 2006; Ito et al. 2009; Michel and Ache 1994; Raman et al. 2010; Spors et al. 2006). Figure 11.2a shows examples of olfactory responses recorded from locust ORNs. A single ORN can be responsive to multiple odorants and be insensitive to others. And, a given ORN can generate different responses for different odorants: some odorants elicit a transient burst of spiking, some elicit tonic excitation outlasting the stimulus, some inhibit spiking to a level below that of spontaneous activity, and some drive the ORN in different combinations of the above. The latency, duration, and intensity of the response can vary with the odorant. Further, different concentrations of the same odorant can elicit different patterns of spiking from an ORN.

Visual and auditory receptor neurons have two-dimensional tuning profiles, corresponding to the frequencies and intensities of light or sound. By contrast, the tuning profiles of ORNs are more complicated; ORNs respond to chemicals that vary along multiple dimensions including size, shape, charge, polarity, mechanical flexibility, functional group, and concentration (Kaupp 2010). Because the dimensionality of the olfactory world is so high, it is challenging for researchers to fully explore the range and diversity of an ORN's sensitivity. This diversity also contributes to the difficulty of using the known response profile of an ORN to accurately predict its response to a new odor.

With this difficulty in mind, comparisons of the tuning widths of ORNs can be approximated using large panels of odorants delivered at several concentrations.

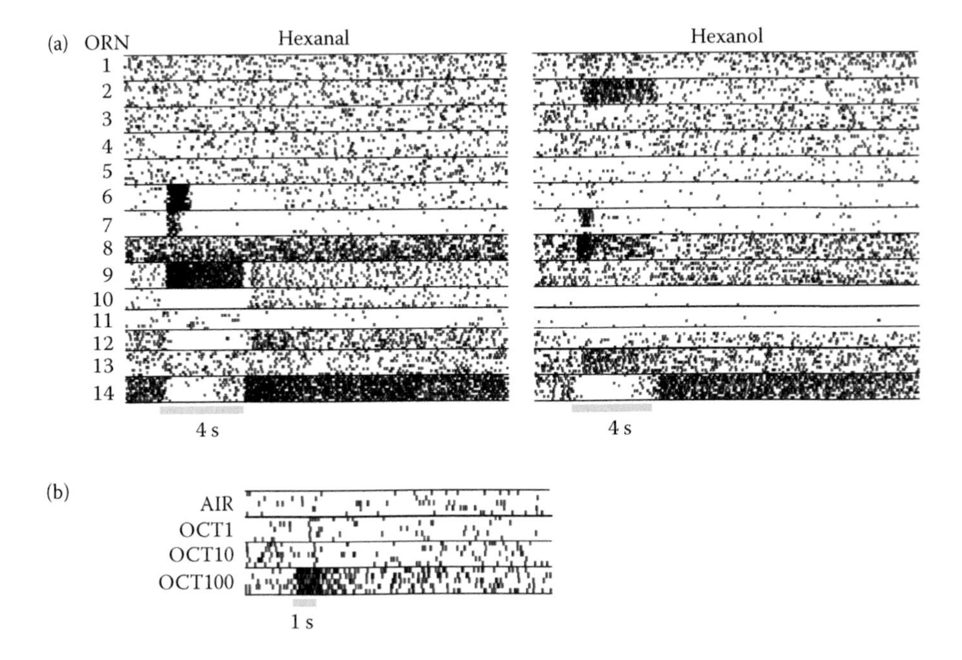

FIGURE 11.2 Responses of locust olfactory receptor neurons (ORNs) to odors. (a) Raster plots show spiking responses of 14 ORNs, recorded sequentially, to repeated 4 s presentations of two odorants, hexanal and hexanol; each raster row shows results from one trial. Responses include excitation and inhibition, and tonic and transient spiking. (b) ORNs respond with reduced specificity as the odor concentration increases. This example ORN does not respond to clean air, or to low and medium concentrations of octanol (OCT1, OCT10), but does respond to a high concentration of octanol (OCT100). Odor solutions were prepared in bottles as order-of-magnitude dilutions in mineral oil. The headspace above the odor solution was pulsed into a constantly flowing stream of air directed at the locust's antenna. (Adapted from Raman, B. et al. 2010. *The Journal of Neuroscience* 30, 1994–2006.)

In most animals, the tuning widths of ORNs fall along a continuum. Some ORNs are narrowly tuned and have been shown to respond very selectively to a single, or just a few odorants, but some ORNs are broadly tuned and respond to many and diverse odorants. Similarly, ORNs exhibit a spectrum of sensitivity to different concentrations of an odorant; some ORNs respond to a wide range of concentrations of a given odorant, while some ORNs respond only to high concentrations of that odorant (Figure 11.2b) (Duchamp-Viret 1999; Firestein et al. 1993; Hallem et al. 2006; Ito et al. 2009).

Although ORNs tend to fire more spikes when the concentrations of odors increase, their maximum firing rates are sharply constrained. Exposure to an odorant for more than a fraction of a second causes responsive ORNs to undergo sensory adaptation, a process that decreases the rate of spiking over the duration of the response. And independent of adaptation, ORNs that are precisely tuned to an odorant, and thus may be highly responsive even to low concentrations, saturate when higher concentrations of that odorant are presented. As a result, as concentrations of the odorant continue to increase, precisely tuned ORNs may continue to generate spikes at rates similar to those evoked by lower concentrations (de Bruyne et al. 2001; Firestein et al. 1993; Ito et al. 2009).

Given these constraints on firing rate, how can ORNs encode the concentrations of odors? Like all biological receptors, ORNs become less selective as the concentration of ligand increases. Thus, more types of ORNs will respond to a higher concentration of an odor (reviewed in Buck, 1996). In fact, it has been shown in the moth that most of the olfactory system's dynamic range is encoded by the size of the responsive population of ORNs rather than by the firing rates of those ORNs (Ito et al. 2009).

11.3 RESPONSE PROPERTIES OF OLFACTORY RECEPTOR NEURONS

The tuning properties of ORNs are best understood in terms of the underlying molecular biology. A series of elegant experiments has shown that in vertebrates as well as in invertebrates, the tuning characteristics of an ORN are specified by the OR expressed by the neuron (Bozza et al. 2002; Dobritsa et al. 2003; Malnic et al. 1999). In vertebrates, ORNs usually express only one OR (but see Mombaerts 2004). In insects, ORNs usually express two ORs, one of which is responsible for the tuning properties of the ORN, while the other, an olfactory receptor coreceptor (Orco), assists in membrane trafficking and signal transduction of odorant binding events (Clyne et al. 1999; Vosshall et al. 1999).

The sensitivity of an OR to odorants is almost certainly determined by the shape and charge distribution of its ligand-binding pocket. Crystal structures have yet to be solved for any OR, so the exact shapes of odor receptors remain unknown. However, an analysis by Araneda et al. (2000) provides an intriguing preview. In this study, the authors transfected an overabundance of a single type of receptor (I7) responsive to a known ligand (octanal) into olfactory epithelia. The authors then constructed a large number of chemical variants on octanal, each sporting a different functional group, or a different carbon backbone, and so on. The authors then measured the ORN population's responses to this panel of odorants, comparing the chemical properties of each odorant to the electrophysiological responses it elicited from the ORNs, and then used these results to constrain a model of the physical characteristics of the OR's odorant binding pocket. According to this model, odorants that fit most neatly into the pocket elicited the strongest spiking responses. Odorants that fit less well elicited less intense responses. And at least one odorant appeared to compete for binding space, leading to inhibitory responses (Araneda et al. 2000). Strikingly, an assortment of differently structured odorants elicited responses of one sort or another from the ORN.

The concentration range of an odorant that will elicit a response from an ORN can most likely be attributed directly to OR binding pocket structure. An odorant that fits snugly into an OR will drive its ORN to modulate its firing of spikes even when the odorant is present at low concentrations.

What mechanisms unleash the diversity of temporal patterning in the spiking responses of ORNs? Several factors are likely to contribute to the timing of spikes. When exploring the environment, animals often engage in active sampling behaviors, such as moving their bodies, sweeping their antennae through an odor field (insects), or sniffing (vertebrates). As these behaviors are performed at varying frequencies, they likely contribute to the variety of temporal structure of spiking in ORNs (Cury and Uchida 2010; Rinberg et al. 2006; Tripathy et al. 2010; Verhagen et al. 2007; Wachowiak and Cohen 2001).

And even before an odorant can reach an OR expressed in the membrane of an ORN, it must navigate a potentially chromatographic pass through the protein-filled layer of liquid (sensillar lymph in insects, mucosa in vertebrates) surrounding the ORs. In insects, sensillar lymph has been shown to contain odorant binding proteins (OBPs), which selectively speed odorant molecules, which can be hydrophobic, along to their OR targets (Jacquin-Joly and Merlin 2004; Rützler and Zwiebel 2005; Vogt and Riddiford 1981). Other proteins include an assortment known collectively as odorant-degrading enzymes, which prevent a constant state of saturation (Rützler et al. 2005). In some cases, the partially digested components of an odorant may themselves bind to ORs. All together, this extracellular apparatus appears to provide many opportunities for the olfactory system to regulate the timing with which different odorants reach the ORs. These variations in timing seem likely to underpin some of the temporal diversity in the odor-elicited spiking patterns of ORNs.

When an odorant finally binds to an OR, it sets an elaborate transduction process into action. In vertebrates, the ORs are types of G-protein coupled receptors that communicate signals through a cyclic adenosine monophosphate (cAMP)-mediated cascade. This cascade results in the efflux of chloride, thus depolarizing the ORN and triggering spikes (reviewed by Kleene, 2008). In insects, the precise intracellular transduction mechanisms of ORs are still under debate; it is unclear whether odorants cause ion channels to open directly or indirectly (Sato et al. 2008; Wicher et al. 2008). Although questions remain about the earliest stages of olfaction, it is apparent that numerous steps along the way could generate the odor-specific variations seen in the temporal patterning of spikes in ORNs.

11.4 COMBINATORIAL CODING OF OLFACTORY INFORMATION

Patterns of spikes in ORNs provide the first neural representations of odorants. What features of odor responses in ORNs are important for processing olfactory information? For many years, a "labeled line" model for olfactory coding was prominent. This model emerged from observations that a single ORN generally receives its odor response profile from a single OR and projects very specifically to only one or two stereotyped glomeruli, and from the assumption that ORNs are narrowly tuned to specific odors. If odors activate only one or a few types of ORNs, and those ORNs project to highly specific locations, then downstream neurons need only know which ORNs are active to know the identity of an odor in the environment (Ressler et al. 1993; Vassar et al. 1993; Vosshall et al. 2000). According to this model, the precise timing of spikes in the ORNs might contribute little, if any, additional information about odorants.

This model was attractive for a number of reasons. If ORs are narrowly tuned, each odor would activate a nonoverlapping subset of the total population of ORNs, leading to very reliable classification of odors. This model also provided an intuitive explanation for why some species have large numbers of OR genes: to reliably detect N odors in a labeled line system, N ORs would be needed. In practice, this model works well to explain the encoding of a few critical species-specific odorants, such as pheromones (Lei et al. 2002).

However, limits to the labeled line model have emerged from studies of general, nonpheromonal, olfactory processing. Many ORs have been shown to be broadly tuned (Hallem and Carlson 2006; Raman et al. 2010). Further, tuning inevitably becomes less specific as the concentration of an odorant increases; even an OR that is relatively well tuned for a given odorant may respond to other odorants presented in sufficiently high concentrations. Given these experimental observations, the labeled-line model predicts that high concentrations of odors should saturate the olfactory system, reducing an animal's ability to identify or classify an odor as a function of its concentration. Yet, psychophysical studies show just the opposite: high concentrations of odors are easiest to discriminate (Hummel et al. 1997; Wright et al. 2005). Further, although the pathway formed by an ORN to a glomerulus and then to a projection neuron or mitral cell may appear to be a direct and insulated channel, in fact, excitatory and inhibitory interneurons link multiple glomeruli, leading to a sharing and reorganization of information across "channels."

The alternate model of olfactory information processing, combinatorial coding, posits that most odors are represented by spiking in unique but overlapping populations of multiple types of ORNs. As ORNs possess a diversity of both tuning width and dynamic range, a combinatorial coding process that includes many ORNs can represent both the chemical identities and concentrations of odors (Bhandawat et al. 2007; Malnic et al. 1999; Uchida et al. 2000). Combinatorial codes can represent vast numbers of odors with a relatively small number of ORs: given N ORs, a combinatorial code can represent ~2^N odors. Although in some species, a small number of critical odorants such as pheromones may be encoded by labeled lines, most odorants are encoded combinatorially.

11.5 VISUALIZING THE RESPONSES OF OLFACTORY RECEPTOR NEURONS OVER TIME

The identities of ORNs responding to odorants provide a coding dimension often called the spatial aspect: one can readily envision a spatially distributed map of neurons either responding to or ignoring a given odor. But the responses of olfactory neurons have been shown to consist of time-varying patterns of spikes that provide an additional coding dimension, the temporal aspect. Thus, olfactory responses consist of spatiotemporal patterns of activity. How can one analyze these time-varying patterns of spiking distributed across a population of neurons?

One simple way (Stopfer et al., 2003) requires four steps, as illustrated in Figure 11.3. Step one: record odor-elicited spike trains from a diverse sample of N ORNs. Step two: select a timescale for dividing the spike trains into bins. This timescale can be arbitrary, but in some cases it is possible to identify a particularly meaningful one, such as the integration time window of downstream olfactory neurons—this would allow viewing the population response as a downstream neural population would view it. For each ORN, count the spikes in each time bin. Step three: assign each ORN to a number line, and assemble these lines orthogonally to create an N-dimensional space. Using the spike count values, plot one point for each time bin (in the example shown in Figure 11.3a, in the first bin the spike count for each neuron is zero, so the first point is 0,0,0). Step four: connect the resulting series of points to illustrate the trajectory taken by the ORN population through the response space. If you included more than two or three ORNs,

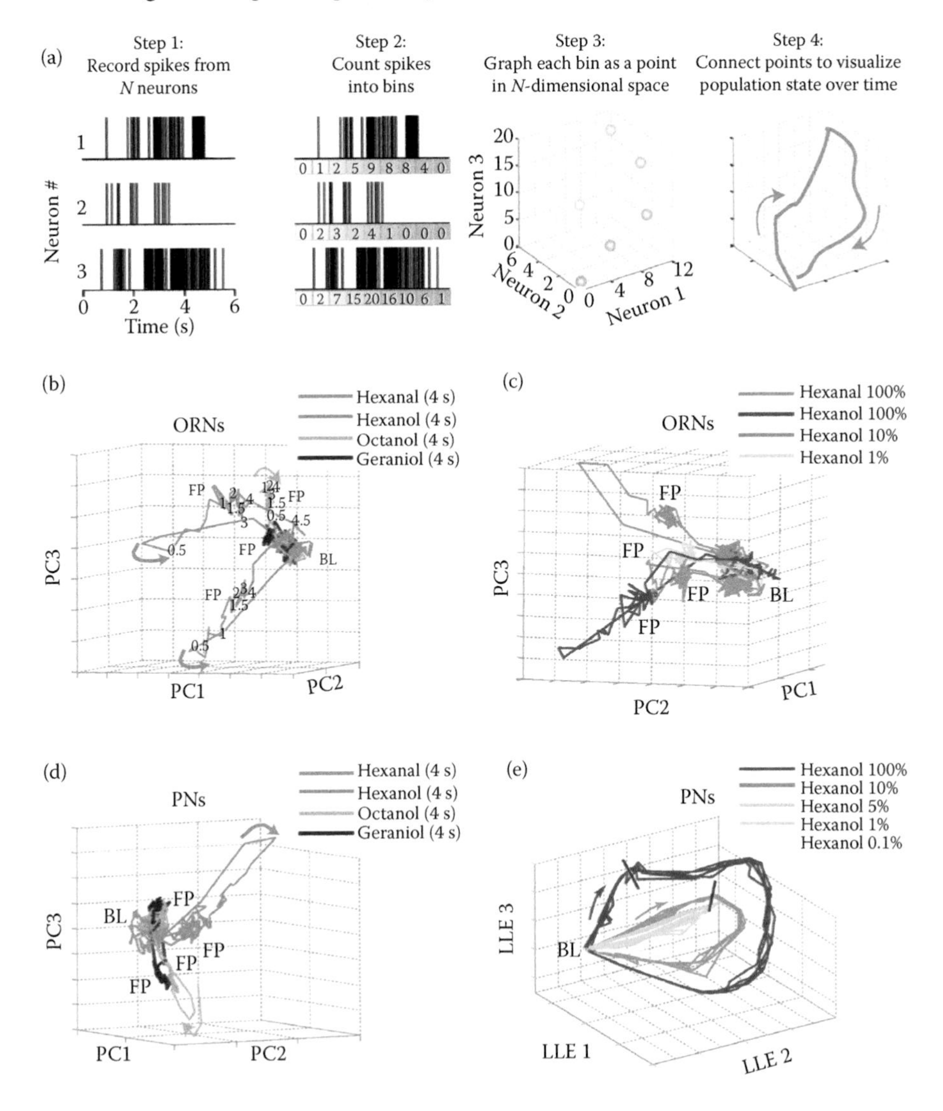

FIGURE 11.3

the high-dimensional result might be difficult to graph on paper, but a practical way to visualize this information is to use a dimensionality reduction technique (such as principle component analysis) to capture as much of the variance in the population response as possible in an easily graphed two or three dimensions (Figure 11.3b–e).

When considered this way, the odor-elicited spiking responses of groups of neurons form looping trajectories within the ORN population space (Figure 11.3b). The starting point reflects spontaneous activity in the ORNs (Joseph et al. 2012) before the odor is presented. Changing the concentration of an odor leads to small changes in a trajectory, but changing the chemical identity of the odor leads to much larger changes (Figure 11.3c). In the vocabulary of dynamical systems theory, odor-specific manifolds (surfaces) exist in ORN population space, and concentration-specific trajectories lie on these manifolds. The fact that ORN population responses to different odors form different manifolds and trajectories demonstrates that their distributed spiking patterns contain information about both odor identity and intensity, even when the patterns are considered over short timescales. Indeed, simple classification algorithms require only 50 ms to accurately identify the odor and its concentration eliciting these responses; providing the algorithms a succession of these time windows further improves classification success rates (Raman et al. 2010).

The odor-elicited responses of ORNs can occur in several motifs. Some responses consist of brief, transient patterns of spiking that, within 1500 ms or so, return to levels of spontaneous activity even if the odor is still present (see Figure 11.2a, Hexanal, ORN7). Other responses consist of tonic changes in the ORN spike rate throughout the odor's duration (see Figure 11.2a, Hexanol, ORN2, and ORN14). And when the odorant is withdrawn, both transient and tonic kinds of responses can erupt with fresh patterns of spikes (see Figure 11.2a, Hexanol, ORN14). Trajectory

---

FIGURE 11.3 **(See color insert.)** Trajectory plots illustrate spiking patterns of populations of ORNs and projection neurons (PNs) over time. (a) Method used to construct these plots (see text). (b) Spiking responses of 53 ORNs to four odorants each presented for 4 s. The first three principal components (PC1–3, see text) reveal that the population response departs from baseline activity (BL), undergoes an odor specific trajectory, settles into a fixed point (FP), and then returns to baseline via another pathway when the odor is removed. Numbers show time (seconds) after the odor's onset. (c) Trajectory responses of 43 ORNs responding to the odorant Hexanal, and to three different concentrations of a different odorant with a similar structure, Hexanol. Increasing concentrations of an odor elicit responses within the same area of response space (within the same manifold) but with trajectories of monotonically increasing length. (d) Trajectory responses of 94 PNs tested with the same odorants as in panel (b) reveal broadly similar time-varying response features. (e) Trajectory responses of 110 PNs responding to 1 s presentations of five different concentrations of Hexanol. Here, the trajectories were visualized using local linear embedding (LLE, see Roweis and Saul 2000), a dimensionality-reduction technique. Like the responses of ORNs shown in panel (c), PN trajectory magnitude increases monotonically within a manifold with odor concentration. (Panels (a) and (e): adapted from Stopfer, M., Jayaraman, V., and Laurent, G. 2003. *Neuron* 39, 991–1004.; Panels (b) through (d): adapted from Raman, B. et al. 2010. *The Journal of Neuroscience* 30, 1994–2006.)

plots illustrate these time-varying, dynamic features of spiking responses distributed across populations of neurons.

When presented with a brief pulse of odor, the ORN population response sets out on an odor- and concentration-specific trajectory, and then smoothly swerves back to the start point. But if the duration of the odor pulse exceeds 1500 ms or so, the trajectory appears to seize up; it settles into a spot (in the language of dynamical systems, a "fixed point attractor") in the population space away from the origin (Figure 11.3a and b). This point reflects the underlying tonic firing or inhibition sustained by some ORNs after the conclusion of the transient patterning elicited by the odor's onset. The trajectory returns to the origin by a different path when the odor is withdrawn (Figure 11.3) (Mazor and Laurent 2005; Raman et al. 2010).

Thus, it is possible to analyze and visualize information contained not only in the identity of ORNs responsive to an odor but also in the time-varying unfurling of that response. Neurons downstream using this information will have the benefit of increased coding capacity because a spatiotemporal code can represent odors not only combinatorially as partially overlapping subsets of active ORNs but also in completely overlapping subsets of ORNs that fire spikes in different timing patterns (Laurent 2002).

11.6 REFORMATTING INFORMATION IN THE ANTENNAL LOBE

Does the AL simply relay the spatiotemporal responses of ORNs to deeper brain areas, or does it perform additional information processing? Historically, physiological support for the view that the AL serves essentially as a relay came from studies employing techniques that used genetically encoded optical sensors of neural activity; these studies concluded that PNs largely share the response profiles of their presynaptic ORNs (Ng et al. 2002; Wang et al. 2003). But such studies must be interpreted cautiously because genetically encoded optical sensors, at present, provide relatively low temporal resolution, highly nonlinear relationships with electrical activity, and sensitivities that may vary across neurons expressing different genetic drivers of the sensor protein, making it difficult to compare the responses arising from different types of neurons (Jayaraman and Laurent 2007).

In fact, direct electrophysiological recordings of ORNs and their immediate postsynaptic PNs in *Drosophila* have shown that the AL does not simply pass along the responses of ORNs, but rather restructures the information carried by the ORNs: the PNs are less selectively tuned than their presynaptic ORNs (Wilson et al. 2004). This reformatting consists, in part, of a reorganization of the population of neurons activated by an odor. In *Drosophila*, this likely occurs because, within the AL, excitatory LNs distribute information about odors across glomeruli (Huang et al. 2010; Olsen et al. 2007; Shang et al. 2007; Yaksi and Wilson 2010). In locusts and other animals, widely branching PNs likely contribute to this broad distribution as well. The circuitry of the AL is often compared to that of the retina, in which interneurons serve to narrow the response profiles of output neurons as a form of contrast enhancement (Hartline 1956; Kuffler 1953; Werblin and Dowling 1968). But, notably, the circuitry of the AL actually performs the opposite function (Laurent 2002).

The circuitry of the AL also substantially reformats the timing of spikes. When odor pulses are presented repeatedly, for example, the trial-to-trial timing of spiking is more reliable in PNs than in the ORNs presynaptic to them (Bhandawat et al. 2007). This occurs because many ORNs, each expressing the same OR set, converge upon each PN; in *Drosophila*, for example, ~30 ORNs converge onto each PN (Stocker 1994). Because ORNs of the same type generate similar firing patterns (Dobritsa et al. 2003), the convergence of many copies of a spike train onto one PN allows for signal averaging, reducing noise (Pouget et al. 2000).

Another way the AL reformats spike timing is by generating an odor-elicited oscillatory rhythm that biases the PNs to spike in synchrony (Bazhenov et al. 2001a; Ito et al. 2009; Laurent and Davidowitz 1994; MacLeod and Laurent 1996). The synchronization of odor-activated PNs can be observed directly in simultaneous recordings of multiple PNs, and indirectly as the oscillating local field potential (LFP) generated by the spiking output of ensembles of PNs, recorded from target regions of the PNs, such as the MB. Throughout the animal kingdom, odor-elicited oscillations like these are ubiquitous, observed in species ranging from *Drosophila* (Tanaka et al. 2009) to humans (Sobel et al. 1998). In honeybees, these oscillations have been shown to be important for the neural representation of odors: pharmacologically disrupting the oscillations impairs odor discrimination (Hosler and Smith 2000; Stopfer et al. 1997).

The mechanism responsible for the oscillatory synchronization of spiking has been studied in some detail in locusts, moths, and *Drosophila* (Bazhenov et al. 2001a; MacLeod et al. 1996; Ito et al. 2008, 2009; Tanaka et al. 2009). Odor-elicited spikes from ORNs depolarize PNs and inhibitory LNs. These LNs, which synapse directly upon PNs, are driven to release the neurotransmitter γ-aminobutyric acid (GABA), which then activates $GABA_A$-type receptors on PNs and on other LNs, causing them to briefly hyperpolarize. Because some of the LNs branch very extensively throughout the AL, they broadcast this "pause" signal throughout the population of PNs. After each burst of fast hyperpolarization wears off (in the locust, these inhibitory postsynaptic potentials, IPSPs, last ~50 ms), groups of odor-driven PNs are more likely to spike together. And reciprocal connections between PNs and LNs reinforce the oscillatory cycle. As a result of this clocklike synchronizing circuitry, spikes in PNs tend to occur during the rising phase of the oscillatory LFP cycle recorded in the MB, and are organized into ~20 Hz waves (Bazhenov et al. 2001a).

But it is not the case that every spike in a PN is synchronized with the spikes of every other PN. Although the oscillatory mechanism biases spiking to occur during the rising phase of the oscillatory cycle, spikes can turn up in any phase position. However, spikes in a given PN typically phase-lock with the ensemble of PNs at specific, reliable times during the response to an odor (Wehr and Laurent 1996). For example, a PN may respond to a 1 s pulse of odor with a 1.5 s long pattern of spiking, but within this PN, only those spikes occurring 0.4–0.7 s into the response will reliably synchronize with the ensemble. Other PNs, of course, may synchronize at other times during the response, contributing to the population oscillatory pattern. These bouts of synchronization reliably vary with, and thus carry information about, the odorant eliciting the response.

11.7 ENCODING INFORMATION ABOUT ODOR CONCENTRATION

Earlier theoretical work (Hopfield 1995) suggested that odor concentration could be encoded by the phase position of PN spikes relative to the ensemble oscillation. However, a series of electrophysiological recordings directly testing this idea established that changes in odor concentration have no effect on the preferred phase position of spikes in PNs. In fact, all tested odorants and concentrations evoked spikes in PNs at the same point of the rising phase of the oscillatory cycle, making it unlikely that anything is encoded in the oscillatory phase of spikes in PNs (Stopfer et al. 2003).

The frequency of the oscillations serves to encode neither the odor's identity (Laurent and Davidowitz 1994) nor concentration (Ito et al. 2009; Stopfer et al. 2003). However, oscillation frequency can change: several species exhibit a reduction in odor-elicited oscillation frequency over the course of responding to a lengthy (>0.5 s or so) pulse of odor. This change in frequency has been shown to reflect not the concentration of the odor but rather the sensory adaptation in the ORNs that provide input to the circuitry generating oscillations (Ito et al. 2009). When the concentration of an odor decreases, fewer types of ORNs respond; the size of the responsive ORN population decreases but the responsive ORNs continue to fire at roughly the same rate. In contrast, when the ORNs undergo sensory adaptation while responding to a lengthy odor presentation, the size of the responsive ORN population remains the same but the firing rate decreases. This decrease in excitatory drive leads to a shift to slower oscillation frequencies (Ito et al. 2009).

Although the frequency of oscillation reflects nothing about the concentration of an odor, the strength of oscillatory synchrony can vary with concentration. As concentrations of an odorant increase, spiking within the population of PNs becomes more precisely synchronized, evident in intracellular recordings as increases in the amplitude of IPSPSs in LNs and PNs, and in LFP recordings as increases in the amplitude of the oscillations. This happens because higher concentrations of odors result in increased drive to LNs, which can then more effectively coordinate spiking across the AL network (Ito et al. 2009; Stopfer et al. 2003).

However, independent of this phenomenon, changes in the extent of oscillatory synchrony can also occur when odors are encountered repeatedly. Interestingly, oscillatory synchronization of spiking does not occur when an odor is presented for the first time—only repeated encounters with an odorant come to evoke synchrony. Repeated encounters with odorants are common; they occur within odor plumes, and as the result of iterative sampling behaviors. The mechanism responsible for this odor-specific buildup of synchronous spiking has been shown to reside within the AL and likely requires activity-dependent facilitation of synapses between LNs and PNs (Stopfer and Laurent 1999). A computational analysis showed this "fast learning" process may contribute to reducing noise and increasing the reliability of spiking responses to odors that are encountered repeatedly (Bazhenov et al. 2005). Because the degree of spike synchrony can shift independently of changes in odor concentration, this response parameter cannot be used by the olfactory system as an unambiguous code for odor concentration.

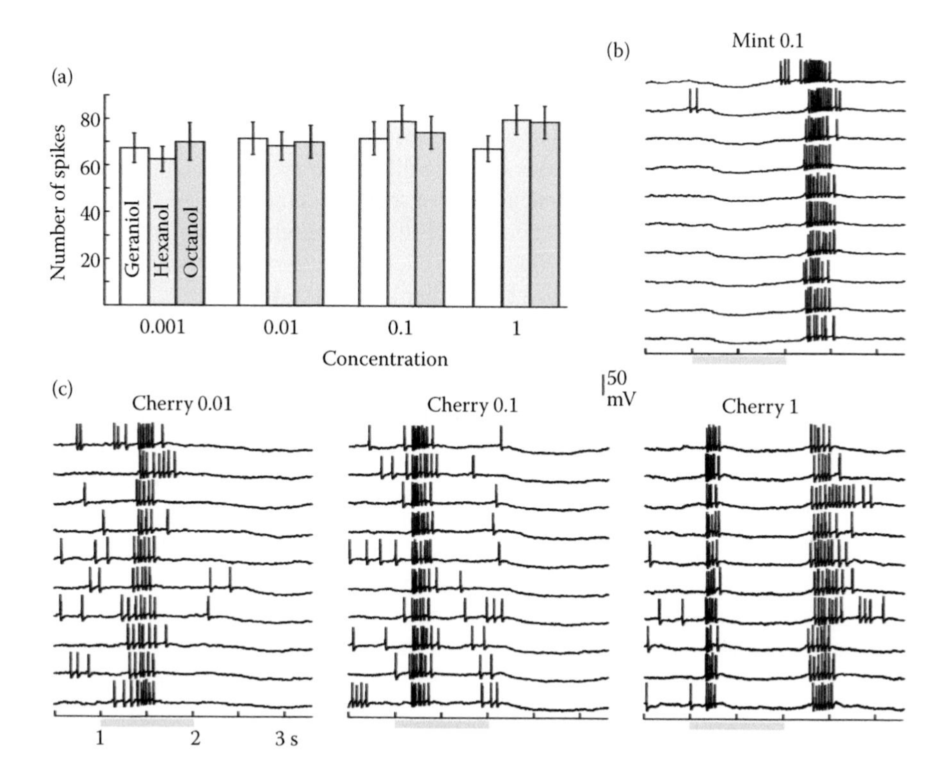

FIGURE 11.4 The timing of spikes, but not the number of spikes, varies in PNs with odor identity and concentration. (a) The average (+/– SEM) numbers of spikes in 110 PNs were about the same when elicited by 1 s presentations of four orders of magnitude concentrations of geraniol, hexanol, and octanol. (b) Intracellular traces show a single locust PN responding to ten 1 s presentations of mint odor (gray bar). (c) The same PN responding to three orders-of-magnitude concentrations of another odor, cherry. This PN responds reliably to both odors and all concentrations tested with similar numbers, but different temporal patterns, of spikes. (Panel (a) adapted from Stopfer, M., Jayaraman, V., and Laurent, G. 2003. *Neuron* 39, 991–1004.)

The population-wide firing rates of PNs are also unlikely to encode odor concentration, as these firing rates are remarkably stable across 1000-fold changes in concentration of chemically diverse odorants. (Figure 11.4a, Stopfer et al. 2003). This is because ORNs activate both excitatory PNs and inhibitory LNs, which then interact within the AL. Thus, when an odor, presented at a high concentration, strongly drives ORN input to the AL, PNs can be strongly excited, strongly inhibited, or both in sequence. Depending upon the specific identity, connectivity, and activity levels of the neurons presynaptic to it, any given PN may respond with more or fewer spikes to any given change in concentration.

How then do PNs encode information about odor identity and concentration? In PNs, as in ORNs, different odors elicit responses with different patterns of spiking. Thus, the timing of the spikes contains information about the identity of the odorant. Figure 11.4b shows an example of a PN responding to 10 successive presentations of a diluted mint odor; the response reliably consists of a period of inhibition followed by a burst of spikes. The same example PN shown in Figure 11.4b also responds to the same dilution of cherry odor (Figure 11.4c, center panel), but with a pattern of spike timing different from that evoked by mint: a burst of spikes followed by a period of inhibition. Presentations of more dilute cherry odor (Figure 11.4c, left panel) elicit approximately the same number of spikes, but the timing is different: the latency is longer, and the interspike interval is shorter. Presentations of more concentrated cherry odor (Figure 11.4c, right panel) elicit a brief burst of spikes followed by a period of inhibition, then followed by another burst of spikes.

Thus, changes in either the odor's identity or concentration can result in changes in the spiking patterns of PNs. How can the olfactory system disambiguate the identity and concentration of odors? To resolve this conundrum, it is useful to consider how the output of the population of PNs is read by its follower neurons, the KCs (see Figure 11.1). In locusts, for example, anatomical and physiological studies show that each KC receives direct synaptic input from hundreds of PNs (Jortner et al. 2007). Further, the oscillatory output of the PN population causes a cyclical, staccato series of excitatory post-synaptic potentials (EPSPs) observable in the KC membrane potential. GGN, the giant inhibitory neuron receiving input from all KCs, counters this periodic excitation with feedback inhibition, causing the membrane potential to return to baseline every cycle of the PN oscillation (Gupta and Stopfer 2012; Papadopoulou et al. 2011). These cycles therefore establish the integration time windows of KCs, and provide a biologically meaningful definition of "synchrony." Further, this perspective can be used to construct a biologically meaningful method of analysis: odor-elicited spikes from large numbers of PNs can be segmented into a series of time bins with widths corresponding to the oscillatory cycle.

So analyzed, the responses of the PN population can be shown to follow odor- and concentration-specific trajectories (Figure 11.3d; Stopfer et al. 2003). This analysis shows that changing the concentration of an odor results in a relatively small change in the population-wide spiking patterns of PNs. But changing the chemical identity of the odor results in a much greater change in spiking patterns (Figure 11.3e). Although the odor-elicited responses of individual PNs are ambiguous with respect to the identity and concentration of an odor, when responses distributed across the population of PNs and over time are considered from the perspective of the KCs, both the identity and concentration of an odor are readily extracted from the timing of the spikes. The existence of this information, made visible by trajectory graphs, can be more rigorously demonstrated by classification analyses (Stopfer et al. 2003). By enabling the use of timing as a coding dimension, the circuitry of the AL allows PNs to represent an enormous range of odor concentrations without saturating, as might a system in which increasingly intense stimuli evoke ever greater numbers of spikes rather than differently timed patterns of spikes. Although odors elicit lengthy firing patterns from the AL, it is important to note that the entire sequence need not be analyzed downstream; enough information exists within a single oscillatory cycle "snapshot" of activity to classify an odor's identity and concentration (Brown et al., 2005; Stopfer et al., 2003). In fact, no evidence suggests follower neurons (such as KCs) integrate these responses over lengthy periods of time. (See Chapter 15 for a more general discussion of this kind of coding in the brain.)

In broad strokes, the distributed, time-varying spiking responses of the PN population resemble those of the population of ORNs (compare Figures 11.3b and c, and 11.3d and e). In both ORN and PN populations, brief pulses of odor evoke rapid changes in firing patterns that can be graphed as population trajectories that rapidly loop away from and then back to baseline points. When long odor pulses are presented, both population responses rapidly undergo odor onset trajectories, then settle into a fixed point, and then return to baseline activity through a different offset path (Mazor and Laurent 2005; Raman et al. 2010). PNs do not simply inherit their temporal dynamics from the ORNs; the odor-elicited responses of PNs, which can include a half-dozen distinct epochs of excitation and inhibition, are much more elaborate and complex than those of ORNs. This complexity results from interactions within the AL among ORNs, PNs, and LNs (Bazhenov et al. 2001b). However, the AL requires variety in the timing of spiking patterns in the ORN population to generate information-rich temporally structured spiking patterns in PNs (Raman et al. 2010).

To date, the coding of odor concentration has not been extensively investigated in the vertebrate, but, despite some significant differences (i.e., vertebrates sniff while insects do not), several lines of evidence point to a direct analogy between coding in the AL and the OB. MCs receive multiglomerular input via local interneurons (Egger and Urban 2006; Wachowiak and Shipley 2006) and respond to odors with reliable temporally structured patterns of spikes (Cury et al. 2010; Dhawale et al. 2010; Friedrich and Laurent 2001; Spors and Grinvald 2002). These spikes are synchronized into odor-induced oscillations (reviewed in Kay et al. 2009). Further, the responses of MCs, viewed as a population, have been shown to traverse odor- and concentration-specific trajectories similar to those described in the insect (Bathellier et al. 2008; Cury et al. 2010).

As the AL reformats information about the identity and concentration of odors, it also adjusts the timing of spikes to more effectively differentiate the neural representations of different odors. Odorants that are chemically similar elicit similar patterns of spikes from overlapping subsets of ORNs, potentially making it difficult to discriminate one odor from another. The circuitry of the zebrafish OB has been shown to push similar responses apart in a process that plays out over the duration of an odor response. When the response to a given odor begins, the firing patterns of MCs, like those of ORNs, are similar to those evoked by other, chemically related odors: the responses to such odors are highly correlated. But as the response progresses, the firing patterns evoked by the given odor become increasingly different, or decorrelated, from those evoked by other odors (Friedrich et al. 2001). By adjusting the spiking patterns of MCs, the process of decorrelation allows a single olfactory circuit to achieve apparently opposed coding goals. At the response's onset, the circuitry favors a coarse kind of classification: it groups together odorants with similar chemical features (e.g., "it smells like a type of fruit"). But over the course of the response, the circuitry begins to favor more precise identification: it distinguishes those odorants from one another (e.g., "it smells like apple"). Analogous experiments have not been reported in insects, but computational models suggest interactions between LNs and PNs in the AL should achieve the same sort of decorrelation as the olfactory bulb (Assisi et al. 2012; Friedrich and Laurent 2001; Raman et al. 2010). It seems likely that responses elicited by different concentrations of the same odorant also decorrelate.

Thus, the AL does not simply relay input from the ORNs. Rather, in several ways, the AL restructures the spatial distribution and timing of spiking it receives. It functions to reduce noise, organize its spiking output into waves of oscillatory synchrony, disambiguate odor identity from intensity, and distinguish odor representations through decorrelation.

11.8 AFTER THE ANTENNAL LOBE

Olfactory signals arise in the periphery as relatively simple patterns of spiking distributed across a population of ORNs. The format of this information is transformed by the AL into much more elaborate patterns of synchronized spikes broadly distributed across a population of PNs. These output neurons, which provide the only olfactory pathway beyond the AL, carry their oscillatory patterns of spikes directly to the mushroom body, where they synapse upon KCs, which also respond to odors. Recordings from KCs reveal yet another dramatic transformation in the format of olfactory information.

The properties of KCs differ from those of PNs in three important ways: a given KC responds to many fewer odorants and concentrations than a given PN; it responds to odors with many fewer action potentials; and it is nearly silent in the absence of odor stimulation (Figure 11.5). Therefore, in several respects, odor representations

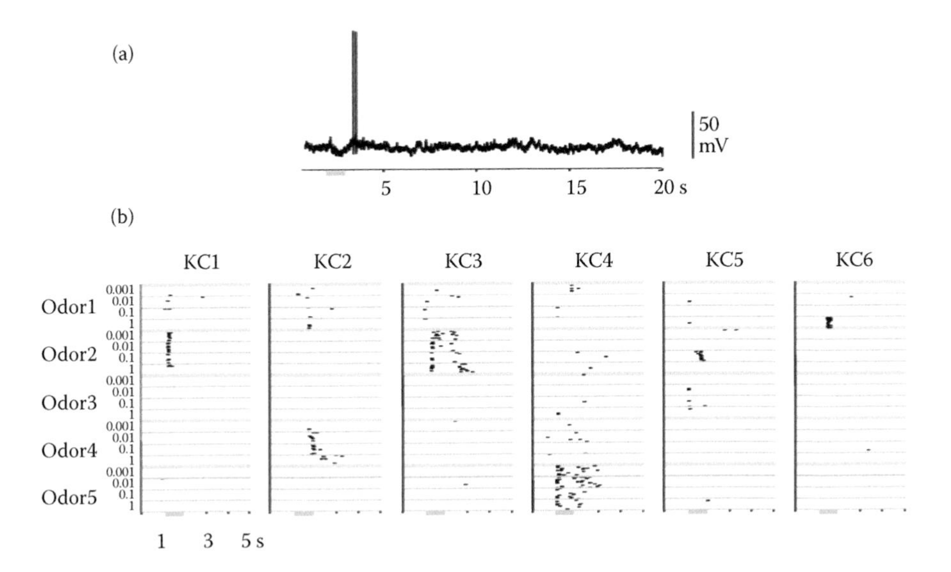

FIGURE 11.5 Sparse spiking responses of Kenyon cells (KCs). (a) A characteristic intracellular recording from a KC; the neuron is silent except for two spikes elicited by a 1 s presentation of odor (hexanol, gray bar). (b) Rasters show spikes recorded from six KCs responding to 1 s presentations of five odorants in four order-of-magnitude concentrations. KCs 1–4 respond to multiple concentrations of an odor; KCs 5–6 respond only to a single concentration. (Panel (b) adapted from Stopfer, M., Jayaraman, V., and Laurent, G. 2003. *Neuron* 39, 991–1004.)

are more "sparse" in the MB than in the AL (Ito et al. 2008; Perez-Orive et al. 2002; Stopfer et al. 2003; Turner et al. 2008). Sparse responses are evoked by every odor tested to date, including mixtures of odorants, which may more closely approximate odors an animal would actually encounter in the wild (Broome et al. 2006). What mechanisms underlie the timing and reformatting of odor-elicited spiking in the KCs?

Intracellular recordings made from locusts show that, although any given KC is rarely driven to spike by any given odor, it will often exhibit subthreshold membrane potential oscillations. The subthreshold activity and the rare spikes observed in KCs are strongly phase-locked to the LFP oscillations originating in the AL (Laurent and Naraghi 1994), suggesting that the AL's rhythmic output strongly influences the timing of spiking in KCs.

The strong coherence observed between odor-elicited membrane potential oscillations in KCs and the odor-elicited, AL-generated LFP oscillation is consistent with results of anatomical studies and electrophysiological paired recordings of PNs and KCs showing the two types of neurons are densely connected. In the locust, each of the ~50,000 KCs receives input from about half of the population of ~830 PNs (Jortner et al. 2007). The firing thresholds of KCs are unusually high (Demmer and Kloppenburg 2009; Perez-Orive et al. 2002): it has been estimated that ~100 nearly coincident spikes from PNs are necessary to drive a KC to fire a single action potential (Mazor et al. 2005). In the absence of odors, spontaneous spiking within the population of PNs rarely reaches the degree of synchrony needed to cause a KC to spike (Joseph et al. 2012), and so KCs show the nearly nonexistent baseline firing rate of about 0.025 Hz (Perez-Orive et al. 2002). But during odor presentations, ~150 PNs are driven to spike during each oscillation cycle (Mazor et al. 2005), driving a small subset of KCs to spike. Because the odor-elicited spiking of PNs is sculpted into slowly evolving temporal patterns of excitation and inhibition, the composition of the set of PNs spiking coincidently changes as the population glides along an odor- and concentration-specific trajectory (see Figure 11.3). Consequentially, the sparse set of KCs stimulated sufficiently to spike also evolves over the course of its response to an odor (Stopfer et al. 2003).

Although all PNs in locusts are thought to be excitatory, electrophysiological recordings revealed rhythmic IPSPs in the membrane potentials of KCs (Laurent and Naraghi 1994). In the locust, this inhibition appears to arise completely from a single broadly branched neuron, the GGN, which has been shown to provide feedback inhibition to the entire KC population (Papadopoulou et al. 2011). (Earlier work suggested this inhibition arose from a different population of interneurons in the lateral horn; see Perez-Orive et al. 2002.) Because GGN also receives convergent input from the entire KC population, it is extremely broadly tuned, likely responding to every odor that activates the olfactory system (Gupta and Stopfer 2012; Papadopoulou et al. 2011).

GGN-mediated feedback inhibition plays a role in the sparsening of odor responses in KCs. During each oscillatory cycle, PNs tend to spike at a characteristic phase position, driving KCs to likewise spike during a narrow window of time. GGN-mediated inhibition regulates the duration of this temporal window, resulting in sparse spiking precisely timed in relation to the oscillatory cycle. The contribution of this mechanism to sparse spiking in KCs is significant; inhibiting the effect of GGN by hyperpolarizing it reduces KC sparseness, exciting GGN by depolarizing it increases KC sparseness (Papadopoulou et al. 2011). (See Chapter 7 for an analysis of input synchrony on a neuron's spiking.)

The role of GGN becomes particularly interesting in the context of concentration coding. As noted earlier, increasing concentrations of odor cause PN synchrony to increase. Because intrinsic and circuit properties make KCs sensitive to synchronous input (Laurent and Naraghi 1994; Papadopoulou et al. 2011; Perez-Orive et al. 2002), one would expect higher odor concentrations to evoke less sparse odor representation across the KC population. However, it has been shown that KC spiking remains equally sparse across a broad range of odor concentrations (Stopfer et al. 2003). This maintenance of sparseness is due to GGN; higher-concentration odorants increase the coincidence of input to KCs, which in turn cause KCs to become more excited, which in turn produce stronger responses in GGN, which in turn provides stronger inhibition to the KCs. By providing feedback inhibition proportional to the level of excitation across the KC population, KC firing sparseness is maintained across odor concentrations (Assisi et al. 2007; Papadopoulou et al. 2011).

Comparisons of insects and vertebrates reveal many similarities between the MB and the piriform cortex (PC). Like PNs in the locust, MCs of the OB project to inhibitory interneurons as well as pyramidal cells in the PC (Neville and Haberly 2003, Stokes and Isaacson 2010). Like the GGN, some of these interneurons send feedback inhibition to the pyramidal cells (Stokes et al. 2010). Pyramidal cell spiking responses are sparse, and are only seen at certain phases of the oscillation generated in the OB (Poo and Isaacson 2009).

11.9 VIRTUES OF SPARSENESS

Why are precisely timed, sparse spiking representations of odors employed by the olfactory systems of so many species? One attractive feature of sparse representations is that individual action potentials can carry a great deal of information. The broad input connectivity and sparse responses of KCs allow them to summarize the current state of a large portion of the PN population with individual spikes. Further, odor representations in KCs are decorrelated relative to odor representations in PNs: KCs vastly outnumber PNs and thus provide a higher-dimensional coding space, and the narrower tuning of KCs means that odor representations share fewer dimensions within the KC coding space than that of the PNs. This additional separation of odor representations may be utilized by neurons downstream from KCs, allowing for fine discrimination of similar odors (Laurent 2002).

Sparse representations may also be optimal for memory formation and recall. Indeed, many studies have shown that the MB plays an important role in learning and memory (Dubnau et al. 2001; Erber et al. 1980; Heisenberg et al. 1985; Krashes et al. 2007, 2009; McGuire et al. 2001; Schwaerzel et al. 2002). In the MB, odors are represented by relatively small numbers of neurons, and so a relatively small number of responses would need to be compared or modified to recall or form memories of odor representations (Heisenberg 2003; Laurent 2002).

11.10 AFTER THE KENYON CELLS

Knowledge of olfactory coding downstream from the KC population remains very limited. KCs form synapses with a range of cells projecting from the MB and into many regions throughout the brain. The best-studied region is probably the β-lobe, which contains a small population of cells called β-lobe neurons. β-lobe neurons are broadly responsive to odorants, likely because each of these cells appears to receive input from large numbers of KCs. Although the information-processing logic behind the massive convergence of KCs to β-lobe neurons is unclear, the pattern of connectivity suggests that individual β-lobe neurons are sensitive to distributed features of upstream cell populations, including the precise timing of spikes. Consistent with this, injection of the $GABA_A$ blocker picrotoxin into the AL, which desynchronizes PN spiking, drastically changes the responses of individual β-lobe neurons to odors (MacLeod et al. 1996).

The odor-evoked spiking responses of β-lobe neurons maintain very precise timing, like those of their upstream inputs, with spikes occurring preferentially at a particular phase of the LFP oscillatory cycle. At first glance, this is surprising, as one might expect variability in the timing of spikes across many KCs to compound in β-lobe neurons, degrading their own ability to respond in phase. However, Cassenaer and Laurent (2007), by recording simultaneously from KCs and β-lobe neurons, discovered that a mechanism employing spike timing-dependent plasticity (STDP) serves to offset this variability, ensuring stable spike timing in β-lobe neurons. If a KC causes a β-lobe neuron to spike at a later-than-usual phase position, the synapse between them is strengthened, causing subsequent spikes to occur at an earlier phase position. Conversely, if a KC causes a β-lobe neuron to spike earlier than usual, the synapse between them is weakened, causing future spikes to occur later. Thus, by means of an active, corrective process, neural synchrony originating in the AL is maintained downstream from the KCs. It remains unknown what role this precise timing of spiking in β-lobe neurons plays in the coding or decoding of odor and concentration, but its active maintenance across layers of neurons suggests it may be important.

11.11 CONCLUSIONS

In recent years, it has become increasingly clear that the neural representations of odors are dramatically transformed as they traverse successive populations of neurons. In the periphery, ORNs with different tuning specificities respond to odorants with patterns of spikes that vary with the odorant, and include periods of excitation and inhibition. Because the firing rates of ORNs are constrained by sensory adaptation and saturation mechanisms, changes in odor concentration are reported to downstream neurons mostly as changes in the size of the population of ORNs responding to an odorant, rather than as changes in the net amount of spiking in those ORNs. In insects, ORNs activate the AL circuitry to create patterns of spiking in PNs that are more broadly distributed, and more temporally complex, than in the periphery. Considered at the population level, the AL circuitry further disambiguates odor identity from concentration, increases the difference between responses elicited by different odors, and imposes oscillatory dynamics that tend to transiently synchronize populations of volubly spiking PNs in an odor-dependent manner. PNs project onward to the MB, where the odor-elicited responses of KCs are remarkably specific with respect to an odor's identity and concentration, and remarkably sparse, consisting of just a few well-timed spikes upon a nearly silent background. The precision of spike timing carries beyond the KCs, at least as far as the β-lobe. Many features of olfactory processing gleaned from work in insects appear to hold true in vertebrates, as well. Although the format of information about odors changes dramatically along the olfactory pathway, one consistent property is that the timing of spikes matters a great deal.

REFERENCES

Araneda, R. C., Kini, A. D., and Firestein, S. 2000. The molecular receptive range of an odorant receptor. *Nature Neuroscience* 3, 1248–55.

Assisi, C., Stopfer, M., and Bazhenov, M. 2012. Balanced excitation and inhibition optimize odor processing in insect olfactory system. *PLoS Computational Biology* 8(7), e1002563.

Assisi, C., Stopfer, M., Laurent, G., and Bazhenov, M. 2007. Adaptive regulation of sparseness by feedforward inhibition. *Nature Neuroscience* 10, 1176–84.

Bathellier, B., Buhl, D. L., Accolla, R., and Carleton, A. 2008. Dynamic ensemble odor coding in the mammalian olfactory bulb: Sensory information at different timescales. *Neuron* 57, 586–98.

Bazhenov, M., Stopfer, M., Rabinovich, M., Huerta, R., Abarbanel, H. D., Sejnowski, T. J., and Laurent, G. 2001a. Model of transient oscillatory synchronization in the locust antennal lobe. *Neuron* 30, 553–67.

Bazhenov, M., Stopfer, M., Rabinovich, M., Huerta, R., Abarbanel, H. D., Sejnowski, T. J., and Laurent, G. 2001b. Model of cellular and network mechanisms for odor-evoked temporal patterning in the locust antennal lobe. *Neuron* 30, 569–81.

Bazhenov, M., Stopfer, M., Sejnowski, T. J., and Laurent, G. 2005. Fast odor learning improves reliability of odor responses in the locust antennal lobe. *Neuron* 46, 483–92.

Bhandawat, V., Olsen, S. R., Gouwens, N. W., Schlief, M. L., and Wilson, R. I. 2007. Sensory processing in the *Drosophila* antennal lobe increases reliability and separability of ensemble odor representations. *Nature Neuroscience* 10, 1474–82.

Bozza, T., Feinstein, P., Zheng, C., and Mombaerts, P. 2002. Odorant receptor expression defines functional units in the mouse olfactory system. *Journal of Neuroscience* 22, 3033–43.

Broome, B. M., Jayaraman, V., and Laurent, G. 2006. Encoding and decoding of overlapping odor sequences. *Neuron* 51(4), 467-82.

Brown, S. L., Joseph, J., and Stopfer, M. 2005. Encoding a temporally structured stimulus with a temporally structured neural representation. *Nature Neuroscience* 8, 1568–76.

Buck, L. B. 1996. Information coding in the vertebrate olfactory system. *Annual Reviews Neuroscience* 19, 517–44.

Cassenaer, S. and Laurent, G. 2007. Hebbian STDP in mushroom bodies facilitates the synchronous flow of olfactory information in locusts. *Nature* 448, 709–13.

Cassenaer, S. and Laurent, G. 2012. Conditional modulation of spike-timing-dependent plasticity for olfactory learning. *Nature* 482, 47–52.

Clyne, P. J., Warr, C. G., Freeman, M. R., Lessing, D., Kim, J., and Carlson, J. R. 1999. A novel family of divergent seven-transmembrane proteins: Candidate odorant receptors in *Drosophila*. *Neuron* 22, 327–38.

Couto, A., Alenius, M., and Dickson, B. J. 2005. Molecular, anatomical, and functional organization of the *Drosophila* olfactory system. *Current Biology* 15, 1535–47.

Cury, K. M. and Uchida, N. 2010. Robust odor coding via inhalation-coupled transient activity in the mammalian olfactory bulb. *Neuron* 68, 570–85.

de Bruyne, M., Foster, K., and Carlson, J. R. 2001. Odor coding in the *Drosophila* antenna. *Neuron* 30, 537–52.

Demmer H. and Kloppenburg, P. 2009. Intrinsic membrane properties and inhibitory synaptic input of Kenyon cells as mechanisms for sparse coding? *Journal of Neurophysiology* 102.3, 1538–50.

Dhawale, A. K., Hagiwara, A., Bhalla, U. S., Murthy, V. N., and Albeanu, D. F. 2010. Non-redundant odor coding by sister mitral cells revealed by light addressable glomeruli in the mouse. *Nature Neuroscience* 13, 1404–12.

Dobritsa, A. A., Der Goes Van Naters, W. van, Warr, Coral G., Steinbrecht, R. A., and Carlson, J. R. 2003. Integrating the molecular and cellular basis of odor coding in the *Drosophila* antenna. *Neuron* 37, 827–41.

Dubnau, J., Grady, L., Kitamoto, T., and Tully, T. 2001. Disruption of neurotransmission in *Drosophila* mushroom body blocks retrieval but not acquisition of memory. *Nature* 411, 476–80.

Duchamp-Viret, P. 1999. Odor response properties of rat olfactory receptor neurons. *Science* 284, 2171–4.

Egger, V. and Urban, N. N. 2006. Dynamic connectivity in the mitral cell-granule cell microcircuit. *Seminars in Cell & Developmental Biology* 17, 424–32.

Erber, J., Masuhr, T., and Menzel, R. 1980. Localization of short-term memory in the brain of the bee, *Apis mellifera*. *Physiological Entomology* 5, 343–58.

Esslen, J. and Kaissling, K. E. 1976. Zahl und Verteilung antennaler Sensillen bei der Honigbiene (*Apis mellifera* L.). *Zoomorphologie* 83, 227–51.

Firestein, S., Picco, C., and Menini, A. 1993. The relation between stimulus and response in olfactory receptor cells of the tiger salamander. *The Journal of Physiology* 468, 1–10.

Fishilevich, E. and Vosshall, L. B. 2005. Genetic and functional subdivision of the *Drosophila* antennal lobe. *Current Biology* 15, 1548–53.

Friedrich, R. W. and Laurent, G. 2001. Dynamic optimization of odor representations by slow temporal patterning of mitral cell activity. *Science* 291, 889–94.

Gupta, N. and Stopfer, M. 2012. Functional analysis of a higher olfactory center, the lateral horn. *Journal of Neuroscience* 32, 8138–48.

Hallem, E. A. and Carlson, J. R. 2006. Coding of odors by a receptor repertoire. *Cell* 125, 143–60.

Hartline, H. K. 1956. Inhibition in the eye of the limulus. *The Journal of General Physiology* 39, 651–73.

Heisenberg, M. 2003. Mushroom body memoir: From maps to models. *Nature Reviews. Neuroscience* 4, 266–75.

Heisenberg, M., Borst, A., Wagner, S., and Byers, D. 1985. *Drosophila* mushroom body mutants are deficient in olfactory learning. *Journal of Neurogenetics* 2, 1–30.

Hildebrand, J. G. and Shepherd, G. M. 1997. Mechanisms of olfactory discrimination: Converging evidence for common principles across phyla. *Annual Review of Neuroscience* 20, 595–631.

Homberg, U., Christensen, T. A., and Hildebrand, J. G. 1989. Structure and function of the deutocerebrum in insects. *Annual Review of Entomology* 34, 477–501.

Hopfield, J. J. 1995. Pattern recognition computation using action potential timing for stimulus representation. *Nature* 376, 33–6.

Hosler, J. S. and Smith, B. H. 2000. Blocking and the detection of odor components in blends. *The Journal of Experimental Biology* 203, 2797–806.

Huang, J., Zhang, W., Qiao, W., Hu, A., and Wang, Z. 2010. Functional connectivity and selective odor responses of excitatory local interneurons in *Drosophila* antennal lobe. *Neuron* 67, 1021–33.

Hummel, T., Sekinger, B., Wolf, S. R., Pauli, E., and Kobal, G. 1997. "Sniffin" sticks: Olfactory performance assessed by the combined testing of odour identification, odor discrimination and olfactory threshold. *Chemical Senses* 22, 39–52.

Ito, I., Bazhenov, M., Ong, R. C., Raman, B., and Stopfer, M. 2009. Frequency transitions in odor-evoked neural oscillations. *Neuron* 64, 692–706.

Ito, I., Ong, R. C.-Y., Raman, B., and Stopfer, M. 2008. Sparse odor representation and olfactory learning. *Nature Neuroscience* 11, 1177–84.

Jacquin-Joly, E. and Merlin, C. 2004. Insect olfactory receptors: Contributions of molecular biology to chemical ecology. *Journal of Chemical Ecology* 30, 2359–97.

Jayaraman, V. and Laurent, G. 2007. Evaluating a genetically encoded optical sensor of neural activity using electrophysiology in intact adult fruit flies. *Frontiers in Neural Circuits* 1, 3.

Jortner, R. A., Farivar, S. S., and Laurent, G. 2007. A simple connectivity scheme for sparse coding in an olfactory system. *The Journal of Neuroscience* 27, 1659–69.

Joseph, J., Dunn, F. A., and Stopfer, M. 2012. Spontaneous olfactory receptor neuron activity determines follower cell response properties. *The Journal of Neuroscience* 32, 2900–10.

Kaupp, U. B. 2010. Olfactory signalling in vertebrates and insects: Differences and commonalities. *Nature Reviews. Neuroscience* 11, 188–200.

Kay, L. M., Beshel, J., Brea, J., Martin, C., Rojas-Líbano, D., and Kopell, N. 2009. Olfactory oscillations: The what, how and what for. *Trends in Neurosciences* 32, 207–14.

Kay, L. M. and Stopfer, M. 2006. Information processing in the olfactory systems of insects and vertebrates. *Seminars in Cell & Developmental Biology* 17, 433–42.

Kleene, S. J. 2008. The electrochemical basis of odor transduction in vertebrate olfactory cilia. *Chemical Senses* 33, 839–59.

Krashes, M. J., DasGupta, S., Vreede, A., White, B., Armstrong, J. D., and Waddell, S. 2009. A neural circuit mechanism integrating motivational state with memory expression in *Drosophila. Cell* 139, 416–27.

Krashes, M. J., Keene, A. C., Leung, B., Armstrong, J. D., and Waddell, S. 2007. Sequential use of mushroom body neuron subsets during *Drosophila* odor memory processing. *Neuron* 53, 103–15.

Kuffler, S. W. 1953. Discharge patterns and functional organization of mammalian retina. *Journal of Neurophysiology* 16, 37–68.

Laurent, G. 2002. Olfactory network dynamics and the coding of multidimensional signals. *Nature Reviews. Neuroscience* 3, 884–95.

Laurent, G. and Davidowitz, H. 1994. Encoding of olfactory information with oscillating neural assemblies. *Science* 265, 1872–5.

Laurent, G. and Naraghi, M. 1994. Odorant-induced oscillations in the mushroom bodies of the locust. *The Journal of Neuroscience* 14, 2993–3004.

Lei, H., Christensen, T. A., and Hildebrand, J. G. 2002. Local inhibition modulates odor-evoked synchronization of glomerulus-specific output neurons. *Nature Neuroscience* 5, 557–65.

MacLeod, K. M. and Carr, C. E. 2007. Beyond timing in the auditory brainstem: Intensity coding in the avian cochlear nucleus angularis. In: *Computational Neuroscience: Theoretical Insights Into Brain Function.* Eds: P. Cisek, T. Drew, and J.F. Kalaska. Elsevier, Amsterdam.

MacLeod, K. and Laurent, G. 1996. Distinct mechanisms for synchronization and temporal patterning of odor-encoding neural assemblies. *Science* 274, 976–9.

Malnic, B., Hirono, J., Sato, T., and Buck, L. B. 1999. Combinatorial receptor codes for odors. *Cell* 96, 713–23.

Mazor, O. and Laurent, G. 2005. Transient dynamics versus fixed points in odor representations by locust antennal lobe projection neurons. *Neuron* 48, 661–73.

McGuire, S. E., Le, P. T., and Davis, R. L. 2001. The role of *Drosophila* mushroom body signaling in olfactory memory. *Science* 293, 1330–3.
Michel, W. C. and Ache, B. W. 1994. Odor-evoked inhibition in primary olfactory receptor neurons. *Chemical Senses* 19, 11–24.
Mombaerts, P. 2004. Odorant receptor gene choice in olfactory sensory neurons: The one receptor-one neuron hypothesis revisited. *Current Opinion in Neurobiology* 14, 31–6.
Mombaerts, P., Wang, F., Dulac, C., Chao, S. K., Nemes, A., Mendelsohn, M., Edmondson, J., and Axel, R. 1996. Visualizing an olfactory sensory map. *Cell* 87, 675–86.
Neville, K. R. and Haberly, L. B. 2003. Beta and gamma oscillations in the olfactory system of the urethane-anesthetized rat. *Journal of Neurophysiology*, 90(6), 3921–30.
Ng, M., Roorda, R. D., Lima, S. Q., Zemelman, B. V., Morcillo, P., and Miesenböck, G. 2002. Transmission of olfactory information between three populations of neurons in the antennal lobe of the fly. *Neuron* 36, 463–74.
Olsen, S. R., Bhandawat, V., and Wilson, R. I. 2007. Excitatory interactions between olfactory processing channels in the *Drosophila* antennal lobe. *Neuron* 54, 89–103.
Papadopoulou, M., Cassenaer, S., Nowotny, T., and Laurent, G. 2011. Normalization for sparse encoding of odors by a wide field interneuron. *Science* 332, 721–5.
Perez-Orive, J., Mazor, O., Turner, G. C., Cassenaer, S., Wilson, R. I., and Laurent, G. 2002. Oscillations and sparsening of odor representations in the mushroom body. *Science* 297, 359–65.
Pomeroy, S., LaMantia, A., and Purves, D. 1990. Postnatal construction of neural circuitry in the mouse olfactory bulb. *The Journal of Neuroscience* 10, 1952–66.
Poo, C. and Isaacson, J. S. 2009. Odor representations in olfactory cortex: "Sparse" coding, global inhibition, and oscillations. *Neuron* 62, 850–61.
Pouget, A., Dayan, P., and Zemel, R. 2000. Information processing with population codes. *Nature Reviews. Neuroscience* 1, 125–32.
Raman, B., Joseph, J., Tang, J., and Stopfer, M. 2010. Temporally diverse firing patterns in olfactory receptor neurons underlie spatiotemporal neural codes for odors. *The Journal of Neuroscience* 30, 1994–2006.
Ressler, K. 1993. A zonal organization of odorant receptor gene expression in the olfactory epithelium. *Cell* 73, 597–609.
Rinberg, D., Koulakov, A., and Gelperin, A. 2006. Speed-accuracy tradeoff in olfaction. *Neuron* 51, 351–8.
Roweis, S. T. and Saul, L. K. 2000. Nonlinear dimensionality reduction by locally linear embedding. *Science* 290, 2323–6.
Rützler, M. and Zwiebel, L. J. 2005. Molecular biology of insect olfaction: Recent progress and conceptual models. *Journal of Comparative Physiology*. A 191, 777–90.
Sato, K., Pellegrino, M., Nakagawa, T., Nakagawa, T., Vosshall, L. B., and Touhara, K. 2008. Insect olfactory receptors are heteromeric ligand-gated ion channels. *Nature* 452, 1002–6.
Schoppa, N. 2003. Dendritic processing within olfactory bulb circuits. *Trends in Neurosciences* 26, 501–6.
Schwaerzel, M., Heisenberg, M., and Zars, T. 2002. Extinction antagonizes olfactory memory at the subcellular level. *Neuron* 35, 951–60.
Shang, Y., Claridge-Chang, A., Sjulson, L., Pypaert, M., and Miesenböck, G. 2007. Excitatory local circuits and their implications for olfactory processing in the fly antennal lobe. *Cell* 128, 601–12.
Sobel, N., Prabhakaran, V., Desmond, J. E., Glover, G. H., Goode, R. L., Sullivan, E. V., and Gabrieli, J. D. 1998. Sniffing and smelling: Separate subsystems in the human olfactory cortex. *Nature* 392, 282–6.
Spors, H. and Grinvald, A. 2002. Spatio-temporal dynamics of odor representations in the mammalian olfactory bulb. *Neuron* 34, 301–15.

Spors, H., Wachowiak, M., Cohen, L. B., and Friedrich, R. W. 2006. Temporal dynamics and latency patterns of receptor neuron input to the olfactory bulb. *The Journal of Neuroscience* 26, 1247–59.

Stocker, R. F. 1994. The organization of the chemosensory system in *Drosophila melanogaster*: A review. *Cell and Tissue Research* 275, 3–26.

Stokes, C. C. A. and Isaacson, J. S. 2010. From dendrite to soma: Dynamic routing of inhibition by complementary interneuron microcircuits in olfactory cortex. *Neuron* 67, 452–65.

Stopfer, M., Bhagavan, S., Smith, B. H., and Laurent, G. 1997. Impaired odour discrimination on desynchronization of odour-encoding neural assemblies. *Nature* 390, 70–4.

Stopfer, M., Jayaraman, V., and Laurent, G. 2003. Intensity versus identity coding in an olfactory system. *Neuron* 39, 991–1004.

Stopfer, M. and Laurent, G. 1999. Short-term memory in olfactory network dynamics. *Nature* 402, 664–8.

Su, C.-Y., Menuz, K., and Carlson, J. R. 2009. Olfactory perception: Receptors, cells, and circuits. *Cell* 139, 45–59.

Tanaka, N. K., Ito, K., and Stopfer, M. 2009. Odor-evoked neural oscillations in *Drosophila* are mediated by widely branching interneurons. *The Journal of Neuroscience* 29, 8595–603.

Tripathy, S. J., Peters, O. J., Staudacher, E. M., Kalwar, F. R., Hatfield, M. N., and Daly, K. C. 2010. Odors pulsed at wing beat frequencies are tracked by primary olfactory networks and enhance odor detection. *Frontiers in Cellular Neuroscience* 4, 1.

Turner, G. C., Bazhenov, M., and Laurent, G. 2008. Olfactory representations by *Drosophila* mushroom body neurons. *Journal of Neurophysiology* 99, 734–46.

Uchida, N., Takahashi, Y. K., Tanifuji, M., and Mori, K. 2000. Odor maps in the mammalian olfactory bulb: Domain organization and odorant structural features. *Nature Neuroscience* 3, 1035–43.

Vassar, R., Ngai, J., and Axel, R. 1993. Spatial segregation of odorant receptor expression in the mammalian olfactory epithelium. *Cell* 74, 309–18.

Verhagen, J. V., Wesson, D. W., Netoff, T. I., White, J. A., and Wachowiak, M. 2007. Sniffing controls an adaptive filter of sensory input to the olfactory bulb. *Nature Neuroscience* 10, 631–9.

Vogt, R. G. and Riddiford, L. M. 1981. Pheromone binding and inactivation by moth antennae. *Nature* 293, 161–3.

Vosshall, L. B., Amrein, H., Morozov, P. S., Rzhetsky, A., and Axel, R. 1999. A spatial map of olfactory receptor expression in the *Drosophila* antenna. *Cell* 96, 725–36.

Vosshall, L. B., Wong, A. M., and Axel, R. 2000. An olfactory sensory map in the fly brain. *Cell* 102, 147–59.

Wachowiak, M. and Cohen, L. B. 2001. Representation of odorants by receptor neuron input to the mouse olfactory bulb. *Neuron* 32, 723–35.

Wachowiak, M. and Shipley, M. T. 2006. Coding and synaptic processing of sensory information in the glomerular layer of the olfactory bulb. *Seminars in Cell & Developmental Biology* 17, 411–23.

Waldrop, B., Christensen, T. A., and Hildebrand, J. G. 1987. GABA-mediated synaptic inhibition of projection neurons in the antennal lobes of the sphinx moth, *Manduca sexta*. *Journal of Comparative Physiology A* 161, 23–32.

Wang, J. W., Wong, A. M., Flores, J., Vosshall, L. B., and Axel, R. 2003. Two-photon calcium imaging reveals an odor-evoked map of activity in the fly brain. *Cell* 112, 271–82.

Wehr, M. and Laurent, G. 1996. Odour encoding by temporal sequences of firing in oscillating neural assemblies. *Nature* 384, 162–6.

Werblin, S. and Dowling, E. 1969. Organization of the retina of the mudpuppy, recording necturus macubsus. II. Intracellular. *Journal of Neurophysiology* 32, 339–55.

Wicher, D., Schäfer, R., Bauernfeind, R., Stensmyr, M. C., Heller, R., Heinemann, S. H., and Hansson, B. S. 2008. *Drosophila* odorant receptors are both ligand-gated and cyclic-nucleotide-activated cation channels. *Nature* 452, 1007–11.

Wilson, R. I. and Mainen, Z. F. 2006. Early events in olfactory processing. *Annual Review of Neuroscience* 29, 163–201.

Wilson, R. I., Turner G. C., and Laurent G. 2004. Transformation of olfactory representations in the *Drosophila* antennal lobe. *Science* 303(5656), 366–70.

Wright, G. A., Lutmerding, A., Dudareva, N., and Smith, Brian H 2005. Intensity and the ratios of compounds in the scent of snapdragon flowers affect scent discrimination by honeybees (*Apis mellifera*). *Journal of Comparative Physiology. A* 191, 105–14.

Yaksi, E. and Wilson, R. I. 2010. Electrical coupling between olfactory glomeruli. *Neuron* 67, 1034–47.

12 Spike Timing as a Mechanism for Taste Coding in the Brainstem

Patricia M. Di Lorenzo

CONTENTS

12.1 INTRODUCTION

When a neuron is effectively stimulated by its inputs, it responds by increasing or decreasing its firing rate. For many students of neural coding, these global changes in the output of a neuron, measured as the spike count over a defined response interval, are a sufficient description of how a stimulus is represented by the nervous system. Some incorporate the spike counts observed across the population of neurons in a given structure or system, or some sample thereof, into neural coding theories, but the basic unit of data is nearly always the same (see Di Lorenzo and Lemon, 2001). More recently, investigators have focused on the *dynamics* of evoked spike trains as a potential means of communication among neurons. In both sensory and motor systems, it has been shown that various temporal characteristics of neural responses may be just as, if not more, informative as spike count.

The evaluation of the temporal characteristics of spike trains poses several challenges, each of which has been addressed in different ways by different investigators. For example, it is not entirely obvious how the changes in the firing rate of a neuron over time are quantified. While spike count over a defined interval clearly reflects the most basic form of temporal coding, that is, firing rate change, there are more subtle dynamics that may also be important. Even if the temporal characteristics of neural responses can be shown to be informative, both the neural mechanisms that underlie the generation of these temporal patterns and the effectiveness of conveying information to target neurons that read the temporal patterns are serious issues that warrant investigation. Last, and perhaps most importantly, is the issue of whether spike timing can actually be used to guide behavior, that is, its functional relevance.

The focus of the present chapter is the neural coding of gustatory stimuli in the brainstem, with a particular emphasis on the temporal characteristics of taste-evoked spike trains, that is, spike timing. We will show that spike timing enables many cells in the central gustatory system to multiplex different types of information about taste stimuli by taking advantage of different timescales. The nature of the neural mechanism(s) involved in giving rise to the temporal patterning of taste responses will also be examined. Finally, we will present evidence that spike timing associated with specific taste stimuli can generate specific and appropriate gustatory sensations. Other aspects of taste coding, including other types of temporal coding, for example, rate envelope, synchronicity among ensembles, and so on, have been covered in a previous review (Hallock and Di Lorenzo, 2006) and will not be discussed here. In sum, we will present evidence for the presence, mechanism, and function of temporal coding in the brainstem gustatory system.

There are several compelling reasons that make the taste system an attractive model for studying temporal coding. For one, there are relatively few stimulus dimensions, represented by the "basic" taste qualities. Taste qualities can be described as groups of water-soluble chemicals that taste alike. In mammals, these are sweet, sour, salty, bitter, and umami (savory). Also, and perhaps most importantly, the successful encoding of taste stimuli can often be gauged by both innate and learned behavioral reactivity, that is, stereotypical patterns of behaviors. These attributes facilitate the exploration of neural coding with stimuli that span the domain of the gustatory world and provide a basis for demonstrating whether and how temporal coding is actually used. On both counts, there is accumulating evidence that the temporal characteristics of taste-evoked spike trains can convey at least as much, and often much more, information as spike count alone and so can be considered a fundamental mechanism of information transmission in the gustatory system.

12.2 NEURAL CODING OF SENSORY INFORMATION: GENERAL CONCEPTS

The action potential was first described in the late nineteenth century (see Schuetze, 1983, for a review) but it was not until the mid-twentieth century that its role as the basic unit of information transmission in the nervous system was fully appreciated. As the infant field of neuroscience was developing, theorists in computer science and engineering seized upon the deceptive simplicity of neuronal communication to devise models that might emulate the enormous computational power of the brain. The properties of the action potential made this venture particularly attractive. For example, the action potential is an all-or-none event. That is, a neuron either "fires," that is, generates and action potential, or not. In addition, any given neuron integrates information from many sources and makes the singular decision of whether to generate its own action potential. In effect, the action potential can be thought of as a binary event: it is either a one or a zero, and represents the results of the computed weighted influence of many inputs. This analogy was brilliantly exploited by McCulloch and Pitts (1943) in their treatise on how the properties of the nervous system could inform the construction of simple yet powerful computational devices (see Kay, 2001, for a description of the historical context).

Modeling computers after the nervous system has been fruitful for the field of computer science (e.g., the concept of parallel process was gleaned from neuroscience) but has also enabled a deeper understanding of how the nervous system works. For example, just as information can be represented as patterns of ones and zeros within a computer, so too does the nervous system represent information in the distributed activity of action potentials across many neurons. But unlike computers where the components that signal information are essentially all the same, neurons are diverse in their morphology and connectivity, and there is much processing that is graded rather than all-or-none. This affords the nervous system a richness and complexity that is not paralleled by any man-made device.

The neural representation of the ambient energy that surrounds an organism in its environment is one of the most basic functions of the brain. This energy, defined as a sensory stimulus once it is fully processed by the brain, comes in many forms, not all of which are sensed by any one species. That is, the sensory organs and brains afforded to members of each species are exquisitely well suited to process the signals in its environment that enable it to survive. In the case of the gustatory system, one of the evolutionarily oldest sensory systems, the detection and evaluation of chemical candidates for ingestion is a critical function. But how is this accomplished by the nervous system? The first step is the transduction of exposure to a chemical (tastant) into a biological signal in the sensory receptor cells. These graded potentials are then converted into the stream of action potentials that constitute what we call the neural code.

12.3 CONSTRAINTS ON ANALYSES OF SPIKE TIMING IN NEURAL CODE FOR TASTE

For any neural code of a sensory stimulus, there are two aspects of the stimulus–organism interaction that are necessary to know. The first is the essential interval over which information is conveyed. For example, if an organism can identify taste quality within say 200 ms, as has been suggested (Gutierrez et al., 2010; Halpern and Tapper, 1971; Halpern, 1985; MacDonald et al., 2009), then presumably all the critical information *about taste quality* is contained and conveyed within this interval. However, a full evaluation of a taste stimulus may take much longer (e.g., Halpern, 1997) depending on its intensity (weak tastants require more time for identification; Saito et al., 1998) and the degree of similarity to other tastants of the same quality (subtle distinctions among tastants may require more time to accomplish; Scott, 1974). Recently, we have shown that although rats can use the information contained in a single lick to make a decision about the identity of a taste stimulus, learning, motivation, and stimulus concentration can alter the speed with which the behavioral reaction occurs (Weiss and Di Lorenzo, 2012). In fact, studies by Stapleton et al.

(2006, 2007) have shown that in the gustatory cortex of awake rats, electrophysiological responses to a single lick can discriminate among different taste qualities. On the other hand, Katz has argued that stimulus-specific taste responses in the gustatory cortex do not occur until the first 0.2 s of response (Katz et al., 2001). Nevertheless, data showing the rapid identification of taste stimuli at least under some conditions are compelling and underscore the capacity of the taste system to react quickly to a brief input (see Lemon and Katz, 2007). The second aspect of the response that is important to specify is the point of reference. That is, the question of what signals the beginning of a response is critical, whether spike count, spike timing, or any other aspect of the response is studied. In the taste system, most investigators studying anesthetized animals use the point of contact of the tastant with the tongue. In the awake, freely licking rat, the lick may be a natural point of reference (Roussin et al., 2012; Gutierrez et al., 2010). In fact, in many sensory systems, there is a motor act that could serve as the reference point (either the act itself or a burst of sensory activity due to the act). In addition to licking, these include sniffing, whisking, or a saccade, for example.

12.4 ANATOMY OF THE TASTE SYSTEM: A BRIEF SUMMARY

Taste transduction occurs in taste buds located in the oropharyngeal area. There are about 10,000 taste buds in the human tongue, soft palate, and epiglottis (Miller, 1995). Each taste bud contains 50–150 taste receptor cells arranged like the sections of an orange. Taste receptor cells are replaced about every 12 days, necessitating nearly constant reinnervation. At the base of the taste bud are 3–14 sensory nerve endings that receive the transduced signals from the receptor cells and send action potentials to the brain (reviewed in Chaudhari and Roper, 2010). Taste buds located in the anterior two-thirds of the tongue, as well as those in the soft palate, are innervated by two branches of the facial nerve (cranial nerve VII): the chorda tympani (CT) nerve and the greater superficial petrosal (GSP) nerve, respectively. Taste buds in the posterior third of the tongue are innervated by the lingual tonsillar branch of the glossopharyngeal nerve (cranial nerve IX). The superior laryngeal branch of the vagus nerve (cranial nerve X) innervates the taste buds in the larynx, pharynx, and epiglottis.

Gustatory coding becomes more complex and multifaceted at the level of the brainstem. (See Figure 12.1 for a diagram of the anatomy of the central gustatory pathways.) The facial, glossopharyngeal, and vagus nerves that innervate the taste buds in the mouth all send their fibers to the rostral nucleus of the solitary tract (NTS) in an orderly, topographical organization, with some overlap. In the rodent, the NTS sends fibers to the parabrachial nucleus of the pons (PbN) as well as to the intermediate and lateral parvocellular divisions of the medullary reticular formation (Halsell et al., 1996). Ascending (to the PbN) and descending (to the reticular formation) NTS fibers, however, originate from nonoverlapping cell populations. The PbN provides ascending input to the insular cortex via the parvocellular subdivision of the ventral posteromedial nucleus of the thalamus (VPMpc) while a second pathway from the PbN sends information to the lateral hypothalamus, central nucleus of the amygdala, bed nucleus of the stria terminalis, and the substantia innominata

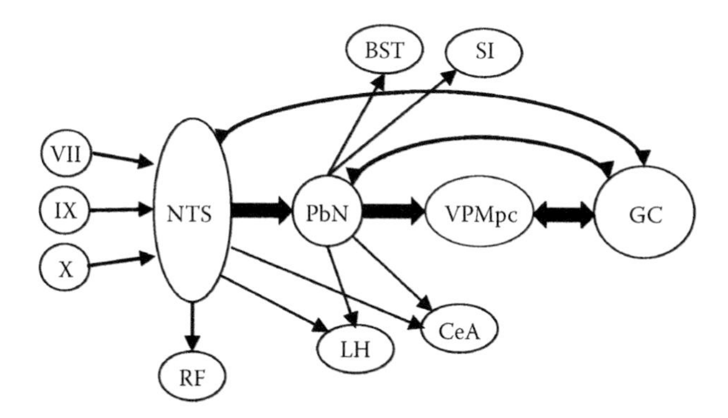

FIGURE 12.1 Diagram of the anatomical structures and pathways of the gustatory system in the rodent. VII, facial nerve; IX, glossopharyngeal nerve; X, vagus nerve; NTS, nucleus tractus solitarius; PbN, parabrachial nucleus of the pons; VPMpc, parvicellular ventroposteromedial nucleus of the thalamus; GC, gustatory cortex; BST, bed nucleus of the stria terminalis; SI, substantia inominata; LH, lateral hypothalamus; CeA, central nucleus of the amygdala; RF, reticular formation.

(reviewed in Lundy and Norgren, 2004). The two pathways from the PbN, that is, the dorsal pathway ending in the cortex and the ventral pathway to the limbic system, are thought to mediate different taste functions. The dorsal path coordinates taste discrimination while the ventral path supports hedonic evaluation of taste (pleasure or disgust). In humans and Old World monkeys, the central gustatory pathway does not have a synapse in the PbN; instead the NTS projects directly to the thalamus (Beckstead et al., 1980) and there is no known distinction between the dorsal and ventral pathways (see Scott and Small, 2009).

12.5 SPATIAL MODELS OF PERIPHERAL GUSTATORY CODING

Historically, there have been two primary hypotheses related to gustatory coding in the periphery that have dominated the literature: the labeled line (LL) and the across-fiber pattern (AFP) theories. Both of these are spatial codes that use spike count as their starting point. Although there is some variability in the actual response interval that is analyzed (usually varying between 3 and 5 s), there is no other consideration of time.

For the LL theory, the central hypothesis is that there are groups or types of nerve fibers that convey information exclusively about a single taste quality. In a typical experiment, one or more exemplars of the four or five* basic taste qualities are bathed over the tongue and the electrophysiological responses are recorded. Prototypical stimuli representing the basic taste qualities are NaCl for salty, HCl or citric acid for sour, sucrose for sweet, quinine HCl for bitter, and monosodium glutamate (MSG) for umami. Peripheral nerve fibers are known to respond to tastants of more than

* It has only been a last decade since many experimenters have included umami as a basic taste. Prior to that, only sweet, sour, salty, and bitter stimuli were tested in the typical taste experiment. Some experimenters continue that practice even now.

one taste quality, that is, they are broadly tuned; however, the prototypical taste stimulus that evokes the largest response, the so-called "best stimulus," predicts the relative response magnitude evoked by the remaining tastants representing the other taste qualities (see Frank, 1973). For example, if a neuron was classified as NaCl best, the stimulus that evoked the next most vigorous response would predictably be HCl, followed by quinine and sucrose; a sucrose best neuron's next best stimulus would be NaCl, and so on. In the last decade, the LL theory has garnered a good deal of support studies using genetic techniques. In general, the conclusion from these investigations is that each taste quality can be associated with a dedicated receptor molecule and that genetic deletion of this molecule selectively eliminates behavioral reactivity* to that taste quality (Chandrashekar et al., 2010; Huang et al, 2006; Mueller et al., 2005; Zhao et al., 2003). On the basis of these observations, the contention is that separate information channels connect peripheral receptors to the central nervous system.

In contrast, the AFP theory emphasizes the nearly ubiquitous observation that taste receptor cells and peripheral nerve fibers are broadly tuned across taste qualities. If a cell responds to more than one taste quality, any response in that cell to a particular tastant would necessarily convey an ambiguous message. The AFP theory solves this conundrum by proposing that the code for any given tastant is the pattern of neural activity distributed across the population of taste-responsive cells. With that conceptualization, the patterns of activity generated by stimuli that taste alike should be more similar (highly correlated) to each other than to patterns generated by stimuli that taste very different. Such is the case (see Doetsch and Erickson, 1970, for an example of early work and Smith et al., 2000, for a more recent review).

12.6 TEMPORAL CODING OF TASTE IN THE PERIPHERY

Temporal coding at the level of the peripheral nerves has, until recently, been restricted to analyses of the rate envelope, often called the time course, of the response. Electrophysiological responses to taste in peripheral nerves are most often described as showing an initial phasic burst of firing followed by a more prolonged tonic and elevated firing rate. This general feature of the rate envelope has also been noted in recordings from the NTS (Nakamura and Norgren, 1991, 1993) and PbN (Nishijo and Norgren, 1990, 1991, 1997) in awake rats under conditions of passive infusion of taste stimuli into the mouth and in the PbN when the animal is free to acquire tastes by licking (Nishijo and Norgren, 1991). Several early reports noted that stimuli of different taste qualities evoked reliably different time courses (rate envelopes) of response (Covey, 1980; Di Lorenzo and Schwartzbaum, 1982; Fishman, 1957; Funakoshi and Ninomiya, 1977; Nagai and Ueda, 1981; Ogawa et al., 1973; Ogawa et al., 1974; Perrotto and Scott, 1976). More recent work by Lawhern et al. (2011) has examined the contribution of both spike count and spike timing to the information conveyed by single neurons in the geniculate ganglion.

* The phrase "behavioral reactivity" is used here to mean the relative intake of a taste stimulus indicative of the animal's preference or aversion to it, and by inference, its detection.

Spike timing was analyzed using the van Rossum metric (van Rossum, 2001).* By determining mutual information contained in the responses, they concluded that the combination of spike rate and spike timing provides more information about taste quality than spike count alone. This is the first study to apply an information-theoretic approach to the analysis of data from the peripheral gustatory system.

12.7 SPIKE TIMING IN THE BRAINSTEM

Gustatory coding becomes more complex and multifaceted at the level of the brainstem. There is evidence of the convergence of peripheral nerve fibers in the NTS (e.g., Doetsch and Erickson, 1970) as well as consistent reports of a broadening of tuning at each successive level of the gustatory neuraxis (see Spector and Travers, 2005). Collectively, these data suggest that spatial codes may not be sufficient to account for the ability to discriminate among taste qualities, intensity (concentration), and mixtures. For example, in cells that respond to more than one taste quality (the majority of cells), there are concentrations at which stimuli of different taste qualities will evoke similar firing rates. In that case, it is obvious that spike count would convey an ambiguous message about stimulus identity. Similarly, in the case of taste mixtures, information from spike count alone would be imprecise in all but the most narrowly tuned cells.

In a series of investigations of taste responses in the NTS of anesthetized rats, we applied metric space analyses (Victor and Purpura, 1996, 1997)† to the spike trains evoked by a variety of taste stimuli. Briefly, this method quantifies the information conveyed by spike timing, by the rate envelope of the response, and by spike count alone. We began our studies by analyzing responses to repeated presentations of exemplars of four of the five basic taste qualities: sweet, sour, salty, and bitter. Results showed that in about half of the taste-responsive NTS cells, spike timing conveyed

---

* The van Rossum metric can be used to quantify the similarity or dissimilarity of two spike trains (van Rossum, 2001). Briefly, each spike train, consisting of a series of discrete time points, is first transformed into a continuous function by adding a "tail" to every spike, reminiscent of a postsynaptic potential. Next, two spike trains are compared by calculating the area of the difference between the corresponding continuous functions. By varying the length of the tail, one can vary the temporal precision at which the interspike train distance is measured.

† In metric space analysis (Victor and Purpura, 1996, 1997), the similarity of two spike trains are gauged by the "cost" of adding or deleting spikes as necessary, and by moving spikes in time, so that the two spike trains become identical. The cost of adding or deleting a spike is an arbitrary value of "1." If the number of spikes in each spike train is made equivalent, they result in a metric, called "D^{count}," that is the arithmetic difference in the number of spikes between the two spike trains. If spikes are moved in time, the cost will depend on the level of temporal precision, called "q." In general, moving a spike in time costs "tq," where q is in units of 1/s and t is the amount of time that the spike is moved. The resulting metric, called "$D^{spike}[q]$," is calculated at several levels of q. Note that when $q = 0$, D^{spike} is the same as D^{count}. That is, the timing of spikes in the response conveys no information beyond what the number of spikes conveys. When applied to taste-evoked spike trains, responses to the various taste stimuli are compared using $D^{spike}[q]$ as a measure of dissimilarity. High values of D^{spike} imply that the two spike trains are very different in terms of both spike count and spike timing and low values imply the opposite. A similar analysis can be applied to interspike intervals, resulting in $D^{interval}$. From the resulting confusion matrix, the information conveyed by spike timing and spike count can be calculated. Generally, the maximum amount of information, H_{max}, conveyed by spike timing and/or spike count is compared to the information conveyed by spike count alone, H_{count}.

significantly more information than either the rate envelope of the response or the spike count alone (Di Lorenzo and Victor, 2003, 2007). Interestingly, we reported that those cells with the most highly variable responses across stimulus presentation replications were those that were most likely to convey information with spike timing. In subsequent studies, we confirmed that spike timing could also be used to convey information that might be used to discriminate between tastants of the same quality but different chemical composition (Roussin et al., 2008) and among different concentrations of the same taste stimulus (Chen et al., 2011). Moreover, we showed that spike timing was also informative about taste quality when tastants of different qualities were presented as a mixture (Di Lorenzo et al., 2009a).

There were several overarching conclusions that resulted from the series of studies just described. First, it was apparent that the contribution of spike timing to the information necessary to discriminate among tastants along any of several dimensions was most prominent when stimulus-evoked spike count provided an ambiguous message. That is, when two taste stimuli evoked similar firing rates, the temporal characteristics of the response could still be used to distinguish between them. Figure 12.2 shows an example of this point in one cell. Here, we show the results of a multidimensional scaling (MDS) analysis where the temporal characteristics of the responses were used as a measure of similarity. Responses to each of the 10 taste stimuli (four basic taste qualities and six binary mixtures) were placed in the response space such that tastants that evoked responses that were similar both in spike count and in the temporal characteristics of the response were placed close together and those that were dissimilar were paced far apart. This figure shows that when only spike count is used to assess similarity among taste responses (Figure 12.2b), there is a good deal of overlap, indicating ambiguity. In contrast, responses to the various taste stimuli in the response space using the temporal characteristics of the responses as a measure of similarity (Figure 12.2a) are separated from each other, suggesting discriminability. Second, and perhaps not surprisingly, the most broadly tuned taste-responsive cells tended to convey information predominantly through the temporal aspects of their response while narrowly tuned cells conveyed information predominantly through spike count. Third, the temporal pattern of the taste-evoked spike train for a given taste quality was consistent across cells even when the stimulus was presented in varying concentrations in a mixture with a stimulus of a different taste quality (see Figure 12.3). To show this, we applied principal component analyses to the rate envelopes of the responses. The results revealed temporal pattern "signatures" associated with each of the four basic taste qualities. Finally, the results of Chen et al. (2011) suggested that the information from spike timing increased as successively longer response intervals were considered. These results are consistent with the recent study of taste responses in the NTS of awake, freely licking rats (Roussin et al., 2012).

Spike timing was also informative in the PbN of anesthetized rats (Rosen et al. 2011). Comparison of data from the PbN with previous data from the NTS (Di Lorenzo and Victor, 2003, 2007) showed that trial-to-trial variability was higher in the PbN than in the NTS; however, temporal coding in both nuclei occurred in a similar proportion of cells and contributed a similar fraction of the total information.

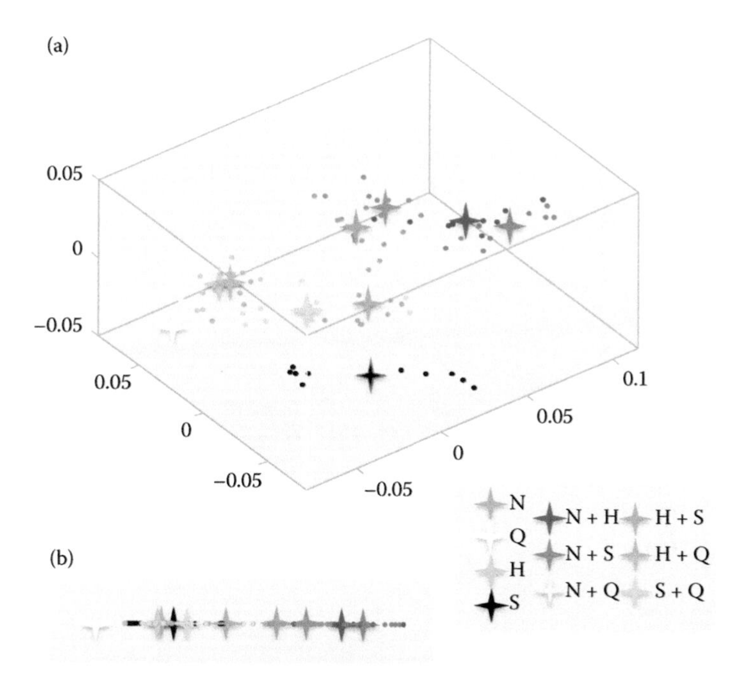

FIGURE 12.2 **(See color insert.)** Multidimensional scaling (MDS) using differences in spike timing as a measure of similarity among responses. Dots indicate the location of individual responses and asterisks indicate the centroid of the clusters of responses to a given taste stimulus. Axes are labeled in arbitrary units. (a) The three-dimensional response space created by MDS of the spike time distances $D^{spike}[q_{max}]$, that is, the difference in spike timing calculated at the level of temporal precision (q) that conveys the maximum amount of information. For this cell, $q_{max} = 2.83$. (b) The one-dimensional response space created by MDS of the spike count distances, D_{count}. N, 0.1 M NaCl; Q, 0.01 M quinine HCl; H, 0.01 M HCl; S, 0.5 M sucrose; N + H, 0.1 M NaCl plus 0.01 M HCl; N + S, 0.1 M NaCl plus 0.5 M sucrose; N + Q, 0.1 M NaCl plus 0.01 M quinine HCl; H + S, 0.01 M HCl plus 0.5 M sucrose; H + Q, 0.01 M HCl plus 0.01 M quinine HCl; S + Q, 0.5 M sucrose plus 0.01 quinine HCl. (Reprinted from Di Lorenzo PM, Chen JY, and Victor JD. *J Neurosci* 29(29): 9227–38, 2009a.)

Moreover, information from spike timing peaked at the same average level of temporal precision. These data suggest that information about taste quality conveyed by the temporal characteristics of evoked responses is transmitted with high fidelity from the NTS to the PbN. In fact, in a study of pairs of simultaneously recorded NTS–PbN cells (Di Lorenzo and Monroe, 1997; Di Lorenzo et al., 2009c), it was shown that taste-evoked spikes in the PbN follow those in the NTS in a roughly rhythmic fashion for the first ~3 s of the response, most predominantly for sucrose (see Figure 12.4). Taste-evoked activity in the PbN gradually becomes more independent from NTS activity as the response develops over time. Even though this rhythm was several times slower than a lick rhythm (~1.5 Hz vs. licks at 6–7 Hz), these observations are consistent with the idea that sensory transmission is pulsatile in nature (see Schroeder et al., 2010). In addition, it is interesting to note that the

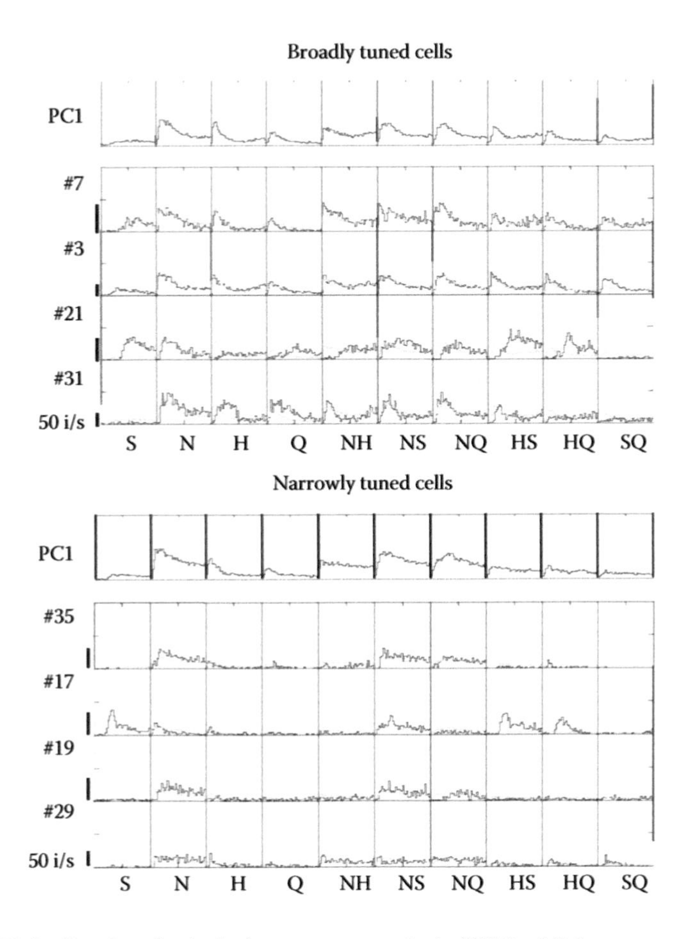

FIGURE 12.3 Results of principal component analysis (PCA) of firing rate envelope for broadly and narrowly tuned cells. Separate analyses were applied to the more broadly tuned (top) and the more narrowly tuned (bottom) cells. Spike rates were normalized according to overall firing rates within each cell. For each group of cells, the top row shows the first principal component for all tastants. The additional rows of plots show the predicted rate envelope (in gray) superimposed on the actual rate envelope (in black) for all stimuli in several individual cells. The line on the bottom left indicates 2 s. (Reprinted from Di Lorenzo PM, Chen JY, and Victor JD. *J Neurosci* 29(29): 9227–38, 2009a.)

typical optimal precision of temporal encoding in the NTS, that is, 150–500 ms (Di Lorenzo et al., 2009a), is in close correspondence to the half-period of this ~1.5 Hz rhythm.

In the awake freely licking rat, taste-responsive cells in the NTS (Roussin et al., 2012) and PbN (unpublished results) also convey information about taste quality using spike timing. NTS and PbN cells were recorded in moderately water-deprived rats as they licked various taste stimuli from a spout. The spout could deliver ~12 μL of fluid (tastant or water rinse) whenever the rat licked. Since licking is by nature episodic, it is not surprising that most taste responses followed the lick pattern and

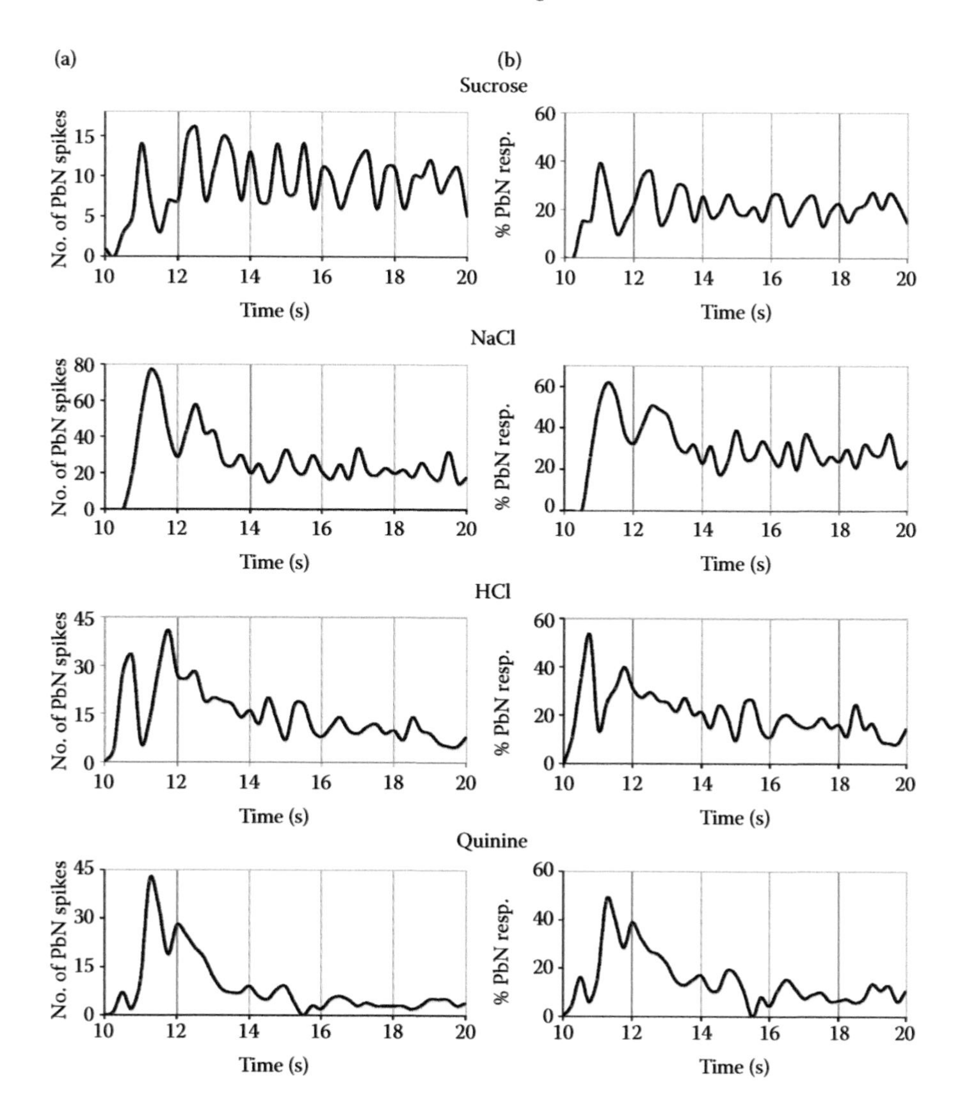

FIGURE 12.4 Time course of NTS input to the PbN. (a) Plot of the average number of PbN spikes that were preceded within 3 ms by an NTS spike in pairs of NTS and PbN neurons that were recorded simultaneously. For each stimulus, $n = 10$. (b) Plot of the proportion of the total number of PbN spikes that were preceded within 3 ms by an NTS spike over time in pairs of NTS and PbN neurons that were recorded simultaneously. (Reprinted from Di Lorenzo PM, Platt D, and Victor JD. *Ann NY Acad Sci* 1170: 365–71, 2009c.)

thus the activity pattern appeared pulsatile. Many cells showed a gradual increase in taste-evoked activity over several licks, presumably as liquid filled the mouth. Overall, in the NTS, the information conveyed by spike timing and spike count increased as the response interval considered increased, but the information from spike timing increased at a faster rate than that from spike count alone. However, the total amount of information conveyed by single cells was much less than that conveyed by cells in anesthetized rats. By itself, this result suggests that cooperative activity across ensembles of NTS cells may be essential to identify and discriminate among taste stimuli, given that NTS cells in the awake rat are almost all broadly tuned.

To add to the complexity of taste coding in the brainstem, we have found that taste-responsive cells in the awake rat are found alongside lick-related cells and cells that stop firing during licking. Lick-related cells were found in the taste-responsive portion of both the NTS (Roussin et al., 2012) and the PbN (unpublished results). Some of these showed differential firing rates to the various taste stimuli, in addition to closely following the lick pattern. Thus, there may be a continuum of cell/response types in the taste-related nuclei in the brainstem, extending from purely taste-related to purely lick-related activity, with most cells showing some measure of both types of activity. The presence of strongly driven lick-related spike activity may buttress the taste-related activity coming from the periphery.* Similar lick-related cells have also been reported in several other structures in the gustatory pathway (Gutierrez et al., 2010; Stapleton et al., 2006; Chapter 13 of this book) and these may serve the same function there. A third cell type that is present in the NTS stops firing when the animal is licking. We called these cells "antilick" cells. Most interestingly, these cells show a surge in firing rate just before a lick is about to begin and just after it ends. This property suggests that these cells provide the initial conditions for the network of taste cells to acquire sensory information.

While reciprocal activation of antilick and taste-responsive cells suggests some degree of connectivity, it is likely that antilick cells are driven by input from more central structures. Krause et al. (2010) have described a subpopulation of cells in the nucleus accumbens, called type 1 cells, which show similar firing characteristics to the antilick cells in the NTS, that is, they are silent during licking but active otherwise. Like the antilick cells described here, these cells in the accumbens show a surge just before and just after the lick bout. Such a surge is reminiscent of a postinhibitory rebound but the circuitry that would underlie this effect has not been described. Krause et al. (2010) proposed that these type 1 cells actively inhibit feeding and suggested that this may occur through an interaction with the lateral hypothalamus. Since the lateral hypothalamus is known to send projections to the NTS (Lundy and Norgren, 2004) and to modulate the activity in taste-responsive neurons in the NTS (Cho et al., 2002, 2003; Matsuo et al., 1984), these may, in part, modulate activity in antilick cells. Alternatively, modulation of antilick cells may arise from other structures or even from within the NTS through interneurons. In any case, the transition of the antilick cells to quiescence may signal the initiation of a sensory acquisition mode wherein NTS would actively process the taste stimuli as the animal licks. The reciprocal activation of antilick and taste-responsive cells directly marries taste processing in the NTS to behaviors that result in taste perception.

* Parenthetically, it should be noted that we identified lick-related cells because our paradigm involved licking when no fluid was delivered, called "dry licks." In nature, it is unlikely that an animal would lick without stimulating the taste system in some way. The signals from lick-related cells would thus be available to reinforce the message from incoming taste-driven fibers/cells.

12.8 MECHANISMS FOR MODULATING SPIKE TIMING IN THE GUSTATORY BRAINSTEM

Although there is now ample evidence to support the idea that information conveyed by spike timing contributes to the representation of taste quality in the brainstem, the question is of how and whether the neural circuitry could support the generation of precisely timed sequences of spikes. Our experiments in this area were based on the idea that the interplay of inhibition and excitation might offer a mechanism that could modulate the timing of taste-evoked spike trains. While it is known that the inputs to the NTS from the periphery are all glutaminergic, and thus exclusively excitatory (Li and Smith, 1997; Smith and Li, 1998), once the taste signal reaches the NTS, there are extensive intranuclear connections that serve as an abundant source of gamma-amino-butyric acid-ergic (GABAergic) inhibition (Lasiter and Kachele, 1988). Additionally, feedback connections from the forebrain also provide inhibitory input to the NTS (Smith et al., 1998). Indeed, about two-thirds of the taste-responsive cells in the NTS receive some form of inhibitory influence (King, 2003; Smith and Li, 1998; Wang and Bradley, 1993) and inhibition is known to affect the breadth of tuning (Lemon and Di Lorenzo, 2002; Rosen and Di Lorenzo, 2009; Smith and Li, 1998). Since the breadth of tuning appears to predict the importance of spike timing in information processing about taste, a further examination of the role of inhibition in the NTS seemed like a fruitful strategy to uncover the neural circuitry that might underlie reliable taste-evoked spike timing.

Our first set of experiments was aimed at confirming a role for inhibition in modulating taste responses in general. These were based on a report by Grabauskas and Bradley (1999) showing that tetanic stimulation of NTS cells *in vitro* results in the potentiation of inhibition. To explore the possible effects of this phenomenon on actual taste responses in the intact animal, we delivered tetanic stimulation to the CT nerve in anesthetized rats just prior to bathing the tongue with various taste stimuli (Lemon and Di Lorenzo, 2002). Results showed that many taste responses were indeed altered following CT stimulation. Perhaps, the most interesting result with respect to spike timing was that tetanic trains preceding some taste stimuli in a subset of cells disrupted the temporal organization of the taste-evoked spike trains by increasing the variability of interspike intervals (ISIs) in the initial portion of the response. To demonstrate this effect, we calculated the standard deviation (SD) of a sliding window of 10 sequential ISIs across the first 300 ISIs. Our reasoning was that, when a cell fires at regular intervals, the SD of its ISIs across time is small, but if this regularity is disrupted, the SD will increase. In the affected cells, tetanic stimulation increased the SDs in the first 2 s of the response without changing the overall firing rate of the cell for the same interval (see Figure 12.5). One possible explanation for this result may be that the tetanic stimulation may have produced a biphasic inhibitory post synaptic potential (IPSP), that is, an IPSP followed by a burst of spikes, similar to that observed in NTS cells *in vitro* (Grabauskas and Bradley 1998). Following tetanic stimulation of the CT, then, these postinhibitory bursts of spikes may have intruded into the orderly progression of the taste-evoked spike train. Alternatively, by inducing long-lasting inhibitory potentials, tetanic stimulation may have disrupted the ability of inhibitory interneurons to regulate the timing of response-related firing patterns.

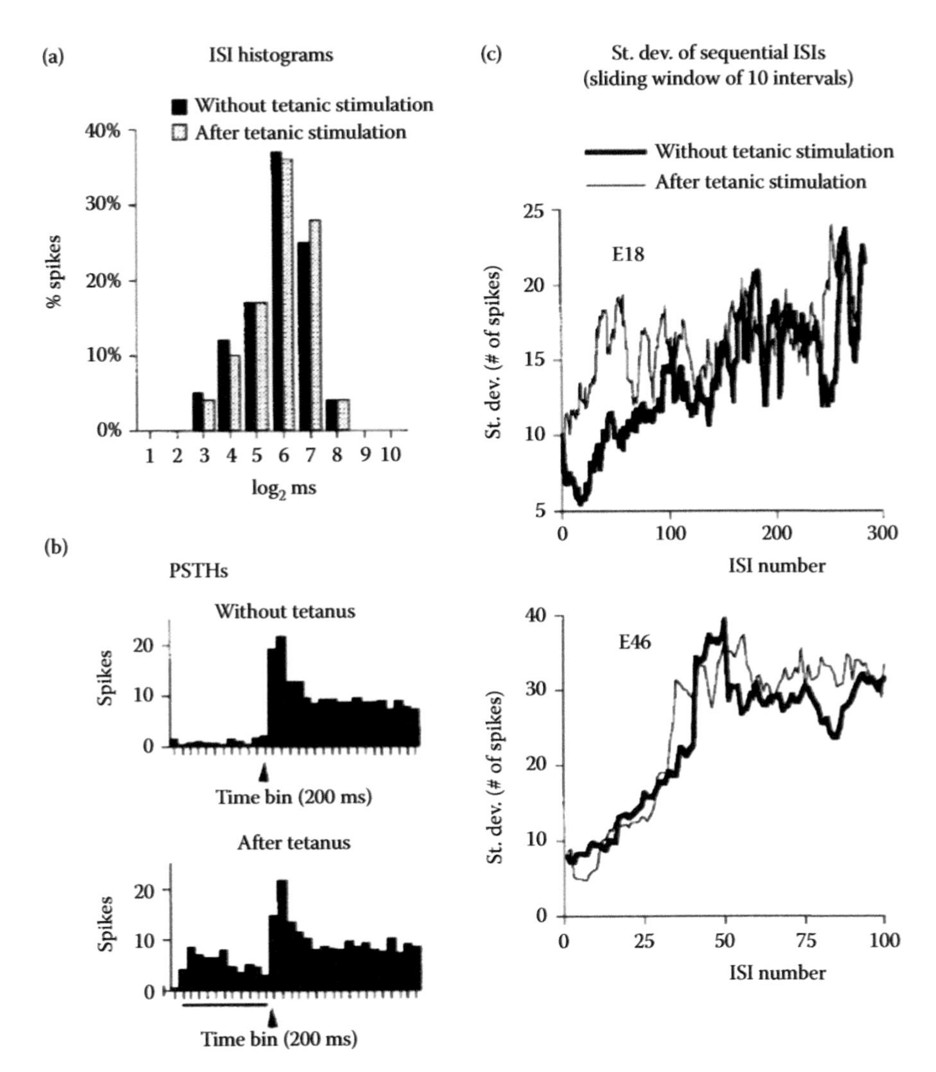

FIGURE 12.5 Tetanic stimulation of the CT nerve changes the temporal organization of spikes in the initial segment of a taste response. (a) Interspike interval histogram showing that the distribution of interspike intervals in unit #E18 during NaCl responses without ($n = 4$) and following tetanic stimulation ($n = 4$) of the CT nerve are nearly identical. (b) Peristimulus-time histograms (PSTHs) of a single NaCl response without (top) and after (bottom) tetanic CT stimulation in unit #E18. The bar below the histograms shows the presentation of tetanus. The arrow indicates the onset of a 5 s taste stimulus. Tetanic stimulation of the CT nerve had no effect on the PSTH. (c) Plot of the standard deviation of a sliding window of 10 sequential interspike intervals during the response to NaCl without and following the tetanic stimulation of the CT in unit #E18 (top, $n = 4$ without and $n = 3$ following tetanus) and #E46 (bottom, $n = 4$ without and $n = 3$ following tetanus). Tetanic stimulation of the CT nerve increased the variability in the interspike intervals in the first ~2 s of the response in unit #E18 but not in unit #E46. (Reprinted from Lemon CH and Di Lorenzo PM. *J Neurophysiol* 88: 2477–89, 2002.)

To further explore the role of inhibition in regulating spike timing in the NTS, our next set of experiments also employed electrical stimulation of the CT nerve. In these experiments (Rosen and Di Lorenzo, 2009; Rosen et al., 2010), taste-responsive NTS cells with CT-evoked responses were tested with paired pulses (10–2000 ms interpulse interval; blocks of 100 trials) delivered to the CT nerve. Results showed that CT stimulation produced two types of inhibitory effects on NTS cells, each with its own time course. One type peaked at ~10 ms and decayed rapidly (by ~100 ms) and a second type peaked at ~50 ms and decayed more slowly (by ~500 ms) (Rosen and Di Lorenzo 2009; Rosen et al., 2010). Cells that showed a short-time course of inhibition were either narrowly (Rosen and Di Lorenzo, 2009) or broadly (Rosen et al., 2010) tuned across taste qualities and had a short latency of CT-evoked response while cells that showed a longer time course of inhibition were generally more broadly tuned and showed longer latency CT-evoked responses. Considering these results, we hypothesized that the longer time course of inhibition to CT stimulation could refine and stabilize the temporal patterns of response by filtering high-frequency, noisy signals.

To demonstrate how NTS cells might show sensitivity to the temporal pattern of input, we presented triads of CT stimulation and recorded the responses in taste-responsive NTS cells in anesthetized rats (Rosen and Di Lorenzo, unpublished results). The first and third pulses of each triad were always 100 ms apart, but the second pulse followed the first at 25, 50, or 75 ms. The experimental design and method of analysis for the triad stimulation were based on those described in a similar investigation of neurons in *Aplysia* by Segundo et al. (1963). Our reasoning was that each triad of electrical pulses could be thought of as a temporal "word" in the sense that it represented an elementary temporal pattern that was one step beyond a pair of pulses. To summarize the results, the great majority of NTS cells with CT-evoked responses showed sensitivity to the temporal pattern of pulses delivered within a 100 ms interval. That is, they showed different responses to the triads based on the timing of the second pulse, and typically favored only one triad pattern. In most cases, responses could be predicted by the time course of inhibitory influences as revealed by the effects of paired pulses. Thus, the upshot of this experiment was that most NTS cells filter CT input by their particular sensitivity to the fine-grain temporal arrangement of incoming pulses.

12.9 MECHANISMS FOR FUNCTIONALITY OF SPIKE TIMING: CAN THE ANIMAL USE THE INFORMATION?

In the final analysis, studies of electrophysiological responses to taste stimuli in either anesthetized or awake behaving animals can only produce correlational evidence that temporal coding plays a role in taste processing. That is, because electrophysiological responses obviously consist of a combination of changes in firing rate, spike timing, and coherent firing among ensembles of neurons, it is difficult to assess the contribution of any single mechanism or to demonstrate its sufficiency. One strategy has been to correlate distinctive temporal patterns of sensory responses with behaviors evoked by specific sensory stimuli. For example, in their study of the *Manduca sexta* caterpillar, Glendinning et al. (2006) showed that two stimuli that evoked responses in the primary taste cells that differed only in whether the firing rate accelerated or decelerated over time-evoked behaviors that were completely different. Furthermore, the pattern of generalization of habituation to a particular stimulus could be predicted based solely on the temporal pattern of the evoked spike train.

Another strategy is based on the idea of replaying stimulus-specific temporal patterns of neural responses with electrical stimulation and examining the evoked behavior. Perhaps, the first use of this tack was an experiment by Covey (1980). She first recorded taste responses from the rat CT nerve and then used the temporal patterns of those responses to drive the activity in the CT of decerebrate rats. Remarkably, each of the temporal patterns of electrical stimulation associated with the various taste stimuli produced appropriate orofacial behaviors, suggesting that different patterns of electrical stimulation were perceived as different taste stimuli.

Building on the basic idea of Covey's work, our previous work has shown that taste-like sensations of an identifiable quality were evoked when the temporal patterns of NTS responses to either sucrose or quinine were played back (as electrical pulse trains) into the NTS in awake rats (Di Lorenzo and Hecht, 1993; Di Lorenzo et al., 2003a, 2009b). Specifically, rats with electrodes implanted in the rostrocentral NTS learned to avoid licking water when licking produced an electrical pulse train with the temporal characteristics of a sucrose or quinine (Di Lorenzo et al., 2003a, 2009b) response in a conditioned aversion paradigm. That is, lick-contingent electrical stimulation sessions were followed by injections of LiCl, an emetic agent. Trained animals actually generalized the conditioned aversions *specifically* to real sucrose or quinine, respectively. Figure 12.6 shows these results for sucrose. These data suggest that the patterned electrical stimulation can evoke a specific gustatory precept, but they do not provide evidence that the

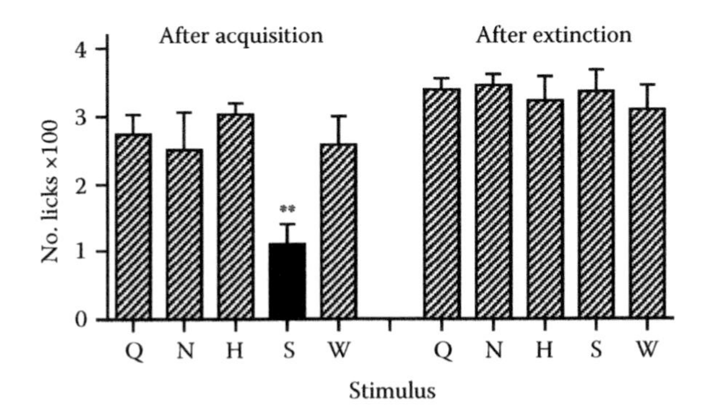

FIGURE 12.6 Mean (±SEM) number of licks to 1-min presentations of natural taste stimuli following acquisition of a conditioned aversion to the sucrose simulation pattern of electrical stimulation and following extinction ($n = 7$). Q, 0.025 mM quinine; N, 25 mM NaCl; H, 2.5 mM HCl; S, 125 mM sucrose; W, water. Asterisks denote a statistically significant difference ($p < 0.01$) between the number of licks for sucrose after acquisition and the number of licks for sucrose after extinction in the experimental group. (Reprinted from Di Lorenzo PM, Lemon CH, and Reich CG. *J Neurosci* 23: 8893–902, 2003b.)

temporal pattern per se is critical to this effect. To study this issue, we took advantage of the fact that rats will avoid quinine without training. We first showed that lick-contingent patterns of electrical pulses based on quinine responses predictably produced significantly lower lick rates than plain water (Di Lorenzo et al., 2003a, 2009b). The critical control patterns of pulses were constructed by randomly reordering the interpulse intervals in the original effective pattern. These randomized control patterns had the same complement of interpulse intervals and the same number of pulses as the pulse patterns that mimicked an actual quinine response. These patterns produced lick rates not different from plain water presented without electrical stimulation (Di Lorenzo and Hecht, 1993; Di Lorenzo et al., 2003a). Collectively, these data imply that the information about taste conveyed by temporal coding in NTS responses (Di Lorenzo and Victor, 2003, 2007; Roussin et al., 2008) is indeed functionally relevant. These results also emphasize that animals make substantial use of intrinsic firing patterns in the generation of taste-related behavior.

12.10 CONCLUSIONS AND FUTURE DIRECTIONS

Neurons are clearly capable of conveying information about sensory stimuli in many ways. In the gustatory system, and especially in the brainstem, studies of the extent to which the timing of spikes contributes to this process have begun to yield fruit. For example, spike timing of taste-evoked responses in the brainstem in both anesthetized and awake rodents has been shown to contribute a significant amount of information about taste quality, intensity, and taste mixtures above and beyond that contributed by spike count alone. Further, the neural circuitry that may produce precisely timed spike trains has begun to be fleshed out. Perhaps most importantly, the demonstration that the temporal patterns of brainstem taste responses can evoke identifiable taste sensations when used to drive neural activity in awake animals speaks directly to the heart of the question of whether these seemingly idiosyncratic temporal patterns of neural activity can be interpreted by the animal. In sum, evidence for the presence, mechanism, and function of temporal coding in the brainstem gustatory system has been presented.

So, where do we go from here? There are several issues that remain unanswered in the field of temporal coding in gustation. One that seems particularly pressing is the question of how temporal patterns of spikes are transmitted and/or transformed as they are relayed to other neural structures. Do temporal patterns in one structure resemble those in upstream nuclei? If so, what neural circuitry and mechanisms support that? Another important issue is the role of synchrony among neurons as another expression of temporal coding. This is beginning to be addressed in studies of the gustatory cortex (Katz et al., 2002; Stapleton et al., 2007); hopefully in the near future, similar data will be available in the brainstem. In fact, this mechanism may be particularly important in awake animals that are actively seeking sensation, as our recent data in the NTS (Roussin et al., 2012) have suggested. Finally, there is much left to discover about how neurons interpret the timing of their input and how that interpretation results in perception. The data presented here only provide the slightest glimpse of how that might happen.

REFERENCES

Beckstead RM, Morse JR, and Norgren R. The nucleus of the solitary tract in the monkey: Projections to the thalamus and brain stem nuclei. *J Comp Neurol* 190: 259–82, 1980.

Chandrashekar J, Kuhn C, Oka Y, Yarmolinsky DA, Hummler E, Ryba NJ, and Zuker CS. The cells and peripheral representation of sodium taste in mice. *Nature* 464(7286): 297–301, 2010.

Chaudhari N and Roper SD. The cell biology of taste. *J Cell Biol* 190(3): 285–96, 2010.

Chen J-Y, Victor JD, and Di Lorenzo PM. Temporal coding of intensity of NaCl and HCl in the nucleus of the solitary tract of the rat. *J Neurophysiol* 105(2): 697–711, 2011.

Cho YK, Li CS, and Smith DV. Taste responses of neurons of the hamster solitary nucleus are enhanced by lateral hypothalamic stimulation. *J Neurophysiol* 87(4): 1981–92, 2002.

Cho YK, Li CS, and Smith DV. Descending influences from the lateral hypothalamus and amygdala converge onto medullary taste neurons. *Chem Senses* 28(2): 155–71, 2003.

Covey E. *Temporal Coding in Gustation*. Unpublished doctoral dissertation. Durham, NC: Duke University, 1980.

Di Lorenzo PM and Hecht GS. Perceptual consequences of electrical stimulation in the gustatory system. *Behav Neurosci* 107: 130–8, 1993.

Di Lorenzo PM and Lemon CH. Methodological considerations for electrophysiological recording and analysis of taste-responsive cells in the brain stem of the rat. In: *Methods in Chemosensory Research*, Simon SA and Nicolelis MAL, eds., New York: CRC Press, pp. 293–324, 2001.

Di Lorenzo PM, Chen JY, and Victor JD. Quality time: Representation of a multidimensional sensory domain through temporal coding. *J Neurosci* 29(29): 9227–38, 2009a.

Di Lorenzo PM, Hallock RM, and Kennedy DP. Temporal coding of sensation: Mimicking taste quality with electrical stimulation of the brain. *Behav Neurosci* 117(6): 1423–33, 2003a.

Di Lorenzo PM, Lemon CH, and Reich CG. Dynamic coding of taste stimuli in the brain stem: Effects of brief pulses of taste stimuli on subsequent taste responses. *J Neurosci* 23: 8893–902, 2003b.

Di Lorenzo PM, Leshchinskiy S, Moroney DN, and Ozdoba JM. Making time count: Functional evidence for temporal coding of taste sensation. *Behav Neurosci* 123(1): 14–25, 2009b.

Di Lorenzo PM and Monroe S. Transfer of information about taste from the nucleus of the solitary tract to the parabrachial nucleus of the pons. *Brain Res* 763: 167–81, 1997.

Di Lorenzo PM, Platt D, and Victor JD. Information processing in the parabrachial nucleus of the pons: Temporal relationships of input and output. *Ann NY Acad Sci* 1170: 365–71, 2009c.

Di Lorenzo PM and Schwartzbaum JS. Coding of gustatory information in the pontine parabrachial nuclei of the rabbit: Magnitude of neural response. *Brain Res* 251: 229–44, 1982.

Di Lorenzo PM and Victor JD. Taste response variability and temporal coding in the nucleus of the solitary tract of the rat. *J Neurophysiol* 90: 1418–31, 2003.

Di Lorenzo PM and Victor JD. Neural coding mechanisms for flow rate in taste-responsive cells in the nucleus of the solitary tract of the rat. *J Neurophysiol* 97(2): 1857–61, 2007.

Doetsch GS and Erickson RP. Synaptic processing of taste-quality information in the nucleus tractus solitarius of the rat. *J Neurophysiol* 33(4): 490–507, 1970.

Fishman IY. Single fiber gustatory impulses in rat and hamster. *J Cell Comp Physiol* 49: 319–34, 1957.

Frank M. An analysis of hamster afferent taste nerve response functions. *J Gen Physiol* 61(5): 588–618, 1973.

Funakoshi M and Ninomiya Y. Neural code for taste quality in the thalamus of the dog. In *Food Intake and the Chemical Senses*, Katsuki Y, Sato M, Takagi SF, and Oomura Y, eds., Tokyo: University Park Press, pp. 223–32, 1977.

Glendinning JI, Davis A, and Rai M. Temporal coding mediates discrimination of "bitter" taste stimuli by an insect. *J Neurosci* 26(35): 8900–8, 2006.

Grabauskas G and Bradley RM. Ionic mechanism of GABAA biphasic synaptic potentials in gustatory nucleus of the solitary tract. *Ann NY Acad Sci* 855: 486–7, 1998.

Grabauskas G and Bradley RM. Potentiation of GABAergic synaptic transmission in the rostral nucleus of the solitary tract. *Neuroscience* 94(4): 1173–82, 1999.

Gutierrez R, Simon SA, and Nicolelis MA. Licking-induced synchrony in the taste-reward circuit improves cue discrimination during learning. *J Neurosci* 30(1): 287–303, 2010.

Hallock RM and Di Lorenzo PM. Temporal coding in the gustatory system. *Neurosci Biobehav Rev* 30: 1145–60, 2006.

Halpern BP. Time as a factor in gustation: Temporal patterns of stimulation and response. In: *Taste, Olfaction and the Central Nervous System*, Pfaff DW, ed., New York: Rockefeller University Press, pp. 181–209, 1985.

Halpern BP. Psychophysics of taste. In: *Tasting and Smelling. Handbook of Perception and Cognition* (2nd ed.), Beauchamp GK and Bartoshuk L, eds., San Diego, CA: Academic Press, pp. 77–123, 1997.

Halpern BP and Tapper DN. Taste stimuli: Quality coding time. *Science* 171(977): 1256–8, 1971.

Halsell CB, Travers SP, and Travers JB. Ascending and descending projections from the rostral nucleus of the solitary tract originate from separate neuronal populations. *Neuroscience* 72(1): 185–97, 1996.

Huang A, Chen X, Hoon MA, Chandrashekar J, Guo W, Tränkner D, Ryba NJ, and Zuker CS. The cells and logic for mammalian sour taste detection. *Nature* 442: 934–8, 2006.

Katz DB, Simon SA, and Nicolelis MA. Dynamic and multimodal responses of gustatory cortical neurons in awake rats. *J Neurosci* 21(12): 4478–89, 2001.

Katz DB, Simon SA, and Nicolelis MA. Taste-specific neuronal ensembles in the gustatory cortex of awake rats. *J Neurosci* 22(5): 1850–7, 2002.

Kay L. From logical neurons to poetic embodiments of mind: Warren S McCulloch's project in neuroscience. *Sci Context* 14(4): 591–615, 2001.

King MS. Distribution of immunoreactive GABA and glutamate receptors in the gustatory portion of the nucleus of the solitary tract in rat. *Brain Res Bull* 60(3): 241–54, 2003.

Krause M, German PW, Taha SA, and Fields HL. A pause in nucleus accumbens neuron firing is required to initiate and maintain feeding. *J Neurosci* 30(13): 4746–56, 2010.

Lasiter PS and Kachele DL. Organization of GABA and GABA-transaminase containing neurons in the gustatory zone of the nucleus of the solitary tract. *Brain Res Bull* 21: 23–36, 1988.

Lawhern V, Nikonov AA, Wu W, and Contreras RJ. Spike rate and spike timing contributions to coding taste quality information in rat periphery. *Front Integr Neurosci* 5: 18, 2011.

Lemon CH and Katz DB. The neural processing of taste. *BMC Neurosci* 8(Suppl 3): S5, 2007.

Lemon CH and Di Lorenzo PM. Effects of electrical stimulation of the chorda tympani nerve on taste responses in the nucleus of the solitary tract of the rat. *J Neurophysiol* 88: 2477–89, 2002.

Li CS and Smith DV. Glutamate receptor antagonists block gustatory afferent input to the nucleus of the solitary tract. *J Neurophysiol* 77(3): 1514–25, 1997.

Lundy RF and Norgren R. Gustatory system. In *The Rat Nervous System*. Paxinos G, ed. San Diego, CA: Elsevier Academic Press, pp. 891–921, 2004.

MacDonald CJ, Meck WH, Simon SA, and Nicolelis MA. Taste-guided decisions differentially engage neuronal ensembles across gustatory cortices. *J Neurosci* 29(36): 11271–82, 2009.

Matsuo R, Shimizu N, and Kusano K. Lateral hypothalamic modulation of oral sensory afferent activity in nucleus tractus solitarius neurons of rats. *J Neurosci* 4(5): 1201–7, 1984.

Mc Culloch WS and Pitts W. A logical calculus of the ideas immanent in nervous activity. *Bull Math Biophys* 5: 115–33, 1943.
Mueller KL, Hoon MA, Erlenbach I, Chandrashekar J, Zuker CS, and Ryba NJ. The receptors and coding logic for bitter taste. *Nature* 434(7030): 225–9, 2005.
Miller IJ. Anatomy of the peripheral taste system. In *Handbook of Olfaction and Gustation.* Doty RL, ed., New York: Marcel Dekker, pp. 521–47, 1995.
Nagai T and Ueda K. Stochastic properties of gustatory impulse discharges in rat chorda tympani fibers. *J Neurophysiol* 45(3): 574–92, 1981.
Nakamura K and Norgren R. Gustatory responses of neurons in the nucleus of the solitary tract of behaving rats. *J Neurophysiol* 66(4): 1232–48, 1991.
Nakamura K and Norgren R. Taste responses of neurons in the nucleus of the solitary tract of awake rats: An extended stimulus array. *J Neurophysiol* 70(3): 879–91, 1993.
Nishijo H and Norgren R. Responses from parabrachial gustatory neurons in behaving rats. *J Neurophysiol* 63(4): 707–24, 1990.
Nishijo H and Norgren R. Parabrachial gustatory neural activity during licking by rats. *J Neurophysiol* 66(3): 974–85, 1991.
Nishijo H and Norgren R. Parabrachial neural coding of taste stimuli in awake rats. *J Neurophysiol* 78(5): 2254–68, 1997.
Ogawa H, Sato M, and Yamashita S. Variability in impulse discharges in rat chorda tympani fibers in response to repeated gustatory stimulations. *Physiol Behav* 11(4): 469–79, 1973.
Ogawa H, Yamashita S, and Sato M. Variation in gustatory nerve fiber discharge pattern with change in stimulus concentration and quality. *J Neurophysiol* 37(3): 443–57, 1974.
Perrotto RS and Scott TR. Gustatory neural coding in the pons. *Brain Res* 110: 283–300, 1976.
Rosen AM and Di Lorenzo PM. Two types of inhibitory influences target different groups of taste-responsive cells in the nucleus of the solitary tract of the rat. *Brain Res* 1275: 24–32, 2009.
Rosen AM, Victor JD, and Di Lorenzo PM. Temporal coding of taste in the parabrachial nucleus of the pons of the rat. *J Neurophysiol* 105(4): 1889–96, 2011.
Rosen AM, Sichtig H, Schaffer JD, and Di Lorenzo PM. Taste-specific cell assemblies in a biologically informed model of the nucleus of the solitary tract. *J Neurophysiol* 104(1): 4–17, 2010.
Roussin AT, Victor JD, Chen J-Y, and Di Lorenzo PM. Variability in responses and temporal coding of tastants of similar quality in the nucleus of the solitary tract of the rat. *J Neurophysiol* 99(2): 644–55, 2008.
Roussin AT, D'Agostino AE, Fooden AM, Victor JD, and Di Lorenzo PM. Taste coding in the nucleus of the solitary tract of the awake, freely licking rat. *J Neurosci* 32(31): 10494–506, 2012.
Saito S, Endo H, Kobayakawa T, Ayabe-Kanamura S, Kikuchi Y, Takeda T, and Ogawa H. Temporal process from receptors to higher brain in taste detection studied by gustatory-evoked magnetic fields and reaction times. *Ann NY Acad Sci* 855: 493–7, 1998.
Schuetze SM. The discovery of the action potential. *Trends Neurosci* 6: 164–8, 1983.
Schroeder CE, Wilson DA, Radman T, Scharfman H, and Lakatos P. Dynamics of active sensing and perceptual selection. *Curr Opin Neurobiol* 20: 172–6, 2010.
Scott TR. Behavioral support for a neural taste theory. *Physiol Behav* 12(3): 413–17, 1974.
Scott TR and Small DM. The role of the parabrachial nucleus in taste processing and feeding. *Ann NY Acad Sci* 1170: 372–7, 2009.
Segundo JP, Moore GP, Stensaas LJ, and Bullock TH. Sensitivity of neurons in *Aplysia* to temporal pattern of arriving impulses. *J Exp Biol* 40: 643–67, 1963.
Smith DV and Li CS. Tonic GABAergic inhibition of taste-responsive neurons in the nucleus of the solitary tract. *Chem Senses* 23: 159–69, 1998.

Smith DV, Li CS, and Davis BJ. Excitatory and inhibitory modulation of taste responses in the hamster brainstem. *Ann NY Acad Sci* 855: 450–6, 1998.
Smith DV, John SJ, and Boughter JD. Neuronal cell types and taste quality coding. *Physiol Behav* 69(1–2): 77–85, 2000.
Spector AC and Travers SP. The representation of taste quality in the mammalian nervous system. *Behav Cog Neurosci Rev* 4(3): 143–91, 2005.
Stapleton JR, Lavine ML, Wolpert RL, Nicolelis MA, and Simon SA. Rapid taste responses in the gustatory cortex during licking. *J Neurosci* 26(15): 4126–38, 2006.
Stapleton JR, Lavine ML, Nicolelis MA, and Simon SA. Ensembles of gustatory cortical neurons anticipate and discriminate between tastants in a single lick. *Front Neurosci* 1(1): 161–74, 2007.
van Rossum MCW. A novel spike distance. *Neural Comput* 13: 751–63, 2001.
Victor JD and Purpura KP. Nature and precision of temporal coding in visual cortex: A metric-space analysis. *J Neurophysiol* 76: 1310–26, 1996.
Victor JD and Purpura KP. Metric-space analysis of spike trains: Theory, algorithms, and application. *Network* 8: 127–64, 1997.
Wang L and Bradley RM. Influence of GABA on neurons of the gustatory zone of the rat nucleus of the solitary tract. *Brain Res* 616(1–2): 144–53, 1993.
Weiss MS and Di Lorenzo PM. Not so fast: Taste coding time in the rat revisited. *Front Integr Neurosci* 6: 27, 2012.
Zhao GQ, Zhang Y, Hoon MA, Chandrashekar J, Erlenbach I, Ryba NJ, and Zuker CS. The receptors for mammalian sweet and umami taste. *Cell* 115: 255–66, 2003.

13 Increases in Spike Timing Precision Improves Gustatory Discrimination upon Learning

Ranier Gutierrez and Sidney A. Simon

CONTENTS

13.1 INTRODUCTION

A primary goal of neuroscience is to identify the neural computations that underlie behavior. In a recent commentary, Carandini stated that "to understand neural computations, we must record from a myriad of neurons in multiple brain regions (Carandini, 2012)," and in a similar vein, György Buzsáki noted, "that transiently active ensembles of neurons, known as 'cell assemblies,' underlie numerous operations of the brain, from encoding memories to reasoning (Buzsaki, 2010)." These two statements illustrate the approach we have undertaken in obtaining ensembles of neurons across four different brain areas to obtain the neural events associated with animal's learning to distinguish among tastants or, in a broader sense, to associate neural activity with behavior.

At a more fundamental level, scientists have inquired how inputs from the physical world such as an object like a tree or a sensation arising from a tastant is represented in the brain. It is generally thought that its representation is likely to be some high-dimensional depiction of the object (or taste) and it will include information about one's familiarity with it and also if it is liked or disliked. It follows that the number of neurons involved in the perception of an object will increase with distance (time) from the periphery. Moreover, since the perception will include identity and other characteristics as well as feelings, its representation will be distributed over several brain areas. Thus, to fully understand the neural correlates of perception, it is necessary to record the activity of neuronal ensembles in multiple brain regions. Neuronal ensembles are obtained by simultaneously recording groups of them for an unspecified interval. This is achieved either optically (e.g., using voltage-sensitive dye imaging techniques) in a manner that can cover one or several surface areas or with arrays of electrodes that could be implanted in one or more deep areas (Nicolelis et al., 1995; Komiyama et al., 2010). Recording from multiple areas has the advantage of determining which area provides the most information in relation to a task and also if the prediction increases by combining information from different areas (Carmena et al., 2003). Multiarea recordings may be used to determine the temporal sequence of information between areas (Nicolelis et al., 1995). Another reason for using ensembles or population activity, as opposed to individual neurons (that obviously contain information), is that single neurons are inherently noisy whereas, in many cases, the noise will be averaged out in an ensemble and thus increase, in a single trial, the probability of transmitting the correct message (Laubach et al., 2000). Another positive aspect of ensembles is reliability in the sense that if each neuron contains a fraction of the total information content (i.e., information is distributed over the population), it follows that if some neurons are lost or inactive during a trial, the output (message) will remain relatively unchanged (Carmena et al., 2005). This of course does not imply that all the neurons in a population contain the same amount of information or that each area will produce equal amounts of information. Despite these advantages, to understand the neural underpinning of behavior, several issues arise that pertain to the use of ensembles. It has often been found that the predictive value of networks increases with the ensemble size (Laubach et al., 2000). It is important to inquire for the presence of correlations between neurons in an ensemble. Is information conveyed by the precise timing of spikes ("spike timing code") or by the number of spikes occurring over an arbitrary time window (rate code)? In the latter case, each neuron of the ensemble simply contributes to the population in proportion to its firing rate (Theunissen and Miller, 1995). It is also important to determine whether all the neurons in an ensemble are of the same type (i.e., pyramidal cells). Although this would be desirable, in most ensembles, at least those obtained from extracellular recordings, the cell types have usually not been ascertained. Thus, these ensembles are likely to contain several types of neurons, which likely include both projection neurons and interneurons that may respond differently during a behavioral task (Woloszyn and Sheinberg,

2012). The effect of having a heterogeneous neuronal population on the ensemble's predictive value is not known. Finally, and perhaps the most difficult issue regarding ensembles is to determine what they are actually computing that is relevant to the behavioral task.

In this chapter, we will focus on the gustatory sensory system. The first half will discuss in great detail the motor and preoromotor substrates underlying rhythmic licking behavior. As the topic of the book is about spike timing, if the reader is interested only in this aspect, we recommend that they go to Section 13.3. In that section, we provide evidence that the taste system, in rodents, is assisted by a rhythmic motor output (licking) that entrains and synchronizes large parts of the distributed gustatory system, and as a mechanism that increases the spike timing precision (and reliability) across neurons of the gustatory system, which enhances taste discrimination (Gutierrez et al., 2010).

13.2 GENERALITIES OF INGESTIVE BEHAVIOR

Although some foods such as saccharides and bitter tastants are innately accepted or rejected, the vast majority of food preferences are acquired through associative learning. Food intake commences first by identifying whether it is palatable, a process that could arise by tasting or through experience. Once in the mouth, animals need to rapidly decide whether to ingest or reject it. This decision is based on many factors besides its palatability such as taste expectations and the animals' state of hunger or satiety (Rolls, 2007; Yoshida and Katz, 2011; Samuelsen et al., 2012). Another factor in gustatory processing arises when animals *actively* seek food to make perceptual decisions as to ingest it (Kleinfeld et al., 2006; Uchida et al., 2006).

13.2.1 Oscillations and Chunking Information

For many sensory modalities, an external stimulus is processed via rhythmic motor behaviors such as whisking, sniffing, and licking, in the somatosensory, olfactory, and gustatory sensory systems, respectively (Kepecs et al., 2006). Interestingly, all these rhythmic behaviors occur at 4–12 Hz, that is, in the theta frequency band. Although the actual function for processing sensory stimuli in a rhythmic manner is still a matter of debate (Wesson et al., 2009), in the olfactory system (and presumably in other sensory systems), it has been proposed that sensing at this theta rhythm can fragment continuous sensory stimuli into, more tractable, chunks of information (Uchida and Mainen, 2003).

Under this scheme, each rhythmic cycle would represent a functional sensory processing unit, where specific phases of the cycle can be used to decode sensory information differentially (Bathellier et al., 2008), thus allowing the use of spike-phase coding algorithms (Kayser et al., 2009). Regarding olfaction, these data indicate that olfactory information is quickly encoded, in one or two sniffs, (~250 ms) (Uchida and Mainen, 2003; Wachowiak, 2011). Regarding taste, despite the fact that rodents can also discriminate among tastants in one or two licks, the role that rhythmic licking plays in gustatory processing is currently unknown (Halpern and Tapper, 1971; Halpern and Marowitz, 1973; MacDonald et al., 2009; Gutierrez et al., 2010). Nevertheless, processing sensory inputs in a rhythmic manner, such as in sniffing, whisking, and licking, is attractive as it allows, in addition to firing rates, the use of spike timing as an extra source to encode sensory information (Flight, 2009). Moreover, in this scheme, each rhythmic cycle can open a window of opportunities ("the duty phase") where ensembles of neurons can synchronize their spikes, and can thus potentially transfer (and coordinate) information between distant brain regions (Nicolelis et al., 1995; Buzsaki and Draguhn, 2004).

13.2.2 Oromotor and Preoromotor Substrates of Ingestive Behavior

Video analysis of ingestive behavior in the rat reveals that mouth opening does not occur during the approach phase of eating or drinking but follows a prolonged period (50–800 ms) of perioral contact with the food (Zeigler et al., 1984). This shows the importance of somatosensory stimulation for eating and drinking (Zaidi et al., 2008). The consummatory phase begins after the rat opens its mouth in the form of stereotypical bouts of rhythmic tongue movements at 6–7 Hz (Figure 13.1a; mice at 8–11 Hz, humans chew at 0.3–4 Hz). Rhythmic licking is generated by a central pattern generator (CPG) located at the medullary reticular formation (Figure 13.1b and c) that is under the descending control of several cortical and subcortical regions (Figure 13.1d, see dashed lines).

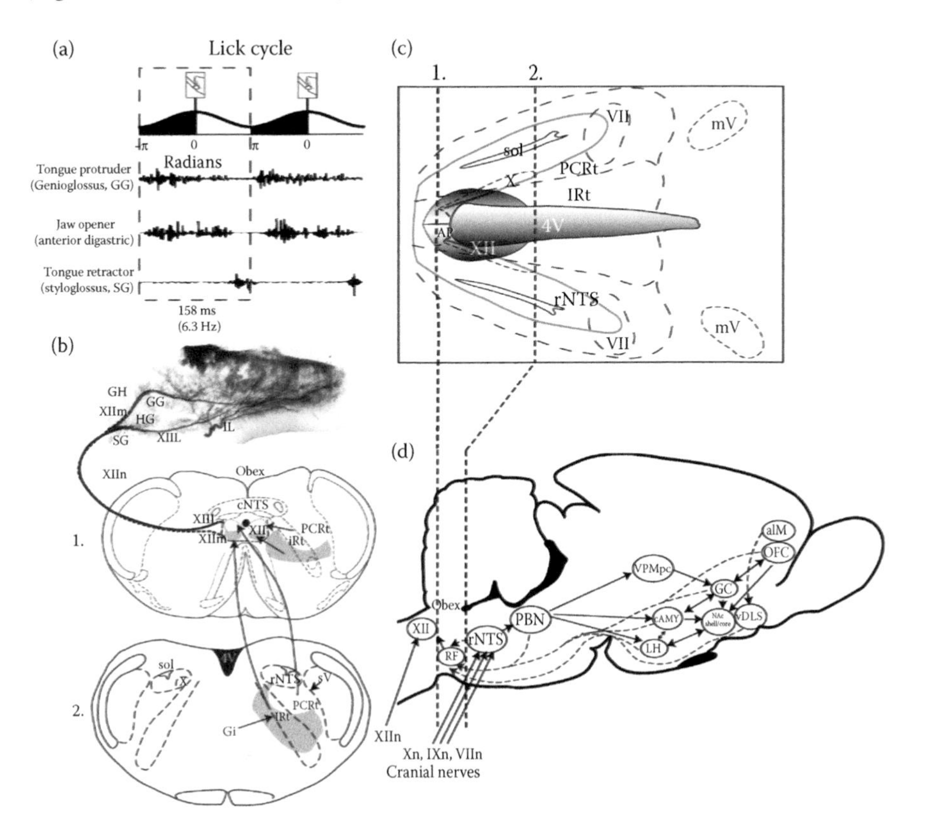

FIGURE 13.1

FIGURE 13.1 Diagram showing the central pattern generator (CPG) for licking and components of the taste–reward pathway. (a) Rodents ingest tastants by licking in a quasiperiodic cycle (from $-\pi$ to π) at 6–7 Hz. At 0 radians, the tongue contacts the sipper; from 0 to $-\pi$ radians, the jaw is opened and the tongue protrudes; and from π to 0 radians, the tongue retracts and the jaw closes. Electromyography (EMG) analysis indicates that a lick cycle encompasses two alternating phases in sequence: first, activation of the protruder muscles (genioglossus (GG) and geniohyoid (GH), not shown), accompanied by jaw opening (anterior digastrics muscle; see EMG traces), whereas in the second phase, the tongue is rapidly retracted (activation of retrusor muscles styloglossus (SG) and hyoglossus (HG), not shown). (Reprinted with permission from Travers JB, Dinardo LA, Karimnamazi H. 1997. *Neurosci Biobehav Rev* 21:631–647.) (b) The selection of a rat's several motor nerves associated with licking. The tongue was stained using Sihler's technique. (Reprinted with permission from McClung JR, Goldberg SJ 2000. *Anat Rec* 260:378–386.) The intrinsic (not shown) and extrinsic (GG, GH, SG, and HG) tongue muscles are innervated by the hypoglossal nerve (XIIn), which contains axons of motoneurons. Close to the muscles, XIIn bifurcates to form medial (Xllm) and lateral (XllL) branches. The hypoglossal nuclei, XII, contain, at least, two types of motoneuron, one that fires in phase with lingual protrusion (dark-shaded first half of the licking cycle in a), located more ventrally (dark-shaded area of the nucleus in b), and neurons that fire in phase with tongue retraction (light-shaded second half of the licking cycle in a) distributed more dorsally (light-shaded area of the nucleus in b) (Travers and Jackson, 1992). The medullary reticular formation (the intermediate, iRt and the parvocellular part, PCRt) contains a large portion of preoromotor neurons that sends second-order axons to XII motoneurons to generate the characteristic 6.3 Hz rhythmic pattern of licking in rats. The gigantocellularis nucleus, Gi, contains the CPG related to mastication. The motor root of the trigeminal cranial nerve (mV) receives projections from the jaw-closer (masseter and temporalis) and jaw-opener muscles (anterior, digastric, and mylohyoid), and the trigeminal nerve deafferentation attenuates jaw opening. (c) The upper panel shows a horizontal view of brainstem structures involved in feeding behavior; for visualization purposes, the dorsoventral dimension was collapsed. Jaw-closing (masseter and temporalis) and jaw-opening (anterior, digastric, and mylohyoid) muscles are innervated by the motor part of the trigeminal cranial nerve, mV. (d) The taste–reward pathway: Three cranial nerves VIIn, IXn, and Xn innervate different parts of the oral cavity and convey taste information to the rostral part of the nucleus of the solitary tract (rNTS). Input from the sensory trigeminal complex (sV) also contributes to gustatory processing. Taste information then projects to the pontine parabrachial nucleus (PBN), for rodents, which in turn distributes gustatory information via a thalamocortical pathway or throughout a ventral forebrain pathway. The PBN projects to the parvicellular part of the ventroposterior medial nucleus of the thalamus (VPMpc). Then the VPMpc projects to the primary gustatory cortex (GC), which in turn sends axon projections to the orbitofrontal cortex (OFC). Brain structures in the ventral forebrain that receive input from the PBN include the central nucleus of the amygdala (cAMY) and the lateral hypothalamus (LH). All four brain regions—LH, cAMY, GC, and OFC—send unidirectional axons to medium spiny neurons of the nucleus accumbens (NAc). The NAc shell sends heavy projections to LH. The dashed line shows the brain regions that send descending projections to the medullary reticular formation, and thus can modulate voluntary licking behavior. Note: Projections from these brain regions do not necessarily synapse on the same set of brainstem neurons. alM, anterolateral motor cortex; vDLS, ventro dorsolateral striatum.

The extrinsic tongue muscles include the genioglossus and geniohyoid muscles that contribute to tongue protrusion, whereas the styloglossus and hyoglossus muscles retract the tongue. All these muscles are directly innervated by the hypoglossal nerve (XIIn). Close to the muscles, XIIn bifurcates to form medial (XIlm) and lateral (XIlL) branches (Figure 13.1b) (McClung and Goldberg, 2000). Electrical stimulation of the medial or lateral branch yields tongue protrusion or retraction, respectively. Extracellular recordings in the hypoglossal nuclei (XII) revealed motoneurons that fired in phase with lingual protrusion (Figure 13.1b, XII darker shadow), and in phase with tongue retraction (XII, light shadow) (Travers and Jackson, 1992). Both neuronal types are topographically distributed in the ventral (protruder) and dorsal (retrusor) regions of the hypoglossal nuclei, respectively (Dobbins and Feldman, 1995). Anatomical studies using the transsynaptic pseudorabies virus injected in the medial or lateral branches of the XIIn have also identified preoromotor neurons that project to hypoglossal motoneurons (XII). These preoromotor neurons are primarily located in the intermediate (iRt) and parvocellular (PCRt) region of the medullary reticular formation (Figure 13.1b) (Dobbins and Feldman, 1995). Both the iRt and PCRt contain neurons that are rhythmically active during licking (Travers et al., 1997). Importantly, in decerebrated rats, intraoral infusions of palatable tastants can elicit rhythmic licking (Grill and Norgren, 1978), suggesting that the CPG for licking is located in the brainstem. The medullary reticular formation is also the reservoir of many other oropharyngeal motor patterns, including mastication (Figure 13.1b; gigantocellular nucleus (Gi)), swallowing, coughing, and breathing. However, a detailed description of these other CPGs is beyond the scope of this chapter, and thus we focus only on the CPG for licking (Dobbins and Feldman, 1994; Gestreau et al., 2005).

13.2.3 Subcortical Inputs to the CPG for Licking

As noted, the CPG for licking is influenced by several subcortical brain regions. Figure 13.1d shows some of the regions that send direct or indirect projections to reticular formation (iRt or PCRt—see dashed lines). Briefly, the rostral area of the nucleus tractus solitarius (rNTS) projects to preoromotor, iRt and PCRt, neurons, and consequently the electrical stimulation of rNTS elicits oromotor behaviors such as licking or gaping (Kinzeler and Travers, 2008). Di Lorenzo has recently identified neurons with rhythmic licking activity in the rNTS (Roussin et al., 2012). As the rNTS projects to the taste region of the parabrachial nucleus (PBN) in rodents, it was not surprising to find that electrical stimulation of the PBN produces rhythmic licking (Galvin et al., 2004). The central amygdala (AMY) and the lateral hypothalamus (LH), and other components of the taste pathway, also contain neurons that fire rhythmically with licking (Kaku, 1984; Yamamoto et al., 1989).

Having discussed the CPG for licking, we now briefly describe the taste pathways and its intrinsic relationship with rhythmic licking. From taste cells that contain receptors and signal transduction pathways throughout the oral cavity, electrical signals from cranial nerves VIIn, IXn, and Xn that contain information on the chemical properties of tastants (as well as somatosensory inputs via cranial nerve V) are conveyed to rNTS of the medulla (Figure 13.1d). Recently, Whitehead and colleagues used a transsynaptic pseudorabies virus to label a few geniculate ganglion neurons (the cells of the cranial nerve VIIn that innervate the fungiform papillae on the anterior tongue and the anterior foliate papillae on the lateral tongue) (Zaidi et al., 2008). They identified two basic taste circuits: one composed of geniculate ganglion cells that project to 10 or more rNTS neurons and, in turn, to the taste region of the PBN (see below), and a second population of geniculate ganglion cells that project to the rNTS neurons and then to preoromotor neurons of the reticular formation, modulating licking and gaping behaviors. Therefore, it follows that rhythmic licking and gustatory processing are intrinsically interrelated. On the other hand, the ventral division of the nucleus tractus solitarius (vNTS) receives vagal input and projects to the dorsal motor nucleus of X and the nucleus ambiguus where they regulate, among other things, the release of appetite-regulating hormones (Berthoud and Morrison, 2008). In rodents, rNTS efferents project to gustatory centers in the PBN that synapse with neurons in the parvicellular part of the ventroposterior medial nucleus of the thalamus (VPMpc). A ventral PBN pathway has been shown to project to the central nucleus of the amygdala (cAMY) and LH. In primates, the rNTS projection fibers bypass the PBN to synapse directly into the VPMpc, whereas the ventral pontine parabrachial nuclei (vPBN) conveys the general visceral information to specialized thalamic nuclei. In either case, thalamic afferents project to the primary gustatory cortex (GC), which has both chemosensitive and visceral areas. In turn, the GC projects to the cAMY, from where gustatory information reaches the LH. The GC also sends projections to the orbitofrontal cortex (OFC), which is sometimes referred to as the secondary taste cortex.

13.2.4 Cortical Modulation of Voluntary Licking

Several cortical regions have been shown to be involved with the licking behavior. In particular, electrical stimulation of the GC can induce tongue and jaw movements (Sasamoto et al., 1990), probably as a consequence of the projection of GC to the medullary reticular formation (Zhang and Sasamoto, 1990). In the rat, the OFC receives strong inputs from the GC and also contains neurons that fire rhythmically with licking (Gutierrez et al., 2006). Electrical stimulation of the OFC was shown to induce rhythmical jaw and tongue movements (Neafsey et al., 1986) and ablation of the OFC results in deficits in tongue protrusion and in the use of the mouth and tongue (Whishaw and Kolb, 1983). Anatomical studies show that OFC sends direct projections to the lower brainstem reticular formation where the CPG for licking is located (vanEden and Buijs, 2000).

As expected, the motor cortex is involved with licking. In this regard, repetitive electrical microstimulation of the anterior–lateral motor cortex (alM) elicits distinct rhythmic jaw movements. Moreover, lesions of the GC did not alter alM's electrically induced jaw movements, suggesting that these two cortical regions work in parallel and independently to each other (Sasamoto et al., 1990). In rodents, transsynaptic tracing found that alM projects to neurons in the ventral part of the dorsolateral striatum (vDLS) (Komiyama et al., 2010). In a licking-related task, neurons in the vDLS were found to covary with licking. However, across days, as the rats improved their performance in the task, these neurons become less active (Tang et al., 2009).

This suggests that cortico-striatal inputs during licking depend on the level of training. vDLS neurons that receive input from alM (Figure 13.1d) project to preoromotor neurons located on the Gi, an area that is involved in mastication. They also project the iRt where the CPG for licking is located (Komiyama et al., 2010). In mice, the temporal inactivation of alM severely impairs rhythmic licking, indicating that alM plays a crucial role in voluntary licking (Komiyama et al., 2010).

In summary, all these results indicate that rhythmic licking behavior activates a widespread and distributed neural network, comprising cortical and subcortical sensorimotor brain regions and major components of the taste–reward pathway (see below).

13.2.5 Taste–Reward Pathway

We will now delve into more detail on four important components of the taste–reward pathways that are involved in the experiments that will be discussed in the next section. As paraphrased in Gutierrez et al. (2010), the taste–reward circuit consists of a highly interconnected neural network that is involved with multiple aspects of ingestive behavior, associative learning, and reward expectation. The insular cortex (IC), which contains the primary GC, is a multimodal area that processes taste, visceral, somatosensory, and hedonic information. The OFC integrates information from several primary sensory systems and assays the relative reward value of sensory stimuli, including those associated with foods. The OFC encodes the economic value of goods, which might be important to choose among different food options, based on the subject's assigned value. OFC and AMY neurons fire selectively to sensory cues according to their predictive value of a reward (Schoenbaum et al., 1998). Neurons from the IC, OFC, and AMY project to the nucleus accumbens (NAc), defining a circuit that processes information about gustatory cues, their predictive value of rewards, and motivational significance (Pecina and Berridge, 2005). Consequently, the NAc is considered as a limbic–motor interface that transforms motivational information generated in limbic regions into movements to achieve a goal, such as eating (Mogenson and Yang, 1991). In fact, the shell part of the NAc contains heavy projections to the LH (Luiten et al., 1987), and such projections are thought to be important in the regulation of feeding behavior (Saper et al., 2002; Kelley, 2004).

13.3 SPIKE TIMING AND GUSTATORY PROCESSING

An exhaustive review of the taste coding is beyond the scope of this chapter, but the following reviews are recommended: Zhao et al. (2003), Simon et al. (2006), Breslin and Spector (2008), Carleton et al. (2010), and Chandrashekar et al. (2010). Briefly, some data support the idea that the taste system utilizes a "labeled-(cell)" followed by "labeled line" coding scheme from the periphery to the GC (Chen et al., 2011). That is, taste responsive cells convey information of one and only one taste modality (bitter, sweet, salt, sour, or umami). However, other studies throughout the taste system found that although there are taste-selective neurons, many of them are broadly tuned in the sense that they are responsive to more than one tastant (Lemon and Smith, 2005; Stapleton et al., 2006; Di Lorenzo et al., 2009), suggesting that ensembles of broadly tuned neurons throughout the gustatory system encode taste qualities ("across-neurons pattern") (Erickson, 2001; Katz et al., 2001). Although many questions remain regarding the coding of taste stimuli, it is clear that spike timing adds additional information that can be used to distinguish among tastants (Gutierrez et al., 2010).

The gustatory system is ideal to investigate the role of spike timing as a neural code, since a code based on spike counts alone cannot unambiguously discriminate among tastants (Lemon and Smith, 2005; Roussin et al., 2008). Katz et al. (2001) were one of the first to analyze the temporal dynamics of taste responses by showing that a single neuron may respond best (i.e., exhibit the greatest activity) to more than one tastant, albeit at different times, during the tastant-evoked activity. This fact rules out the use of spike count as a complete neural code for taste quality (Katz et al., 2001). As noted, the hallmark of taste responsive neurons in the CNS is that they exhibit broadly tuned profiles. These are important since the more broadly tuned the neuron, the greater the taste information it conveys (Di Lorenzo et al., 2009). In particular, a neuron's temporal pattern, for example, the spike timing and envelope of their responses allow a broadly tuned neuron to represent the entire taste quality domain. Thus, taste quality may be encoded by population of neurons with different broadly tuned profiles.

A seminal study using electrical microstimulation of the nucleus tractus solitarius (NTS) revealed that spike timing also plays a causal and not only a correlative role for taste coding (Di Lorenzo and Hecht, 1993; Di Lorenzo et al., 2003). This group first recorded the neural activity of rNTS neurons while rats drank sucrose or quinine, creating two separate templates of firing patterns for each tastant. Then, while a rat rhythmically licked for water, they microstimulated the rNTS by replaying the spike timing activity previously recorded for one of the tastants. When they played back the quinine pattern, the rats avoided water as if it was a bitter-tasting compound. Two other aspects of this experiment are worthy of comment. First, to work the playback of the quinine template, it necessarily involves the coupling with rhythmic licking for water and second, shuffling the quinine-stimulation template (but maintaining the same number of electrical pulses) did not result in avoiding water, thereby indicating that specific electrical stimulation patterns can trigger a "bitter or aversive taste sensation." Moreover, and perhaps even more impressive, was the pairing of the sucrose-stimulation pattern (also triggered when the animal was licking for water) with a gastric malaise agent (LiCl)-induced conditioned taste aversion to real sucrose. In this experiment, the playback of the sucrose-stimulation pattern, in the absence of gastric malaise, extinguished the aversion generalized to real sucrose. These data provide compelling evidence of the important role that spike timing may have for taste processing. However, beyond the NTS, the role that both rhythmic licking and spike timing play on the taste–reward pathway and gustatory processing is currently unknown. Thus, owing to the paucity of information, we decided to use a taste-discrimination task and multichannel recordings to address the following two fundamental questions: that is, it remains to be known whether rhythmic licking may function as an internal temporal frame that synchronizes spiking across brain regions and equally untouched is the question whether spike timing would enhance the ability of the taste–reward pathway to discriminate among gustatory cues.

13.3.1 Go/No-Go Taste-Discrimination Task

The study of taste discrimination in freely moving rats is usually accomplished by employing a taste-guided task. In a discrimination task, the taste stimuli serve as cues for a reward or a punishment. This ensures that neural and behavioral responses are not driven by the animal's natural preference for a particular taste stimulus. Instead, taste-guided tasks are used to study how an arbitrary taste cue becomes predictive of a reward or a punishment. It also allows one to uncover the neuronal correlates of taste learning. In a recent study using a novel taste-discrimination go/no-go task, Gutierrez et al. (2010) simultaneously recorded neuronal activity from four areas of the taste–reward circuit (OFC, AMY, GC, and NAc) to obtain the neural activity while rats learn to discriminate between two taste cues (Figure 13.2a and b).

The task is as follows (see Figure 13.2b): The trials were performed in an operant box that contained two widely spaced (~17 cm) drinking compartments. Each compartment contained a photobeam lickometer that was used to register each lick. The sipper tubes consisted of independent bundles of stainless-steel tubes connected to solenoid valves (Stapleton et al., 2006). Each lick delivered ~20 μL drop of liquid (in 10 ms). The behavioral task and details about the quantification of learning are shown in Figure 13.2b. Briefly, prior to surgery, all rats were trained on the taste-discrimination task. A trial began when the door opened allowing access to the cue compartment. The rats were then allowed to lick the sipper and at the fourth lick they received a drop of water. The rats were required to continue licking the dry sipper five additional times and on the sixth lick, they received a cue consisting of a drop of either 0.1 M NaCl or 0.1 M monopotassium-L-glutamate (MPG). These two tastants were chosen because rodents can distinguish between them independently of the sodium content (Maruyama et al., 2006). In the first session, one of these tastants was randomly chosen as either the positive (C+) or negative (C–) cue. After cue delivery, access to the outcome compartment was granted and, although no more licks were required, the rats continued licking, on average, 1.2 s before moving to the outcome port. The rats had 10 s after cue delivery to move from the cue compartment to the outcome compartment, where after three empty licks, they could receive the signaled outcome. If no response was observed after 10 s, both doors closed, thereby terminating the trial. In the outcome compartment, the positive cue (C+) was associated with three deliveries of 0.4 M sucrose and the negative cue (C–) signaled the availability of up to three deliveries of 1 mM QHCl (Figure 13.2b). Learning was quantified using a state-space algorithm to determine within a single session the "learning trial" (Smith et al., 2004), although we acknowledge that learning is a gradual process. Therefore, the "learning trial" does not mean that learning occurred in that particular trial; instead it reflects the point at which there is sufficient evidence that the subject began performing above the chance level and its correct performance remains in that way for the rest of the session.

13.3.2 Rhythmic Licking as a Global Internal "Clock" That Entrains the Taste–Reward Circuit

We first analyzed the level in which the firing of each of the neurons recorded was coherent with rhythmic licking. Coherence is defined as a measure of the

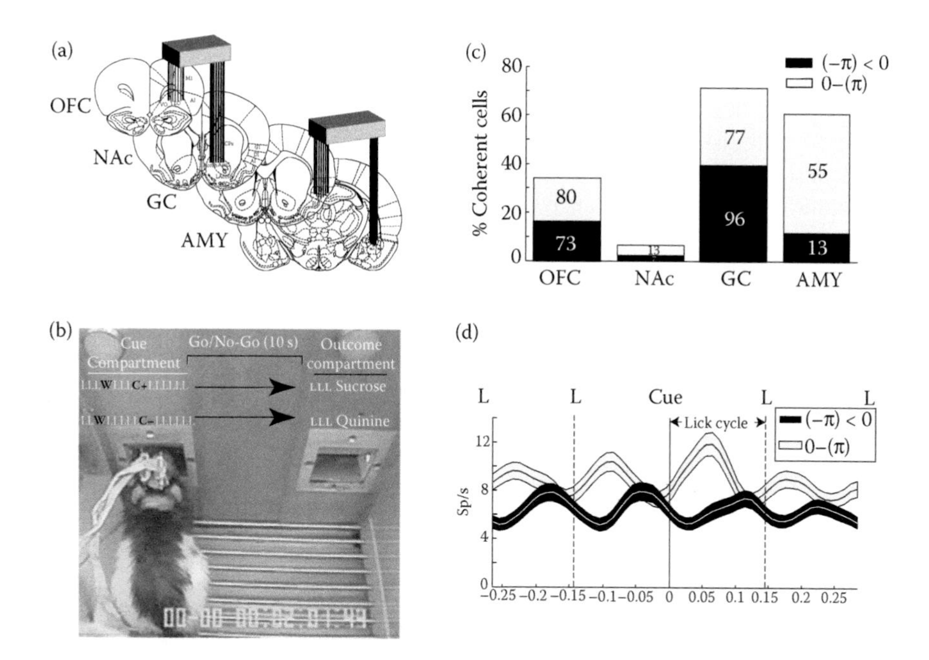

FIGURE 13.2 Rhythmic licking entrains neural spike timing across the taste–reward pathway. (a) Schematic representation of multielectrode recording sites in four components of the rat taste–reward circuit. OFC, orbitofrontal cortex; NAc, nucleus accumbens; GC, gustatory cortex; AMY, amygdala. (b) A rat implanted with four microarray electrodes while performing a taste-discrimination go/no-go task. In each trial, rats approached a sipper in the cue compartment, and then licked (L) an empty sipper 3 times, wherein on the fourth lick, they received a water drop (W). Then they licked the empty sipper another 5 times and on the 10th lick, they received a taste cue (a 20 μL drop of either 0.1 M NaCl or 0.1 M monopotassium glutamate (MPG). In positive cue trials, C+, one of the arbitrary taste cues signaled the availability of sucrose, whereas in negative cue trials, C–, the other taste cue signaled the availability of quinine. After cue delivery, subjects had 10 s to leave the cue compartment and move to the outcome compartment where after three additional empty licks, they received either sucrose or quinine. (c) Histogram showing the percentage of neurons with significant coherence for each brain area: percentage of neurons with positive (0–π, gray; retraction) and negative (−π < 0, black; protrusion) phase coherence. (d) Population PSTH of 414 neurons that fired coherence with rhythmic licking, in the OFC–NAc–GC–AMY, sorted by their phase relationship with licking. Black: activity of neurons phase locked from −π < 0 radians (tongue protrusion). Gray: population PSTH of neurons, which spikes are phased locked from 0 to π radians (tongue retraction). Activity is aligned to cue delivery (time 0 s). The interval from the solid vertical line to the right dotted line indicates the first lick cycle after cue delivery. Note that the neuronal activity from these two subpopulations of neurons reset in each lick cycle.

interdependence of licking and neural activity in the relevant frequency domain. A coherence of 0 or 1 means that two signals are completely uncorrelated, or completely correlated in frequency and phase, respectively. The phase of the coherence ranges between ±π radians and indicates the extent to which a neural discharge follows or precedes a lick (tongue contacting the sipper tube). Specifically, at phase 0 radians, the tongue contacts the sipper; from −π to 0 radians, the jaw is opened and the tongue protrudes; and from 0 to π radians, the tongue retracts and the jaw closes (see Figure 13.1a).

We found that rhythmic licking entrains the spike timing of neurons located in disparate and distant brain regions (Figure 13.2c). As might be expected from anatomical studies and previous studies, the GC > AMY > OFC contained the majority of neurons that fired in coherence with rhythmic licking (Figure 13.1d). Thus, despite the fact that individual neurons do not necessarily spike in every lick cycle, the population peristimulus time histogram (PSTH) combining spike activity recorded in all the rhythmic licking-coherent neurons represents every phase of the lick cycle (Figure 13.2d). This suggests that in several regions of the taste–reward pathway, rhythmic licking can be reliably reconstructed by a population code.

In contrast, the NAc was the region with fewer rhythmic neurons (Figure 13.2c). On the basis of anatomical grounds and since all GC–AMY–OFC projects to the NAc, it would be expected that neurons in the NAc would in fact oscillate with licking even more; however, this was not the case. Currently, there is no clear explanation for this result, but it could be attributed to the low-input resistance of medium spiny neurons (MSN), which renders them less excitable. What is clear is that MSNs in the NAc are not linearly integrating glutamatergic and rhythmic inputs from the GC, AMY, and OFC evoked during licking (Yuste, 2011). Nevertheless, what is known is that during licking, a large majority of NAc neurons are strongly inhibited during feeding (Krause et al., 2010; Tellez et al., 2012), which can rationalize why only a few NAc neurons covaried with licking.

13.3.3 Rhythmic Licking Coordinates Spike Timing between Brain Regions

As seen in Figure 13.3a, after a taste cue has been delivered and during the delay epoch when the rat had to make a decision whether to "go" or "not go" to obtain a reward, the activity of two simultaneously recorded neurons, one in AMY and the other in the OFC, become synchronous. In other words, both neurons fired synchronously (Figure 13.3a, right panel) and coherently with licking. It is obvious that these neurons do not reflect somatosensory input derived from licking for; otherwise, they would fire through the entire time the animal was licking in the cue compartment (Gutierrez et al., 2010).

Indeed, we found that the vast majority of neurons whose spiking activity was phase locked with licking *do not* arise from pure oromotor inputs. The reason being that the neuronal activity was usually found to covary with licking only during specific lick cycles in the trial (e.g., see Figure 13.3a; in raster from "W" to "C+"). One important result of this study was that the proportion of neurons that fire in coherence with licking dramatically increased after learning (not shown, see Gutierrez et al., 2010). These results suggest that rhythmic licking may have a more dynamic function than previously thought. Under this scheme, we proposed that each lick may serve as an internal "clock" signal (a global oscillatory drive) against which neurons from distant brain regions can synchronize their spiking activity. Moreover, it was found that synchronous firing among pairs of neurons recorded simultaneously also increased after learning and importantly, this effect was stronger in neurons that also fired in coherence with licking (Figure 13.3a, right panels).

Throughout the CNS, numerous brain oscillations are recorded in the local field potential (LFP), and have been found to be distributed over frequencies that range from about 1 to 100 Hz. Analysis of these LFP's oscillations can provide information of the internal state of the animal, such as whether they are awake, attending, or in various states of sleep (Gervasoni et al., 2004). On the other hand, global oscillations, whether they be internally generated, as we noted may occur through CPG, such as saccades, sniffing, whisking, and licking, or be driven by external stimuli can all act to synchronize ensembles, and by doing so, they provide additional information relevant to the behavioral state. However, LFP's oscillations can vary from brain region to region (Canolty et al., 2010).

We will now discuss a recent paper that is relevant to the above discussion of spike timing (van Wingerden et al., 2010). In a two-*odor* go/no-go discrimination task, it was found that theta oscillations, at 6 Hz, in the OFC LFP were always and exclusively observed during rhythmic licking. Specifically, theta power increased during

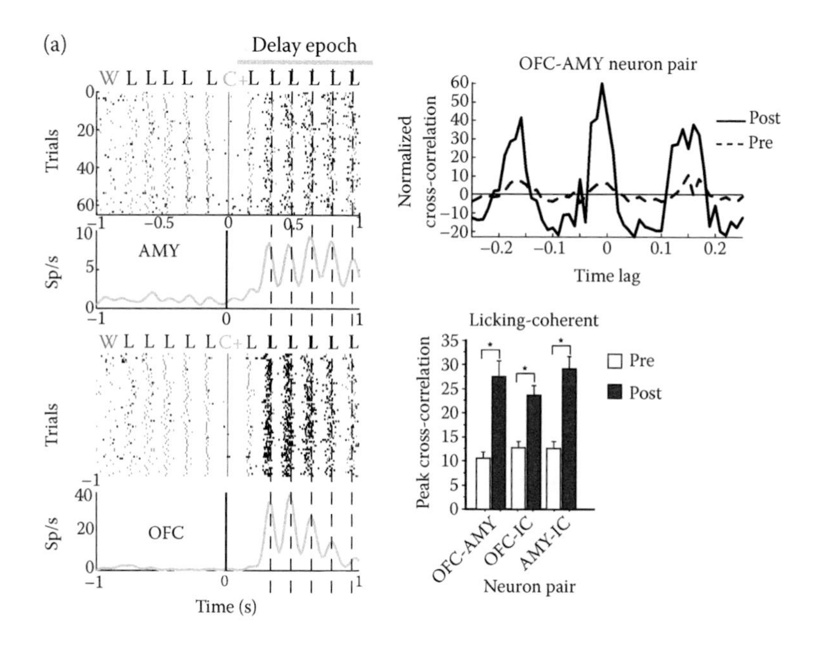

FIGURE 13.3 Rhythmic licking serves as an internal "clock" that dynamically coordinates spike timing between brain regions and improves cue-taste discrimination as a function of learning. (a) Rasters and PSTHs showing the responses of two neurons simultaneously recorded from AMY (top) and OFC (bottom). Note that their action potentials were phase locked to the lick cycle and were in synchrony only after cue delivery (vertical dashed lines). Gray marks represent the tongue's contact with the sipper tube. The black tick marks indicate the occurrence of action potentials. The panel at the right top shows the cross-correlation between the same two neurons recorded in AMY–OFC ("a"). It is seen that after learning (post), this neuron pair became more synchronous. Bottom: The histogram shows that synchronous spiking increased as a function of learning for pairs of OFC–AMY, OFC–IC, and AMY–IC neurons. As few NAc neurons fired in coherence with licking, they were not included in this analysis (Figure 13.2c).

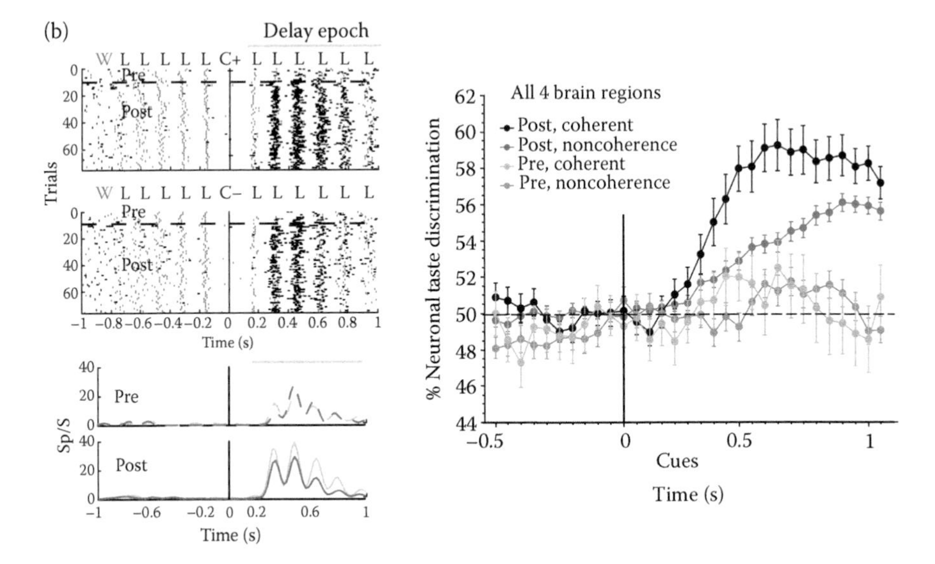

FIGURE 13.3 (Continued.) (b) The rasters (top) and PSTHs (bottom) of the same OFC neuron plotted in "a." After learning, this neuron developed a stronger firing rate in C+ trials. In the raster plot, the horizontal dashed lines ("the learning trial") separate trials in the prelearning and postlearning of the go/no-go taste-discrimination task. At the right, the graph shows that, after learning (postlearning), licking-coherent neurons (black dots) were better at discriminating the taste cues than noncoherent neurons (dark gray dots). (Modified with permission from Gutierrez R, Simon SA, Nicolelis MA 2010. *J Neurosci* 30:287–303.)

anticipatory licking to get the sucrose reward or during fluid licking for sucrose. Accordingly, in false alarm trials, where quinine was delivered, rats did not rhythmically lick, and in those trials, no increase in theta power was observed. Furthermore, OFC neurons also fired in phase to LFP's theta oscillation, in both anticipatory licking and during licking for sucrose solution. Despite these results, they reported that the "vigor" of licking does not correlate with the degree of spike phase locking to LFP oscillations. However, they did not explore whether the same neurons that fire in phase with LFP theta-band also fire in phase with rhythmic licking. Therefore, it is unclear whether rhythmic licking (5–7 Hz, a common oscillatory drive) and/or LFP theta oscillations (at 6 Hz, a local oscillation source) represent two independent parts or are both (two) complementary parts of the same "clock" signal that OFC neurons used to synchronize its spiking activity during feeding. Nevertheless, their data are highly consistent with our proposal, and suggests that rhythmic licking boosts LFP's theta oscillations and phase locking of OFC neurons.

13.3.4 Spike Timing Precision (and Reliability) Improves Taste Discrimination upon Learning

To solve this go/no-go task, rats needed to determine which of the taste cues predicted the reward (C+) and which predicted the aversive outcome (C−). Rats initially

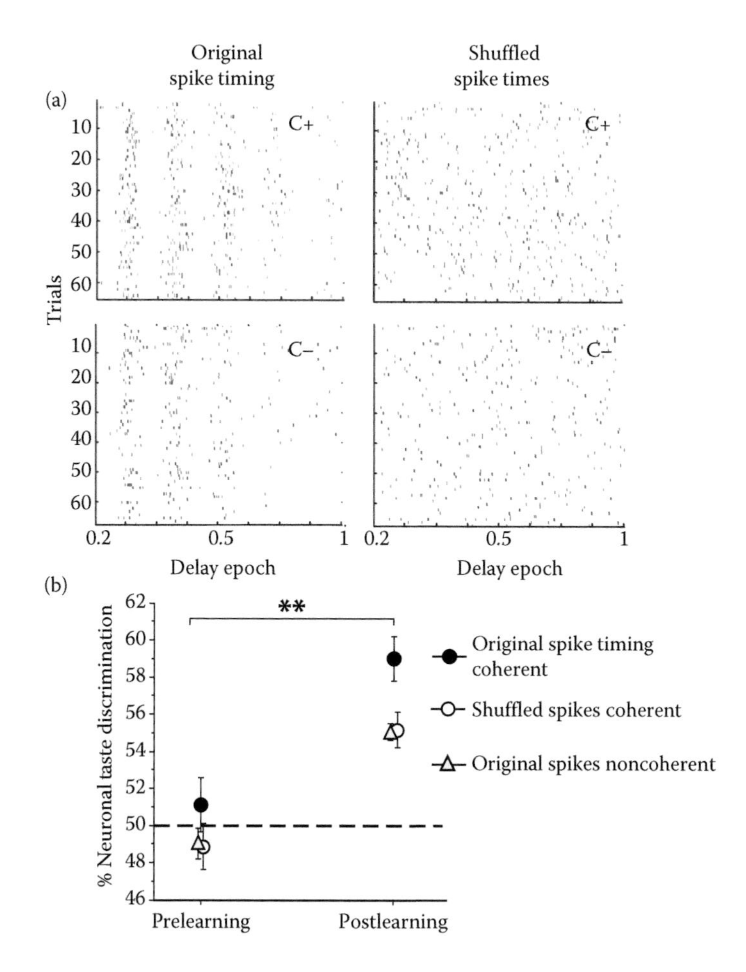

FIGURE 13.4 Spike timing conveys the extra taste information, not present in spike count alone. (a) Top left: raster plot of spikes of a licking-coherent neuron in the delay epoch, C+ trials in the postlearning phase. The taste cue was delivered at time 0 s (data not shown). Top right: to determine whether there is more information contained in the spike timing than in spike counts, we shuffled the spikes, such that each trial had the same number of spikes as in the original raster plot but random spike timing. It is clear that shuffling removed the spike-timing information and retained the mean firing rate. To quantify cue discrimination, the same spike shuffling procedure was performed for C– trials (bottom panels), and the neural taste discrimination was recalculated on these shuffle spikes. In this example, the cue discrimination (C+ vs. C–) in the postlearning phase was 71% (trials correct) using the original raster plots, whereas the cue discrimination dropped to 60% after removing spike-timing information. (b) Plots of the amount of discrimination (percentage of trials correctly classified) of all cue-selective neurons in the delay epoch as a function of learning phase (solid circles). It is seen that when the spike-timing information is removed (by shuffling the spikes), the licking-coherent cue-selective neurons showed the same level of discrimination as non-coherent cue-selective cells (compare open circles vs. triangles). (Modified with permission from Gutierrez R, Simon SA, Nicolelis MA 2010. *J Neurosci* 30:287–303.)

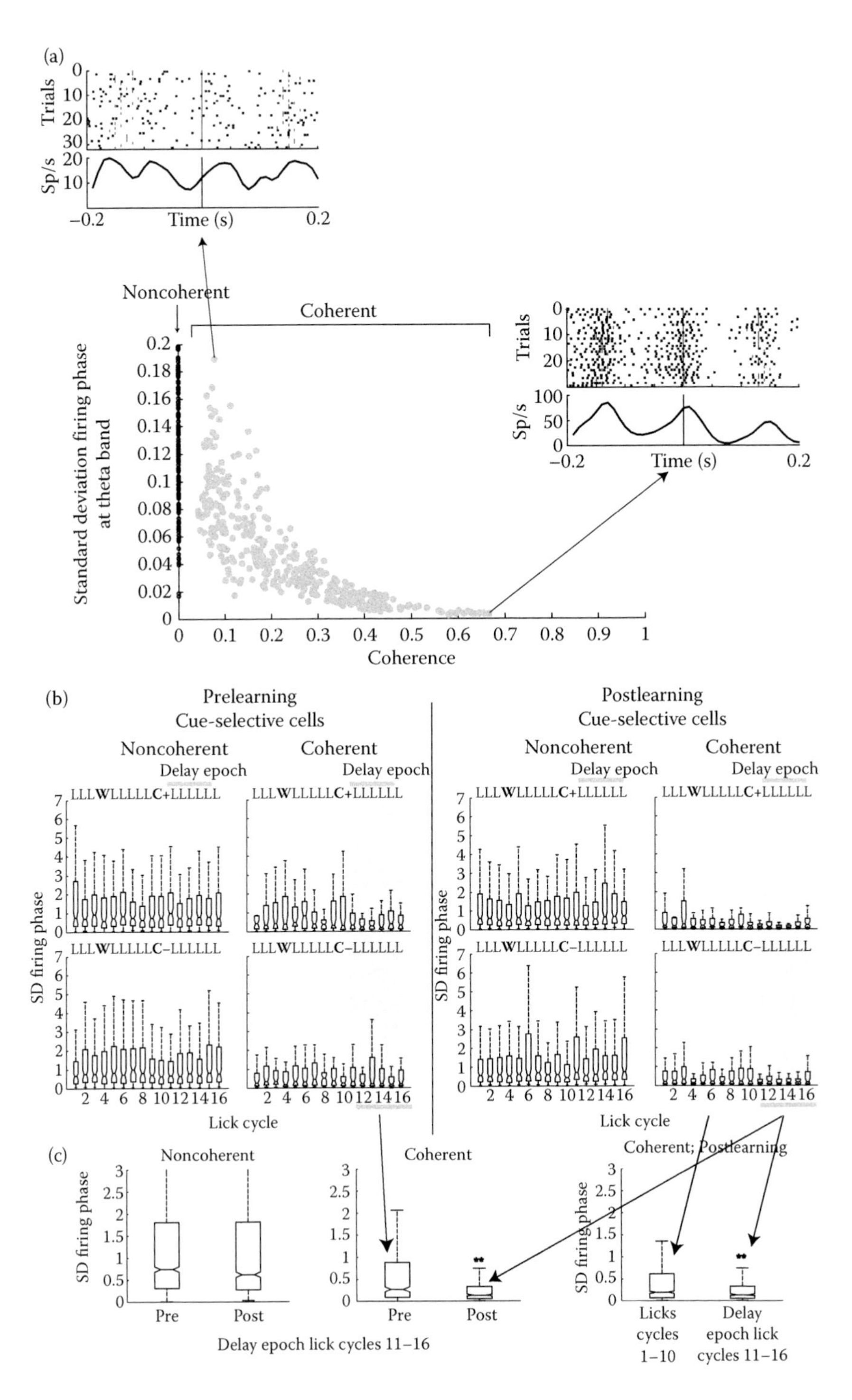
(a)
Trials
Sp/s
Time (s)
Noncoherent
Coherent
Standard deviation firing phase at theta band
Coherence
(b)
Prelearning
Cue-selective cells
Postlearning
Cue-selective cells
Noncoherent
Coherent
Delay epoch
LLLWLLLLLC+LLLLLL
LLLWLLLLLC–LLLLLL
SD firing phase
Lick cycle
(c)
Noncoherent
Coherent
Coherent; Postlearning
SD firing phase
Pre
Post
Delay epoch lick cycles 11–16
Licks cycles 1–10
Delay epoch lick cycles 11–16

FIGURE 13.5

responded to each cue by making a go response and received the corresponding outcome until the cues acquired a predictive value. At this point, rats learned to avoid quinine, while continuing to respond after the positive cue to obtain sucrose. Thus, as rats learned to perform this task, neuronal responses to initially nonpredictive taste cues became more distinct and predictable in the four brain areas recorded, suggesting that learning induces a significant functional reorganization of neural activity throughout the major components of the taste–reward circuit.

We identified two types of neurons that developed cue selectivity with learning: one that fires in coherence with some phase of the licking cycle (Figure 13.3b) and one that does not (not shown). In general, cue-selective neurons that synchronized their activity with licking were significantly better at decoding the cue's identity than noncoherent neurons (Figure 13.3b, right panel). A detailed analysis indicated that licking-coherent cells fire with higher spike timing precision than nonlicking-coherent neurons (Figure 13.5a), and that it was *this more precise degree of spike timing* that conveyed the extra cue information of licking-coherent neurons (Figure 13.4). In fact, we found that the hallmark of taste learning was to increase spike timing precision and reliability (Figure 13.5b and c), which in turn may allow licking-induced oscillations to enhance cue discrimination (Gutierrez et al., 2010).

FIGURE 13.5 Licking-induced neural synchronization improves spike timing precision (and reliability) and this effect is potentiated after learning. (a) Plot of the coherence value against the standard deviation (SD) of the phase; it shows an exponential curve in which the larger the coherence, the smaller the SD phase. Shown are two examples from both extremes. One extreme is one in which the cell fired with the largest variability (high SD phase, top raster plot) and a small coherence. This cell fired in coherence with the lick cycle, but since its SD phase is higher, its spikes (black tick marks) are neither reliable (constant) across trials nor precise with regard to the phase of the lick cycle. On the other extreme is a coherent neuron that had a smaller SD phase and a larger coherence (right raster plot). This cell tended to fire with high precision with regard to the lick cycle and greater reliability across trials. Thus, the SD phase is an indicator of how precisely and reliably (spike timing precision) a cell will fire in a lick cycle. Thus, the lower the SD firing phase, the greater the spike timing precision with regard to the lick cycle. (b) The SD phase of coherent and noncoherent cue-selective cells for the first 16 licks in the cue compartment. The first four panels at the left are for the prelearning phase, and the right four panels are for the postlearning phase. Note that, in all 16 licks, coherent cells tend to fire with a smaller SD phase than noncoherent cells, but especially during the delay epoch (lick cycles 11–16; gray bands; see (c)). In this regard, noncoherent cells most frequently displayed a high SD phase. (c) Plots of the SD phase during the delay epoch as a function of learning. For visualization purposes, error bars (left panel) are truncated. The noncoherent neurons showed no significant difference between prelearning and postlearning phases (Wilcoxon's rank sum test, p-value > 0.05, NS). In contrast, the coherent neurons, in the postlearning phase, further improved their spike timing precision by significantly reducing the SD phase (Wilcoxon's rank sum test, p-value < 0.001). The panel at the right compared the SD of the firing phase of licking-coherent neurons during lick cycles 1–10 versus lick cycles in the delay epoch (11–16) and found a significant difference between epochs (p-value < 0.0001). This suggests that, during the delay epoch, in which these coherent neurons discriminated the cues also corresponded to the epoch with the smallest spike timing variability. (Reprinted from Gutierrez R, Simon SA, Nicolelis MA. 2010. *J Neurosci* 30:287–303.)

Consistent with these results, it has been shown that sniffing-induced neuronal oscillations (at theta rhythm) enhance stimulus discrimination in the olfactory bulb by ensuring action potential precision (Schaefer et al., 2006). Therefore, rhythmic behaviors associated with gustation and olfaction seems to share several neuronal coding mechanisms. Moreover, modeling of spike trains provides theoretical evidence that a common oscillatory drive, such as that observed in saccades eye movements (3–5 Hz) and sniffing (6–12 Hz), improves the reliability of spike timing, indicating that global oscillations may serve as a reference point to phase the firing of neurons and to facilitate spike timing-dependent plasticity (STDP) (Masquelier et al., 2009). STDP is a learning rule in which synaptic strength is modified as a function of the relative timing of pre- and postsynaptic spikes (Flight, 2009). Masquelier et al. (2009) recently demonstrated the superiority of 6–12 Hz oscillations for both STDP-based learning and the speed of decoding. That is, in each oscillatory cycle, oscillations may facilitate learning by increasing spike timing reliability and by helping a postsynaptic neuron to decode the spike input pattern from its afferents. The fact that rhythmic licking enhances the spike timing precision of neurons in the GC, AMY, and OFC strongly suggests that licking-induced oscillations can also facilitate STDP, among neurons in the taste–reward pathway.

In summary, our data is in agreement with a recent comment of György Buzsáki suggesting that rodents actively explore the world by "dedicated motor outputs, such as whisking, sniffing or licking because they assist their specific sensory systems by 'resetting' or synchronizing spiking activity in large parts of the corresponding sensory system and/or creating transient gains, which enhance the sensory system's ability to process the inputs (Buzsaki, 2010)." In agreement, with this line of ideas, our results indicate that rhythmic licking acts as a global oscillatory signal that increases spike timing precision and synchronizes neural activity across multiple brain structures throughout the taste–reward circuit, enhancing taste discrimination upon learning.

13.4 FUTURE ISSUES

The next step in gustatory physiology is now restrained by limitations in our available technologies but in the future this will no longer be a problem. A crucial challenge is, no doubt, to be able to identify taste-related neurons, from taste receptor cells all the way through GC, and then to simultaneously record them. In mammals, this experiment is not yet feasible, but once accomplished, it will definitely answer the question of how taste information is encoded throughout the brain. In the near future, however, the next step is to use optogenetic tools to express light-sensitive proteins, opsins, for example, in neurons responsible for the CPG for licking. This will help us to trigger or to stop at wish the rhythmic licking machinery and then to causally determine the functional role that rhythmic licking behavior actually plays on gustatory processing and feeding behavior in general. Similarly, if we were able to activate, for example, with light pulses, specific sweet (bitter)-sensitive taste receptor cells (or geniculate ganglion neurons or in the GC taste "hot spots" (Chen et al., 2011)), this will give us the ability to control taste activation at the millisecond scale, which among other things will definitely answer the question of how much time does a rat need to identify a tastant and to decipher the neural code (spike timing or firing rates) that the brain actually uses to encode taste information. No doubt, the use of optogenetic techniques foretells the next level of gustatory physiology and will pave the way to investigate the sweet taste of light.

ACKNOWLEDGMENTS

We thank Professor Miguel Nicolelis for many of his scientific insights. This work was supported by NIH grant DC-01065 to SAS and CONACYT grant 179484, Salud2010-02-151001, ICYTDF-PICSA12-126 and Productos Medix 652 to R.G.

REFERENCES

Bathellier B, Buhl DL, Accolla R, Carleton A. 2008. Dynamic ensemble odor coding in the mammalian olfactory bulb: Sensory information at different timescales. *Neuron* 57:586–598.

Berthoud HR, Morrison C. 2008. The brain, appetite, and obesity. *Annu Rev Psychol* 59:55–92.

Breslin PA, Spector AC. 2008. Mammalian taste perception. *Curr Biol* 18:R148–R155.

Buzsaki G. 2010. Neural syntax: Cell assemblies, synapsembles, and readers. *Neuron* 68:362–385.

Buzsaki G, Draguhn A. 2004. Neuronal oscillations in cortical networks. *Science* 304:1926–1929.

Canolty RT, Ganguly K, Kennerley SW, Cadieu CF, Koepsell K, Wallis JD, Carmena JM. 2010. Oscillatory phase coupling coordinates anatomically dispersed functional cell assemblies. *Proc Natl Acad Sci U S A* 107:17356–17361.

Carandini M. 2012 From circuits to behavior: A bridge too far? *Nat Neurosci* 15:507–509.

Carleton A, Accolla R, Simon SA. 2010. Coding in the mammalian gustatory system. *Trends Neurosci* 33:326–334.

Carmena JM, Lebedev MA, Henriquez CS, Nicolelis MA. 2005. Stable ensemble performance with single-neuron variability during reaching movements in primates. *J Neurosci* 25:10712–10716.

Carmena JM, Lebedev MA, Crist RE, O'Doherty JE, Santucci DM, Dimitrov DF, Patil PG, Henriquez CS, Nicolelis MA. 2003. Learning to control a brain-machine interface for reaching and grasping by primates. *PLoS Biol* 1:E42.

Chandrashekar J, Kuhn C, Oka Y, Yarmolinsky DA, Hummler E, Ryba NJ, Zuker CS. 2010. The cells and peripheral representation of sodium taste in mice. *Nature* 464:297–301.

Chen X, Gabitto M, Peng Y, Ryba NJ, Zuker CS. 2011. A gustotopic map of taste qualities in the mammalian brain. *Science* 333:1262–1266.

Di Lorenzo PM, Hecht GS. 1993. Perceptual consequences of electrical stimulation in the gustatory system. *Behav Neurosci* 107:130–138.

Di Lorenzo PM, Hallock RM, Kennedy DP. 2003. Temporal coding of sensation: Mimicking taste quality with electrical stimulation of the brain. *Behav Neurosci* 117:1423–1433.

Di Lorenzo PM, Chen JY, Victor JD. 2009. Quality time: Representation of a multidimensional sensory domain through temporal coding. *J Neurosci* 29:9227–9238.

Dobbins EG, Feldman JL. 1994. Brainstem network controlling descending drive to phrenic motoneurons in rat. *J Comp Neurol* 347:64–86.

Dobbins EG, Feldman JL. 1995. Differential innervation of protruder and retractor muscles of the tongue in rat. *J Comp Neurol* 357:376–394.

Erickson RP. 2001. The evolution and implications of population and modular neural coding ideas. *Prog Brain Res* 130:9–29.

Flight MH. 2009. Neural coding: Oscillations help to decode spike patterns. *Nat Rev Neurosci* 10:834–835.

Galvin KE, King CT, King MS. 2004. Stimulation of specific regions of the parabrachial nucleus elicits ingestive oromotor behaviors in conscious rats. *Behav Neurosci* 118:163–172.

Gervasoni D, Lin SC, Ribeiro S, Soares ES, Pantoja J, Nicolelis MA. 2004. Global forebrain dynamics predict rat behavioral states and their transitions. *J Neurosci* 24:11137–11147.

Gestreau C, Dutschmann M, Obled S, Bianchi AL. 2005. Activation of XII motoneurons and premotor neurons during various oropharyngeal behaviors. *Respir Physiol Neurobiol* 147:159–176.

Grill HJ, Norgren R. 1978. Chronically decerebrate rats demonstrate satiation but not bait shyness. *Science* 201:267–269.

Gutierrez R, Carmena JM, Nicolelis MA, Simon SA. 2006. Orbitofrontal ensemble activity monitors licking and distinguishes among natural rewards. *J Neurophysiol* 95:119–133.

Gutierrez R, Simon SA, Nicolelis MA. 2010. Licking-induced synchrony in the taste-reward circuit improves cue discrimination during learning. *J Neurosci* 30:287–303.

Halpern BP, Tapper DN. 1971. Taste stimuli: Quality coding time. *Science* 171:1256–1258.

Halpern BP, Marowitz LA. 1973. Taste responses to lick-duration stimuli. *Brain Res* 57:473–478.

Kaku T. 1984. Functional differentiation of hypoglossal motoneurons during the amygdaloid or cortically induced rhythmical jaw and tongue movements in the rat. *Brain Res Bull* 13:147–154.

Katz DB, Simon SA, Nicolelis MA. 2001. Dynamic and multimodal responses of gustatory cortical neurons in awake rats. *J Neurosci* 21:4478–4489.

Kayser C, Montemurro MA, Logothetis NK, Panzeri S. 2009. Spike-phase coding boosts and stabilizes information carried by spatial and temporal spike patterns. *Neuron* 61:597–608.

Kelley AE. 2004. Ventral striatal control of appetitive motivation: Role in ingestive behavior and reward-related learning. *Neurosci Biobehav Rev* 27:765–776.

Kepecs A, Uchida N, Mainen ZF. 2006. The sniff as a unit of olfactory processing. *Chem Senses* 31:167–179.

Kinzeler NR, Travers SP. 2008. Licking and gaping elicited by microstimulation of the nucleus of the solitary tract. *Am J Physiol Regul Integr Comp Physiol* 295:R436–R448.

Kleinfeld D, Ahissar E, Diamond ME. 2006. Active sensation: Insights from the rodent vibrissa sensorimotor system. *Curr Opin Neurobiol* 16:435–444.

Komiyama T, Sato TR, O'Connor DH, Zhang YX, Huber D, Hooks BM, Gabitto M, Svoboda K. 2010. Learning-related fine-scale specificity imaged in motor cortex circuits of behaving mice. *Nature* 464:1182–1186.

Krause M, German PW, Taha SA, Fields HL. 2010. A pause in nucleus accumbens neuron firing is required to initiate and maintain feeding. *J Neurosci* 30:4746–4756.

Laubach M, Wessberg J, Nicolelis MA. 2000. Cortical ensemble activity increasingly predicts behaviour outcomes during learning of a motor task. *Nature* 405:567–571.

Lemon CH, Smith DV. 2005. Neural representation of bitter taste in the nucleus of the solitary tract. *J Neurophysiol* 94:3719–3729.

Luiten PG, ter Horst GJ, Steffens AB. 1987. The hypothalamus, intrinsic connections and outflow pathways to the endocrine system in relation to the control of feeding and metabolism. *Prog Neurobiol* 28:1–54.

MacDonald CJ, Meck WH, Simon SA, Nicolelis MA. 2009. Taste-guided decisions differentially engage neuronal ensembles across gustatory cortices. *J Neurosci* 29:11271–11282.

Maruyama Y, Pereira E, Margolskee RF, Chaudhari N, Roper SD. 2006. Umami responses in mouse taste cells indicate more than one receptor. *J Neurosci* 26:2227–2234.

Masquelier T, Hugues E, Deco G, Thorpe SJ. 2009. Oscillations, phase-of-firing coding, and spike timing-dependent plasticity: An efficient learning scheme. *J Neurosci* 29:13484–13493.

McClung JR, Goldberg SJ. 2000. Functional anatomy of the hypoglossal innervated muscles of the rat tongue: A model for elongation and protrusion of the mammalian tongue. *Anat Rec* 260:378–386.

Mogenson GJ, Yang CR 1991. The contribution of basal forebrain to limbic-motor integration and the mediation of motivation to action. *Adv Exp Med Biol* 295:267–290.

Neafsey EJ, Bold EL, Haas G, Hurley-Gius KM, Quirk G, Sievert CF, Terreberry RR. 1986. The organization of the rat motor cortex: A microstimulation mapping study. *Brain Res* 396:77–96.

Nicolelis MA, Baccala LA, Lin RC, Chapin JK. 1995. Sensorimotor encoding by synchronous neural ensemble activity at multiple levels of the somatosensory system. *Science* 268:1353–1358.

Pecina S, Berridge KC. 2005. Hedonic hot spot in nucleus accumbens shell: Where do mu-opioids cause increased hedonic impact of sweetness? *J Neurosci* 25:11777–11786.

Rolls ET. 2007. Understanding the mechanisms of food intake and obesity. *Obes Rev* 8(Suppl 1):67–72.

Roussin AT, Victor JD, Chen JY, Di Lorenzo PM. 2008. Variability in responses and temporal coding of tastants of similar quality in the nucleus of the solitary tract of the rat. *J Neurophysiol* 99:644–655.

Roussin AT, D'Agostino AE, Fooden AM, Victor JD, Di Lorenzo PM. 2012. Taste coding in the nucleus of the solitary tract of the awake, freely licking rat. *J Neurosci* 32:10494–10506.

Samuelsen CL, Gardner MP, Fontanini A. 2012. Effects of cue-triggered expectation on cortical processing of taste. *Neuron* 74:410–422.

Saper CB, Chou TC, Elmquist JK. 2002. The need to feed: Homeostatic and hedonic control of eating. *Neuron* 36:199–211.

Sasamoto K, Zhang G, Iwasaki M. 1990. Two types of rhythmical jaw movements evoked by stimulation of the rat cortex. *Shika Kiso Igakkai Zasshi* 32:57–68.

Schaefer AT, Angelo K, Spors H, Margrie TW. 2006. Neuronal oscillations enhance stimulus discrimination by ensuring action potential precision. *PLoS Biol* 4:e163.

Schoenbaum G, Chiba AA, Gallagher M. 1998. Orbitofrontal cortex and basolateral amygdala encode expected outcomes during learning. *Nat Neurosci* 1:155–159.

Simon SA, de Araujo IE, Gutierrez R, Nicolelis MA. 2006. The neural mechanisms of gustation: A distributed processing code. *Nat Rev Neurosci* 7:890–901.

Smith AC, Frank LM, Wirth S, Yanike M, Hu D, Kubota Y, Graybiel AM, Suzuki WA, Brown EN. 2004. Dynamic analysis of learning in behavioral experiments. *J Neurosci* 24:447–461.

Stapleton JR, Lavine ML, Wolpert RL, Nicolelis MA, Simon SA. 2006. Rapid taste responses in the gustatory cortex during licking. *J Neurosci* 26:4126–4138.

Tang CC, Root DH, Duke DC, Zhu Y, Teixeria K, Ma S, Barker DJ, West MO. 2009. Decreased firing of striatal neurons related to licking during acquisition and overtraining of a licking task. *J Neurosci* 29:13952–13961.

Tellez LA, Perez IO, Simon SA, Gutierrez R. 2012. Transitions between sleep and feeding states in rat ventral striatum neurons. *J Neurophysiol* 108:1739–1751.

Theunissen F, Miller JP. 1995. Temporal encoding in nervous systems: A rigorous definition. *J Comput Neurosci* 2:149–162.

Travers JB, Jackson LM. 1992. Hypoglossal neural activity during licking and swallowing in the awake rat. *J Neurophysiol* 67:1171–1184.

Travers JB, Dinardo LA, Karimnamazi H. 1997. Motor and premotor mechanisms of licking. *Neurosci Biobehav Rev* 21:631–647.

Uchida N, Mainen ZF. 2003. Speed and accuracy of olfactory discrimination in the rat. *Nat Neurosci* 6:1224–1229.

Uchida N, Kepecs A, Mainen ZF. 2006. Seeing at a glance, smelling in a whiff: Rapid forms of perceptual decision making. *Nat Rev Neurosci* 7:485–491.

van Eden CG, Buijs RR. 2000. Functional neuroanatomy of the prefrontal cortex: Autonomic interactions. In: *Cognition, Emotion and Autonomic Responses: The Integrative Role of the Prefrontal Cortex and Limbic Structures* (Uylings HBM, VanEden CG, DeBruin JPC, Feenstra MGP, Pennartz CMA, eds), pp. 49–62. Amsterdam: Elsevier Science Bv.

van Wingerden M, Vinck M, Lankelma J, Pennartz CM. 2010. Theta-band phase locking of orbitofrontal neurons during reward expectancy. *J Neurosci* 30:7078–7087.

Wachowiak M. 2011. All in a sniff: Olfaction as a model for active sensing. *Neuron* 71:962–973.

Wesson DW, Verhagen JV, Wachowiak M. 2009. Why sniff fast? The relationship between sniff frequency, odor discrimination, and receptor neuron activation in the rat. *J Neurophysiol* 101:1089–1102.

Whishaw IQ, Kolb B. 1983. “Stick out your tongue”: Tongue protrusion in neocortex and hypothalamic damaged rats. *Physiol Behav* 30:471–480.

Woloszyn L, Sheinberg DL. 2012. Effects of long-term visual experience on responses of distinct classes of single units in inferior temporal cortex. *Neuron* 74:193–205.

Yamamoto T, Matsuo R, Kiyomitsu Y, Kitamura R. 1989. Response properties of lateral hypothalamic neurons during ingestive behavior with special reference to licking of various taste solutions. *Brain Res* 481:286–297.

Yoshida T, Katz DB. 2011. Control of prestimulus activity related to improved sensory coding within a discrimination task. *J Neurosci* 31:4101–4112.

Yuste R. 2011. Dendritic spines and distributed circuits. *Neuron* 71:772–781.

Zaidi FN, Todd K, Enquist L, Whitehead MC. 2008. Types of taste circuits synaptically linked to a few geniculate ganglion neurons. *J Comp Neurol* 511:753–772.

Zeigler HP, Jacquin MF, Miller MG. 1984. Trigeminal sensorimotor mechanisms and ingestive behavior. *Neurosci Biobehav Rev* 8:415–423.

Zhang GX, Sasamoto K. 1990. Projections of two separate cortical areas for rhythmical jaw movements in the rat. *Brain Res Bull* 24:221–230.

Zhao GQ, Zhang Y, Hoon MA, Chandrashekar J, Erlenbach I, Ryba NJ, Zuker CS. 2003. The receptors for mammalian sweet and umami taste. *Cell* 115:255–266.

14 Spike Timing in Early Stages of Visual Processing

Paul R. Martin and Samuel G. Solomon

CONTENTS

14.1 BACKGROUND AND OVERVIEW

This chapter introduces the basic principles of spike timing in the mammalian subcortical visual system. The main components of this system are: (1) the eye, which transduces light energy into nerve signals; (2) the optic nerve and optic tract, along which trains of action potentials from the retinal ganglion cells are transmitted to the brain; (3) the brain centers where the optic-tract fibers terminate. Some of these targets of optic-tract fibers are involved in the control of controlled reflex eye movements and aspects of visual function that do not reach consciousness normally. Other brain targets of optic fibers relay signals directly to the primary visual cortex, and they are required for conscious visual perception (Rodieck 1998).

To a first approximation, ganglion cell spike trains can be understood as a code of local light intensity variation, represented by spike rate (Kuffler 1953; Troy and Shou 2002; Solomon and Lennie 2007; Lee et al. 2010). At this level of approximation, spike timing conveys information only inasmuch as the occurrence of individual spikes allows for an estimation of the time-varying spike rate. However, while a time-varying rate code accounts for a large fraction of the information that the retinal output carries, the timing of individual spikes is a source of information that can be extracted by subsequent processing levels (McClurkin et al. 1991; Reinagel and Reid 2002; Pillow et al. 2008; Jacobs et al. 2009). With these considerations in mind, our main focus is the physiology and circuitry underlying the control of time-varying firing rate of retinal ganglion cells; second (e.g., Section 14.4.5.5), we will consider the mechanisms that control the timing of individual spikes.

We examine the dynamics of retinal ganglion cell inputs from several points of view. First, the early stages of visual processing involve parallel pathways that transmit information about different aspects of the visual world (Stone 1983; Rodieck 1998; Wässle 2004). The nerve circuits that give rise to these specializations are one of the main topics of this chapter. A second principal topic of the chapter is how retinal circuits for light adaptation can influence spike timing in retinal outputs. Third, we consider the consequences of response timing across groups of visual cells with distinct functional properties, and the potential advantages and disadvantages of correlated activity within and across cell populations.

Since our special expertise is in the primate visual system, we will give greater weigh to this subject. In primates, the two best-characterized divisions of the afferent visual pathway pass through the visual dorsal thalamic relay nucleus (lateral geniculate nucleus, LGN). They are named magnocellular and parvocellular pathways for the distinctive layers of the LGN though which their signals pass. The magnocellular layers contain large relay cells and the parvocellular layers contain small relay cells (the anatomical organization of parvocellular and magnocellular pathways is also summarized in Section 14.4.4). The cells in these two pathways show distinct but partly overlapping physiological properties in spatial, temporal, and spectral domains (Shapley and Perry 1986; Kaplan et al. 1989; Casagrande and Xu 2003; Lee et al. 2010), and provide distinct contributions to submodalities of visual perception such as color, form, and motion vision. Understanding spike timing in magnocellular and parvocellular pathways is thus of importance to understanding the functional properties of human vision, and is also of medical relevance for detecting and treating visual dysfunctions.

14.2 GENERATION OF SPIKE TRAINS IN THE OPTIC NERVE

14.2.1 EXCITATORY ("THROUGH") RETINAL PATHWAYS

A schematic anatomical view of the monkey retina (Figure 14.1a) shows the "through" (excitatory) pathway comprising photoreceptors, interneurons called bipolar cells, and retinal output neurons (retinal ganglion cells). Ganglion cells are spiking neurons; their activity can be assessed by extracellular recording from their cell bodies in the retina or their axons in the optic nerve and in the optic tract. Most retinal neurons communicate by graded neurotransmitter release, which in turn is based on graded voltage changes and membrane conductances (Dowling 1987; Wässle and Boycott 1991; Baccus 2007; Oesch et al. 2011). Signal timing in these neurons is therefore studied by measuring changes in the membrane potentials, current flows across cell membranes, and surrogates such as changes in intracellular calcium (Ca^{2+}) concentration. Responses of retinal cell types in the through pathway are shown schematically in Figure 14.1b; responses are represented as membrane potential for photoreceptors and bipolar cells and as trains of action potentials in ganglion cells.

14.2.1.1 Photoreceptors

The light-collecting components of the eye bring an image of the external world into focus near the outer border of the retina, where the photoreceptors (rods and cones) form a monolayer array in the outer nuclear layer (ONL, Figure 14.1a). In whole-animal recordings with intact optics, a visual stimulus is typically projected onto an external screen while ganglion cell responses are recorded in the eye or from axons in the optic nerve. Responses of preganglion cell retinal elements are more commonly studied by focusing a light stimulus directly onto an excised piece of retina. In either case, the stimulus is defined as a change in the local photon flux over a small area of retina, and signal spread in the receptors and postreceptoral neurons is measured. The stimulus in Figure 14.1b is sinusoidal intensity modulation at 5 Hz (i.e., 5 stimulus cycles/s). Cone photoreceptors and rod photoreceptors (discussed in a later section), are hyperpolarized by an increase in light intensity and are depolarized by a decrease in light intensity (Toyoda et al. 1969; Werblin and Dowling 1969; Baylor and Fuortes 1970; Schnapf et al. 1990). The response of photoreceptors is delayed relative to the light intensity changes because the process of phototransduction takes time (the delay constitutes some tens of milliseconds at indoor lighting levels, and

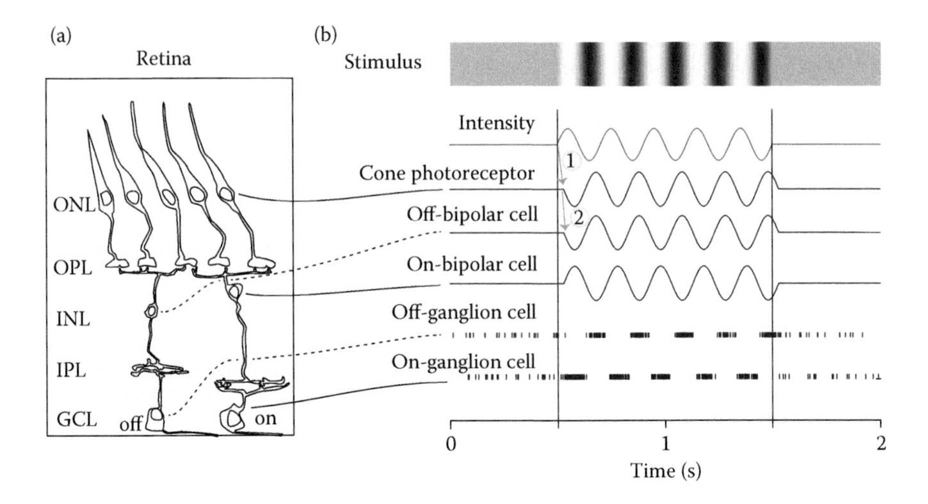

FIGURE 14.1 Response timing in excitatory ("through") retinal pathways. (a) Schematic radial view through the layers of a primate (monkey) retina showing the connections between cone photoreceptors, bipolar cells, and ganglion cells. Cone photoreceptors have cell bodies in the outer nuclear layer (ONL) and make synaptic contact with on-type and off-type bipolar cells in the outer plexiform layer (OPL). Bipolar cells have cell bodies in the inner nuclear layer (INL) and their axonal processes make contact with the dendrites of ganglion cells in the inner plexiform layer (IPL). Ganglion cells have cell bodies in the ganglion cell layers (GCL); their axons leave the retina at the optic disk to form the optic nerve. (b) Schematic representation of signal timing in the retina in response to 1 s of sinusoidal intensity variation at 5 Hz. The stimulus intensity is represented by the upper (intensity) curve. The cone photoreceptor membrane potential is delayed (1) by ~30 ms and is inverted relative to stimulus intensity. The off-bipolar and on-bipolar responses show additional synaptic delay (2) of ~5 ms and the on- and off-types show opposite polarity of membrane potential variation. The membrane potential of bipolar cells is translated to a rectified pulse–frequency code at the synapse between bipolar cells and ganglion cells where trains of action potentials are generated and transmitted to the central visual pathways through the optic nerve.

decreases with the increasing intensity). Thus, under low-to-moderate image contrast, the photoreceptor membrane potential is an inverted and delayed representation of photon absorption rate.

14.2.1.2 Bipolar Cells

The second key step of visual processing occurs at the synapse between photoreceptors and bipolar cells in the outer plexiform layer (OPL, Figure 14.1a). Here, a fundamental dichotomy of response polarity emerges, with distinct types of bipolar cells showing "On-type" or "Off-type" responses. On-bipolar cells are depolarized by light increments and off-bipolar cells are depolarized by light decrements (Werblin and Dowling 1969; Kaneko 1970; Nelson and Kolb 2003; Schiller 2010). This dichotomy is preserved at many subsequent processing levels (Nelson and Kolb 2003; Westheimer 2007). The distinct on- and off-responses arise because on- and off-bipolar cells express different types of neurotransmitter receptors, on the postsynaptic side of the photoreceptor–bipolar synapse. Photoreceptors release the amino acid neurotransmitter glutamate. On-bipolar cells express metabotropic glutamate receptor subtype 6 (mGluR6); the activation of these receptors leads to the closing of cGMP (cyclic guanosine monophosphate)-gated cation channels and bipolar cell hyperpolarization. By contrast, off-bipolar cells express ionotropic glutamate (AMPA)* and kainate receptors, activation of these receptors leads to increased sodium permeability, and bipolar cell depolarization (Slaughter and Miller 1981; Kaneko 1983; Nakajima et al. 1993; Euler et al. 1996; DeVries and Schwartz 1999; Nelson and Kolb 2003). These complementary processes involve quite distinct (ionotropic and metabotropic) biochemical pathways, but their timing must be very similar because the responses of on-type and off-type cells to periodic stimuli appear phase shifted close to 180° (Lee et al. 1989; Lankheet et al. 1998; Burkhardt et al. 2004). We return to the issue of timing symmetry and its importance in a later section (Section 14.4.4). For now, we note that mGluR6 transduction kinetics are much more rapid than those of other metabotropic glutamate receptor classes (Nakanishi 1992), which implies that rapid signaling is an important requirement for a useful visual system. Bipolar cells transmit their signals to the inner plexiform layer (IPL, Figure 14.1a) where they release glutamate at points of synaptic contact with ganglion cells (Massey and Redburn 1987; Mittman et al. 1990; Grünert et al. 2002).

14.2.1.3 Ganglion Cells

All visual signals reaching the brain are transmitted as a pattern of spike trains in ganglion cell axons; thus, understanding how ganglion cells respond to light stimuli is of major importance for understanding visual processing. The receptive field of ganglion cells is customarily defined by the area of retina (or equivalently, the area of visual space imaged on that part of the retina) where presentation of an appropriate stimulus leads to a change in the frequency of action potentials. Ganglion cells express AMPA glutamate receptors, and are therefore depolarized by glutamate released from bipolar cell–axon terminals in the IPL (Figure 14.1a). Ganglion cells thus inherit their on- or off-response signature from the bipolar cells that they contact (Nelson et al. 1993; Wässle 2004; Oesch et al. 2011). Off-bipolar cells contact off-ganglion cells in the outer- (sclerad) half of the IPL and on-bipolar cells contact on-ganglion cells in the inner- (vitread) half of the IPL. Thus, signals about photon flux are first carried as membrane potentials in bipolar cells, then they are converted into pulse trains in ganglion cell bodies, and axons (Figure 14.1b).

Superimposed on the main division into off- and on-type responses are subvarieties of response types that will be described in a later section of this chapter. Additionally, "non-standard" ganglion cell types with complex properties also form important outputs of the retina. In the retina of humans and nonhuman diurnal primates, these nonstandard cells make up only about 10% of ganglion cells but it is worth noting that their total number (~100,000) is close to the total number of ganglion cells in the retina of the cat (Wässle 2004). The attributes of nonstandard cells have been reviewed elsewhere (Masland and Martin 2007) and will not be covered in detail in this chapter.

* Alpha-amino-3-hydroxy-5-methyl-4-isoxazole-propionic acid.

14.2.2 Inhibitory ("Lateral") Retinal Pathways

14.2.2.1 Horizontal Cells

The majority of bipolar cells and retinal ganglion cells show antagonistic "center-surround" receptive fields. The "center" of the receptive field is the region of the retina where excitatory responses (membrane depolarization and/or increased frequency of the action potential) are elicited by light increments (for on-type cells) or decrements (for off-type cells). The center response is generated by the excitatory glutamatergic pathway described above. The antagonistic surround covers a larger, approximately concentric area of the retina and reacts in the opposite way to a light stimulus (Kuffler 1953). The surround is the product of inhibitory retinal processes, which chiefly increases in chloride permeability following the binding of GABA* and glycine to receptors on bipolar cells and retinal ganglion cells (see Kamermans and Spekreijse 1999; Flores-Herr et al. 2001; Wilson 2003; Wässle 2004; Duebel et al. 2006).

In the OPL, horizontal cells receive input from multiple cone photoreceptors and send reciprocal output to cones and feed-forward output to cone bipolar cells (Figure 14.2a). Horizontal cells are also joined by gap junctions to form a syncytial network. Hence, changes in illumination at the position of any given cone can lead to widespread influences on other cones and bipolar cells via horizontal cells (Wilson 2003; Wässle 2004). It is now known that the principle shown in Figure 14.2b and c for off-bipolar cells applies for on-bipolar cells, that is, inhibition by horizontal cells counteracts the depolarization by light increments. This is likely because GABA release causes outward (depolarizing) chloride currents in on-bipolar dendrites (Duebel et al. 2006).

Since the spatial extent of surround inhibition is greater than the spatial extent of center excitation, the receptive fields of downstream retinal neurons become selective for the size of a stimulus: the most effective stimulus becomes one which fills the center but does not encroach on the surround. When responses to spatial intensity modulation are measured, the center–surround organization of bipolar and ganglion cell receptive fields is manifested as band-pass tuning for spatial frequency (Enroth-Cugell and Robson 1966; Derrington and Lennie 1984; Frishman et al. 1987).

Surround inhibition has an important consequence for spike timing in ganglion cells. In the OPL, the inhibitory (cone → horizontal cell → bipolar cell) surround circuit introduces a synaptic delay of 5–10 ms compared to the excitatory (cone → bipolar cell) center circuit (Frishman et al. 1987; Perlman et al. 2003; Crook et al. 2011). The consequence of this additional delay is shown schematically in Figure 14.2b and c, which shows the effects of intensity modulation at low (5 Hz, Figure 14.2b) and high (50 Hz, Figure 14.2c) temporal frequency. A 10 ms synaptic delay represents a phase lag of only 18° at 5 Hz, so that the phase of the surround inhibition is close to the phase of center excitation (shaded rectangle, Figure 14.2b), and the center and surround signals largely cancel each other. But now, consider more rapid temporal modulation. The same 10 ms delay in the surround circuit represents a phase lag of 180° at 50 Hz; thus at high frequency, the surround inhibition is minimal when center excitation is maximal, and the horizontal signals no longer subtract from (or, alternatively, no longer "shunt") the center excitation (shaded rectangle, Figure 14.2c).

* Gamma-aminobutyric acid.

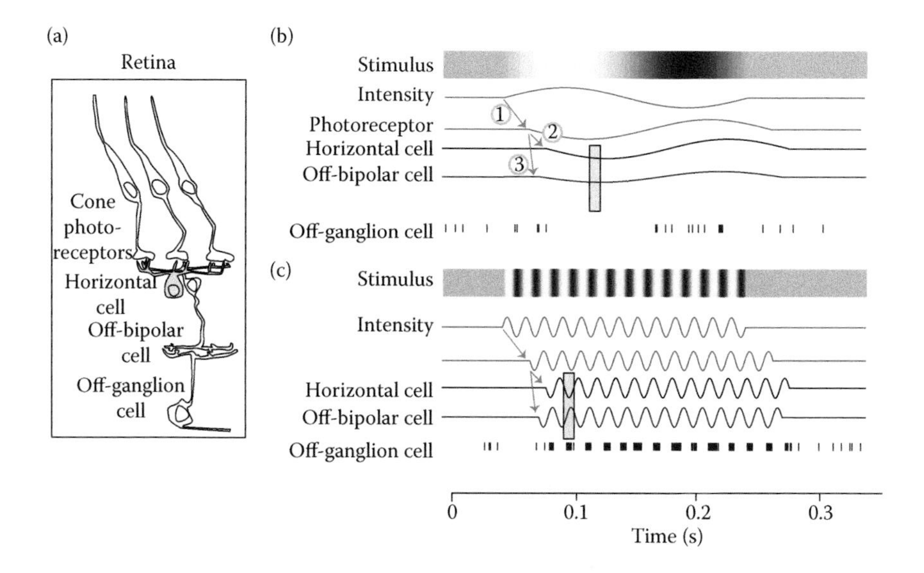

FIGURE 14.2 Response timing in lateral inhibitory retinal pathways. For simplicity, only the lateral connections in OPL are illustrated. (a) Inhibitory interneurons (horizontal cells) contact bipolar cells and cone photoreceptors. The horizontal cells are depolarized by glutamate released from photoreceptors and produce widespread inhibition to generate the antagonistic receptive field surround. (b) Schematic representation of signal timing in the same format as Figure 14.1. In addition to (1) phototransduction delay, the inhibitory effect of horizontal cells is delayed (2) by 5–10 ms relative to the direct photoreceptor excitation of (3) bipolar cells. At low temporal frequencies, the phase delay of inhibition relative to excitation is small (shaded box); so ganglion cell response is weak. (c) At high temporal frequencies, the phase delay represents close to half the stimulus cycle (shaded region) so that the inhibition becomes synergistic to direct excitation and ganglion cell responses are boosted.

In sum, lateral inhibition by horizontal cells makes the receptive fields of bipolar and ganglion cells selective for spatial and temporal frequency. At low spatial or temporal frequencies, the surround reduces the center's effectiveness, but at high spatial or temporal frequency, the surround is ineffective or (if the inhibition is truly subtractive), it can even act to enhance the center response.

For simplicity, many important details of horizontal cell organization are not shown in Figure 14.2. For example, most vertebrate retinas studied so far contain different subtypes of horizontal cells that can be distinguished by their patterns of connectivity with different spectral classes of cone photoreceptors (Kolb et al. 1980; Dacey et al. 1996; Perlman et al. 2003). Furthermore, horizontal cell–axon terminals make contact with rod photoreceptors and specialized rod bipolar cells in the scotopic (night vision) retinal pathways (Boycott and Kolb 1973; Perlman et al. 2003). The fact that surround inhibition is substantially weaker under scotopic than under photopic conditions (Barlow et al. 1957; Enroth-Cugell and Lennie 1975; Troy et al. 1999); this fact presents a problem for interpreting the practically identical anatomy of rod and cone contacts to horizontal cells.

14.2.2.2 Amacrine Cells

Vertebrate retinas contain multiple subclasses of amacrine cells, which make contact with the axon terminals of bipolar cells and dendrites of ganglion cells in the IPL (Vaney 2002). The function of some of these amacrine cell types is well described, but for most, we have at best a rudimentary understanding. Two well-understood circuits are (1) the "day-night switch" circuit whereby the AII class of amacrine cell communicates signals from rod bipolar cells to cone bipolar cells in scotopic conditions (Bloomfield and Dacheux 2001; Wässle 2004) and (2) the starburst amacrine circuit, that generates selectivity for image motion in some classes of ganglion cells (reviewed by Vaney and Taylor 2002). As should be obvious from these two contrasting examples, generalizations about the role of amacrine cells in controlling spike timing in ganglion cells are not easy to make. However, the following broad principles apply (Wässle and Boycott 1991; Vaney 2002). First, the great majority of amacrine cells release the inhibitory neurotransmitters glycine or GABA. Second, glycinergic amacrine cells typically show "narrow field" morphology, meaning an individual cell integrates signals over a more restricted region of the retina than that covered by a typical GABAergic amacrine cell. Third, the inhibitory outputs of amacrine cells are commonly arranged as feed-forward inhibition at "dyad" synapses, meaning they can act to sharpen temporal tuning both by inhibiting ganglion cells directly and also by cutting-off bipolar excitation to ganglion cells (Nirenberg and Meister 1997). In other words, amacrine cells can high-pass filter the excitatory input to ganglion cells. The amacrine/bipolar/ganglion cell microcircuit may also generate oscillatory potentials in the IPL and may also impose resonance on the signals of ganglion cells (Lachapelle 2006). Such resonances can become apparent at high temporal frequencies, as described below.

14.3 ADAPTIVE PROCESSES AND RESPONSE TIMING

14.3.1 Light Adaptation and Ganglion Cell Temporal Response

14.3.1.1 Rod and Cone pathways

Visual environments show wide variation and vertebrate retinas have adaptive mechanisms to extract information from different images (Walraven et al. 1990). Consider a dim-light environment, such as starlight. Only a few photons are reflected from an object to the eye and each photon may provide an important source of information about the outside world. In brighter environments, visual images are largely redundant. Across time and space, the contrast of retinal images is roughly invariant with the intensity of the illumination so that it is more useful to signal change (contrast) than absolute light level. At low light levels, the signals of photoreceptors are summed across time and space to maximize sensitivity; at high light levels, the receptor signals are optimized by adaptation processes to signal contrast, as explained below.

Evolution has dealt effectively with different light environments by parceling out light detection to different circuits at different light levels, and by providing adaptive mechanisms at receptor and postreceptor levels. The eyes of most mammals contain two major classes of photoreceptors (rods and cones) with different light sensitivities (Walraven et al. 1990; Rodieck 1998; Lamb et al. 2007). The presence of two receptor classes, plus dedicated retinal circuits to amplify small signals from rod receptors, allow vision over a wide range of light environments from starlight (where rods can act as single photon detectors) to high albedo desert sunlight (where rods are saturated but cones can provide modulated signals). The details of the retinal pathways by which the rod and cone signals reach ganglion cells are reviewed elsewhere (Rodieck 1998; Bloomfield and Dacheux 2001; Wässle 2004); our focus here is on the influence that ambient light level has on optic nerve spike trains.

On transition from low background intensity (starlight or moonlight) to high background intensity (bright indoor lighting or daylight), the main changes in ganglion cell functional properties are reduction in absolute increment sensitivity, increased sensitivity to rapid intensity changes, and increased spatial acuity. The temporal sensitivity changes result from variation in the response kinetics of rods and cones at low-to-intermediate light levels. Most importantly, with increasing photopic light levels, the phototransduction shutoff time in cones is decreased, which truncates the receptor response and increases temporal resolution (Pugh et al. 1999; Lamb 2011). Under intense daylight levels, cone receptor sensitivity is also lowered by reduced availability of activatable (unbleached) pigment (Lamb 2011). Increased spatial acuity at photopic versus scotopic levels of illumination is attributable to reduced spatial pooling in cone pathways compared to rod pathways within the retina (Rodieck 1998; Bloomfield and Dacheux 2001; Wässle 2004). Changes in temporal frequency sensitivity are considered here in detail because they have influences on response timing in optic nerve axons.

14.3.1.2 Timing of Ganglion Cell Responses in Day and Night Vision

Figure 14.3 shows temporal modulation transfer functions (tMTF) of ganglion cells recorded at comparable pairs of background intensities under different experimental conditions. Figure 14.3a shows the average tMTFs of phasic (parasol–magnocellular pathway) ganglion cells recorded *in vivo* from macaque retina (Lee et al. 1990). The background light level is expressed in trolands (Td), a photometric unit that incorporates the effect of pupil size on retinal image intensity. At 2 Td (moonlight) intensity, the tMTF shows low-pass characteristics with peak near 3 Hz and negligible sensitivity above 10 Hz. At 2000 Td (daylight) intensity, the response has much stronger band-pass characteristic with a peak near 20 Hz and extending to frequencies above 80 Hz. These changes have two main causes. First, compared to the responses of cones, responses of rods have slow onset and decay slowly, making them less able to follow rapid intensity changes. At 2 Td, the cone receptors are near the lower sensitivity limit whereas rods are in the middle of their sensitivity range. The tMTF at 2 Td, therefore, is mostly determined by the sluggish temporal characteristic of rod photoreceptors. Second, with increasing intensity, postreceptoral circuits play an increasingly important role in light adaptation. The inhibition from horizontal and amacrine cells reduces low temporal frequency sensitivity, and differential delays in these circuits create resonances that boost sensitivity at high temporal frequencies.

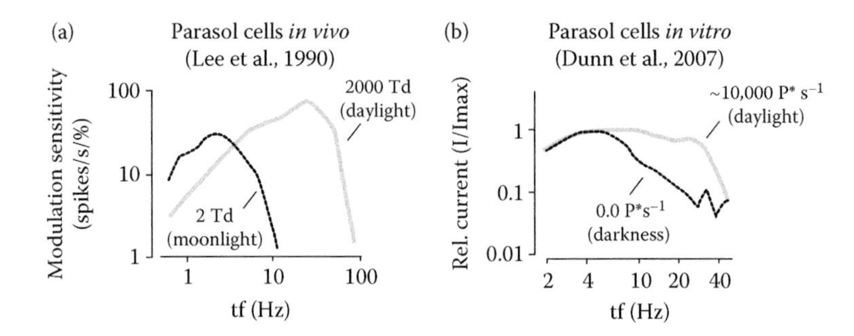

FIGURE 14.3 Response timing in day and night vision. (a) Temporal modulation transfer of parasol–magnocellular pathway ganglion cells under moonlight (scotopic) and daylight (photopic) background intensity. Note the increased response and sharper band-pass character of ganglion cell responses under photopic conditions; these changes are consistent with different temporal properties of rod and cone receptors. (Modified from Lee, BB et al. 1990. *Journal of the Optical Society of America A* 7, 2223–36.) (b) Temporal modulation transfer of transmembrane current in parasol cells recorded *in vivo* under conditions where rod photoreceptor activity is eliminated. This result shows that additional temporal filtering enhances cone circuit responses to high temporal frequencies at high background intensities. (Modified from Dunn, FA et al. 2007. *Nature* 449, 603–07.)

Complementary to Figure 14.3a, Figure 14.3b shows transmembrane currents recorded from a parasol–magnocellular pathway ganglion cell *in vitro,* under conditions where rod activity was eliminated by pharmacological blockade and/or selective adaptation (Dunn et al. 2007). Here, the natural optics are not present and the background light level is expressed as photon absorption rate per receptor (P*) per second. In normal intact human eye, 1 scotopic troland represents ~8.6 P* s^{-1} per rod (Breton et al. 1994). Consistent with the *in vitro* recordings, the difference between the tMTF in darkness (0.0 P* s^{-1}) and sunlight (10,000 P* s^{-1}) is greatest at temporal frequencies above 10 Hz, again likely reflecting temporal tuning in postreceptoral circuits. Pandarinath et al. (2010) showed that low-frequency shift in temporal tuning in scotopic conditions is more marked in on-type than in off-type cells. This asymmetry would in principle compensate for the reduced detectability of increments when the photon catch is low (Pandarinath et al. 2010), but the retinal circuits underlying the difference are not established.

14.3.2 Tracing Adaptation Pathways in the Retina

Further evidence for distinct receptoral and postreceptoral influences on timing in retinal circuits is shown in Figure 14.4 (modified from Dunn et al. 2007). The recordings here were made from midget–parvocellular pathway neurons (also see Section 14.4.4) as shown schematically in Figure 14.4a, under conditions that eliminate activity in rod pathways (as in Figure 14.3b). Between darkness and indoor light levels (Figure 14.3b, left and center columns), there is little change in the sensitivity of individual cone receptors (top row) or midget–parvocellular pathway bipolar cells (center row), but sensitivity in the ganglion cells is reduced (lower row) indicating signal

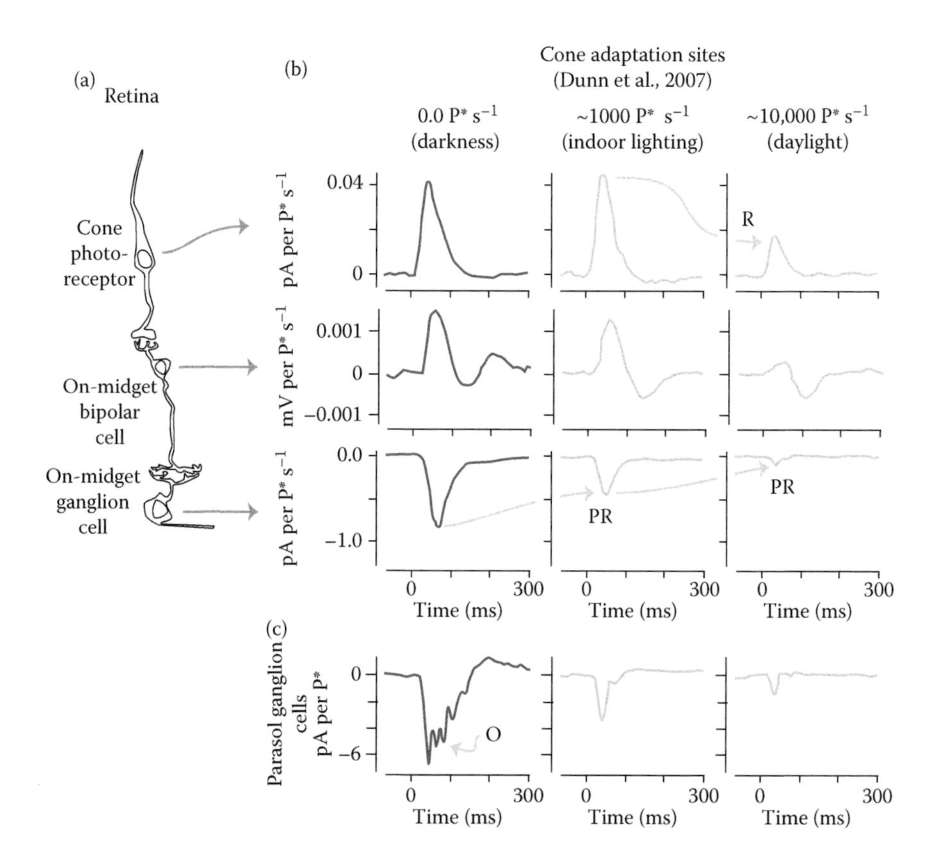

FIGURE 14.4 **(See color insert.)** Influence of receptor and postreceptor circuits on response timing in the retina. (a) Schematic of midget–parasol pathway connections in monkey retina. The cone photoreceptors connect via midget–bipolar cells to midget–ganglion cells. (b) Light-evoked responses (membrane current flow or membrane potential changes) in these three cell classes. Each row shows the responses of one cell type at three different background light intensities. The sensitivity of individual cone receptors (upper row) shows little change between darkness and indoor light levels but receptor adaptation (R) is apparent with increases to outdoor light levels. Sensitivity of midget–bipolar cells (center row), parallels cone sensitivity, indicating that bipolar adaptation is inherited from the receptors. By contrast, ganglion cells show signs of postreceptor adaptation (PR) on transition from darkness to indoor levels (center column) and from indoor to outdoor lighting levels (right column). (c) The recordings from parasol cells show oscillatory potentials (O) attributable to reciprocal and feed-forward synaptic circuits in the IPL. (Modified from Dunn, FA et al. 2007. *Nature* 449, 603–07.)

transfer between bipolar and ganglion cells as a site of adaptation across this range of intensity. Further increase of background intensity (to a level roughly comparable with indirect sunlight, Figure 14.4b, right column) reduces sensitivity both at the receptor (R) and postreceptoral (PR) stages of retinal processing. The likely involvement of amacrine cell inhibitory circuits is supported by the recordings of parasol cells (Figure 14.3b, lower row) that show oscillatory potentials (O); these oscillations are considered to arise in the inner nuclear layer from reciprocal and feed-forward amacrine cell synapses to bipolar and ganglion cells (Boycott and Dowling 1969; Koontz and Hendrickson 1987; Raviola and Dacheux 1987). The oscillatory potentials are most prominent at low light levels, consistent with resonance in circuits amplifying small signals under these conditions.

14.4 RESPONSE TIMING IN PARALLEL VISUAL PATHWAYS

14.4.1 Origin of Sustained and Transient Channels

The ganglion cell outputs of the retina consist multiple parallel pathways tuned to different parts of the time, space, and wavelength spectrum of the visual image. In Section 14.2.1.2, we saw how the first synapse in the retina, between photoreceptors and bipolar cells, is where the fundamental dichotomy on on-type and off-type responses is generated. Overlaid on this dichotomy is bipolar population showing more low-pass ("sustained") or band-pass ("transient") temporal characteristics. Just as on-type and off-type responses are based on the expression of distinct glutamate receptors by bipolar cells, the sustained and transient channels are thought to be initiated by subtle differences within the on-type and off-type glutamate receptor classes.

Figure 14.5a–c shows how sustained and transient channels emerge in the retina of a terrestrial diurnal mammal (13-lined ground squirrel *Spermophilus tridecemlineatus*). The distinct morphology of two of the three subclasses of off-type bipolar cells identified in this species is shown in Figure 14.5a (modified from DeVries 2000). Subtype b2 expresses AMPA-sensitive glutamate receptors at the cone synapse whereas subtype b3 expresses kainate-sensitive receptors. The b2 cell type shows more transient responses to glutamate released by cone photoreceptors than does the b3 type (Figure 14.5b), and the b2 type shows more rapid recovery from sustained cone depolarization (Figure 14.5c). These distinct response kinetics tune the cells to different rates of glutamate release, which are in turn produced by different temporal frequencies in the visual image. The b2 subtype is tuned to transient intensity changes whereas the b3 subtype is tuned to slower (sustained) intensity changes. An analysis of receptor expression at the cone-to-bipolar synapse in monkey retinas suggests that different bipolar classes express different complements of AMPA and kainate receptors (Morigiwa and Vardi 1999; Haverkamp et al. 2000, 2001), but the necessary details of which bipolar cells express which receptor classes have not yet been filled in. It is puzzling that the only specifically tested bipolar type (flat-midget–bipolar cell) expresses rapidly adapting AMPA receptors (Puller et al. 2007), which is inconsistent with the prediction that midget–bipolar cells would show sustained responses (as they are part of the parvocellular pathway; also see Section 14.4.4). Resolution of this inconsistency is obviously an important goal for interpreting the function from anatomical studies of the human and monkey retina.

14.4.2 Sustained and Transient Ganglion Cell Responses

14.4.2.1 Origins of Sustained and Transient Responses

The best-understood parallel ganglion cell systems are in the cat (where *X*-class cells show sustained responses, and *Y*-class cells show transient responses (Enroth-Cugell

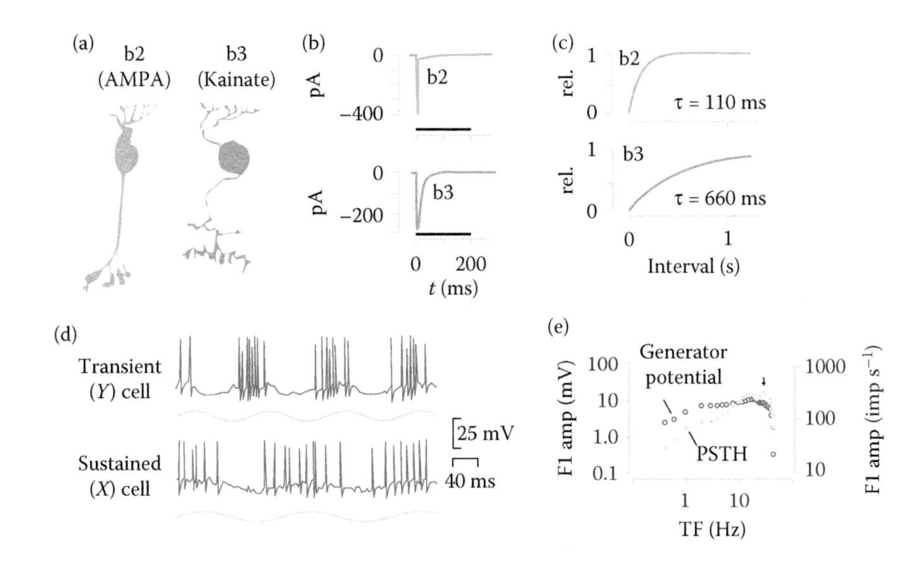

FIGURE 14.5 Sustained and transient channels in the retina. (a) Examples of bipolar cells drawn following intracellular injection of fluorescent dye in ground squirrel retina. Subtype b2 expresses AMPA-sensitive glutamate receptors at the cone synapse whereas subtype b3 expresses kainate-sensitive receptors. (b) The responses (transmembrane currents) to 200 ms depolarization (black bar) of presynaptic cone photoreceptors show that b2 cell responses are more transient than b3 cell responses. (c) The response recovery time (normalized responses to the second of two cone-depolarizing pulses) of the b2 cell is more rapid than the response recovery time of the b3 cell. These distinct response kinetics tune the b2 cells to transient intensity changes and tune b3 cells to sustained intensity changes. (d) Intracellular recordings of transient (*Y*) cell and sustained (*X*) cat ganglion cell responses to sinusoidal intensity modulation (gray traces). The *Y*-cell response is more transient and is phase advanced relative to the *X*-cell response. (e) Comparison of the average generator potential and spike frequency measured from peristimulus time histograms (PSTH) shows the enhancement of high temporal frequencies in the PSTH relative to the generator potential. (Panels (a)–(c) modified from DeVries, DH. 2000. *Neuron* 28(3), 847–56; and panels (d) and (e) modified from Lankheet, MJM et al. 1989b. *Vision Research* 29(5), 505–17.)

and Robson 1966; Cleland et al. 1971)) and in the monkey (where parvocellular-projecting cells show sustained responses, and magnocellular-projecting cells show transient responses (Wiesel and Hubel 1966; Dreher et al. 1976; Derrington and Lennie 1984)).

The different temporal channels that most likely emerge at the first synapse are further tuned at the second synaptic layer of the retina, where processes of amacrine cells, bipolar cells, and ganglion cells make contact. The principles here are established but many details remain poorly understood. Reciprocal and feed-forward synapses between amacrine and bipolar cells should truncate sustained glutamate release at bipolar terminals, and might be expected to predominate in circuits serving transient rather than sustained channels (Baccus 2007). But the anatomical and functional evidence are not consistent. Anatomical analyses of inputs to parallel ganglion cell pathways (in cat and in the monkey) show generic, rather than specialized, microcircuitry of synaptic inputs to sustained versus transient ganglion cells (Kolb 1979; Koontz and Hendrickson 1987; Kolb and Dekorver 1991; Calkins and Sterling 1996; Grünert et al. 2002). On the other hand, it could be argued that functional specificity is shown by the presence of oscillations in light-induced synaptic currents of parasol cells but not in midget cells (Figure 14.4b and c) and high-frequency resonance in parasol tMTFs (Figure 14.3). These differences may depend on yet-to-be-discovered fine differences in the inner retinal circuitry, but may also result from the activation of generic circuits by bipolar cells with distinct temporal properties.

14.4.2.2 Generation of Sustained and Transient Spike Trains

The spike rate code of ganglion cells is inherently nonlinear. Most obviously, ganglion cell firing rates cannot fall below zero, and cannot exceed the maximum set by the absolute and relative refractory periods of the action potential (typically 1–2 ms). More subtle nonlinear processes also shape the temporal properties of ganglion cell spike trains. Figure 14.5d compares the transient (*Y*) cell and sustained (*X*) cell responses from intracellular recordings in cat retina (Lankheet et al. 1989a). The stimulus was 6.25 Hz sinusoidal modulation of a circular light spot covering the receptive field center. The transient (*Y*) cell response appears as a train of narrowly spaced spikes that encompass ~25% of the stimulus phase cycle. The sustained (*X*) cell pulse train is more widely spaced (i.e., the spike is of lower frequency) and the pulse train fills the majority of the stimulus phase cycle. Thus, at a crude level, the spike rate of the sustained cell gives a more faithful representation of the input than does the spike rate of the transient cell. Nevertheless, for both sustained and transient cells, the spike-generating mechanism acts as a high-pass filter, by attenuating low-frequency components in the underlying generator potential (Figure 14.5e). In summary, the temporal integration required to reach the spiking threshold acts as a high-pass temporal filter; detailed treatments of the spike-generating process have been given elsewhere (Lankheet et al. 1989b; Meister and Berry 1999).

14.4.3 Temporal Receptive Fields

14.4.3.1 Goal of Temporal Receptive Field Description

We have introduced the mechanisms that shape temporal response properties in the retina, and we have illustrated how different response properties emerge. But we would also like to predict how different stages of the retinal pathway will respond to arbitrary patterns of light, and for this, we need simple mathematical descriptions of these processes. For example, we now have good biophysical models of the photon-triggered cascade of events within a photoreceptor (Lamb and Pugh 2006), of horizontal cell feedback to cone photoreceptors (Smith et al. 2001), and of the spike-generating mechanism in ganglion cells (Lankheet et al. 1989b; Meister and Berry 1999). The overarching goal here is to predict how imaged photons are transformed into action potentials in the optic nerve; in other words, the goal is to develop a capable model of the receptive fields of ganglion cells.

14.4.3.2 Linear Models

The visual response of many ganglion cells can be well described by describing the pathway from receptors to ganglion cells as a series of linear filters, and by describing the ganglion cell response as a linear sum of these inputs. The convolution of a visual image with the receptive field is thus represented in the rate of ganglion cell action potentials. The standard model for ganglion cell receptive fields consists of (1) a spatial function specifying the strength and sign of inputs to the ganglion cells as a function of position along the retinal sheet, (2) a temporal function specifying how the ganglion cell inputs change in time, and (3) a function to convert the ganglion cell membrane potential into spike rate. Under sinusoidal intensity modulation, the discharge of a linear cell is modulated at the fundamental frequency of the stimulus, and for low contrast and low-to-intermediate spatial frequencies, the temporal frequency limits of spike generation can be safely ignored. These facts mean that a histogram of ganglion cell spike rates, created from an average of many stimulus cycles, allows reasonable inference of the underlying analog signal. Simple models of this kind are sufficient to describe most of the visual response of *X*-cells in cat retina, and of parvocellular- and magnocellular-projecting cells in monkey retina (Derrington et al. 1984; Frishman et al. 1987; Lee et al. 2010).

Early work used the above-described linear systems approach to measure the spatial and temporal receptive fields of ganglion cells, and showed that most ganglion cells are bandpass in spatial and temporal frequency. However, the spatial- and temporal-frequency tuning functions are not independent because the temporal frequency response becomes more bandpass at low spatial frequency. This interdependence arises largely because of the delay between center-and-surround signals to the receptive field discussed above (see Section 14.2.2). At high spatial frequencies, the surround is largely ineffective; so center–surround interactions are much weaker. Further, resonant circuits in the inner retina are less effective at high spatial frequency.

14.4.3.3 Nonlinear (*Y*)-Receptive Fields in Cat Retina

An important early discovery was that some ganglion cells do not show linear spatial summation (Enroth-Cugell and Robson 1966). Nonlinear spatial summation is most obvious in *Y*-class cells of cat retina. For counterphase modulation of gratings of high spatial frequency, the spike rate of *Y*-cells increases in both the on-phase and off-phase of modulation, meaning that the response tracks the second harmonic of the temporal modulation frequency (Enroth-Cugell and Freeman 1987). Further, on exposure to a high spatial frequency-drifting grating, *Y*-cells respond with an unmodulated increase in the discharge rate. The simplest explanation for these properties is that bipolar signals feeding different parts of the receptive field are half-wave rectified (clipped), either by a threshold or a saturating nonlinearity (Hochstein and Shapley 1976; Victor and Shapley 1979). The rectification means that the signals from a bipolar cell that "sees" the bright phase of the stimulus are not canceled by a bipolar cell that "sees" the dark phase of the stimulus, causing the ganglion cell to be depolarized and to respond to both phases of modulation. This hypothesis is supported by the observation that the second harmonic response can be elicited with gratings that are much finer than can be resolved by the linear *Y*-cell center mechanism (Hochstein and Shapley 1976). A likely site for rectification is at the synapse between bipolar cells and amacrine or ganglion cells (Werblin and Dowling 1969; Nelson and Kolb 2003).

14.4.3.4 Nonlinear (*Y*)-Receptive Fields in Monkey Retina

Early work suggested that second harmonic responses were much less prevalent in monkey retinal ganglion cells, and true "Y-class" behavior is rarely observed only in *in vivo* recordings from monkey retina or LGN (Kaplan and Shapley 1986; Kaplan 2003). More recent recordings made *in vitro* from peripheral regions of monkey retina suggest that *Y*-class responses are more common than first thought, and can be elicited in both magnocellular-projecting ganglion cells and sparser ganglion cell classes (Crook et al. 2008). One possibility is that the nonlinear behavior is at best weakly manifest *in vivo* because the eye's natural optics attenuate high spatial frequencies, and so it is not possible to construct stimuli capable of driving second harmonic response. This explanation is supported by the fact that midget pathway cells, which show highly linear behavior under natural optics, show much more complex receptive field substructure when stimuli produced by interference fringes (bypassing natural optics) are used (McMahon et al. 2000). An alternative explanation is that the rectification is malleable, and for whatever reason, it is more pronounced in the *in vitro* preparation.

14.4.3.5 Contrast Gain Control

In addition to the strong nonlinear behavior defining *Y*-cells, more subtle nonlinearities are common in both cat and monkey retina. The best studied is the so-called contrast gain control, which is prominent in the receptive fields of both *X*- and *Y*-class cells of cat retina, and magnocellular-projecting cells in monkey retina (Shapley and Victor 1978, 1981; Kaplan and Benardete 2001). These cell classes are most sensitive to changes in contrast when the overall contrast of the retinal image is low, and are less sensitive to changes in contrast when the overall image contrast is higher. The temporal frequency tuning of these cells also depends on contrast, shifting from lower-to-higher-temporal frequencies as image contrast increases; this shift is accompanied by a reduction in latency to the response peak. One explanation for these changes is that transmission at the bipolar–ganglion cell synapse depends on the overall amplitude of bipolar cell response—becoming faster and less sensitive when bipolar cell responses are large. This explanation, however, does not account for the fact that contrast gain can be reduced by a stimulus located outside the classical receptive field, that is, a stimulus that does not by itself change the ganglion cell firing rate. Thus, contrast gain control cannot simply be explained by activity-sensitive mechanisms such as synaptic depression or spike generation. An alternative (albeit less specific) proposal is that a separate pathway calculates the image contrast over a large area of space (much larger than the receptive field) and over a long time window (tens or hundreds of milliseconds). The resultant signal is fed back either to bipolar cells or to ganglion cells via some form of inhibition.

A model of retinal output that accurately predicts the ganglion cell responses to random-noise stimuli must include the kinds of temporal nonlinearities described in the preceding paragraphs, and must also incorporate the details of the spike-generating process. As described above, the impact of light intensity and contrast on the temporal response are largely independent and can be mathematically described by resistor–capacitor and simple-oscillator circuits. Recent work has also provided equations that can describe the spike generation process, including the impact of voltage-gated currents (Pillow et al. 2008). Thus, it is now possible to take the response of any ganglion cell to arbitrary visual stimuli, and to predict the probability that a spike will be generated. An important limitation to this statement is discussed below (Section 14.4.5.5), that is, if ganglion cells show correlated activity, then the source and structure of the correlations between cells needs to be known.

14.4.4 Spike Timing On and Off Ganglion Cells

Given that the photoreceptor signals are transferred to on- and off-bipolar cells by different types of glutamate receptors, and that these pathways are served by largely independent mechanisms in the inner retina, it seems no accident that under daylight conditions, the temporal signals of on- and off-ganglion cells are almost exactly symmetric (see Section 14.2.1.2). It is true that there may be subtle asymmetries in the response of on- and off-ganglion cells at photopic light levels (also see Chichilnisky and Kalmar 2002). Yet, without knowledge of the stimulus light polarity, an experimenter would find it impossible to distinguish the responses of an on-cell from the responses of an off-cell. Furthermore, variance in the response latency between on- and off-cells within a given ganglion cell class is much less than variance in the response latency between cell classes (e.g., between *X* and *Y* classes; Lu et al. 1995). Thus, it is fair to ask: why are signals of on- and off-pathways largely symmetric, and why do they have almost identical latency?

We give here two general hypotheses to address this question. The first hypothesis is that for some reason, it is important that on- and off-signals arrive simultaneously for later stages of visual processing in the brain. This hypothesis is attractive, but has two main problems. First, visual response latency in the brain depends on how far the action potential had to travel; for example, in unmyelinated axons in the retina, propagation speed is slow and signals of ganglion cells further from the optic disk take longer to get to the rest of the brain. Second, it is usually thought that the signals of different pathways (e.g., *X* and *Y*, or *P* and *M*) are recombined at some stage in the visual cortex; yet, their signals have very different response latencies. We propose here an alternative hypothesis based on the requirements of intraretinal processing. One very important function of the retina requires tight temporal synchrony of on- and off-pathways; that is, the retinal circuits that create direction-selective cells (Barlow and Levick 1965; Vaney and Taylor 2002). Direction-selective cells support the critical "pursuit" eye movements for tracking moving objects, and other ocular reflexes. If on- and off-type inputs to these cells had different latencies, their responses would confound contrast polarity with retinal slip velocity. In other words, our alternative hypothesis is that timing symmetry in on- and off-pathways is the outcome of evolutionary drive to improve retinal computations, and not a constraint imposed by high-level visual processes. The partial breakdown of on–off symmetry under scotopic conditions (Pandarinath et al. 2010, see Section 14.3.1.2) is consistent with this alternative hypothesis, on the assumption that under scotopic conditions, retinal computations are optimized for maximum sensitivity. This topic is taken up elsewhere in this book (Chapter 1).

14.4.5 Spike Timing in Magnocellular and Parvocellular Pathways

14.4.5.1 Anatomical Organization

Diurnal (day-active) primates are specialized for visually guided food foraging. At the center of the field of vision, where visual acuity is the highest, is a specialized region of the retina called the fovea. Here, cone photoreceptors are tightly packed and make one-to-one connections with midget–bipolar cells and midget–ganglion cells (Wässle et al. 1989; Kolb and Dekorver 1991). The axons of midget–ganglion cells terminate in the parvocellular layers of the LGN. The low convergence in the midget system and the high density of cone photoreceptors mean that midget–parvocellular pathway cells serving foveal vision (the central-most 2° of the visual field) form more than half of all axons in the optic nerve. Midget–parvocellular cells show sustained responses to maintained contrast, and in primates with trichromatic ("normal") color vision, most midget–parvocellular cells in the fovea show red–green color opponent properties.

The second main set of thalamic projecting ganglion cells consists parasol morphology ganglion cells, which in the fovea receive convergent input from ~50 cones via diffuse bipolar cells (Goodchild et al. 1996; Silveira et al. 2004). Parasol cells project to the magnocellular layers of the LGN. Parasol cells show transient responses to maintained contrast and in trichromatic cells, primates get excitatory input from medium- ("green") and long-wavelength ("red") sensitive cones.

With increasing distance from the fovea, the number of cones feeding midget–bipolar cells increases as does the number of midget–bipolar cells converging onto midget–parvocellular ganglion cells; likewise, convergence increases with eccentricity for parasol–magnocellular cells. The increased convergence leads to systematic eccentricity-dependent changes in the receptive field properties in both cell classes.

14.4.5.2 Summary of Response Properties

Figure 14.6 summarizes a typical experimental configuration for *in vivo* assessment of spike timing in magnocellular and parvocellular pathways, and compares the spatial- and temporal-tuning properties in these pathways. An anesthetized animal is placed before an apparatus that can produce visual stimuli. Nowadays, a computer monitor is the most commonly used visual stimulation apparatus. The image of the computer monitor is focused through the eye's optics onto the retina while the extracellular recorded activity of a single neuron in the LGN is monitored. The stimulus (in this example, a drifting sine grating) is moved to find the part of the visual field that causes changes in spike rate of the recorded neuron; this part of the visual field defines the classical receptive field of the neuron. The majority of neurons in the primate LGN is "relay cells," which get dominant input from one retinal ganglion cell and have an axonal projection to the primary visual cortex (V1, Figure 14.6a). With suitable recording conditions, it is possible to distinguish excitatory postsynaptic potentials (arising from the retinal input) in the recorded waveform (Figure 14.6b);

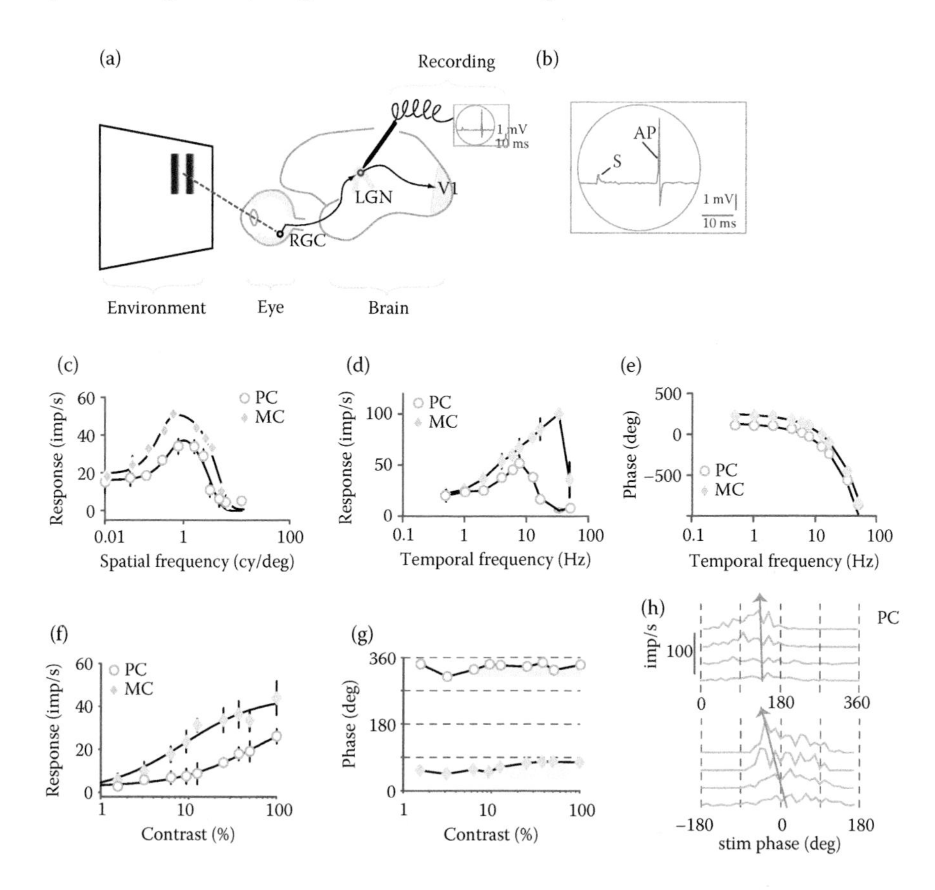

FIGURE 14.6 Spike timing in magnocellular and parvocellular pathways in the visual system of monkeys. (a) Schematic view of extracellular electrode recording in the lateral geniculate nucleus (LGN). Visual stimuli in the environment are placed in the receptive field of retinal ganglion cells (RGC) that project to the LGN recording site. The extracellular potentials from relay cells projecting to the primary visual cortex (V1) are monitored. (b) Schematic oscilloscope recording showing two kinds of potentials typically recorded under such conditions: synaptic potentials (S) from RGC afferent to the LGN cell normally (but not always) produce relay cell action potentials (AP). (Modified from Kaplan, E, and R Shapley. 1984. *Experimental Brain Research* 55, 111–16.) (c) Spatial frequency-tuning functions of one parvocellular (PC) and one magnocellular (MC) cell recorded at comparable visual field positions (PC, 8.6°; MC, 8.4°). The responses show comparable receptive field dimensions and band-pass characteristic indicating center–surround antagonism. (d) Temporal-tuning functions of the same two cells. The MC cell is much more responsive than the PC cell for frequencies between 10 and 50 Hz. (e) The MC cell responses are phase advanced relative to PC cell responses. (f) Contrast–response amplitude-tuning curves show that the MC cell has higher contrast gain and saturated responses compared to the PC cell. (g) Contrast–response phase-tuning curves show the phase advance of MC cell responses to contrast above 10% (shaded area) but show negligible phase advance of PC cell over the same contrast range (shaded area). (h) Phase advance is not evident in PC cell PSTH (upper traces, traces shifted vertically for clarity) but is prominent in MC cell histograms (lower traces, arrow). (Data for panels (c)–(h) are from Martin P, S Solomon, K Cheong, S Pietersen, unpublished.)

the data from such experiments shows that in monkeys, the great majority of LGN spikes is driven from the retina by activity originating in a single ganglion cell. The timing of LGN spikes is thus dominated by the timing of spikes from a single retinal ganglion cell. The arrangement is similar in cat LGN, but in cats, each LGN cell probably receives a small input from additional retinal afferents.

It has been known for some time that LGN neurons sometimes produce two or more action potentials in a "burst" of firing. Bursts make up a small percentage of all spikes emitted by LGN neurons (1–4%) and the frequency of bursts depends on the brain state. Bursts gained prominence because they are more often seen in deep sleep or anesthesia than in waking animals, and because they provide a potential mechanism for increasing the efficacy of LGN signals (Sherman 1996). However, the contribution of retinal afferent activity to bursts remains controversial. An important recent study on macaque LGN (Sincich et al. 2007) indicates that the retinal input spikes contribute equally to rapidly emitted LGN spike trains ("burst mode") and to spikes with longer interspike intervals ("tonic mode"). This indicates that burst firing is a property of retinal ganglion cells and the temporal high-pass filtering on spiking activity in LGN provides an increased probability of observing such events.

Responses of one parvocellular and one magnocellular cell recorded in the LGN of a marmoset monkey are compared in Figure 14.6c–h. The receptive fields of these cells were located at comparable visual field eccentricity. Both cells show center–surround organization; the surround antagonism is manifested as low-frequency roll-off in the spatial frequency-tuning curves (Figure 14.6c) and in the temporal frequency-tuning curves (Figure 14.6d). The magnocellular cell shows greater low temporal frequency roll-off and the tMTF shows higher peak frequency and extends to higher frequencies than the tMTF of the parvocellular cell; this is the temporal frequency–domain characteristic of the sustained/transient distinction in these cells' response to sustained contrast steps. These tMTFs were measured at high contrast (~95%); the response phase of the magnocellular cell is advanced relative to the response phase of the parvocellular cell (Figure 14.6e). This phase advance is consistent with contrast gain control acting on the magnocellular cell at high contrast. The typical spike timing effect of contrast gain control is most obvious in the contrast–response functions (Figure 14.6f–h), where the magnocellular cell shows greater response saturation (Figure 14.6f) than the parvocellular cell, and the response at high-contrast levels is phase advanced relative to the response at low contrast levels (Figure 14.6g and h).

14.4.5.3 Stimulus Specificity and Response Redundancy

As is clear from Figure 14.6, the spatial and temporal tuning of magnocellular and parvocellular cells overlap substantially. The signals provided by a magnocellular cell and a parvocellular cell that "see" the same part of visual space are, therefore, substantially redundant. In trichromatic primates (where parvocellular cells multiplex color and luminance signals), this redundancy is reduced. Not withstanding this detail, the signals of magnocellular and parvocellular cells are certainly more redundant than the signals of their respective on- and off-subdivisions. Evidence for this assertion is shown in Figure 14.7.

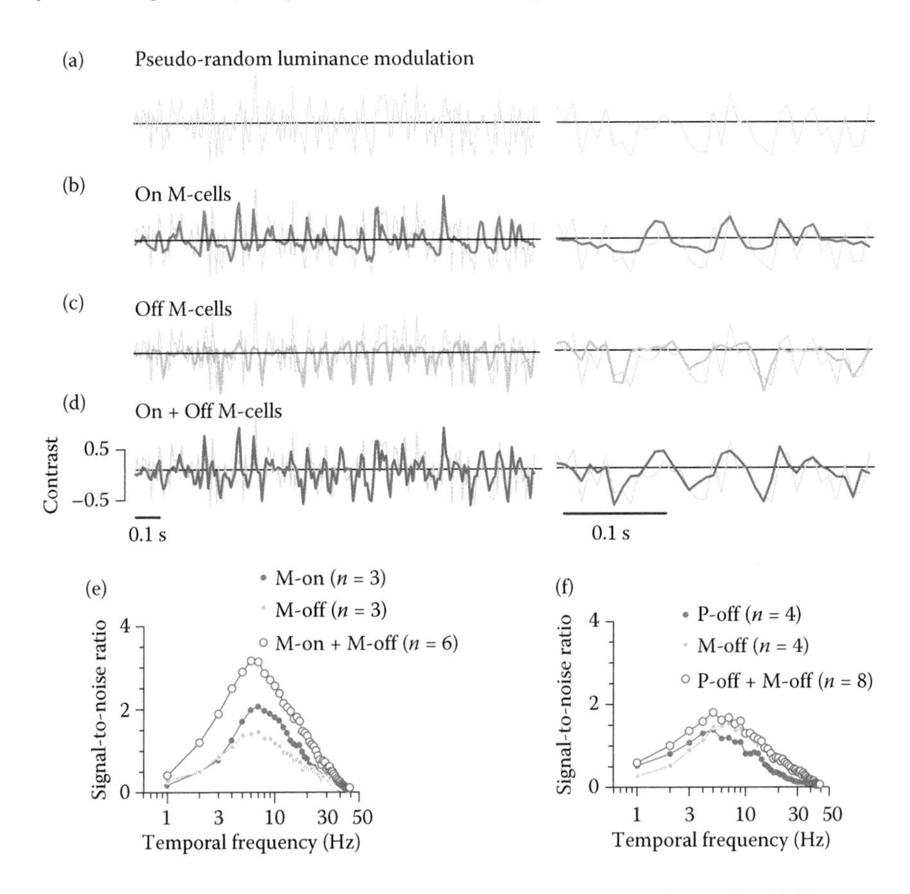

FIGURE 14.7 Redundancy and independence in the signals of LGN cells. (a) The luminance of a spatially uniform field was varied in a pseudorandom fashion with Gaussian intensity distribution. Left, luminance variation over 1 s; right, 250 ms on an expanded timescale. (b) Stimulus profile superimposed with predictions of an optimal linear decoder based on spike trains of 3 MC-on cells recorded in marmoset LGN. The reconstruction is more accurate for luminance increments than for luminance decrements. (c) When the decoder is based on the spike trains of three MC-off cells, the reconstruction is better for luminance decrements than for luminance increments. (d) When the decoder is based on spike trains of three MC-on and three MC-off, responses to increments and decrements are well predicted. (e) Base-2 logarithm of signal-to-noise ratio of reconstructions b–d as a function of temporal frequency. The decoder performs better when based on both MC-on and MC-off responses, suggesting that their signals are largely independent. (f) When the decoder is based on PC-off and MC-off cells, performance tracks that of the most informative cell class, suggesting that PC-off and MC-off signals are largely redundant. (Data reanalyzed from Solomon, SG et al. 2010. *Journal of Neurophysiology* 104(4), 1884–98.)

Figure 14.7a shows the temporal profile of luminance of a small circular disk displayed on a computer monitor and centered on the receptive field of an LGN neuron. The luminance was changed to a new random value (drawn from a Gaussian distribution) every 8.3 ms. The responses of three magnocellular-on cells were recorded and the linear transformation that best predicted the luminance sequence on the basis of these responses was calculated (Warland et al. 1997; Solomon et al. 2010). The model was then exercised by a pseudorandom sequence that was not included in the training set. The model predictions and luminance sequence are shown in Figure 14.7b: the model reconstructs well the amplitude of luminance increments in the stimulus but not the amplitude of luminance decrements. A complementary pattern is seen after training the model with responses of three magnocellular-off cells (Figure 14.7c); here, the model reconstructs well the amplitude of luminance decrements in the stimulus but not the amplitude of luminance increments. When the model was trained with responses of both on- and off-cells, it was capable of recovering both increases and decreases in luminance (Figure 14.7d).

Figure 14.7e quantifies the capacity of neuronal signals to reconstruct the luminance sequence. Here, the signal-to-noise ratio (SNR) is shown as a function of temporal frequency (Solomon et al. 2010). The SNR is the power in the predicted luminance sequence divided by the difference between the predicted and actual sequences. By this metric, the magnocellular-on cells provide better signals than the magnocellular-off cells, and the signal provided by combining on- and off-cells (open symbols, Figure 14.7e) is close to that predicted by adding the signals of the two subclasses. Thus, the signals of magnocellular-on and magnocellular-off cells are not redundant. Figure 14.7f shows the result of training the model with responses of four parvocellular-off cells, four magnocellular-off cells, or a mixture of magnocellular-off and parvocellular-off cells. When the model is trained on the mixture of parvocellular and magnocellular responses, the SNR is not substantially greater than the SNR of the populations in isolation. Rather, the SNR tracks the most informative neurons (parvocellular cells at low temporal frequencies and magnocellular cells at high temporal frequencies). These data and analyses provide evidence that the temporal signals of magnocellular and parvocellular cells are largely redundant over most temporal frequencies.

14.4.5.4 Lagged and Nonlagged Populations in LGN

About 30% of X and Y relay cells in the cat LGN show a "lagged" pattern of spike timing under appropriate stimulus conditions (Mastronarde 1987; Humphrey and Weller 1988; Saul and Humphrey 1990; Hartveit 1992; Mastronarde 1992; Wang et al. 1994). In response to square-wave temporal modulation of a receptive field center-filling spot, the spike responses of lagged cells appear complementary to those of conventional cells, that is, the responses of lagged cells appear with longer latency and decay less rapidly than those of conventional cells. The spike responses of lagged cells to sinusoidal temporal modulation appear phase lagged by about 90° relative to conventional cells at low temporal frequencies. The combined "quadrature" pattern of spiking in lagged and nonlagged cells could in principle be decoded to reconstruct the high temporal frequency (edge) components of contrast borders or could be contributed to cortical direction selectivity at low temporal frequencies (Saul and Humphrey 1992). More recent recordings from waking monkey LGN also revealed a small proportion of cells with lagged timing signature (Saul 2008). Lagged cells are likely generated by LGN circuitry rather than from distinct retinal inputs, and overall, they appear at the end of the range of response timing exhibited by LGN populations. Whether lagged cells play a specific functional role in cortical afferent streams is still not clear. One difficulty is that lagged cells do not bear a chromatic or temporal response signature that could be unambiguously interpreted in the responses of cortical cells. Nevertheless, the accumulated evidence of a distinct timing stream generated in the LGN is important and deserves further study.

14.4.5.5 Sources of Correlated Spiking in Magnocellular and Parvocellular Pathways

In and around the fovea, the on- and off-subclasses of parvocellular-projecting (midget) tile the retina with little or no overlap, and as explained above (see Section 14.4.5.1), the precise one-to-one connections in the retina and LGN mean that the output of each foveal bipolar cell is conveyed to the cortex by a single retinal ganglion cell. However, this pattern of connections is rather the exception than the rule. For magnocellular-projecting (parasol) ganglion cells in the fovea, and for all cells in the peripheral primate retina, and throughout the retina of most nonprimate mammals studied so far, the output of bipolar cells diverges to influence more than one ganglion cell. The consequence of this divergence is that noise in photoreceptors, or bipolar cells, can be inherited by more than one ganglion cell. It follows that the probability of an action potential in neighboring ganglion cells is not always independent. Additional correlations in the ganglion cell activity arise because many classes of retinal neurons are electrically connected by gap junctions. Furthermore, in scotopic vision, additional correlations are generated by convergence of amacrine cells into bipolar cells. Finally, correlations can arise in the LGN if a single retinal ganglion cell drives more than one geniculocortical relay cell.

The result of all this connectivity is that in many situations, the signals of ganglion cells or LGN cells are not independent estimates of the retinal image (Mastronarde 1983; Greschner et al. 2011). For closely neighboring neurons in the peripheral primate retina, knowing the structure of correlations between neurons in a population can improve the information theoretically available by 10–20% (Pillow et al. 2008). This improvement arises because some of the trial-to-trial variability in singleneuron responses is shared by other neurons in the population, and so, the population is less noisy than predicted by the measurements of a single neuron. This redundancy may help in detecting weak signals, where the responses of a single neuron could otherwise be lost in noise. Yet, it remains a future challenge to establish whether the increased information is important (does the brain care about the correlation structure?) or significant (do correlations provide more relevant information than independent signals?). In summary, much remains to be learnt about the impact of correlated activity on visual performance. One route to substantial progress would be to record from populations of LGN cells while an animal is engaged in a behavioral task.

14.4.5.6 Extraclassical Receptive Fields

The classical receptive field of ganglion cells is conventionally defined as the region of visual space where visual stimuli elicit a change in the firing rate (Hartline 1938; Barlow 1953; Kuffler 1953). Careful measurement reveals in most cells an additional and larger region of visual space where stimuli have modulatory effects on discharge. Such an "extraclassical" receptive field is described in ganglion cells of cats and monkeys (McIlwain 1964; Kruger et al. 1975; Derrington et al. 1979), as well as neurons in LGN (Felisberti and Derrington 2001; Solomon et al. 2002). The extraclassical receptive field can have profound effects on the visual sensitivity of retinal ganglion cells, though less is known about its impact on the response timing.

Rapid movement in the peripheral visual field can either increase or decrease the discharge rate of many ganglion cells: the "periphery-effect" (McIlwain 1964) usually refers to increases in discharge brought about by peripheral stimulation; the "shift effect" (Barlow et al. 1977; Derrington et al. 1979; Fischer and Krüger 1980) usually refers to a suppression of discharge. Both effects are generally weaker in primate ganglion cells than in cat ganglion cells and in monkeys, the shift effect may be absent in parvocellular cells (Solomon et al. 2006). In cat *Y*-class cells, the periphery-and-shift effects can be functionally dissociated by careful measurement (Passaglia et al. 2001), but how these effects change the signals that are provided by ganglion cells remain poorly understood. The shift effect is reduced in the presence of tetrodotoxin (tetrodotoxin blocks action potentials), suggesting a role of spiking amacrine cells in its generation (Vaney et al. 1988; Demb et al. 1999; Taylor 1999).

A second manifestation of an extraclassical receptive field is the suppression of visual responses by stimuli that have no effect if presented in isolation. This suppression arises from an area of visual field slightly larger than the classical receptive field surround. Along the dimensions that has been studied, the functional impact of this suppressive field is very similar to that of the contrast gain control described above (Shapley and Victor 1979; Solomon et al. 2002, 2006; Bonin et al. 2005), and like the contrast gain control, the suppression is most prominent in the receptive fields of *X*- and *Y*-class cells of cat retina, and of magnocellular cells in primate retina. It is weak or absent in the parvocellular cells of primate retina (Solomon et al. 2006).

14.5 SUMMARY AND UNANSWERED QUESTIONS

We have seen that key factors controlling response timing in the early stages of visual processing are (1) stimulus intensity, (2) stimulus contrast, and (3) derived properties of parallel pathways that carry visual signals. From our perspective as experimentalists, we wish to conclude with three big unanswered questions raised by the data summarized in the foregoing sections. First, there is substantial evidence that the first synapse in the retina between cone photoreceptors and bipolar cells is a key site for splitting receptor signals into parallel visual channels with distinct timing properties. What remains poorly understood is the detailed functional properties of bipolar cells, and how bipolar signals are transferred to ganglion cells; for example, is there convergence of on- and off-type signals ("push-pull") to ganglion cells in the inner retina? Second, why is timing in on- and off-pathways so closely matched within cell classes, despite wider variation of timing between cell classes? We offered in Section 14.4.4 a speculation based on retinal processing requirements, but many other possibilities exist, such as requirements for binocular convergence (Wolpert et al. 1993; DeAngelis et al. 1998). Finally, we have shown that timing in parallel afferent pathways depends on the type of stimulus and on the type of pathway on which a representation of the stimulus is carried to the cerebral cortex. But are timing differences important or irrelevant for cortical processing? A resolution of this question may help us to understand the importance of fine temporal structure in spike trains, a subject that is addressed in the other chapters of this book.

ACKNOWLEDGMENTS

We thank Kenny Cheong and Sander Pietersen for helping to acquire the data shown in Figure 14.6, and Jonathan Victor and Trevor Lamb for helpful discussions and comments on the manuscript.

REFERENCES

Baccus, SA. 2007. Timing and computation in inner retinal circuitry. *Annual Review of Physiology* 69, 271–90.

Barlow, HB. 1953. Summation and inhibition in the frog's retina. *Journal of Physiology* 119, 69–88.

Barlow, HB, AM Derrington, LR Harris, and P Lennie. 1977. The effects of remote retinal stimulation on the responses of cat retinal ganglion cells. *Journal of Physiology* 269, 177–94.

Barlow, HB, R Fitzhugh, and SW Kuffler. 1957. Change of organization in the receptive fields of the cat's retina during dark adaptation. *Journal of Physiology* 137(3), 338–54.

Barlow, HB, and WR Levick. 1965. The mechanism of directionally selective units in rabbit's retina. *Journal of Physiology* 178, 477–504.

Baylor, DA, and MG Fuortes. 1970. Electrical responses of single cones in the retina of the turtle. *Journal of Physiology* 207(1), 77–92.

Bloomfield, SA, and RF Dacheux. 2001. Rod vision: Pathways and processing in the mammalian retina. *Progress in Retinal and Eye Research* 20(3), 351–84.

Bonin, V, V Mante, and M Carandini. 2005. The suppressive field of neurons in lateral geniculate nucleus. *Journal of Neuroscience* 25(47), 10844–56.

Boycott, BB, and JE Dowling. 1969. Organization of the primate retina: Light microscopy. *Philosophical Transactions of the Royal Society of London (B)* 255, 109–76.

Boycott, BB, and H Kolb. 1973. The horizontal cells of the rhesus monkey retina. *Journal of Comparative Neurology* 148, 115–40.

Breton, ME, AW Schueller, TD Lamb, and EN Jr Pugh. 1994. Analysis of erg a-wave amplification and kinetics in terms of the G-protein cascade of phototransduction. *Investigative Ophthalmology and Visual Science* 35(1), 295–309.

Burkhardt, DA, PK Fahey, and MA Sikora. 2004. Retinal bipolar cells: Contrast encoding for sinusoidal modulation and steps of luminance contrast. *Visual Neuroscience* 21(6), 883–93.

Calkins, DJ, and P Sterling. 1996. Absence of spectrally specific lateral inputs to midget ganglion cells in primate retina. *Nature* 381, 613–15.

Casagrande, VA, and X Xu. 2003. Parallel visual pathways: A comparative perspective. In *The Visual Neurosciences*, edited by LM Chalupa, and JS Werner. Cambridge, MA: The MIT press.

Chichilnisky, EJ, and RS Kalmar. 2002. Functional asymmetries in on and off ganglion cells of primate retina. *Journal of Neuroscience* 22(7), 2737–47.

Cleland, BG, MW Dubin, and WR Levick. 1971. Sustained and transient neurones in the cat's retina and lateral geniculate nucleus. *Journal of Physiology* 217, 473–96.

Crook, JD, MB Manookin, OS Packer, and DM Dacey. 2011. Horizontal cell feedback without cone type-selective inhibition mediates "red-green" color opponency in midget ganglion cells of the primate retina. *Journal of Neuroscience* 31(5), 1762–72.

Crook, JD, BB Peterson, OS Packer, FR Robinson, JB Troy, and DM Dacey. 2008. *Y*-cell receptive field and collicular projection of parasol ganglion cells in macaque monkey retina. *Journal of Neuroscience* 28(44), 11277–91.

Dacey, DM, BB Lee, DK Stafford, J Pokorny, and VC Smith. 1996. Horizontal cells of the primate retina: Cone specificity without spectral opponency. *Science* 271 656–59.

DeAngelis, GC, BG Cumming, and WT Newsome. 1998. Cortical area Mt and the perception of stereoscopic depth. *Nature* 394, 677–80.

Demb, JB, L Haarsma, MA Freed, and P Sterling. 1999. Functional circuitry of the retinal ganglion cell's nonlinear receptive field. *Journal of Neuroscience* 15, 9756–67.

Derrington, AM, J Krauskopf, and P Lennie. 1984. Chromatic mechanisms in lateral geniculate nucleus of macaque. *Journal of Physiology* 357, 241–65.

Derrington, AM, and P Lennie. 1984. Spatial and temporal contrast sensitivities of neurons in lateral geniculate nucleus of macaque. *Journal of Physiology* 357, 219–40.

Derrington, AM, P Lennie, and MJ Wright. 1979. The mechanism of peripherally evoked responses in retinal ganglion cells. *Journal of Physiology* 289, 299–310.

DeVries, SH. 2000. Bipolar cells use kainate and AMPA receptors to filter visual information into separate channels. *Neuron* 28(3), 847–56.

DeVries, SH, and EA Schwartz. 1999. Kainate receptors mediate synaptic transmission between cones and "off" bipolar cells in a mammalian retina. *Nature* 397, 157–60.

Dowling, JE. 1987. *The Retina: An Approachable Part of the Brain*. Cambridge, MA: The Belknapp Press of Harvard University Press.

Dreher, B, Y Fukada, and RW Rodieck. 1976. Identification, classification and anatomical segregation of cells with *X*-like and *Y*-like properties in the lateral geniculate nucleus of Old-World primates. *Journal of Physiology* 258, 433–52.

Duebel, J, S Haverkamp, W Schleich, G Feng, GJ Augustine, T Kuner, and T Euler. 2006. Two-photon imaging reveals somatodendritic chloride gradient in retinal on-type bipolar cells expressing the biosensor clomeleon. *Neuron* 49(1), 81–94.

Dunn, FA, MJ Lankheet, and F Rieke. 2007. Light adaptation in cone vision involves switching between receptor and post-receptor sites. *Nature* 449, 603–07.

Enroth-Cugell, C, and AW Freeman. 1987. The receptive-field spatial structure of cat retinal *Y* cells. *Journal of Physiology* 384, 49–47.

Enroth-Cugell, C, and P Lennie. 1975. The control of retinal ganglion cell discharge by receptive field surrounds. *Journal of Physiology* 247, 551–78.

Enroth-Cugell, C, and J Robson. 1966. The contrast sensitivity of retinal ganglion cells of the cat. *Journal of Physiology* 187, 517–52.

Euler, T, H Schneider, and H Wässle. 1996. Glutamate responses of bipolar cells in a slice preparation of the rat retina. *Journal of Neuroscience* 16, 2934–44.

Felisberti, F, and A Derrington. 2001. Long-range interactions in the lateral geniculate nucleus of the New-World monkey, *Callithrix jacchus*. *Visual Neuroscience* 18, 209–18.

Fischer, B, and J Krüger. 1980. Continuous movement of remote patterns and shift-effect of cat retinal ganglion cells. *Experimental Brain Research* 40, 229–32.

Flores-Herr, N, DA Protti, and H Wässle. 2001. Synaptic currents generating the inhibitory surround of ganglion cells in the mammalian retina. *Journal of Neuroscience* 21, 4852–63.

Frishman, LJ, AW Freeman, JB Troy, DE Schweitzer-Tong, and C Enroth-Cugell. 1987. Spatiotemporal frequency responses of cat retinal ganglion cells. *Journal of General Physiology* 89, 599–628.

Goodchild, AK, KK Ghosh, and PR Martin. 1996. Comparison of photoreceptor spatial density and ganglion cell morphology in the retina of human, macaque monkey, cat, and the marmoset *Callithrix jacchus*. *Journal of Comparative Neurology* 366, 55–75.

Greschner, M, J Shlens, C Bakolitsa, GD Field, JL Gauthier, LH Jepson, A Sher, AM Litke, and EJ Chichilnisky. 2011. Correlated firing among major ganglion cell types in primate retina. *Journal of Physiology* 589(1), 75–86.

Grünert, U, S Haverkamp, EL Fletcher, and H Wässle. 2002. Synaptic distribution of ionotropic glutamate receptors in the inner plexiform layer of the primate retina. *Journal of Comparative Neurology* 447, 138–51.

Hartline, HK. 1938. The response of single optic nerve fibers of the vertebrate eye to illumination of the retina. *American Journal of Physiology* 121, 400–15.

Hartveit, E. 1992. Simultaneous recording of lagged and nonlagged cells in the cat dorsal lateral geniculate nucleus. *Experimental Brain Research* 88, 229–32.

Haverkamp, S, U Grünert, and H Wässle. 2000. The cone pedicle, a complex synapse in the retina. *Neuron* 27, 85–95.

Haverkamp, S, U Grünert, and H Wässle. 2001. The synaptic architecture of AMPA receptors at the cone pedicle of the primate retina. *Journal of Neuroscience* 21, 2488–500.

Hochstein, S, and RM Shapley. 1976. Linear and nonlinear spatial subunits in *y* cat retinal ganglion cells. *Journal of Physiology* 262, 265–84.

Humphrey, AL, and RE Weller. 1988. Structural correlates on functionally distinct *X*-cells in the lateral geniculate nucleus of the cat. *Journal of Comparative Neurology* 268, 448–68.

Jacobs, AL, G Fridman, RM Douglas, NM Alam, PE Latham, GT Prusky, and S Nirenberg. 2009. Ruling out and ruling in neural codes. *Proceedings of the National Academy of Sciences of the USA* 106(14), 5936–41.

Kamermans, M, and H Spekreijse. 1999. The feedback pathway from horizontal cells to cones. A mini review with a look ahead. *Vision Research* 39, 2449–68.

Kaneko, A. 1970. Physiological and morphological identification of horizontal, bipolar, and amacrine cells in goldfish retina. *Journal of Physiology* 207, 623–33.

Kaneko, A. 1983. Retinal bipolar cells: Their function and morphology. *Trends in Neurosciences* 6, 219–23.

Kaplan, E. 2003. The M, P, and K pathways of the primate visual system. In *The Visual Neurosciences*, edited by LM Chalupa, and JS Werner. Cambridge, MA: The MIT Press.

Kaplan, E, and E Benardete. 2001. The dynamics of primate retinal ganglion cells. *Progress in Brain Research* 134, 17–34.

Kaplan, E, BB Lee, and RM Shapley. 1989. New views of primate retinal function. In *Progress in Retinal Research*, edited by N Osborne, and J Chader. New York: Pergamon.

Kaplan, E, and R Shapley. 1984. The origin of the *S* (slow) potential in the mammalian lateral geniculate nucleus. *Experimental Brain Research* 55, 111–16.

Kaplan, E, and RM Shapley. 1986. The primate retina contains two types of ganglion cells, with high and low contrast sensitivity. *Proceedings of the National Academy of Sciences of the USA* 83, 2755–57.

Kolb, H. 1979. The inner plexiform layer in the retina of the cat: Electron microscopic observations. *Journal of Neurocytology* 8, 295–329.

Kolb, H, and L Dekorver. 1991. Midget ganglion cells of the parafovea of the human retina: A study by electron microscopy and serial section reconstructions. *Journal of Comparative Neurology* 303, 617–36.

Kolb, H, A Mariani, and A Gallego. 1980. A second type of horizontal cell in the monkey retina. *Journal of Comparative Neurology* 189, 31–44.

Koontz, MA, and AE Hendrickson. 1987. Stratified distribution of synapses in the inner plexiform layer of primate retina. *Journal of Comparative Neurology* 263, 581–92.

Kruger, J, B Fischer, and R Barth. 1975. The shift-effect in retinal ganglion cells of the rhesus monkey. *Experimental Brain Research* 23(4), 443–46.

Kuffler, SW. 1953. Discharge patterns and functional organization of mammalian retina. *Journal of Neurophysiology* 16, 37–68.

Lachapelle, P. 2006. The oscillatory potentials of the electroretinogram. In *Principles and Practice of Clinical Electrophysiology of Vision*, edited by JR Heckenlively, and GB Arden. Cambridge, MA: MIT Press.

Lamb, TD. 2011. Light adaptation in photoreceptors. In *Adler's Physiology of the Eye,* 11th Edition, edited by Levin, LA, SFE Nilsson, J Ver Hoeve, S Wu, PL Kaufman, and A Alm. Edingburg: Elsevier.

Lamb, TD, SP Collin, and EN Jr Pugh. 2007. Evolution of the vertebrate eye: Opsins, photoreceptors, retina, and eye cup. *Nature Reviews Neuroscience* 8(12), 960–76.

Lamb, TD, and EN Jr Pugh. 2006. Phototransduction, dark adaptation, and rhodopsin regeneration of the proctor lecture. *Investigative Ophthalmology and Visual Science* 47(12), 5137–52.

Lankheet, MJ, J Molenaar, and WA van de Grind. 1989a. Frequency transfer properties of the spike generating mechanism of cat retinal ganglion cells. *Vision Research* 29, 1649–61.

Lankheet, MJM, P Lennie, and J Krauskopf. 1998. Distinctive characteristics of subclasses of red-green *P*-cells in LGN of macaque. *Visual Neuroscience* 15, 37–46.

Lankheet, MJM, J Molenaar, and WA van de Grind. 1989b. The spike generating mechanism of cat retinal ganglion cells. *Vision Research* 29(5), 505–17.

Lee, BB, PR Martin, and U Grünert. 2010. Retinal connectivity and primate vision (Review). *Progress in Retinal and Eye Research* 29(6), 622–39.

Lee, BB, PR Martin, and A Valberg. 1989. Amplitude and phase of responses of macaque retinal ganglion cells to flickering stimuli. *Journal of Physiology* 414, 245–63.

Lee, BB, J Pokorny, VC Smith, PR Martin, and A Valberg. 1990. Luminance and chromatic modulation sensitivity of macaque ganglion cells and human observers. *Journal of the Optical Society of America A* 7, 2223–36.

Lu, SM, W Guido, JW Vaughan, and SM Sherman. 1995. Latency variability of responses to visual stimuli in cells of the cat's lateral geniculate nucleus. *Experimental Brain Research* 105, 7–17.

Masland, RH, and PR Martin. 2007. The unsolved mystery of vision (Review). *Current Biology* 17, R577–82.

Massey, SC, and DA Redburn. 1987. Transmitter circuits in the vertebrate retina. *Progress in Neurobiology* 28, 55–96.

Mastronarde, DN. 1983. Correlated firing of cat retinal ganglion cells. I. Spontaneously active inputs to *X*- and *Y*-cells. *Journal of Neurophysiology* 49(2), 303–24.

Mastronarde, DN. 1987. Two classes of single-input *X*-cells in cat lateral geniculate nucleus. I. Receptive-field properties and classification of cells. *Journal of Neurophysiology* 57, 357–80.

Mastronarde, DN. 1992. Nonlagged relay cells and interneurons in the cat lateral geniculate nucleus: Receptive-field properties and retinal inputs. *Visual Neuroscience* 8, 407–41.

McClurkin, JW, TJ Gawne, BJ Richmond, LM Optican, and DL Robinson. 1991. Lateral geniculate neurons in behaving primates. I. Responses to two-dimensional stimuli. *Journal of Neurophysiology* 66(3), 777–93.

McIlwain, JT. 1964. Receptive fields of optic tract axons and lateral geniculate cells: Peripheral extent and barbiturate sensitivity. *Journal of Neurophysiology* 27, 1154–73.

McMahon, MJ, MJM Lankheet, P Lennie, and DR Williams. 2000. Fine structure of parvocellular receptive fields in the primate fovea revealed by laser interferometry. *Journal of Neuroscience* 20, 2043–53.

Meister, M, and MJ Berry. 1999. The neural code of the retina. *Neuron* 22, 435–50.

Mittman, S, RW Taylor, and DR Copenhagen. 1990. Concomitant activation of two types of glutamate receptor mediates excitation of salamander retinal ganglion cells. *Journal of Physiology* 428, 175–97.

Morigiwa, K, and N Vardi. 1999. Differential expression of ionotropic glutamate receptor subunits in the outer retina. *Journal of Comparative Neurology* 405, 173–84.

Nakajima, Y, H Iwakabe, C Akazawa, H Nawa, R Shigemoto, N Mizuno, and S Nakanishi. 1993. Molecular characterization of a novel retinal metabotropic glutamate receptor mGluR6 with a high agonist selectivity for L-2-amino-4-phophonobutyrate. *Journal of Biological Chemistry* 268, 11868–73.

Nakanishi, S. 1992. Molecular diversity of glutamate receptors and implications for brain function. *Science* 258, 597–603.

Nelson, R, and H Kolb. 2003. On and off pathways in the vertebrate retina and visual system. In *The Visual Neurosciences*, edited by LM Chalupa, and JS Werner. Cambridge, MA: The MIT Press.

Nelson, R, H Kolb, and MA Freed. 1993. Off-alpha and off-beta ganglion cells in cat retina. I: Intracellular electrophysiology and Hrp stains. *Journal of Comparative Neurology* 329, 68–84.

Nirenberg, S, and M Meister. 1997. The light response of retinal ganglion cells is truncated by a displaced amacrine circuit. *Neuron* 18, 637–50.

Oesch, NW, WW Kothmann, and JS Diamond. 2011. Illuminating synapses and circuitry in the retina. *Current Opinion in Neurobiology* 21(2), 238–44.

Pandarinath, C, JD Victor, and S Nirenberg. 2010. Symmetry breakdown in the on and off pathways of the retina at night: Functional implications. *Journal of Neuroscience* 30(30), 10006–14.

Passaglia, CL, C Enroth-Cugell, and JB Troy. 2001. Effects of remote stimulation on the mean firing rate of cat retinal ganglion cells. *Journal of Neuroscience* 21, 5794–803.

Perlman, I, H Kolb, and R Nelson. 2003. Anatomy, circuitry, and physiology of vertebrate horizontal cells. In *The Visual Neurosciences*, edited by Chalupa, LM, and JS Werner. Cambridge, MA: The MIT Press.

Pillow, JW, J Shlens, L Paninski, A Sher, AM Litke, EJ Chichilnisky, and EP Simoncelli. 2008. Spatio-temporal correlations and visual signaling in a complete neuronal population. *Nature* 454(7207), 995–99.

Pugh, EN Jr, S Nikonov, and TD Lamb. 1999. Molecular mechanisms of vertebrate photoreceptor light adaptation. *Current Opinion in Neurobiology* 9(4), 410–18.

Puller, C, S Haverkamp, and U Grünert. 2007. Off midget bipolar cells in the retina of the marmoset, *Callithrix jacchus,* express AMPA receptors. *Journal of Comparative Neurology* 502, 442–554.

Raviola, E, and RF Dacheux. 1987. Excitatory dyad synapse in rabbit retina. *Proceedings of the National Academy of Sciences of the USA* 84, 7324–28.

Reinagel, P, and RC Reid. 2002. Precise firing events are conserved across neurons. *Journal of Neuroscience* 22(16), 6837–41.

Rodieck, RW. 1998. *The First Steps in Seeing*. Sunderland: Sinauer.

Saul, AB. 2008. Lagged cells in alert monkey lateral geniculate nucleus. *Visual Neuroscience* 25(5–6), 647–59.

Saul, AB, and AL Humphrey. 1990. Spatial and temporal response properties of lagged and nonlagged cells in cat lateral geniculate nucleus. *Journal of Neurophysiology* 64, 206–24.

Saul, AB, and AL Humphrey. 1992. Evidence of input from lagged cells in the lateral geniculate nucleus to simple cells in cortical area 17 of the cat. *Journal of Neuroscience* 68(4), 1190–208.

Schiller, PH. 2010. Parallel information processing channels created in the retina. *Proceedings of the National Academy of Sciences of the USA* 107(40), 17087–94.

Schnapf, JL, BJ Nunn, M Meister, and DA Baylor. 1990. Visual transduction in cones of the monkey *Macaca fascicularis*. *Journal of Physiology* 427, 681–713.

Shapley, R, and VH Perry. 1986. Cat and monkey retinal ganglion cells and their visual functional roles. *Trends in Neurosciences* 9, 229–35.

Shapley, RM, and JD Victor. 1978. The effect of contrast on the transfer properties of cat retinal ganglion cells. *Journal of Physiology* 285, 275–98.

Shapley, RM, and JD Victor. 1979. Nonlinear spatial summation and the contrast gain control of cat retinal ganglion cells. *Journal of Physiology* 290 (2), 141–61.

Shapley, RM, and JD Victor. 1981. How the contrast gain control modifies the frequency responses of cat retinal ganglion cells. *Journal of Physiology* 318, 161–79.

Sherman, SM 1996. Dual response modes in lateral geniculate neurons: Mechanisms and functions. *Visual Neuroscience* 13, 205–13.

Silveira, LC, CA Saito, BB Lee, J Kremers, M daSilvaFilho, BE Kilavik, ES Yamada, and VH Perry. 2004. Morphology and physiology of primate M- and P-cells. *Progress in Brain Research* 144, 21–46.

Sincich, LC, DL Adams, JR Economides, and JC Horton. 2007. Transmission of spike trains at the retinogeniculate synapse. *Journal of Neuroscience* 27(10), 2683–92.

Slaughter, MM, and RF Miller. 1981. 2-amino-4-phosphonobutyric acid: A new pharmacological tool for retina research. *Science* 211, 182–85.

Smith, VC, J Pokorny, BB Lee, and DM Dacey. 2001. Primate horizontal cell dynamics: An analysis of sensitivity regulation in the outer retina. *Journal of Neurophysiology* 85, 545–58.

Solomon, SG, BB Lee, and H Sun. 2006. Suppressive surrounds and contrast gain in magnocellular-pathway retinal ganglion cells of macaque. *Journal of Neuroscience* 26(34), 8715–26.

Solomon, SG, and P Lennie. 2007. The machinery of color vision. *Nature Reviews Neuroscience* 8, 276–86.

Solomon, SG, C Tailby, SK Cheong, and AJ Camp. 2010. Linear and non-linear contributions to the visual sensitivity of neurons in primate lateral geniculate nucleus. *Journal of Neurophysiology* 104(4), 1884–98.

Solomon, SG, AJR White, and PR Martin. 2002. Extraclassical receptive field properties of parvocellular, magnocellular, and koniocellular cells in the primate lateral geniculate nucleus. *Journal of Neuroscience* 22, 338–49.

Stone, J. 1983. *Parallel Processing in the Visual System*. London: Plenum Press.

Taylor, WR. 1999. Ttx attenuates surround inhibition in rabbit retinal ganglion cells. *Visual Neuroscience* 16, 285–90.

Toyoda, J, H Nosaki, and T Tomita. 1969. Light-induced resistance changes in single photoreceptors of necturus and gekko. *Vision Research* 9(4), 453–63.

Troy, JB, DL Bohnsack, and LC Diller. 1999. Spatial properties of the cat *X*-cell receptive field as a function of mean light level. *Visual Neuroscience* 16(6), 1089–104.

Troy, JB, and T Shou. 2002. The receptive fields of cat retinal ganglion cells in physiological and pathological states: Where we are after half a century of research. *Progress in Retinal and Eye Research* 21, 263–302.

Vaney, DI. 2002. Retinal neurons: Cell types and coupled networks. In EC Azmitia, J DeFelipe, EG Jones, P Rakic, and CE Ribak (Ed.), *Changing Views of Cajal's Neuron* (pp. 239–254). USA: Elsevier Science BV.

Vaney, DI, L Peichl, and BB Boycott. 1988. Neurofibrillar long-range amacrine cells in mammalian retine. *Proceedings of the Royal Society (London) B* 235(1280), 203–19.

Vaney, DI, and WR Taylor. 2002. Direction selectivity in the retina. *Current Opinion in Neurobiology* 12, 405–10.

Victor, JD, and RM Shapley. 1979. Receptive field mechanisms of cat *X* and *Y* retinal ganglion cells. *Journal of General Physiology* 74, (2), 275–98.

Walraven, J, C Enroth-Cugell, DC Hood, DIA MacLeod, and JL Schnapf. 1990. The control of visual sensitivity. In *Visual Perception: The Neurological Foundations*, edited by Spillman, L, and J S Werner. San Diego: Academic Press.

Wang, C, B Dreher, and W Burke. 1994. Non-dominant suppression in the dorsal lateral geniculate nucleus of the cat: Laminar differences and class specificity. *Experimental Brain Research* 97(3), 451–65.

Warland, DK, P Reinagel, and M Meister. 1997. Decoding visual information from a population of retinal ganglion cells. *Journal of Neurophysiology* 78(5), 2336–50.

Wässle, H. 2004. Parallel processing in the mammalian retina. *Nature Reviews Neuroscience* 5(10), 747–57.

Wässle, H, and BB Boycott. 1991. Functional architecture of the mammalian retina. *Physiological Reviews* 71, 447–80.

Wässle, H, U Grünert, J Röhrenbeck, and BB Boycott. 1989. Cortical magnification factor and the ganglion cell density of the primate retina. *Nature* 341, 643–46.

Werblin, FS, and JE Dowling. 1969. Organization of the retina of the mudpuppy, *Necturus maculosus*. Intracellular recording. *Journal of Neurophysiology* 32(3), 339–55.

Westheimer, G. 2007. The on-off dichotomy in visual processing: From receptors to perception. *Progress in Retinal and Eye Research* 26 (6), 636–48.

Wiesel, TN, and D Hubel. 1966. Spatial and chromatic interactions in the lateral geniculate body of the rhesus monkey. *Journal of Neurophysiology* 29, 1115–56.

Wilson, M. 2003. Retinal synapses. In *The Visual Neurosciences*, edited by Chalupa, LM, and JS Werner. Cambridge, MA: The MIT Press.

Wolpert, DM, RC Miall, B Cumming, and SJ Boniface. 1993. Retinal adaptation of visual processing time delays. *Vision Research* 33(10), 1421–30.

15 Cortical Computations Using Relative Spike Timing

Timothy J. Gawne

CONTENTS

15.1 INTRODUCTION

Despite considerable research, there is still no generally held consensus as to what the neural code is in the cerebral cortex. We know from basic principles that, as regards the communication of information from one region of the cortex to another, it is only the pattern and timing of action potentials that is important. Thus, while analog processing plays a strong role in the processing of information for locally connected neurons, as far as the rest of the brain is concerned, it is only the number and timing of action potentials in the projection neurons that matter.

The default hypothesis for the neural code is the rate code, that is, the correct interpretation of the signal carried by a neuron is the number of action potentials per unit time (Shadlen and Newsome, 1994). The rate code has much to commend it, and for cases like primary motor or brainstem oculomotor nucleus neurons, it is likely correct. However, for the cerebral cortex, it is not an entirely satisfactory hypothesis. In particular, the rate code does not fit well with the observed high speed of overall behavior because at least under some conditions there does not appear to be enough time to count spikes (Ghose and Harrison, 2009; VanRullen, 2007). In addition, at the level of the retina, it has been shown that the rate code cannot account for performance on a simple psychophysical task, but adding a dependence on interspike intervals can account for behavior (Jacobs et al., 2009). It should be noted that the bottleneck at the optic nerve makes such studies more practical in the retina than in the cortex. Many possible alternatives to the rate code have been proposed, which use more precise timing changes both between and within single neurons (although there is no reason why the nervous system could not use both a rate and a timing code see Gawne, 2011; Kayser et al., 2009; Kermany et al., 2010).

This chapter concerns the specific hypothesis that much of cortical computation may rely on the *relative* timing of spikes between different neurons in the range of milliseconds to tens of milliseconds. It might seem that the latency of firing of a neuron could not be of any functional use because the nervous system has no general way of determining the absolute latency of a spike relative to any external event. There are exceptions to this, for example, in the visual system, a saccadic eye movement, and in the somatosensory system, a movement of a limb or whisker could be used to determine the absolute latency of an elicited response from a sensory neuron, but the ability of the nervous system to determine an absolute response latency is not general. Thus, it seems likely that it is the *relative* time between spikes that is important. It has long been known that the relative response times at which a sound enters the two ears can be used to determine a bearing to an acoustic source; there is no reason why relative timing could not be of use elsewhere in the nervous system.

It has been proposed that the onset of a population response could be used as a reference point for determining spike timing (Chase and Young, 2007). Evidence from the auditory cortex suggests that there are specialized populations of neurons with response onset latencies that are fixed to external stimulus timing, and that could be used as a reference allowing the brain to decode changing absolute latencies in other populations of neurons (Brasselet et al., 2012). However, it is still the case that the brain only has access to relative latencies between neurons. Often, a study will only quantify the absolute response latency in reference to the time of an external stimulus, and it should be understood that for the nervous system to make use of this information, there must be some sort of internal comparison with the response timing of another population of neurons (which may be a specialist subpopulation of temporal reference neurons, or simply other neurons tuned to different aspects of the external world).

It should be noted that relative spike timing can be defined both between spikes generated by a single neuron and between spikes generated by different neurons. Most physiological experiments concern the former situation, and while certainly important, this is really just temporal filtering of a single signal. What is potentially more interesting is understanding how neurons perform computations on multiple inputs. The nervous system does not use single neurons in isolation, and linear chains from one single neuron to another single neuron can only perform a simple reflex. It is the ability of neurons to combine information from multiple sources that gives the nervous system of complex animals its power. For example, a mechanism that emphasized the first spike in a sequence that was applied to a single input would only perform high-pass filtering, but the same mechanism applied to multiple inputs could perform selection between different inputs. Therefore, most of the emphasis in this chapter is placed on how relative spike timing between different neurons could play a role in neural computation.

Currently, a large proportion of the research on codes using relative spike timing concern synchronous firing (which might or might not be associated with oscillatory behavior), which is the special case of an interspike interval of zero between neurons (Buzsaki, 2006; Eckhorn et al., 1988; Gray and Singer, 1989; Singer, 1999). The idea that the synchronous firing of neurons constitutes a code indicating which patterns of activity should be grouped together has received massive coverage and it is beyond the scope of this review to exhaustively review this literature. It is pointed out that a code that uses synchronous firing and codes that use nonzero time lags between spikes need not be mutually exclusive. Zero interspike intervals could be used to code for activity patterns that need to be grouped together; nonzero interspike intervals could be used to select or gate signals and also provide more positive information on the relations between different signals. For a complex dynamical system like the brain, we should not rule out the possibility that there could be several different operational modes that might come into play under different circumstances or that might interact at the same time.

Because complex visual stimuli can be easily and rapidly presented, and because the visual cortical system is so large and relatively important in primates, the emphasis in this chapter is on the visual cortical system as a model for all of the cerebral cortex (although examples will be used from other sensory modalities when appropriate). Nature tends to be conservative, so it is expected that the neuronal code in one region of the cortex should be the same or similar in other regions.

This chapter has two major components. The first component shows that the prerequisites for coding based on relative spike timing are present at the level of neurons and circuits (Section 15.2), and that there is experimental evidence that relative spike timing influences the responses of cortical circuitry (Section 15.3). The second major component describes possible ways that the nervous system could use relative spike timing to code information, and provides experimental evidence for their plausibility (Section 15.4).

15.2 LOW-LEVEL MECHANISMS SUGGESTING THAT RELATIVE SPIKE TIMING IS IMPORTANT

The rate code remains the default hypothesis; however, studies on reduced preparations routinely find that the relative timing of inputs has extremely strong effects on spike generation. Certainly, cortical circuitry is capable of responding to and generating signals with millisecond-level precision (Mainen and Sejnowski, 1995; Tiesinga et al., 2008). If, at the level of individual neurons and microcircuits, relative spike timing is important, it is hard to see how it could not be important for the nervous system as a whole. It should be noted that many/most of these studies have only considered the role of the relative timing of spikes on a single input to a neuron, but it seems likely that (possibly excepting synaptic depression) similar effects would be found for relative spike timing from multiple inputs. Listed below are three mechanisms demonstrating that relative spike timing should be functionally important. This list is not exhaustive, and complex network properties or active dendrites could compute almost any function of spike timing. The point is that the ubiquity and robustness of circuits that are sensitive to small changes in spike timing is a powerful argument in favor of a functional role of relative spike timing in the cortex.

15.2.1 Feed-Forward Inhibition

It has long been known that, in many areas of the brain, an excitatory input also activates a population of inhibitory interneurons (Shepherd, 1998) (Figure 15.1). This disynaptic, feed-forward inhibition creates a brief temporal window for excitatory input to have an effect. Feed-forward inhibition may be considered to emphasize temporally coincident inputs, to select against inputs with relatively longer latencies, or to temporally sharpen (compute the first derivative of) an input from a single source.

Feed-forward inhibition has been shown to have nontrivial timing effects in somatosensory (Gabernet et al., 2005; Sun et al., 2006) and auditory (Wehr and Zador, 2003) cortex. More generally, disynaptic feed-forward inhibition is both common and powerful in regions of the central nervous system other than the cerebral cortex, including the thalamus (Blitz and Regehr, 2005), cerebellum (Mittmann et al., 2005), hippocampus (Pouille and Scanziani, 2001. See also Kimura et al., 2011 for further evidence of the importance of small relative spike times in hippocampal computations), and retina (Chen et al., 2010). By this one mechanism alone, it seems likely that relative spike timing must have strong functional effects throughout the nervous system.

15.2.2 Synaptic Depression

Even in isolation, single synapses exhibit nontrivial temporal dynamics, which can range in duration from milliseconds to years. Here, we concentrate on short-term

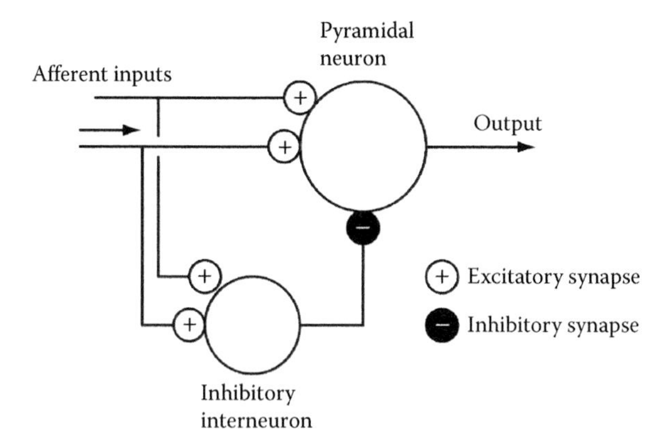

FIGURE 15.1 Whenever excitatory input enters a region of the cortex, there is typically also a feed-forward activation of inhibitory interneurons. Because this disynaptic inhibition must pass through one more synapse than the monosynaptic excitatatory input, this should add temporal delays between the excitation and inhibition. This should result in the relative timing of inputs from different sources being a powerful factor in neural computation.

synaptic depression: the phenomena whereby one input suppresses the effect of a succeeding input for interspike intervals of tens to hundreds of milliseconds (Zucker, 1989). Considerable experimental evidence exists that synaptic depression can affect the temporal response characteristics of cortical neurons and it is therefore strong evidence for the importance of relative spike timing (Carandini et al., 2002; Chance et al., 1998; David et al., 2009; Freeman et al., 2002). However, it should be noted that synaptic depression cannot play a direct role in computing relative spike timing between different inputs at the level of a single neuron because different inputs would use different synapses. Nonetheless, in multineuronal networks, the temporal shaping properties of synaptic depression could be important in calculating spike timing from different sources: consider, for example, if synaptic depression had a role in feed-forward or feedback inhibition (see Diesz and Prince, 1989). Additionally, if synaptic depression resulted in different rules for combining spikes with different relative timings from either the same or different sources, this would be a computation sensitive to relative spike times. For example, synaptic depression could suppress a second spike from the same source, yet allow a second spike from a different source to have an effect.

15.2.3 Feedback/Re-Entrant Activity

Cortical microcircuitry has numerous re-entrant loops that provide both excitatory and inhibitory feedback to cortical neurons (Callaway, 2004; Douglas and Martin, 2004; Douglas et al., 2005). Figure 15.2 illustrates a simplified diagram of this typical cortical microcircuitry. Because synapses are relatively slow, these re-entrant loops must have significant time delays associated with them. In addition, interareal

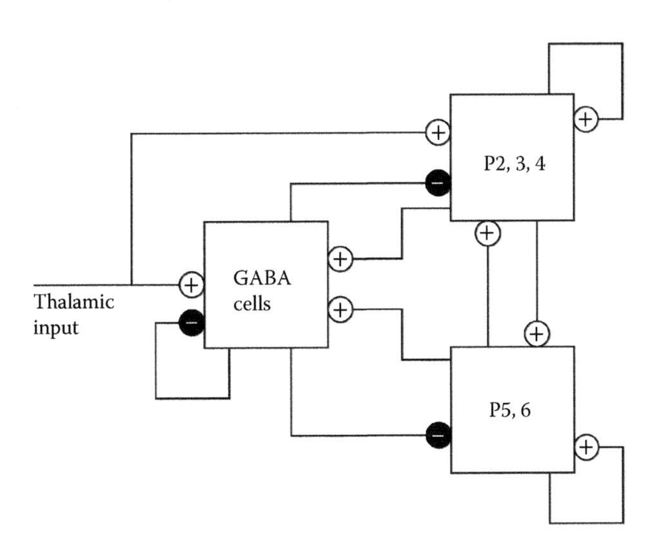

FIGURE 15.2 Simplified schematic of the relatively conserved ("canonical") pattern of cortical microcircuitry. There are numerous re-entrant positive and negative feedback loops, all of which have associated synaptic time delays, and which therefore seems likely to have non-trivial timing effects on cortical computations. (Redrawn from Douglas, R.J., Martin, K.A.C., and Whitteridge, D. *Neural Computation* 1:480–488, 1989.)

feedback connections can also have delays ranging from milliseconds to tens of milliseconds (Schwabe et al., 2006). Therefore, when two inputs arrive at an area of the cortex in sequence, the first input would have activated circuits that will modulate the effect of the second arriving input. The consequences of temporally delayed excitatory and inhibitory feedback on the dynamics of spike train processing has been largely unexplored, probably because the complexity of cortical circuitry makes modeling difficult (but see Haeusler and Maass, 2007; Hang and Dan, 2011), and controlled physiological experiments isolating the effects of these intertwined re-entrant loops is currently nearly impossible, although in simple cases, it has been demonstrated that recurrent inhibition can have strong timing effects in the cortex (Kapfer et al., 2007). Nonetheless, given the organization of cortical circuitry, it is hard to see how it could not be sensitive to differences in relative spike timing.

15.3 EVIDENCE THAT RELATIVE SPIKE TIMING CARRIES INFORMATION *IN VIVO*

In the real world, organisms are exposed to different events that occur at different times, and as these events impact an organism, this will naturally result in spikes that occur at different times. In particular, for auditory stimuli, the nature of the stimulus itself is a sequence of events over time that must translate into a sequence of action potentials distributed across a population of neurons over time. Somatosensory stimuli, as an animal runs either whiskers or a digit across a textured surface, will similarly generate a sequence of events over time. However, there is considerable physiological evidence that even for a single external event, the relative timing of elicited action potentials between cortical neurons can have interesting properties.

In visual cortex, when a single stimulus is flashed on, using the pattern of timing of action potentials in addition to the average spike count approximately doubles the amount of information that a single neuron can transmit about a visual stimulus (Richmond and Optican, 1987; Richmond et al., 1990). It has been demonstrated that it is primarily the response latency that carries this extra information (Oram et al., 2002; Reich et al., 2001; Victor, 2000). Given that cortical circuitry is so organized that differences in input timing should have a strong functional role, and given that neuronal responses to an external event carry significant information that is not present in the spike count, this suggests a role for relative spike timing in cortical computation.

All other things being equal, a stronger response should have a shorter latency than a weaker response. Hence, the first spikes should tend to come from more strongly activated neurons, with later spikes coming from less strongly activated neurons. This would naturally result in a code where relative latency would code for relative saliency. However, the finding that response latency carries information not present in the mean firing rate suggests that latency might do more than just represent response strength in a different manner. In visual cortex, the phase but not the magnitude of the response to a drifting grating is mostly a function of stimulus contrast (Albrecht, 1995; Carandini and Heeger, 1994). For visual stimuli that are abruptly flashed on, the response latency is largely driven by stimulus contrast, and response strength is driven by stimulus form (Gawne et al., 1996; Reich et al., 2001), such that a low-contrast optimal stimulus will have a large-magnitude response and a long latency, and a high-contrast nonoptimal stimulus will have a short latency and a small magnitude. Thus, when an external visual stimulus appears, or when an eye movement changes the visual scene, the responses of visual cortical neurons will be partitioned into different temporal intervals according to contrast.

That spike timing/relative latency codes can add information not present in the spike count has also been demonstrated in auditory (Bizley et al., 2010; Malone et al., 2010) and somatosensory (Johansson and Birznieks, 2004) cortex. Chapter 11 presents evidence for the existence of such codes in the olfactory system of both insects and vertebrates. However, it must be noted that all of these studies are purely correlative, that is, it has not yet been demonstrated that these timing differences are important for cortical computations. This is primarily because it is not currently possible to remove variations in input latency from an intact animal to see how behavior is or is not affected. For now, this is all that we know for certain: neural circuits seem designed to make use of timing differences between inputs, and *in vivo* cortical neurons responding to an external event generate spikes with nontrivial timing properties.

Further evidence that spike timing in the cortex can have potentially important properties comes from evidence that the first spike in a response carries a disproportionate amount of the information, or alternatively that most of the information in a response can be obtained from just the first spike (Arabzadeh et al., 2006; Bale and Petersen, 2009; Foffani et al., 2004; Oram and Perrett, 1992; Petersen et al., 2001; although this has been disputed, see Rolls et al., 2006). Similarly, it has been argued that the speed of visual recognition is so rapid that there is only time for a single spike per neuron, at least for simple perceptual tasks (Delorme and Thorpe, 2001; VanRullen, 2007; VanRullen and Thorpe, 2002). If we only consider the first spike from each neuron, then other than the simple presence or absence of a spike for each neuron, the only parameter that can vary is the relative timing between single spikes.

15.4 SPECIFIC WAYS THAT RELATIVE SPIKE TIMING CAN BE INCORPORATED INTO NEURONAL CODES

There are many possible methods for how the brain could use relative spike timing to perform computations. The following list is not exhaustive but serves only to illustrate some of the more commonly investigated ones.

15.4.1 WINNER-TAKE-ALL CODE

Possibly the simplest mechanism for taking advantage of relative spike times would be the one where the first spikes "win" and lock out the effects of any later-arriving spikes. This could easily be implemented by any of the potential mechanisms listed above, or by network effects. It has been proposed that a "winner-take-all," or "MAX," operator could be computationally powerful and an important aspect of cortical function (Douglas and Martin, 2004; Maas, 2000; Riesenhuber and Poggio, 1999; Yuille and Geiger, 1995), and simulations have suggested plausible biophysical mechanisms whereby differential spike timing could result in one input vetoing the other (Koch et al., 1983).

A recent study attempted to determine if relative spike timing could be responsible for implementing a winner-take-all computation in visual cortex (Gawne, 2008). A rhesus monkey (*Macaca mulatta*) was trained to fixate on a spot of light for juice reward. Isolated single units in visual cortical area V4 were recorded using standard microelectrode techniques. Two visual stimuli were positioned such that each alone elicited a strong response. The stimuli were presented at the widest possible physical separation, so as to minimize interactions in earlier parts of the visual system. The stimuli were presented both separately and in combination, and their contrast and relative onset timing were varied. In general, the response of each neuron to two stimuli was locked to the response to that single stimulus that produced the shortest latency.

Data from this laboratory suggests that this phenomenon is not limited to V4 (Gawne and Nowak, 2007). Figure 15.3 shows example responses from a V1 neuron taken under similar conditions as in Gawne (2008). Panels (a) and (b) are the responses as a function of time for when only one stimulus was presented at a time in the receptive field. Both stimuli elicit strong responses, and in this case the response to the single stimulus in (a) had a longer relative latency than the response to the

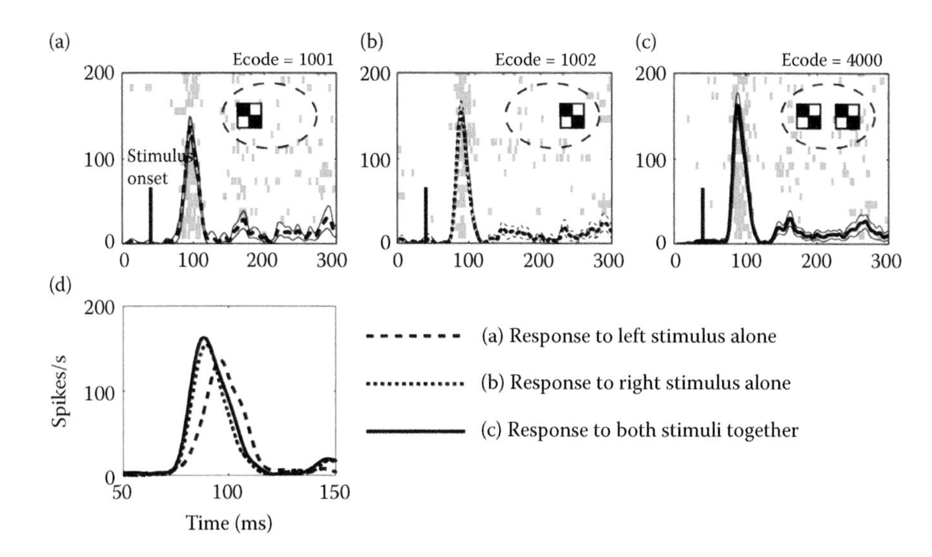

FIGURE 15.3 Example responses from a single primate V1 neuron that exhibited strong temporal gating effects. (a) Response to a single high-contrast stimulus presented alone. The continuous heavy line is the average firing rate as a function of time (smoothed by convolution with a Gaussian kernel; see Gawne 2008), with the standard error of the mean indicated by the thin lines to either side. The rasters (time of single spike occurrences) are shown in light gray. Stimulus configurations within the receptive field are illustrated in the dashed ellipses. (b) Response to a single stimulus at another location. (c) Response to both stimuli presented at the same time. (d) All responses overlaid. The response to the high-contrast stimulus presented alone has a shorter latency than the response to the low-contrast stimulus presented alone. The response to both stimuli presented at the same time tracks the short-latency response with little effect from adding the stimulus that, by itself, elicits a longer latency response.

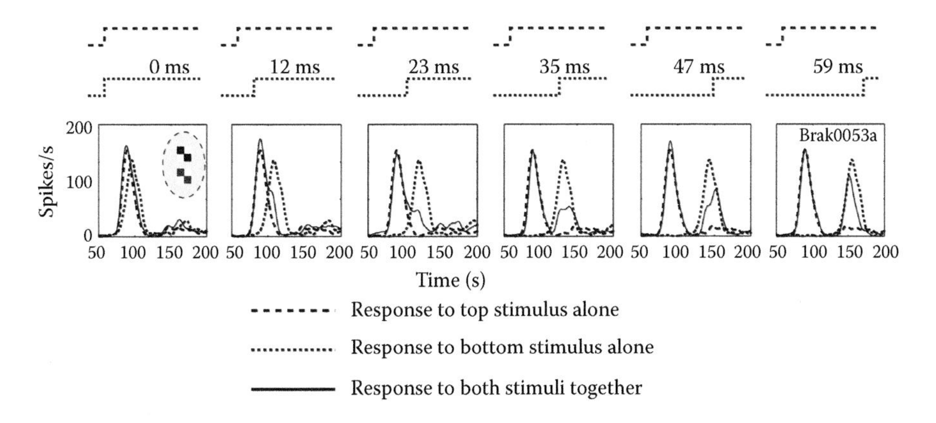

FIGURE 15.4 Data from a single primate V1 cortical neuron responding to two spatially separated stimuli with variable interstimulus timing. When the responses of the stimuli presented separately have a relatively small latency difference, the response with the shortest relative latency effectively suppresses the other response. However, this effect decays with increasing latency differences (rightmost panels).

single stimulus in (b). Panel (c) is the case when both stimuli are presented at the same time. By overlaying and expanding the plots, panel (d) shows that when these two stimuli were presented in the receptive field of a neuron. The response to the stimuli presented together is the same as the response to the stimulus with the earlier latency. It is as if the input from the stimulus that by itself elicited a longer latency had been completely shut out.

Figure 15.4 shows the results of presenting two stimuli in the receptive field of a single V1 neuron but with different temporal offsets. The simultaneous condition is the same as in Figure 15.3c/d. The data in the left-most panel is for when both stimuli were presented at the same time, and the other panels are for when the presentation of one stimulus was delayed relative to that of the other one. The temporal gating effect is strong at short relative latencies, but it is largely gone when the stimuli were presented 59 ms apart.

These data do suggest that relative spike timing could perform a rapid MAX computation in the cortex. A limitation of these sorts of studies is that they do not actually control the separate inputs to a single neuron but only the visual stimuli presented to the entire visual system. The fact that, in the cerebral cortex, it is not possible even in slice preparations to differentially stimulate completely separate inputs is a significant challenge to understanding the nature of cortical computation.

15.4.2 Recruitment Order/Rank Order Code

Another simple and potentially robust way to make use of relative spike timing is to use recruitment order, that is, the pattern at which the first spikes arrive from a population of inputs across an extended temporal interval, and not just the first spikes (Gautrais and Thorpe, 1998; Richmond and Wiener, 2004; Shahaf et al., 2008).

The basic concept is illustrated schematically in Figure 15.5. This would potentially allow for extremely rapid discrimination between different input patterns because you would only need one spike from each of a potentially small number of afferents. As a trivial example, if you had four neurons, and your temporal resolution was 2 ms, then over a 20 ms interval, the population could encode for one of 10,000 possible conditions. Even if the nervous system does not make such precise use of time, it is clear that a rank order code has the potential to rapidly carry a lot of information. It has been argued that our ability to identify visual patterns in approximately 100 ms is so fast that there cannot be enough time for more than one spike from each neuron in the processing system, and that a code based on relative spike times between neurons is the most likely way to solve this problem (VanRullen and Thorpe, 2002).

Experimental evidence for rank order coding has been found in tactile afferents in human peripheral nerve, where it has been demonstrated that the direction and shape of a tactile stimulus can be accurately predicted from the first spike in a modest number of afferents (Johansson and Birznieks, 2004). A variety of computational approaches have shown that, in principle, a recruitment order code could be fast and efficient (Delorme, 2003). Interestingly, both theoretical models and physical electronic devices based on rank order coding have proven to be robust and effective (Furber et al., 2007; Häfliger and Aasebø, 2004; Masmoudi et al., 2010; Qi et al., 2004; Thorpe et al., 2004). Nevertheless, despite the theoretical attractiveness of a

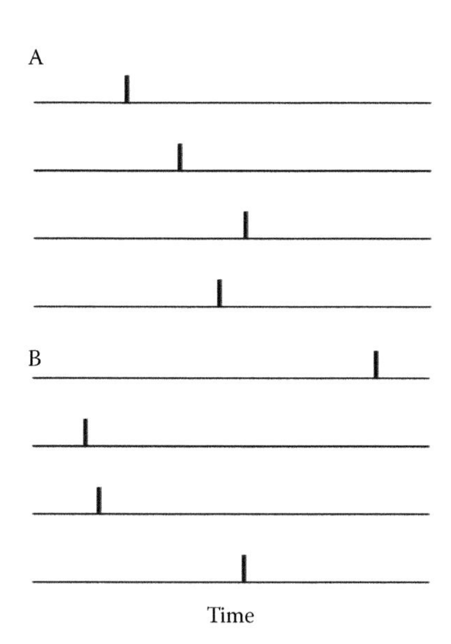

FIGURE 15.5 Schematic of the idea of a rank order code. Each short vertical line indicates the time of occurrence of an action potential from a single neuron. In conditions A and B, each of the four neurons transmits a single spike, but the changes in relative timing between neurons allow the two conditions to be clearly differentiated. In principle, a rank order code could transmit a large amount of data in a very short period of time.

rank order code, and the evidence that cortical circuitry should in general be sensitive to relative spike timing, microcircuits designed specifically to decode rank order codes have not been identified.

15.4.3 Relative Phase of Firing Code

It has been proposed that one method of coding information would be for individual neurons to fire at different phases of some ongoing rhythm. To date, this idea has been most aggressively pursued in the hippocampus (see Buzsaki 2006), but evidence has been found for the presence of such a coding scheme in the cortex as well (Kayser et al., 2009; Vinck et al., 2010). However, there is evidence against the use of spike timing with respect to an oscillation in the olfactory system (see Chapter 11).

Coding via the phase of a spike with respect to a periodic population signal is conceptually similar to the rank order coding discussed in the previous section, except that in phase coding, the "ordering" is with respect to a cyclic signal. The periodic nature of the oscillation could lead to different mathematical properties, and the biophysics of encoding and decoding might be very different. On the other hand, it has been proposed that oscillations simply preserve the initial timing relationships that are created by a stimulus onset (Samonds and Bonds, 2005). Chapters 6 and 7 provide more in-depth discussions of how information could be encoded in the firing times of spikes relative to oscillatory activity.

The rhythms are typically identified using measures of large-scale neuronal activation such as the local field potential (LFP), but the LFP is by itself probably not a functionally useful signal. Rather, when it is reported that a neuron fires with a given phase lag relative to a particular rhythm, what is really being said is that it is firing with a given phase lag relative to another population of neurons, which is conceptually another form of rank order coding.

15.5 CONCLUSION

Experiments on reduced preparations have shown that cortical circuitry is powerfully affected by differences in relative spike timing. At the same time, experiments primarily in sensory cortical areas have shown that the distribution of spike times in response to an external event carries significant information not present in the spike count. A variety of models have suggested how relative spike times could be used in the cortex and have supported the idea that computations involving relative spike times could be fast and efficient. At the same time, there is almost no direct evidence that relative spike timing is actually used by the intact operating brain, and that the interesting timing properties of cortical neurons are not epiphenomena.

To advance, it is suggested that we need to move beyond performing purely correlative experiments and develop new technologies and conceptual approaches that will allow us to directly inject signals into the cortex in a controlled and definable way. Cortical microcircuitry is complex but the apparently modular nature of the cortex, and the ease with which different cortical regions communicate with each other, suggests that this cortical circuitry might operate by set of relatively simple and conserved rules. Determining the dynamical rules by which the cortex operates could be one way to finally crack the neural code, and open the path for a true understanding of the mechanistic basis of the mind.

ACKNOWLEDGMENTS

The author thanks Jonathan D. Victor for editorial assistance. Some of the work presented here was supported by NSF grant IOS 0622318.

REFERENCES

Albrecht, D.G. Visual cortex neurons in monkey and cat: Effect of contrast on the spatial and temporal phase transfer functions. *Visual Neuroscience* 12:1191–1210, 1995.

Arabzadeh, E., Panzeri, S., and Diamond, M.E. Deciphering the spike train of a sensory neuron: Counts and temporal patterns in the rat whisker pathway. *The Journal of Neuroscience* 26:9216–9226, 2006.

Bale, M.R., and Petersen, R.S. Transformation in the neural code for whisker deflection direction along the lemniscal pathway. *Journal of Neurophysiology* 102:2771–2780, 2009.

Bizley, J.K., Walker, K.M.M., King, A.J., and Schnupp, J.W.H. Neural ensemble codes for stimulus periodicity in auditory cortex. *The Journal of Neuroscience* 30:5078–5091, 2010.

Blitz, D.M., and Regehr, W.G. Timing and specificity of feed-forward inhibition within the LGN. *Neuron* 45:917–928, 2005.

Brasselet, R., Panzeri, S., Logothetis, N.K., and Kayser, C. Neurons with stereotyped and rapid responses provide a reference frame for relative temporal coding in primate auditory cortex. *The Journal of Neuroscience* 32:2998–3008, 2012.

Buzsaki, G. *Rhythms of the Brain*. Oxford University Press, New York, 2006.

Callaway, E.M. Feedforward, feedback and inhibitory connections in primate visual cortex. *Neural Networks* 17:625–632, 2004.

Carandini, M., and Heeger, D. Summation and division by neurons in primate visual cortex. *Science* 264:1333–1336, 1994.

Carandini, M., Heeger, D.J., and Senn, W. A synaptic explanation of suppression in visual cortex. *The Journal of Neuroscience* 22:10053–10065, 2002.

Chase, S.M., and Young, E.D. First-spike latency information in single neurons increases when referenced to population onset. *Proceedings of the National Academy of Sciences* 104:5175–5180, 2007.

Chance, F.S., Nelson, S.B., and Abbott, L.F. Synaptic depression and the temporal characteristics of V1 cells. *The Journal of Neuroscience* 18:4785–4799, 1998.

Chen, X., Hsuch, H.A., Greebnerg, K., and Werblin, F.S. Three forms of spatial temporal feedforward inhibition are common to different ganglion cell types in rabbit retina. *Journal of Neurophysiology* 103:2618–2632, 2010.

David, S.V., Mesgarani, N., Fritz, J.B., and Shamma, S.A. Rapid synaptic depression explains nonlinear modulation of spectro-temporal tuning in primary auditory cortex by natural stimuli. *The Journal of Neuroscience* 29:3374–3386, 2009.

Delorme, A. Early cortical orientation selectivity: How fast inhibition decodes the order of spike latencies. *Journal of Computational Neuroscience* 15:357–365, 2003.

Delorme, A., and Thorpe, S.J. Face identification using one spike per neuron: Resistance to image degradation. *Neural Networks* 14:795–803, 2001.

Diesz, R., and Prince, D. Frequency-dependent depression of inhibition in guinea-pig neocortex *in vitro* by GABA B receptor feedback on GABA release. *Journal of Physiology (London)* 412:513, 1989.

Douglas, R.J., and Martin, K.A.C. Neuronal circuits of the neocortex. *Annual Review of Neuroscience* 27:419–451, 2004.

Douglas, R.J., Martin, K.A.C., and Whitteridge, D. A canonical microcircuit for neocortex. *Neural Computation* 1:480–488, 1989.

Douglas, R.J., Koch, C., Mahowald, M., Martin, K., and Suarez, H. Recurrent excitation in neocortical circuits. *Science* 269:981–985, 2005.

Eckhorn, R. Bauer, R., Jordan, W., Brosch, M., Kruse, W., Munk, M., and Reitboeck, H.J. Coherent oscillations: A mechanism of feature linking in the visual cortex? *Biological Cybernetics* 60:121–130, 1988.

Foffani, G., Tutunculer, B., and Moxon, K.A. Role of spike timing in the forelimb somatosensory cortex of the rat. *The Journal of Neuroscience* 24:7266–7271, 2004.

Freeman, T.C.B., Durand, S., Kiper, D.C., and Carandini, M. Suppression without inhibition in visual cortex. *Neuron* 35:759–771, 2002.

Furber, S.B., Brown, G., Bose, J., Cumpstey, J.M., Marshall, P., and Shapiro, J.L. Sparse distributed memory using rank-order neural codes. *IEEE Transactions on Neural Networks* 18:648–659, 2007.

Gabernet, L., Jadhav, S.P., Feldman, D.E., Carandini, M., and Scanziani, M. Somatosensory integration controlled by dynamic thalamocortical feed-forward inhibition. *Neuron* 48:315–327, 2005.

Gautrais, J., and Thorpe, S. Rate coding versus temporal order coding: A theoretical approach. *BioSystems* 48:57–65, 1998.

Gawne, T.J. Stimulus selection via differential response latencies in visual cortical area V4. *Neuroscience Letters* 435:198–203, 2008.

Gawne, T.J. Short-time scale dynamics in the responses to multiple stimuli in visual cortex. *Frontiers in Psychology* 2:1–7, 2011.

Gawne, T.J., and Nowak, P. Temporal dynamics of the responses of visual cortical neurons to multiple stimuli. *Society of Neuroscience Abstracts* 32:615.11, 2007.

Gawne, T.J., Kjaer, T.W., and Richmond, B.J. Latency: Another potential code for feature binding in striate cortex. *Journal of Neurophysiology* 76:1356–1360, 1996.

Ghose, G.M., and Harrison, I.T. Temporal precision of neuronal information in a rapid perceptual judgment. *Journal of Neurophysiology* 101:1480–1493, 2009.

Gray, C.M., and Singer, W. Stimulus-specific neuronal oscillations in orientation columns of cat visual cortex. *Proceedings of the National Academy of Sciences USA* 86:1698–1702, 1989.

Haeusler, S., and Maass, W. A statistical analysis of information-processing properties of lamina-specific cortical microcircuit models. *Cerebral Cortex* 17:149–162, 2007.

Häfliger, P., and Aasebø, E.J. A rank encoder: Adaptive analog to digital conversion exploiting time domain spike signal processing. *Analog Integrated Circuits and Signal Processing* 40:39–51, 2004.

Hang, G.B., and Dan, Y. Asymmetric temporal integration of layer 4 and layer 2/3 inputs in visual cortex. *Journal of Neurophysiology* 105:347–355, 2011.

Jacobs, A.L., Fridman, G., Douglas, R.M., Alam, N.M., Latham, P.E., Prusky, G.T., and Nirenberg, S. Ruling out and ruling in neural codes. *Proceedings of the National Academy of Sciences USA* 106:5936–5941, 2009.

Johansson, R.S., and Birznieks, I. First spikes in ensembles of human tactile afferents code complex spatial fingertip events. *Nature Neuroscience* 7:170–177, 2004.

Kapfer, C., Glickfeld, L.L., Atallah, B.V., and Scanziani, M. Supralinear increase of recurrent inhibition during sparse activity in the somatosensory cortex. *Nature Neuroscience* 10:743–753, 2007.

Kayser, C., Montemurro, M.A., Logothetis, N.K., and Panzeri, S. Spike-phase coding boosts and stabilizes information carried by spatial and temporal spike patterns. *Neuron* 61:597–608, 2009.

Kermany, E., Gal, A., Lyakhav, V., Meir, R., Marom, S., and Eytan, D. Tradeoffs and constraints on neural representation in networks of cortical neurons. *The Journal of Neuroscience* 30:9588–9596, 2010.

Kimura, R., Kang, S., Takahashi, N., Usami, A., Matsuki, N., Fulai, T., and Ikegaya, Y. Hippocampal polysynaptic computation. *The Journal of Neuroscience* 31:13168–13179, 2011.

Koch, C., Poggio, T., and Torre, V. Nonlinear interactions in a dendritic tree: Localization, timing, and role in information processing. *Proceedings of the National Academy of Sciences* 80:2799–2802, 1983.

Maas, W. On the computational power of winner-take-all. *Neural Computation* 12:2519–2535, 2000.

Mainen, Z.F., and Sejnowski, T.J. Reliability of spike timing in neocortical neurons. *Science* 268:1503–1506, 1995.

Malone, B.J., Scott, B.H., and Semple, M.N. Temporal codes for amplitude contrast in auditory cortex. *The Journal of Neuroscience* 30:767–784, 2010.

Masmoudi, K., Antonini, M., and Kornprobst, P. Spike based neural codes: Towards a novel bio-inspired still image coding schema. INRIA Research Report RR-7302, 2010.

Mittmann, W., Koch, U., and Häusser, M. Feed-forward inhibition shapes the spike output of cerebellar Purkinje cells. *Journal of Physiology* 563:369–378, 2005.

Oram, M.W., and Perrett, D.I. Time course of neural responses discriminating different views of the face and head. *Journal of Neurophysiology* 68:70–84, 1992.

Oram, M.W., Xiao, D., Dritschel, B., and Payne, K.R. The temporal resolution of neural codes: Does response latency have a unique role? *Philosophical Transactions of the Royal Society of London. Series B* 357:987–1001, 2002.

Petersen, R.S., Panzeri, S., and Diamond, M.E. Population coding of stimulus location in rat somatosensory cortex. *Neuron* 32:503–514, 2001.

Pouille, F., and Scanziana, M. Enforcement of temporal fidelity in pyramidal cells by somatic feed-forward inhibition. *Science* 293:1159–1163, 2001.

Qi, X., Guo, X., and Harris, J.G. A time-to-first spike CMOS imager. *Proceedings of the International Symposium of Circuits and Systems*, Vancouver, Canada, pp. 23, 2004.

Reich, D.S., Mechler, F., and Victor, J.D. Temporal coding of contrast in primary visual cortex: When, what, and why? *Journal of Neurophysiology* 85:1039–1050, 2001.

Richmond, B.J., and Optican, L.M. Temporal encoding of two-dimensional patterns by single units in primate inferior temporal cortex. II. Quantification of response waveform. *Journal of Neurophysiology* 57:147–161, 1987.

Richmond, B.J., Optican, L.M., and Spitzer, H. Temporal encoding of two-dimensional patterns by single units in primate primary visual cortex. I. Stimulus-response relations. *Journal of Neurophysiology* 64:351–369, 1990.

Richmond, B., and Wiener, M. Recruitment order: A powerful neural ensemble code. *Nature Neuroscience* 7:97–98, 2004.

Riesenhuber, M., and Poggio, T. Hierarchical models of object recognition in cortex. *Nature Neuroscience* 2:1019–1025, 1999.

Rolls, E.T., Franco, L., Aggelopoulos, N.C., and Jerez, J.M. Information in the first spike, the order of spikes, and the number of spikes provided by neurons in the inferior temporal visual cortex. *Vision Research* 46:4193–4205, 2006.

Samonds, J.M., and Bonds, A.B. Gamma oscillation maintains stimulus structure-dependent synchronization in cat visual cortex. *Journal of Neurophysiology* 93:223–236, 2005.

Schwabe, L., Obermayer, K., Angelucci, A., and Bresslof, P.C. The role of feedback in shaping the extra-classical receptive field of cortical neurons: A recurrent network model. *The Journal of Neuroscience* 26:9117–9129, 2006.

Shadlen, M.N., and Newsome, W.T. Noise, neural codes, and cortical organization. *Current Opinion in Neurobiology* 4:569–579, 1994.

Shahaf, G., Eytan, D., Gal, A., Kermany, E., Lyakhov, V., Zrenner, C., and Marom, S. Order-based representation in random networks of cortical neurons. *PLoS Computational Biology* 4: e1000228, 2008.

Shepherd, G.M. *The Synaptic Organization of the Brain*. Oxford University Press, New York, 4th Edition, 1998.

Singer, W. Neuronal synchrony: A versatile code for the definition of relations? *Neuron* 24:49–65, 1999.

Sun, Q.Q., Huguenard, J.R., and Prince, D.A. Barrel cortex microcircuits: Thalamocortical feedforward inhibition in spiny stellate cells is mediated by a small number of fast-spiking interneurons. *The Journal of Neuroscience* 26:1219–1230, 2006.

Thorpe, S.J., Guyonneau, R., Guilbaud, N., Allegrand, J.M., and VanRullen, R. SpikeNet: Real-time visual processing with one spike per neuron. *Neurocomputing* 58:857–864, 2004.

Tiesinga, P., Fellous, J.M., and Sejnowski, T.J. Regulation of spike timing in visual cortical circuits. *Nature Reviews Neuroscience* 9:97–109, 2008.

VanRullen, R. The power of the feed-forward sweep. *Advances in Cognitive Psychology* 3:167–176, 2007.

VanRullen, R., and Thorpe, S.J. Surfing a spike wave down the ventral stream. *Vision Research* 42:2593–2615, 2002.

Victor, J.D. How the brain uses time to represent and process visual information. *Brain Research* 886:33–46, 2000.

Vinck, M., Lima, B., Womelsdorf, T., Oostenveld, R., Singer, W., Neuenschwander, S., and Fries, P. Gamma-phase shifting in awake monkey visual cortex. *The Journal of Neuroscience* 30:1250–1257, 2010.

Wehr, M., and Zador, A.M. Balanced inhibition underlies tuning and sharpens spike timing in auditory cortex. *Nature* 426:442–446, 2003.

Yuille, A.L., and Geiger, D. Winner-take-all mechanisms. In *The Handbook of Brain Theory and Neural Networks*, ed. M.A. Arbib, MIT Press, Cambridge, MA, pp. 1056–1060, 1995.

Zucker, R.S. Short-term synaptic plasticity. *Annual Review of Neuroscience* 12:13–31, 1989.

Index

A

B

C

I

M

N

O

P

R

U

V

W

Z

9780367380106
T - #0135 - 231019 - C444 - 234/156/20 - PB - 9780367380106